Níl sé ceadaithe an leabhar is déanaí atá luaite thíos.
This book must be returned not later than the last date stamped below.

Tropical Pasture and Fodder Plants

TROPICAL AGRICULTURE SERIES
The Tropical Agriculture Series, of which this volume forms part, is published under the editorship of D. Rhind, CMG, OBE, BSc, FLS, FIBiol.

ALREADY PUBLISHED
Tobacco *B. C. Akehurst*
Tropical Pasture and Fodder Plants *A. V. Bogdan*
Coconuts *R. Child*
Yams *D. G. Coursey*
Sorghum *H. Doggett*
Tea *T. Eden*
Rice *D. H. Grist*
Termites *W. V. Harris*
The Oil Palm *C. W. S. Hartley*
Tropical Farming Economics
M. R. Haswell
Sisal *G. W. Lock*
Cattle Production in the Tropics Volume I
W. J. A. Payne
Cotton *A. N. Prentice*
Bananas *N. W. Simmonds*
Tropical Pulses *J. Smartt*
Agriculture in the Tropics
C. C. Webster and P. N. Wilson
An Introduction to Animal Husbandry in the Tropics
G. Williamson and W. J. A. Payne
Cocoa *G. A. R. Wood*

Tropical Pasture and Fodder Plants
(Grasses and Legumes)

A. V. Bogdan FLS

Longman
London and New York

Longman Group Limited London

Associated companies, branches and representatives throughout the world

Published in the United States of America by Longman Inc., New York

© A. V. Bogdan, 1977

All rights reserved. No part of this publication may be reproduced, stored in a retrieval system, or transmitted in any form or by any means, electronic, mechanical, photocopying, recording, or otherwise, without the prior permission of the Copyright owner.

First published 1977

Library of Congress Cataloging in Publication Data
Bogdan, A. V.
  Tropical pasture and fodder plants.
  (Tropical agriculture series)
  Includes index.
  1. Forage plants – Tropics. 2. Pastures – Tropics.
I. Title.
SB208.T7B63 1976   633'.2'00913   76-14977
ISBN 0-582-46676-8

Printed in Great Britain by
Whitstable Litho Ltd., Whitstable, Kent

To the memory of my daughter Natalie Hopewell

Contents

FOREWORD	viii
PLANT NAMES	xi
ABBREVIATIONS	xiii

The Grasses — 1

Introduction — 1

Classification and distribution	1
Environment in relation to some aspects of grass physiology	2
Day length	2
Light intensity	2
Temperature	5
Rainfall	5
Soils	6
The structure of the grass plant	7
Cultivation	11
Establishment	11
Management	12
Association with the legumes	12
Herbage yields	13
Conservation	13
Nutritive value	14
Reproduction	19
The structure of floral parts	19
Flowering and pollination	20
Fertilization	20
Apomixis	21
Seed	23
Seed germination	24
Seed production	25
Introduction and breeding	26

The more important species — 31

The Legumes 302

Introduction 302

Classification and distribution 302
The structure of the leguminous plant 303
Nitrogen fixation by the legumes 304
In mixture with the grasses 307
Cultivation 308
Pests and diseases 310
Yields, utilization 310
Chemical composition, nutritive value 311
Palatability, toxicity 311
Reproduction 312
    The flowers 312
    Flowering, pollination 314
    Seed production and harvesting 315
Breeding 316

The more important species 318

APPENDIX 429

REFERENCES 432

INDEXES OF PLANT NAMES 462

Botanical names 462
Common names 471

Foreword

Fodder and pasture plants attract nowadays more and more attention in the tropics and subtropics, both in advanced and in developing countries. New local fodder and pasture plants are being introduced into cultivation and new species and varieties transferred from the areas and continents rich in fodder and pasture plants to the areas where they are scarce. Methods of establishment, management and seed production are being improved and plant breeders produce new superior varieties and cultivars and develop new forms by hybridization. Pastoralists learn more and more about the value of wild grasses and about the ways they should be managed. Information on these numerous activities and achievements is scattered in innumerable research papers, journals, monographs and sometimes books. Summaries on certain species have been compiled and regional accounts on pasture and fodder plants written and published. This enormous material is difficult to grasp and I am attempting to summarize briefly the work done on individual species in a kind of, perhaps, over-ambitious reference book in which over 300 species are briefly described botanically, the descriptions supplemented by the more essential information on their distribution, cultivation and nutritive value which I have been able to trace and select from the literature available to me.

I hope that this book may be useful to those working in the tropics, especially in places where the libraries may not be so good as in big centres; they can then chose the plants worth trying in their particular country, and I know from my own experience during the first years in the tropics that if such a book had existed it would have saved time, have speeded the work and helped to avoid a number of mistakes. I also hope that this publication may be of use to students training for work in warm countries. In this kind of publication omissions are inevitable and can be considerable and I should be most grateful for any critical comments on the omissions and mistakes, also inevitable.

Grasses and legumes are the two main groups of plants on which the farm animals exist. Both groups are grazed in natural grasslands and range lands and both are cultivated, singly or in mixtures, for fodder or grazing. In mixtures the grasses usually give the main bulk of herbage, whereas the legumes increase the bulk and improve the quality of herbage by enriching it with proteins. The legumes can also improve the nitrogen status of soil by fixing atmospheric nitrogen through a symbiosis with *Rhizobium* bacteria living in root nodules. This ability to fix nitrogen, which is however not so common in the tropics as in temperate countries, can increase the bulk of grass and improve its quality. There are of course plants of botanical families other than Gramineae (grasses) and Leguminosae (legumes) which also form a part of natural vegetation and are browsed or occasionally cultivated, but their role in animal feeding is only secondary to that of the grasses and the legumes and they are not included in the present book. Another reason for their omission is to avoid over-loading the book with material and thus delaying its publication; studies on fodder and pasture plants and their improvement are expanding with ever increasing speed and any substantial delay in publication can result in omissions of new, up-to-date information concerning the two most important plant groups.

Only tropical grasses and legumes are considered and they are understood as plants originating and grown or cultivated in the tropics. The problem of defining tropical pasture and fodder plants is complicated by the topography as at high elevations the climate is far from being tropical but sub-tropical or even temperate, depending on the altitude. If a grass or a legume cultivated at high altitudes is of non-tropical origin and had been introduced from sub-tropical or temperate areas, then its morphology, environment requirements and performance are usually adequately described elsewhere, as e.g. of subterranean clover or cocksfoot, and such plants are only briefly mentioned and their performance at high altitudes of the tropics is only lightly covered; but if a grass or a legume grown at high or medium altitude is of tropical origin, as e.g. *Pennisetum clandestinum* or *Setaria anceps*, then it is dealt with in full.

In this book the tropics are understoood geographically, rather than climatically, as an area between 23°27′ latitude both sides of the equator. A country is considered entirely tropical and is fully covered in the text if its essential portion lies within the geographical tropics, and only if its considerable part extends beyond the tropics is this part not then covered.

In compilation of this book extensive literature has been used with the emphasis on the more recent and more important research papers and on basic textbooks and reviews, especially those recently published. Abstracting journals have been perused, especially the *Herbage Abstracts*, and without this journal the publication of this book would hardly have been possible. In the literature studied, metric and avoirdupois systems have been used for yield records and for various other measurements; these measures have been unified and non-metric data were recalculated and presented as metric and are sometimes slightly simplified by rounding the figures. Some information given in the introductory part is sometimes elementary but it has to be borne in mind that the book is intended for readers with various degrees of technical knowledge. The main emphasis is however on individual species and the space given for each species is in an approximate proportion to its importance either in cultivation or in natural grasslands. They are described in the alphabetic order of botanical names; of the synonyms only the important or better known ones are given. The synonyms and the common names can be found in corresponding indexes. Chromosome numbers of individual species are cited mainly from Bolkhovskikh *et al.* (1969), where references to the original publications and authors can be found.

During the preparation of manuscript I extensively used the library of the Commonwealth Bureau of Pastures and Field Crops, Hurley, United Kingdom, and I am most grateful to its staff, librarians and especially the Director, P. J. Boyle, for their advice and encouragement. I also thank Miss S. Daniels, the librarian of the Commonwealth Mycological Institute, for her invaluable help in my search for literature. The identity of a number of species, especially taxonomically difficult, had to be checked in the Herbarium of the Royal Botanic Gardens, Kew, London, and I am much indebted to Dr W. D. Clayton, Mr S. A. Renvoize and Dr C. E. Hubbard for their advice on the taxonomy of grasses and to Dr B. Verdcourt on the taxonomy of legumes. My thanks are also due to Dr G. W. Burton, L. R. Humphreys, J. O. Green, Dr D. F. Osbourn, C. R. Lonsdale, Professor A. E. Kretschmer, Professor L. V. Crowder, M. G. W. Rodel and Dr C. R. Metcalfe with whom various problems concerning the book were discussed. My special thanks are also due to Mr John Knight, a friend and a colleague who has read a considerable part of the manuscript and made a number of critical comments and corrections, and to my wife for her constant and patient support and encouragement.

Plant Names

Scientific classification of plants is based mainly on their floral morphology although in the last 50 or so years much attention has been given to other characteristics: anatomy, cytology, chemical composition and various aspects of biology. Plants of large botanical families, such as Gramineae (grasses and cereals) or Leguminosae (legumes), are grouped, in a descending order, to **subfamilies**, **tribes**, **genera**, **subgenera**, **species**, **subspecies**, **varietas** and **forms**, and there are some other, less familiar intermediate categories. The names, from the genus downwards, are written in print differing from the text, preferably in italics. Following the binominal system of Linneaus, a botanical name consists of the name of the genus and the specific epithet, e.g. *Chloris gayana* (Rhodes grass). The present usage is to begin *all* specific epithets with a lower-case (not capital) letter. There are some deviations from the binominal system, forced mainly by a pressure of practical requirements, when a subspecies and/or a varietas has to be defined, as e.g. *Andropogon gayanus* var. *bisquamulatus*. To achieve a more exact meaning for the plant name, the name of the author (often abbreviated) who first named the species is added, e.g. *Chloris gayana* Kunth, and the same refers to subspecies and varieties, and actually to all other categories. If, for some reason, the name has been changed, e.g. by transferring a species to another genus or changing its rank, the name of the original author is given in brackets, e.g. *Beckeropsis uniseta* (Nees) Robyns or *Desmodium intortum* (Mill.) Urb. Botanical names are not infrequently changed and the changes often cause confusion and make the taxonomists and nomenclaturists unpopular with the pastoralists and agriculturists. The great majority of changes are however done with a good reason and in accordance with the rules of International Code of Botanical Nomenclature, the last edition of which appeared in 1972, and a number of changes made in the past have now been generally accepted: no one would now apply the name *Panicum brizanthum* to *Brachiaria brizantha* or *Andropogon rufus* to *Hyparrhenia rufa*, but the name of *Panicum purpurascens* is even now often applied to *Brachiaria mutica*. Delaying the acceptance of changed names, such as *Stylosanthes guianensis* instead of *S. gracilis* or *Macroptilium atropurpureum* instead of *Phaseolus atropurpureus* does not help to avoid confusion. In the majority of cases botanical names are applied correctly but wrong identifications are not infrequent. A wrongly named plant recommended for cultivation can be a disappointment and may incur unnecessary expenses. To secure correct naming it is useful to preserve herbarium specimens of all species and varieties under trial and send them for naming or checking to a reputable botanical institution.

There are no rules for common names and although their use is necessary they should be supported by botanical names especially in publications because in different countries the common names can be different; moreover, sometimes the same name can be referred to a different species.

Special rules do however exist for naming bred or selected varieties of cultivated plants. These rules were, in the first instance, intended for ornamental garden plants to clear up the havoc of names that existed in this group, but they are also applicable to all cultivated plants. The term **cultivar** (cv.) has been suggested for selected or bred varieties although **variety** has remained a valid alternate term. The term cultivar has now been widely accepted in a number of countries. To have a valid name every newly developed cultivar should be registered with an appropriate authority. The International Code of Nomenclature for Cultivated Plants formulated by The International Commisson for the Nomenclature of Cultivated Plants of the International Union of Biological Sciences outlines the rules in 57 articles and is a publication most useful for plant breeders and seed traders.

Abbreviations

CF	– crude fibre
CP	– crude protein
cv.	– cultivar
DCF	– digestible crude fibre
DCP	– digestible crude protein
DDM	– digestible dry matter
DE	– digestible energy
DEE	– digestible ether extractives
DM	– dry matter
DNFE	– digestible nitrogen-free extractives
DOM	– digestible organic matter
EE	– ether extractives
FU	– feed unit or fodder unit
f.y.m.	– farm yard manure
GA	– gibberellic acid
ME	– metabolic energy
NFE	– nitrogen-free extractives
OM	– organic matter
PGS	– pure germinable seed
PLS	– pure live seed
PVS	– pure viable seed
SE	– starch equivalent
TDN	– total digestible nutrients
$2n$	– somatic chromosome number (in actual plant)
x	– basic number of chromosomes

The Grasses

Introduction

Classification and distribution

Gramineae (Poaceae) is a large botanical family with about 10,000 species grouped into some 650 genera and the genera into 50–60 tribes; units larger than tribes have also been suggested as sub-families or more loosely defined as groups of tribes. Their number varies from 2 to 12 depending on the views of those working on general classification of the Gramineae. Of these 'groups', festucoid (poaoid), panicoid and chloridoid grasses are of particular interest. The three groups include practically all cultivated, and a large number of valuable wild grasses: the festucoid group – temperate grasses; the panicoid group – tropical and subtropical grasses; and the chloridoid group – a few tropical cultivated grasses and a number of valuable wild grasses of the tropics and of the warmer areas of North America. Festucoid grasses on one side, and panicoid and chloridoid on the other, differ in a number of characters of which leaf anatomy is of particular importance because it seems to be connected with the differences in the process of photosynthesis in the two groups which are adapted to tropical conditions (panicoid and chloridoid grasses) or to temperate climate (festucoid grasses), these differences are discussed in more detail on pp. 2–5.

To fesucoid grasses belong the tribes Triticae (species of *Agropyron*), Festuceae (*Festuca, Dactylis, Lolium, Poa*), Bromeae (*Bromus*), Aveneae (*Avena, Arrhenatherum*), Agrostideae (*Agrostis, Alopecurus, Phleum*); to panicoid group belong Paniceae or Mellinidae (*Panicum, Brachiaria, Digitaria, Melinis, Pennisetum, Cenchrus*), Andropogoneae (*Andropogon, Hyparrhenia, Sorghum, Lasiurus, Themeda*) and Maideae (*Zea, Euchlaena, Tripsacum*), and to chloridoid grasses Chlorideae (*Chloris, Cynodon*) and Eragrosteae (*Eragrostis, Dactyloctenium, Eleusine*). Although panicoid grasses occur mainly in the tropics and subtropics, and festucoid in the temperate zones, panicoid species can sometimes be found in temperate areas and festucoid grasses are not uncommon at high elevations and occasionally in lowlands in the tropics and sub-tropics. Hartley (1958) has shown that the Andropogoneae tribe is connected mainly with the tropics and so is the largest tropical tribe Paniceae. The Chlorideae is less tied with the tropics and its species are common in North America.

Environment in relation to some aspects of grass physiology

The main environmental factors which affect grass plants are climate and soil, and of the climate the day length, light intensity, temperature and rainfall.

Day length

In the equatorial zone, some 5°–10° N. and S. of the equator, photoperiods or day lengths do not vary much and there is only a slight difference in the day length of June and December; further away from the equator these differences are however well felt and at the tropic circles of Cancer and Capricorn the day length changes from 10 h 20 min in winter to 13 h 40 min in summer. The majority of tropical grasses are either indifferent to day length (*Tripsacum dactyloides*, *Acroceras macrum*) or are short-day plants and flower earlier under short than under long photoperiods (*Hyparrhenia hirta*, *Sorghum halepense*); there are, however, tropical species (*Paspalum dilatatum*) which flower easier and earlier under longer than shorter photoperiods (Knight & Bennett, 1953). Herbage production can follow the pattern of flowering (Wang, 1961) but in the majority of grasses long photoperiods stimulate herbage growth and production.

Light intensity

Solar energy in the form of light is the source of energy for the photosynthetic activity of green-plant parts, the activity which results in the synthesis of carbon dioxide (CO_2) of the air and water organic matter – carbohydrates, in the first instance sugars, from which all other organic substances of plants further develop. Only a small proportion of solar energy reaching the earth is utilized in the process of photosynthesis, 1–5 per cent on the yearly basis and 3–10 per cent during maximum active growth (Cooper, 1970). A certain degree of light intensity is needed for photosynthesis to begin and its productivity in terms of amounts of synthesized organic matter increases with the increase of light intensity up to 15,000–25,000 lux and further increases in light intensity do not increase the productivity of photosynthesis of temperate grasses any further. In tropical grasses, however, photosynthetic productivity increases further and reaches its maximum at 50,000–60,000 lux and sometimes even at higher light intensity, and can thus be much greater than in temperate grasses provided that light intensity is sufficiently high. Not all tropical grasses respond, however, to high light intensities, but only those belonging to so-called panicoid and chloridoid groups to which, as it has been mentioned previously, belong the tribes Paniceae, Andropogoneae, and Maideae in the panicoid group and Chlorideae and Eragrosteae tribes in chloridoid grasses. Andropogoneae, Paniceae and Maideae tribes are of tropical origin and they apparently acquired their adaptability to high light intensity during their long history of adaptation to tropical conditions and so did the species of Chlorideae and Eragrosteae although the last named tribes are more widely spread and extend to temperate regions.

Panicoid and chloridoid grasses on one side and festucoid grasses on the other differ in some important details of the photosynthetic process which are in some ways connected with leaf anatomy. In grass-leaf, the mesophyll consists of the smaller, relatively uniform cells containing small chloroplasts and one or two layers of larger cells known as bundle-sheath cells which surround the smaller veins or, in other words, bundles of conductive cells. In chloridoid and in the majority of panicoid grasses these cells have slightly thickened walls and contain large chloroplasts (Fig. 1) often of linear shape, frequently granulated when viewed under electronic microscope (Johnson & Brown, 1973). These cells are known as Kranz-type cells. In festucoid

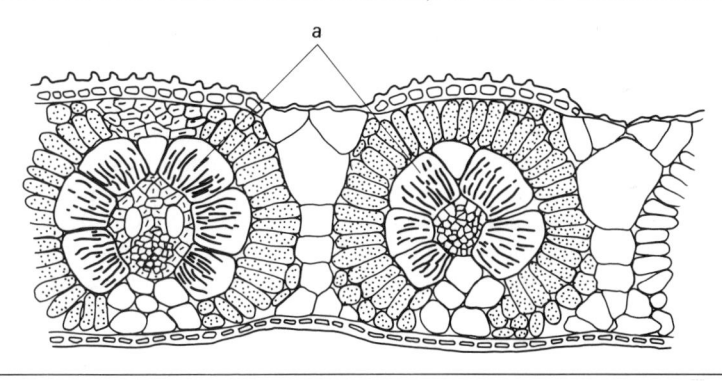

Fig. 1 Leaf cross-section of a chloridoid grass *(Cynodon dactylon)*; (a) bundle sheath of Kranz-type cells. (Diagrammatic drawing from microscopic slides housed at the Jodrell Laboratory, Royal Botanic Gardens, Kew.)

grasses the bundle-sheath cells have thin membraneous walls and usually contain no chloroplasts (Fig. 2) and they are known as non-Kranz type cells. The Kranz cells apparently play a considerable role in photosynthesis and in the passage of the products of carbon assimilation through them to the conductive tissues. Although the Kranz cells are characteristic for panicoid and chloridoid grasses, some genera of

Paniceae have the non-Kranz-type bundle-sheath cells. De Oliveira *et al.* (1973), who examined 72 species of the Andropogoneae, Paniceae, Maideae, Chlorideae and Eragrosteae tribes, have found that they all, except *Acroceras macrum* of Paniceae tribe, have the Kranz-type syndrome, although there can be other species of Paniceae with the non-Kranz-type bundle-sheath cells.

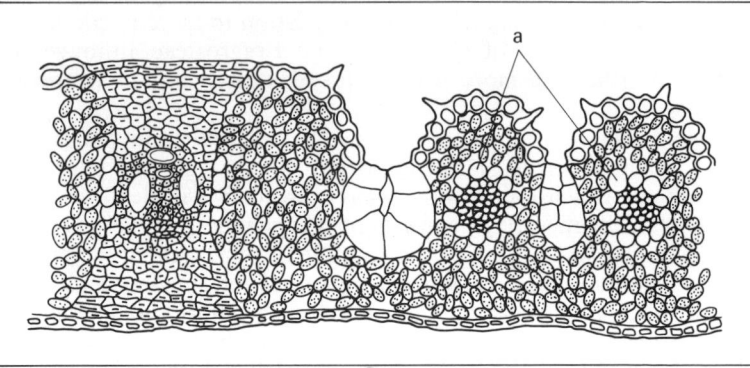

Fig. 2 Leaf cross-section of a festucoid grass *(Bromus fibrosus)*; (a) bundle sheath of non Kranz-type cells. (Drawn from microscope slides housed at the Jodrell Laboratory, Royal Botanic Gardens, Kew.)

In festucoid grasses the photosynthetic process is of the so-called Calvin or C_3 cycle under which the initial products of C assimilation are 3-phosphoglyceric (3-carbon) acids or hexose phosphates further utilized for the formation of carbohydrates (Cooper & Tainton, 1968). In panicoid grasses with the Kranz-type syndrome the initial products of photosynthesis are 4-carbon acids – malate, asparagate, oxalo-acetate – and this photosynthetic cycle is known as 'C_4 pathway', which, at its initial stages is usually combined with the Calvin cycle. Under the Calvin cycle a certain proportion of CO_2 absorbed by the leaf but not utilized is released back into the atmosphere, whereas under the C_4 cycle it penetrates into bundle sheath cells and serves as an additional gaseous source for photosynthesis (Hatch, 1972), and this probably contributes to the efficiency of CO_2 utilization; C_4 pathway of photosynthesis can work under lower concentrations of CO_2 in the atmosphere than the Calvin cycle.

The optimum temperatures for the Calvin cycle photosynthesis are around 15°–20°C, i.e. the optimum air temperatures for temperate grasses. For the C_4 pathway photosynthesis the optimum temperatures are however of the order of 30°–40°C, the temperatures which the leaves of tropical grasses can reach under direct sunlight. The optimum light intensity for the C_4 pathway photosynthesis is about 50,000–60,000 lux and 15,000–30,000 lux for the Calvin cycle of temperate grasses. High efficiency of the C_4 pathway photosynthesis of panicoid grasses can be reached only at high temperatures and high light intensity and then it can be up to 50–70 mg CO_2/dm^2/hour (Cooper, 1970), whereas 20–30 mg CO_2 are the highest photosynthetic rates reached in temperate grasses; this corresponds to about 30–50 g DM/m^2/day (DM = dry matter) and up to about 20 g, respectively. This all shows than under light intensities and high temperature tropical grasses can yield much more than temperate grasses, but tropical grasses do not make use of this advantage under low temperatures or low light intensities. It should be noted that the two photosynthetic cycles can be closely connected and in some grasses, e.g. in some species of *Sorghum* they can interchange at some stages of growth.

Temperature

In tropical grasses nearly all parameters of growth, such as tillering, tiller growth, leaf length, etc., are at their maximum at higher temperatures than in temperate grasses, although leaf width is usually decreased in leaves grown at particularly high temperatures. Chlorophyll formation in young plants does not take place in tropical grasses until the air temperature is above 10°–15°C and the effect of temperature on the productivity of photosynthesis has already been mentioned. Very high natural temperatures – and in leaves exposed to direct sunlight, they can be higher than the air temperatures – seem to produce no harmful effect on tropical grasses provided that the plants are well supplied with water. The effect of low winter or night temperatures in the subtropics or at high elevations result in suppression of plant growth and temperatures of about zero point can kill the leaves of certain tropical grasses but lower temperatures are required to kill a perenial grass. In Queensland, Australia, R. M. Jones (1969) observed on some grass species and varieties the effect of winter in which the lowest temperature was −10°C and found: 100 per cent survival of *Paspalum dilatatum* and *Pennisetum clandestinum*; 97 per cent survival of *Chloris gayana*; 23–91 per cent of *Setaria anceps*, depending on the cultivar; and only 0–6 per cent survival of *Panicum maximum*. Seedlings are killed easier than the adult plants and small and weak seedlings are particularly vulnerable to frost. The herbage affected by frost can loose its palatability and nutritive value.

Rainfall

In the tropics annual rainfall varies within very wide limits from 50–100 mm to well over 3,000 mm, and there are areas, mostly in the equatorial zone, where rain is plentiful and fluctuates little throughout the year. However, in most areas seasonal changes in rainfall occur and they are usually well expressed. Webster & Wilson (1966) distinguish the following seasonally wet, 'monsoon', regions:

1. Areas with 1,000–2,000 mm of annual rainfall with two rainy seasons, and one short and another hardly noticeable dry seasons;
2. Areas with about similar annual rainfall but with well-pronounced two dry seasons;
3. Areas with somewhat lower annual rainfall (750–1,250 mm) and one fairly long dry season;
4. Areas with one very long dry season and a short wet season, and usually with a still lower annual rainfall.

In the areas with low annual rainfall rains can be erratic and the reliability of rainfall should be taken in account when planning to grow certain fodder crops or grasses. In the outer tropics and subtropics most of the rain falls in warmer months, in summer, and winters are cool and dry. The effect of rain on grasses much depends on the relative amounts of rain water absorbed by the soil and of that run off and lost for the grass and this proportion depends to a considerable extent on the density of grass cover. Denuded pastoral lands absorb only very little water, whereas the same land covered with unspoiled grass absorbs infinitely more water and it is remarkable how good the grass can be in the areas which look almost a desert when the grass is sparse or almost completely destroyed. Grass/water relationship in the tropics is in general similar to that of temperate warm areas and is not discussed here.

Soils

Tropical soils can be classified into zonal, intrazonal and azonal (H. Vine in Webster & Wilson, 1966). Zonal soils develop mainly under the effect of zonal factors – climate and vegetation, intrazonal soils mainly under the effect of local factors such as parent rock or waterlogging, which may produce different effects in different zones, and azonal soils, e.g. fresh alluvium, do not depend on either. Zonal soils can be broadly classified into soils of dry areas where leaching of minerals and other soluble substances is restricted or practically non-existent, and those of more moist areas where leaching processes play a considerable part in soil formation. The quality and fertility of such soils depend often on the parent rock and may vary very considerably. In the literature concerning individual grass species, the soils are variously classified or not mentioned at all and soil as an environmental factor is often omitted in grass descriptions. Soil type is however nearly always mentioned when the grass grows on waterlogged soils of flats and depressions; these soils are often heavy in texture and dark or black in colour. Grass flora of such soils is usually quite distinct from that of well-drained habitats. Sandy soils, and especially sands, are also a substrate which suits only certain grass species. The effect of soil fertility, especially when resulting from the application of fertilizers, is briefly mentioned in chapters on grass cultivation, and in more detail when the growth and yields of individual grasses are discussed.

The structure of the grass plant

Similarly with all other flowering plants, the grasses (Gramineae) are built of two main parts: **the shoots** or **tillers** and the **roots**. The shoots in their turn consist of the **stem** (Fig. 3) (**culm, haulm**) and the **leaves**. The leaves of grasses are situated on the stem in two opposite rows, alternately, i.e. the third leaf above the first, the fourth leaf above the second, and so on. The zone of the stem from which the leaf arises, or the **node** (Fig. 3a) can be clearly seen all round the stem and the portion of stem between the nodes is the **internode** (Fig. 3b). The shoots can be upright, ascending or horizontal; the latter can creep on the ground surface and are called the **stolons** – or under the ground, the **rhizomes**.

The leaf consists of the sheath and the blade. The sheath (Fig. 3c) is a cylindrical, or sometimes compressed, lower part of the leaf which clasps the stem and its main function is to protect and support the tender bottom part of stem internode; this is needed because the stem growth is intercalary, each internode elongating in its bottom part for a considerable period of time after the upper part of the internode has ceased to grow. On its top the sheath bears the **blade** (Fig. 3d), a flat, or sometimes folded or rolled in, linear portion of leaf with the main function of photosynthesis. In this book, for brevity, the blade is further referred to

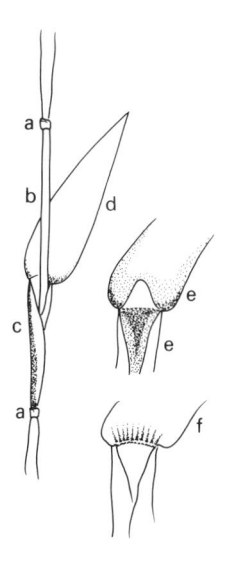

Fig. 3 Grass shoot: (a) node, (b) stem internode, (c) leaf-sheath, (d) leaf-blade, (e) ligule.

as the leaf. At the bottom of the internode, in the leaf axil, there is a bud which may or may not develop into a side tiller, and between the bud and the stem there is a leaf-like outgrowth with two keels – the **prophyll**, which can sometimes be long and even protrude from the leaf sheath. On the borderline between the sheath and the blade is an outgrowth, the **ligule**, which is membraneous or reduced to a rim, often a rim of hairs.

The single tiller which arises from germinating seed soon branches, the branches arising from axillary buds, and the side tillers can branch in their turn near the ground and form the **tuft**. The tillers can retain a certain degree of independence from the main tiller, or a group of tillers, by developing its own root or roots. Some tillers can grow horizontally and form stolons or rhizomes.

The grasses can be annual or perennial. The annual grasses last one season, bear seeds and die out with the onset of winter or a dry season, or even earlier. In temperate areas, or in the areas with well-defined, severe dry seasons annual grasses differ clearly from perennial grasses but in tropical areas with not well expressed dry season the difference may not be very clear as the annuals can survive longer than one season. The annuals can be best distinguished by the absence of remnants of old leaves and stem bases left from the previous seasons which are always present in perennial grasses. With a few exceptions (*Eleusine indica*, *Eragrostis tenuifolia*) the annuals, in contrast with the majority of perennials, can be easily pulled out from the soil. All tillers of the annual grasses can potentially bear seeds and are uniform in structure. Annual grasses occur mainly as weeds of cultivation and they are also common, and often constitute the main grazing, in arid and semi-arid areas where there is not enough moisture in the soil during the dry season to support perennial grass even in a semi-dormant state. Annuals can also occur in appreciable quantities in bush or in light forest in less dry areas and they also occur in overgrazed pastures. Some annual grasses (sorghum, Sudan grass, teosinte, maize, African millet, *Pennisetum pedicellatum*) are cultivated as fodder plants.

Perennial grasses which last for a few to several seasons have normally a more complex structure than the annuals. They usually form tufts which spread in the outer side by repeated tillering or by means of rhizomes with very short internodes and on which new groups of tillers arise close to the older tillers of the tuft. Tufted perennials usually develop two types of tillers or shoots: fertile, seed-producing tillers with elongated internodes and distant leaves and sterile tillers with very short, practically unnoticeable internodes and long crowded leaves often forming the main bulk of grazeable herbage. Sterile tillers can remain as such for years or elongate and flower in the next season. Some perennials, e.g. *Melinis minutiflora* or *Hyparrhenia papillipes*, do not form sterile tillers with short internodes and long leaves but develop elongated sterile tillers; the lower leaves of these tillers are often small, they die out early but the upper leaves develop well and are grazed.

Grasses of this type cannot usually withstand close grazing and are difficult to transplant. In rare cases perennial tufted grasses do not form any sterile shoots. Grass tufts live for several, usually 4–8, years. Tufted grasses usually form the main bulk of herbage in permanent, savanna-type grasslands and a number of them are grown in leys and in sown permanent grasslands, the better known of them being *Panicum maximum, Setaria anceps, Pennisetum purpureum, Paspalum dilatatum, Hyparrhenia rufa, Melinis minutiflora, Andropogon gayanus* and *Cenchrus ciliaris*.

Creeping perennial grasses send out horizontal rhizomes or stolons and form more even stands than tufted grasses. Rhizomatous grasses with underground creeping shoots can grow between tufted perennials or form almost pure stands. They usually occur in more moist areas or on sands and avoid dry areas with hard and dry soil which is difficult to penetrate, and are common in swampy areas but not on seasonally waterlogged soil. Rhizomatous grasses are difficult to eradicate and some of them are noxious weeds of arable land, e.g. *Digitaria scalarum* in Africa or *Cynodon dactylon*, mainly in the subtropics. A number of rhizomatous grasses, including the two named ones, can form both rhizomes and stolons. Stolons usually spread first and occupy the ground and the rhizomes follow later, or simultaneously but they grow slower than the stolons, e.g. Cape Royal grass, a form of *Cynodon dactylon* var. *elegans* much used in Africa as a turf grass, when planted by cuttings covers the ground quickly with stolons which, however, do not form suitable turf. Later on, the rhizomes catch up with the stolons and send up branched tillers which form a desirable lawn. Kikuyu grass (*Pennisetum clandestinum*) is another African example of a rhizomatous/stoloniferous grass.

Purely stoloniferous grasses occur predominantly in the tropics and warm countries. The stolons creep horizontally on the ground surface rooting at the nodes, but in a number of grasses the first internode or two of young plants raised from seed arch slightly and only next internodes grow horizontally. There are, however, some grasses in which the internodes are typically arching, as e.g. in *Cynodon plectostachyus*. In some other species, which are however not numerous (*Sporobolus helvolus, Digitaria macroblephara*), the stolons grow vertically or under an angle and only on reaching a considerable length do they bend down; after having touched the ground they root and then grow horizontally. This can perhaps be regarded as an adaptation to dry conditions: the grass places the new tillers at some distance from the mother plant thus avoiding immediate competition for moisture.

Morphologically, two types of stolons, denoted here as types A and B, can be distinguished. In type A the stolons have a structure normal for the grass shoot: each node has a leaf and the leaves are evenly distributed along the stem. In type B each node (actually a group of nodes with no visible internodes between them) develops two to five leaves and the leaves are grouped on the stem. Type B branches more and earlier than A because the upper leaf sheaths in a group of leaves protects the basal, growing portion of the long internode, whereas side branches are produced from the lower leaf axils when the stolon is still growing. Stolons of A type have been observed mainly in the Paniceae tribe and of B in the Chlorideae and Sporoboleae, e.g. in *Chloris gayana*, species of *Cynodon* and *Sporobolus*.

The speed of stolon growth differs in different grasses. Species with slow-growing stolons (*Paspalum notatum, Sporobolus isoclados*) usually form a dense cover from the early stages of spreading and develop slowly-spreading colonies, but once the ground has been occupied they retain it firmly and do not allow other species to grow within the colonies. In other grasses (*Chloris gayana, Cynodon nlemfuensis*) the stolons grow fast, the colonies spread quickly but they are less aggressive and other grasses can still settle inside the colonies, at least for a period of time.

Stoloniferous, and sometimes rhizomatous grasses as well, are often pioneer species able to occupy quickly the bare ground of denuded pastoral land or arable land when it is left fallow. They can represent a certain phase in plant succession, often as a second stage after the dominance of annual weeds in fallows, and are in their turn eventually replaced by tufted grasses or bush. Stoloniferous grasses in man-established pastures usually last for a shorter time than tufted species unless they are heavily fertilized and closely grazed; they can often become sod-bound. Stoloniferous grasses can usually be relatively easily eradicated and do not normally become serious arable weeds. There are however some exceptions and e.g. *Paspalum notatum* is difficult to plough, and if it is grown in an arable rotation the crop immediately following the grass can be poor. Stoloniferous grasses often produce relatively short sterile tillers on their stolons and form shorter herbage than tufted grasses, but this is not true in regard of all species and stoloniferous *Chloris gayana* or *Cynodon nlemfuensis* develop tall herbage.

From the germinating seed the root appears first followed by the tiller, and the roots that arise from the seed are known as **primary** roots, whereas the **secondary** or **adventitious** roots are those which develop from the nodes of tillers or creeping stems. The number of primary roots is characteristic for the species and the majority of small-seed grasses have only one primary root. Primary roots can branch and they provide the seedlings with water and mineral nutrients during the first stages of growth. Later they are replaced by secondary roots which can be very numerous and play an important role at later stages of plant development and growth and in preventing soil erosion by retaining soil particles between small rootlets. The majority of grass roots are found in the upper soil layers of 0–10 or 0–20 cm but a number of roots can also penetrate deep into the soil, down to 1 or 2 metres or even deeper. Each tiller, or each group of tillers, develop their own roots which makes them independent to a certain degree from other tillers or parts of the tuft in regard to obtaining water and mineral nutrients, although some connection between living tissues of different parts of the tuft can remain for a considerable time. It should be noted that only the young root tips densely covered with root hairs can absorb water and mineral nutrients from the soil; old roots lose this ability.

Cultivation

Establishment

Tropical grasses are established vegetatively or from seed, vegetatively if it is simpler or cheaper, or when the purity and uniformity of hybrid clones or clones of cross-pollinating grasses should be maintained as e.g. in clonal cultivars or hybrids of Bermuda grass (*Cynodon dactylon*). Stoloniferous and rhizomatous species can be propagated by pieces of stolons or rhizomes spread on the ground and buried in by subsequent harrowing or discing. Large tufted grasses, such as elephant grass (*Pennisetum purpureum*), *Tripsacum laxum*, large varieties of *Panicum maximum*, etc., are planted in rows by splits or sprigs. When planting, the top portions of grass should be cut off as the splits with uncut stems and leaves would almost invariably die because of the loss of water through the leaves. Long roots should also be cut short because the old roots would anyhow die out and the plants develop new roots and live on them. It is advisable to pile the splits loosely, water the piles and cover them with sacks for a few days, until the new roots begin to appear. Planting is normally done in rows, the distance between the rows depending on plant size and local custom; the distance between plants in rows is of less importance except in experimental plots. Planting is done by hand or by special machines as those for planting sugar cane or adapted tobacco planters. Elephant grass can be and is normally planted by stem cuttings; the stems should not be very young and cut to pieces, each having about three nodes. The stems are stuck into the soil so that two nodes are underground, and there are other methods of planting by stem cuttings.

Establishment from seed is usually more difficult in the tropics than under temperate conditions: seed is often not in a ready supply, they are mostly small, and drought can kill small weak seedlings although this can also happen under the drier temperate climates. A clean, not too fine seedbed as free from weeds as possible should be prepared; seed is then broadcast or drilled in in rows. Seed rates vary and can be recommended with some certainty only if the quality of seed, i.e. its purity and germination, is known. Seed quality can be expressed as percentages of PGS (pure germinating seed), PVS (pure variable seed) or PLS (pure live seed). It can generally be recommended not to plant seed immediately after harvesting, but after some 6 months of storage, but seed stored for

2 or more years should be re-examined for germinability and recommendations in this respect, if available, are given in the main text dealing with individual species. Rolling after sowing is recommended for nearly all species in order to establish a better contact between the seeds and the moist soil. Fertilizers are given before, during or after sowing; phosphorus, in the form of double or single superphosphate, is usually given before or during the sowing. Nitrogen is given in its nitrate form, ammonium form or as urea. The time of N application is a controversial issue and some authors recommend to apply it at about the sowing time whereas according to Birch & Friend (1956) available N accumulates in the soil during the dry season and is released at the onset of rains, and therefore may be given later in the season when its natural source in the soil has been exhausted. Nitrogen is often applied after each grazing or cutting. If weeds are a problem, and they usually are, the early management may include mowing before the weeds go to flower, at a time when grass plants are still weak, and this may reduce grass-weed competition and result in a better establishment; if the weeds include palatable species early grazing can replace or supplement mowing.

Management

The newly established grass can be first utilized from 2 to 8 months after sowing. For sown grasses rotational or periodical grazing is usually preferred but some species may not respond to rotational grazing and continuous grazing is not always inferior to rotational grazing. Sown grass can be grazed at various intervals between the grazings, i.e. at various stages of growth, and early grazing, before earing, would result in smaller amounts but of better quality herbage than grazing at later stages of growth. The same general consideration is valid when grass is cut for hay, silage or soilage: the early-cut material provides good quality fodder but results in lower yields of herbage. Grasses, especially only lightly fertilized or not fertilized at all, are usually most productive in the second year of growth or, sometimes, in the first year; their productivity decline in the years that follow and on the fourth or fifth year the productivity can fall to a level observed in unimproved local grasslands. Stoloniferous grasses loose their productivity sooner than tufted grasses but if the grass is well fertilized it may remain productive for a longer period of time and stoloniferous and rhizomatous species of old stands can then be as productive as younger stands and live as long as tufted species or even longer.

Association with legumes

Grasses are often grown in mixtures with pasture and fodder legumes. The significance of legumes in mixed swards is that they usually, though not always, fix atmospheric nitrogen, contain more protein than the grasses and thus improve the quality of herbage, especially at the later stages of seasonal growth. Although leguminous plants can fix nitrogen, only seldom do they increase the yields of companion grasses, but often decrease them although total yields of mixed herbage usually increases. In most cases tropical legumes fix nitrogen to a lesser extent than temperate legumes and their beneficial effect is smaller. Nevertheless even small amounts of fixed atmospheric nitrogen can help, especially bearing in mind the high price of nitrogenous fertilizers and relatively low prices of animal production in most of the tropical countries. Legumes used in mixtures with each particular grass are mentioned in the main text and leguminous plants are dealt with in the **Legumes** section.

Herbage yields

Yields of tropical grasses vary enormously and under optimum growth conditions can be very high; Cooper (1970) quotes 85.2 t DM/ha/year obtained from elephant grass well supplied with N and other nutrients, and receiving adequate amounts of water; this is perhaps the highest recorded yield for a fodder grass. Maize can yield more than elephant grass in a single cut but not in terms of the whole year production. On the low side, yields of DM can be well below 1 t/ha in arid and semi-arid areas, especially for natural wild grasses. Under normal circumstances average yields of DM in well fertilized experimental plots can be expected to fluctuate between 20 and 40 t/ha for high-yielding grasses, 10–25 t for medium-yielding grasses and 3–10 t for poor yielders. Yields which are given in the main text for individual species also vary within wide limits depending on climate, weather conditions of the year, water supply, soil fertility, the fertilizers applied and the management, and even under seemingly similar conditions yields of the same species may differ in different countries and in different parts of the same country, and when interpreting experimental data the farmer has to use his own judgement and experience in estimating the expected yields. It should also be borne in mind that under farm conditions herbage yields can be lower than under experiments, especially if the data have been obtained from small-size plots.

Conservation

Grass surplus in the wet season has to be conserved for use in the dry season when grazing is not available or is in short supply. Hay can be prepared in the same way as in temperate countries and the grass is preferably cut at the early flowering stage. At this stage the rains may still continue, making hay curing and drying difficult; the same problem is however encountered in temperate areas. Haymaking can be particularly important in semi-arid and arid areas although various local conditions, including stony or uneven terrain, may not permit it. Grass conservation as 'standing hay', as it is often referred to, is a doubtful way of preserving fodder and should only be done when there are no other ways of conservation. Grass stands left uneaten as a reserve fodder for the dry season consist of dry overmature plants which have seeded, lost nearly all usable nutrients and are of very low feeding value. Good quality standing hay can be obtained only in the years in which the rains stop unexpectedly when the grasses are still in flower; perennial species wilt and dry and the annuals die out before they seed and may produce standing hay of reasonable or even good quality.

Tropical grasses can be ensiled and various types of sorghum and maize produce as good silage as can be obtained in temperate areas. A number of other tropical grasses can also be conserved as silage but ensiling is usually not easy and often results in a product differing from that made of temperate grasses. Catchpoole & Henzell (1971) state in their review of silage making in the tropics that the more difficult aspect of silage preparation from tropical grasses is the exclusion or reduction of aeration which may result in considerable losses of DM and an undesirable type of fermentation. Tropical grasses cut for silage are difficult to compress, possibly because they are more fibrous and more springy than the best temperate grasses of humid areas, such as ryegrass (*Lolium perenne*), and a considerable pressure is required to achieve a satisfactory weight/m^3 of silage. Another difficulty is in directing the fermentation to the lactic-acid route, possibly because of insufficiently high acidity of ensiled material and the pH of silage prepared from tropical grasses and legumes is usually higher than 4.2, regarded as the maximum pH for standard silage in temperate areas. Fermentation is often of the acetic-acid type and volatile acids, including buyric acid, are usually present in larger quantities than is permissible for standard silage as known in temperate countries, and ammonium N is often present in a larger proportion of total N of silage than is desirable. Nevertheless moderate or even good quality silages, with relatively small losses of DM and of N, have been obtained from tropical grasses, legumes and their mixtures, and these are mentioned in the main text on individual species. Grass herbage can also be conserved as haylage which differs from silage by a considerably lower content of water. The herbage, wilted to a water content of about 50 per cent, is stored more or less similarly to silage. This type of conservation has however hardly been reported from the tropics.

Nutritive value

The nutritive value of herbage is estimated by analysing it for the contents of ash, crude protein (CP), crude fibre (CF), ether extractives (EE), nitrogen-free extractives (NFE), and also for the contents of phosphorus (P), calcium (Ca), sometimes also potassium (K), magnesium (Mg), and some microelements, such as Mo, B, Mn, Zn and Cu, and carotene.

Ash constitutes about 8 to 12 per cent of DM; its value is mainly in the contents of mineral P, Ca or K, and grass ash contains large amounts of silica (Si), about 50 per cent of its total weight. Crude protein is a conventional term for all nitrogenous substances: mineral N, amide N, amino acids and proteins, and its determination is based on that of total N; the content of total N is multiplied by a 6.25 factor, e.g. 2.00 per cent N in the herbage corresponds to 12.5 per cent CP; different factors are occasionally used and this is then indicated in the text. Crude protein content usually ranges from 3 to 20 per cent, or even more in very young plants. Its content decreases as the growth of grass progresses and in tropical grasses CP content decreases faster than in temperate species; under water stress it again decreases faster than under the more humid environments. Crude fibre is in fact cell walls consisting of cellulose and lignin; its content ranges from 22 to 25 per cent in young plants and from 30 to 40 per cent in mature plants and is particularly high in tough fibrous grasses; CF content increases with plant age and also depends, to a certain degree, on the temperature under which the grass grows; the higher the temperature the higher is usually the content of CF, and it is normally higher in tropical and subtropical grasses, in which CF content increases with plant age more rapidly than in temperate grasses. Ether extractives contain fats (oil) and etherial oils; their content is usually in the order of 1–3 per cent and higher content, sometimes up to 5–6 per cent, usually indicates the presence of etherial oils. Nitrogen-free extractives contain soluble or near soluble carbohydrates: sugar, starch, etc., and their content ranges from 35 to 55 per cent and remains relatively constant or slightly increases with plant age. Here and throughout this book nutrient contents are expressed as per cent of dry matter; if the percentage is based on fresh fodder or hay this is always mentioned in the text.

In the animal body, CP is utilized for the growth and the replacement of some old cells and tissues, and for the formation of milk and is of particular value for young growing animals and for lactating cows and ewes. In these capacities CP cannot be substituted by any other of the named nutrients. The other nutrients, namely CF, EE and NFE (plus excessive amounts of CP if present) are utilized for walking, mastication and other animal activities, the maintenance of body temperature through the release of energy during the process of respiration, and also for the formation of fat.

Only a part of each chemical component of herbage can be utilized by the animal, the part that is digested. The degree of digestibility of plants and their parts can be established *in vivo*, in the living animal, by determining the contents of nutrients in the consumed herbage and in the feaces (and this is known as 'apparent digestibility'), or *in vitro* by imitating the digestion under laboratory conditions. The *in vivo* digestibility determination has an advantage of obtaining direct results applicable to the animals used in the trial but it has also some disadvantages. *In vivo* trials require considerable work and time, at least a few animals to experiment with, a considerable amount of herbage and are expensive. Moreover, the digestibility of herbage is influenced by the type and breed of the animal, its age, health, the level of herbage intake, and '*in vivo* digestibility is not a constant characteristic of a herbage' (Tilley & Terry, 1963). *In vitro* determination of digestibility is a much more rapid and much less expensive procedure; it requires only small samples of herbage and can therefore be used in grass breeding work, and under the present-day technique it gives reasonably accurate results. A simple method developed by Tilley & Terry (1963) is much in use. It involves incubation of milled herbage with rumen liquid of ruminants and with acid pepsin. At the first stage, which lasts 48 hours, the digestible nutrients, except CP, are completely digested in the rumen liquid and during the next 48 hours the digestion is completed in acid pepsin. The non-digested material is then weighed and related to the weight of the original sample. Total digestible DM is thus determined and expressed as TDN (total digestible nutrients).

In vitro technique can normally be used for determining the digestibility of total DM and digestibilities of single nutrients has to be determined by direct digestibility trials with the animals. Their digestibilities vary, the variation, apart from the animal factors, depends mainly on plant species, age and management. Digestibility of CP ranges from nil to 80 per cent but rarely exceeds 60 per cent; CF digestibility is relatively high and in the majority of cases ranges from 50 to 70 per cent; digestibility of EE ranges from 20 to about 60 per cent and of NFE from 40 to 70 per cent.

Crude protein digestibility is connected with its content in the forage: the higher its CP content the higher is its digestibility. Numerous digestibility trials have lead to the establishment of numerical relationships, a number of equations have been suggested for expressing it in various forages and grass species and a typical equation given by McDonald *et al.* (1973) reads:

$$\%DCP = (\%CP \times 0.9115) - 3.67$$

Osbourn *et al.* (1971) suggest a slightly different equation,

$$DCP = 0.960 CP - 4.210$$

for temperate grasses they studied. At 10 per cent CP the two equations give almost identical contents of DCP, 5.4 per cent DCP, but at CP content of 20 per cent the first equation gives 14.5 per cent DCP and the second equation 15.0 per cent. For quick approximate estimates it can be assumed that per cent DCP = per cent CP − 4 (or perhaps 4.5), which can usually be accurate enough for use in practice, especially bearing in mind a number of other variants affecting digestibility, such as sampling, leaf : stem ratio, etc. These estimates imply that, e.g., a grass with 14 per cent CP contains 10 per cent DCP, with 8 per cent CP – 4 per cent CP, and when the content of CP decreases below 4 per cent the DCP value would be negative and for normal functioning the animal body has to utilize proteins or other nitrogenous substances which have been previously accumulated. Regarding the intricate nature of relationship between the content of CP and its digestibility, recent opinions support an idea that in the digestive tract of the animal CP is digested to almost 100 per cent (or rather 96 per cent), the digested CP enters into metabolic processes in the animal body and a certain, constant amount of N returns back to the digestive tract and is excreted with faeces.

Total digestible nutrients (TDN) are the sources of energy of the animal body. The term TDN, widely used in the USA and in other countries, is now less popular amongst British nutritionists, and the expression digestible energy (DE) is normally used, its value given in calories. The energy value of 1 g of digestible NFE or digestible CF is 4.15 kcal (kilocalories) of digestible EE (fats) 9.40 kcal and of digestible CP 5.65 kcal. Digestible crude protein is however usually utilized to its full extent for body building and/or milk production and should be excluded as a source of energy except when the herbage is rich in protein which cannot be fully utilized as a source of metabolic N; the excess can then serve as a source of energy. Such use of nitrogen is however uneconomical because a pasture grass containing large amounts of N yields less herbage than more mature grass containing less nitrogen. Of real value to the animal is the metabolizable energy, denoted in equations as ME or M, which is the heat of combustion of the feed minus the heat of combustion of faeces, urine and gases produced by digestion. The metabolized energy of forage is expressed in megacalories (Mcal)/kg DM and usually ranges from 1.5 to 2.5 Mcal. In the main bulk of literature concerning animal requirements and the evaluation of herbage in the tropics, the TDN system of expressing the energy of grass and other forages has so far been predominantly applied, whereas the calory system has only relatively recently begun to appear in more advanced publications. In this book the experimental data is therefore expressed predominantly in the TDN system.

Starch equivalent (SE) is another unit expressing the incoming energy which can be denoted as the number of kilograms of starch that would produce in the adult steer the same amount of fat as 100 kg of the feed stuff under consideration. There is still another indicator used for forage evaluation: feed unit or fodder unit (FU) which produces the same nutritive effect as 1 kg of barley grain. This unit, first introduced in Scandinavian countries, is now in use in some other areas, e.g. in the French-speaking tropics.

Animals of different classes and at different stages of their life and activity require different contents of digestible nutrients in the grasses they consume, and young growing animals need herbage of higher nutritive value than the adult animals. Examples of these requirements are given below, mainly from Crampton & Harris (1969). These standards are intended mainly for temperate conditions and can be somewhat different in the tropics.

(a) For maintaining adult cattle at a constant weight without obtaining any production, the herbage should contain about 4.0–4.2 per cent DCP, 50 per cent TDN, 0.12–0.20 per cent Ca and 0.12–0.15 per cent P.

(b) For the normal growth of heifers and steers weighing 180–450 kg, DCP content should be from 4.7 to 7.0 per cent, depending on the animal age and size; TDN content 50–55 per cent, Ca 0.15–0.30 per cent and P 0.15–0.20 per cent.

(c) Finishing yearling cattle need 7.5 per cent DCP, 65 per cent TDN, 0.20–0.25 per cent Ca, and about 0.20 per cent P.

(d) Lactating cows require a minimum of 60 per cent TDN, 0.30 per cent Ca and 0.25 per cent P; DCP requirements are: 4 per cent for maintenance plus extra 40–60 g/kg milk produced by cow; 0.30–0.45 TDN/kg milk are also required.

(e) Non-lactating ewes require during the first 15 weeks of gestation not less than 4.2 per cent DCP, 50 per cent TDN, 0.20–0.30 per cent Ca and 0.16–0.21 per cent P.

(f) Lactating ewes need 4.2–4.8 per cent DCP, 52–58 per cent TDN, 0.24–0.30 per cent Ca and 0.18–0.22 per cent P.

(g) Fattening lambs need 5.2–5.8 per cent DCP, 55–62 per cent TDN, 0.18–0.23 per cent Ca and 0.16–0.21 per cent P.

The data given for individual plant species in the main text would show that for most tropical grasses the contents of TDN and of Ca are adequate for animal requirements but the contents of DCP and of P are adequate only in young or very young grass and that for normal growth and production supplementary feeding or growing grass/legume mixtures are needed.

Animal production depends of course not only on the quality of herbage, i.e. its chemical composition and digestibility, but also on the amounts of consumed herbage. Even if grass as fodder or grazing is freely available, the animals are not necessarily able to consume sufficient amounts of it to satisfy their requirements for digestible nutrients. The amounts consumed depend on the animal appetite, the volume of their stomachs, and also on grass species or cultivar, its age and management and is expressed in grammes of dry matter/kg animal bodyweight/day, normally abbreviated as g/kg W if the weight of the whole body is considered, or as g/kg $W^{0.75}$ or g/kg $W^{0.73}$ when only the metabolic part of the body, accepted as 0.75 or 0.73 of total body weight, is taken into account. The intake of DM usually ranges from 20 to 80 g/kg $W^{0.73}$ for different grasses or for the same grass at different stages of growth or differently managed. The intake is also expressed in kg DM/100 kg W ($W^{0.75}$ or $W^{0.73}$) and then it ranges from 2 to 8.

Reproduction

The structure of floral parts

The grass flower or **floret** (Fig. 4) has two protective scales, the lower – **lemma** (Fig. 4c) – and the upper – **palea** (Fig. 4d). They can be of various shape and texture, from soft membranaceous or herbaceous to hard and crustaceous. In some tribes and genera the lemma has one, but sometimes two to a few, awns (Fig. 4ij), straight or kneed and then usually spirally twisted below the knee (Fig. 4j). The palea (Fig. 4d) has two keels and resembles the prophyll. Between the two scales there are usually three stamens (Fig. 4g), a pistil which consists of an ovary (Fig. 4e) with two, often feathery stigmas (Fig. 4f), free or fused, and two lodicules (Fig. 4h). The lodicules can swell and force apart lemma and palea during the flowering. The ovary contains one ovule the walls of which, the pericarp, is fused with those of the ovary so that the seed cannot be detached from the fruit (**caryopsis** or **grain**) except in species of *Eleusine* and some other genera in which the pericarp can be free. In some grasses the caryopsis is firmly clasped between the lemma and the palea and these are sown as one unit or 'seed'. The florets which have both ovary and stamens are bisexual; those containing only an ovary are female or pistillate florets, those containing only the stamens are male or staminate florets and there can also be florets containing neither the ovary nor the stamens (**empty**, **barren** or **neutral** florets). Both bisexual and female florets can develop a caryopsis and are **fertile** florets.

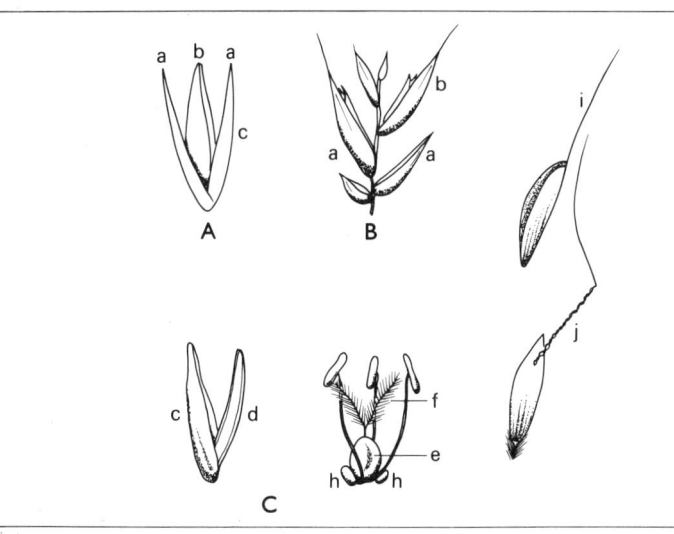

Fig. 4 Spikelets and florets: (A) spikelet with one floret and (B) with several florets; (a) glumes, (b) floret, (C) floret, (c) lemma, (d) palea, (e) ovary, (f) stigmas, (g) stamen, (h) lodicules, (i) and (j) awns.

The florets form a primary inflorescence, the **spikelet** (Fig. 4b), a kind of modified shoot on the axis of which two glumes and one to several florets are arranged. The spikelets are grouped to form more complex inflorescences: the **spikes (ears)** (Fig. 11, 31), in which the spikelets are sessile or nearly sessile on a long axis, the **racemes**, if the spikelets are on clear and not very short pedicels, and **panicles** (Fig. 23, 24) if the inflorescence is branched. Branches of the panicle form a **whorl** (Fig. 14) if they arise from about the same level round the panicle axis and a panicle with a single whorl of spikes on the top of the stem is a **digitate panicle** (Fig. 12, 19). If the branches of the panicle are supported by leaves (often modified) these leaves are known as **spathes** and those supporting ultimate divisions of the panicle as **spatheoles** and the panicle as **spathate panicle**, a common feature in a number of species of the Andropogoneae tribe, e.g. those of *Hyparrhenia*. The branches arising from the spathe axils are termed **rays**.

Flowering and pollination

Flowering, or anthesis, usually begins with the opening of the floral glumes (lemma and palea) helped by swelling of the lodicules (Fig. 4h). Filaments of the stamens elongate and the anthers exert and hang outside the spikelet and dehisc, i.e. they split and release the pollen. Stigmas either bend and protrude sideways from the floret or appear from its top. After the release of pollen the lemma and palea close up again. The process of flowering takes different times to complete, from a few minutes to a few hours, depending on the species and the weather. Stigmas and anthers usually appear more or less simultaneously or they emerge at different times and in species of *Pennisetum* the stigmas appear usually much earlier, often a day or two earlier, than the anthers. Flowering takes place at different times of the day, depending mainly on the species but also on the weather; in the majority of tropical grasses it occurs in the morning, usually beginning at dawn. Grasses are essentially anemophylous plants, i.e. their pollen is transferred from floret to floret and from plant to plant by wind, but insects, mainly pollen-collecting bees (Fig. 5), also transfer the pollen, perhaps more so in tropical than in temperate grasses.

Fertilization

Each grass has in its cell a certain specific number of chromosomes – cell parts carrying genes. The plant of a new generation is initiated by the fusion of male and female sexual cells (**gametes**) but the chromosomes do not fuse and their number is doubled by the fusion in the process of fertilization. However, at a certain stage of plant development the number of chromosomes is reduced by half and cell fusion only restores the number of chromosomes. The reduction takes place during the formation of the embryo sac in the ovule and the formation of pollen in the anthers. The reduced chromosome number is termed haploid (or n) and the full number, after the fusion of gametes, in the embryo sac and in the growing plant diploid or $2n$. In some plants the gametes can have an unreduced number of chromosomes and the plants of the next generation produced sexually would have a tetraploid number of chromosomes; this increased number of chromosomes can be retained in the following generations, e.g. two races of *Chloris gayana* have been found: diploid with $2n$ chromosomes = 20 and tetraploid with $2n$ = 40. There can also be hexaploids and octoploids.

Fig. 5 Honey bee collecting grass pollen.

The basic chromosome number characteristics for each genus or species is denoted as x.

Similarly with other Angiospermae, two male gametes penetrate into the embryo sac with the pollen tube, one of them fuses with an embryo-sac cell (female gamete) which has a reduced number of chromosomes and after the fusion develops into the embryo and later into an adult plant. The other male gamete fuses with an unreduced embryo-sac cell and initiates the endosperm, a body which has $3n$ chromosome number in its cells. The endosperm grows side by side with the embryo and takes part in providing the embryo with nutrients and supplies and other nutrients to young seedlings during seed germination.

Apomixis

In many tropical grasses the number of chromosomes in the embryo-sac cells is not reduced and the embryo can develop and actually develops from an unfertilized cell. This development is known as apomixis or nonsexual reproduction and the plants with apomictic development are apomicts. In the majority of apomicts, pollination is however necessary for the initiation of the endosperm without which seeds cannot fully develop. This phenomenon is termed pseudogamie; it was observed in *Cenchrus ciliaris* and *Panicum maximum* in the 1950s (Warmke, 1954; Snyder *et al.*, 1955) and later in a number of other grasses. Most of the apomicts can, to a limited extent, also form sexual seed (facultative apomicts) by normal fusion of male and female gametes. This occurs very seldom in some species (*C. ciliaris*) and more or less regularly in the others (*P. maximum*). Apomixis has been observed in a large number of species of Paniceae and Andropogoneae, tribes predominantly tropical and apomixis is common in cultivated tropical grasses.

Apomictic grasses breed true to type and retain their characteristics in seed progenies which can be genetically compared with the progenies obtained by vegetative propagation. Amongst the apomicts sexual reproduction occurs usually in diploids rather than in tetraploids or in plants of higher ploidy; some plants with irregular numbers of chromosomes can be propagated by apomictic seed, whereas non-apomictic, sexual species with irregular numbers of chromosomes would be sterile.

The apomicts can show some variability, often considerable, when their occasional sexual plants give sexual progenies and some of the newly shown useful characters can then be fixed in another apomictic generation. Apomicts are valuable plants and Harlan & de Wet (1963) wrote that 'apomictic groups are frequently more variable than sexual ones, more difficult to treat taxonomically, and are often among the most "successful" and adaptable of all plant taxa' and that the apomictic grasses can form 'an efficient genetic system which can cross widely and escape the penalties of sterility'.

Of the more important tropical cultivated grasses, *Panicum maximum*, species of *Brachiaria* and of *Paspalum*, *Cenchrus ciliaris*, *Pennisetum clandestinum*, *Melinis minutiflora* and some forms of *Cynodon* are apomicts, whereas *Setaria*, *Chloris gayana*, *Sorghum* and diploid forms of *Cynodon* are sexually reproduced plants.

To find out whether a grass is an apomict or a cross-fertilized plant, studies on seed formation and progeny behaviour can be useful. Seed setting in isolated plants, or plants with their inflorescences enclosed in paper bags is compared with that of plants grown without any isolation from other plants of the same species. If seed formation fails or is poor in isolated plants and good in non-isolated plants the species can be considered to be cross-fertilized, but if no difference in seed formation has been found the species can be either an apomict or of a self-fertilized type. Variability of progenies of single plants indicates cross-fertilization and their uniformity suggests self-pollination or apomixis. A microscopic cytological investigation differentiates self-pollination and apomixis: unreduced, $2n$ cells of the embryo sac indicate apomixis.

In a number of tropical grasses the apomixic nature of the species has been established by this method only. However, applied alone, the cytological method can sometimes lead to wrong conclusions and testing seed formation and progeny behaviour can help to ascertain the actual breeding behaviour of the species.

Seed

In a botanical sense, seed is a body developed from the ovule. In the grasses 'botanical' seed cannot be separated from the fruit which is a body developed from the ovary. The grass ovary contains only one ovule and the walls of the ovary or fruit (pericarp) and the ovule fuse at early stages of development, and, when ripe, they form a combined body termed the caryopsis or the grain; wheat grain can serve as a model and the caryopses of other grasses can only differ in size and shape. The caryopsis consists of the skin, formed by seed and fruit covers, the embryo and the endosperm. The embryo is the future plant with minute embryonic bud and root. It can be clearly seen at the basal part of caryopsis and, as a rule, is larger (in relation to the caryopsis size) in tropical than in temperate grasses. The endosperm is a kind of a second embryo but it is not differentiated and serves as a source of nutrients for the seedling when seed germinates; it consists mainly of starch but also contains some proteins and free amino acids.

In an agricultural sense, seed is a part of the plant which, when sown, produces a new plant. In some grasses (e.g. *Eragrostis curvula*) seed is in the form of a naked caryopsis. In other grasses (*Chloris gayana, Melinis minutiflora*) in the form of a spikelet containing one to a few caryopses which usually remain enclosed in the spikelet at harvest or cleaning but can be relatively easily separated. In some grasses of the Paniceae tribe (species of *Panicum, Paspalum, Brachiaria, Echinochloa*), the caryopsis is firmly enclosed in hard, often crustaceous lemma and palea and can be separated only with a considerable effort. In most species of the Andropogoneae tribe the hard scales which enclose the floret and the caryopsis are the spikelet scales or glumes. Also in the Andropogoneae the ultimate racemes of the inflorescence often break into joints with two spikelets each, the pair of spikelets usually containing one caryopsis and functions as seed. Seed as harvested, particularly in species of the Andropogoneae tribe (*Hyparrhenia, Andropogon, Bothriochloa*), contains a considerable amount of chaff, much exceeding in weight and volume the amount of true seed. The chaff includes awns, often spirally twisted and stuck together, joints with empty spikelets, and other floral parts. Separating clean seed from chaff is difficult but in the last years seed cleaners and traders have made considerable progress in the technique of seed cleaning.

The size of seed varies and one kg of seed can contain from 75,000 to

100,000 seeds of fodder sorghum to 40 million seeds of *Eragrostis caespitosa*. Of the grasses with intermediate seed size Sudan grass has about 200,000 seeds/kg, *Panicum maximum* 700,000–1,500,000 seeds, depending on the variety, and *Chloris gayana* 2 million of spikelets each with one caryopsis. In temperate grasses the number of seed per kg ranges from 500,000 in *Lolium perenne* to 11 million in *Agrostis alba*.

Seed germination

Before germination begins the seed has to absorb water and swell, the process known as imbibition. Then the sheath of the rootlet breaks through the seed cover and through the lemma or other scales in which the seed is enclosed. This sheath develops numerous fine hairs which hasten water absorption. Then the radicle or the primary root or roots break through the sheath. On the second or third day of germination, or later, appears a coleoptile, the first leaf of a cylindrical shape which later splits and true leaves excert. Secondary roots then appear from the first nodes of the seedling stem, which is very short and can hardly be seen. Germination takes place only under certain temperatures, the minimum temperature for tropical grass germination varies usually from 15° to 20°C, and the optimum temperature from 25° to 35°C, and is considerably higher than the optimum temperature for temperate grass germination. In the majority of trials, alternating temperatures result in faster germination than constant temperature. Some tropical grass seeds germinate better under light and some in darkness, but it seems that the majority of species have no preference. It was also observed that under marginal temperatures seeds germinate better under light than in darkness, even if there was no preference under optimum temperatures. Under optimum conditions seed of some tropical grasses can germinate and the seedlings emerge very fast, the seedlings often emerging on the third–fourth day after sowing; other grasses, e.g. *Panicum maximum*, germinate slowly. Seed germination in some species of the Paniceae tribe, in which the caryopsis is firmly embraced by hard lemma and palea, can be considerably delayed, often because of poor water absorption, and scarification, mechanical or with sulphuric acid, can improve germination. In some grasses floral scales attached to the caryopsis can contain germination inhibitors which can usually be washed off by rain water after which seed germinates. Sowing naked caryopses has been tried with inconsistent results. Under laboratory conditions naked caryopses often show better germination than those of intact spikelets or florets but the position can be reversed under field conditions, when seed germinates in the soil as e.g. is the case with *Cenchrus ciliaris*. In most tropical grasses post-harvest seed maturation is needed and in the first months after harvesting the germination can be poor. The best germination is usually observed in 6 to 12-month-old seeds and then it declines, first slowly, then faster and in 5–6 years it drops to 5–20 per cent and often to nil in 7–8 years time; seeds of some species lose their germinability still earlier. Storage conditions, and especially moisture content of the seed, have a strong effect on the retention of germinability and seeds which contain less than 10 per cent water and are stored in sealed containers can remain viable for a long time. Low temperature can also extend seed viability but to a lesser extent than the low water content. KNO_3 solution often improves germination but is usually applied in seed germination tests only.

Seed production

In the tropics, the production of seed of perennial grasses meets with considerable difficulties. Seed is often poorly formed and a large proportion of spikelets can be empty; ripening seeds shed easily and early, often before maturation; birds eat well developed seeds; the spikelets can be infested with smut or bunt. Moreover, fertile shoots appear gradually, and while some shoots have ripe seed the others only begin to appear and flower. All these factors can reduce potentially high yields to a fraction of their potential. Seed yields of temperate *Phleum pratense* grown in Europe and of *Setaria anceps* grown in Kenya, species with a similar type of growth and inflorescence, were compared (Boonman, 1973). A good crop of *S. anceps* gives 160 kg/ha of 'clean seed' and 50 kg of pure germinating seed (PGS), whereas a good crop of *P. pratense* 800 and 700 kg, respectively. In a variety of *S. anceps* selected for good seed production, seed yields almost trebled and in a hypothetical variety with ideal seed production PGS yields can reach 750 kg/ha, about equalling the yields of *P. pratense*.

Seed is often produced from grazed grass, leaving the stands ungrazed, usually in the second half of the season, every year or in some years only. Seed yields under this system are normally low, especially if grazing is discontinued rather late in the season. Moreover, seed can seldom be harvested from mixed grass/legume crops and grass grown exclusively or mainly for seed production gives higher yields and better quality seed. It has been almost universally proved that fertilizer N considerably increases seed yields. Boonman (1971a) maintains that 100 per cent yield increases are usually obtained in Kenya from the application of 100 kg N/ha over a usual rate of 60 kg and N applied immediately after the onset of rains gives the highest yield increases. Strickland (1970) in his review also gives several examples of considerable seed yield increases from fertilizer N. Spacing of tufted grasses is important and in Kenya narrowing the distance between the rows from the usually accepted 90 cm to 30–50 cm resulted in 30 per cent yield increases. Correct timing of harvesting is another important factor and according to Boonman (1972b) the highest yields are likely to be obtained if seed is harvested when 10–30 per cent of spikelets have shed.

Harvesting is done mechanically or by hand and there are different practices accepted in different countries. The grass can be cut low, either by hand or with a reaper binder, the sheaves are stooked and it is advisable to cover the stook with a turned-up sheaf as a protection against bird damage. The sheaves are then threshed in the fields, on tarpaulins, one or two weeks later. Under other types of mechanical harvesting the grass is cut or stripped high, and practically only the flowering heads are harvested. Harvesting can be done by vehicles fitted with combs which collect easily detachable floral parts of the grass. Sucking vacuum machines have also been used for collecting seed from the ground. Direct combining is at present not much in favour because large amounts of green material is threshed together with seed and considerable losses of seed can occur. After harvesting, seed is dried and, after a preliminary cleaning on the farm, cleaned at special plants in countries where a grass seed industry has been well advanced, e.g. in Australia, USA or Kenya.

Seed yields of some grasses (species of *Brachiaria*, *Hyparrhenia*) can be low, in the order of 30–60 kg/ha, but in some other species, such as *Chloris gayana*, *Panicum maximum*, *Setaria anceps*, *Paspalum dilatatum*, seed yields are considerably higher and can sometimes reach 200–250 kg/ha per harvest. The yields vary, however, and even crops of the same species can give ten-fold higher or lower yields, depending on the variety, but more on the growth conditions. Yields of viable seeds, which actually matter, can only seldom be assessed from published data and even 'clean seed' yields, that are often reported, may contain a high proportion of empty spikelets, which in some species, e.g. *Chloris gayana*, are difficult to separate from those containing a caryopsis. Yields of PGS are usually in the order of 10–50 kg/ha and in experimental plots in Kenya average yields of PGS for seven harvests ranged from 15 to 32 kg for *Setaria anceps*, depending on the variety, 24–44 kg for *Chloris gayana*, 25 kg for a well-seeding variety of *Panicum maximum* and 23 kg for *Brachiaria ruziziensis*, or twice these amounts per year in two harvests (Boonman, 1972b). No wonder that tropical grass seed is expensive and that the cost of seed constitutes a large proportion of the total cost of establishment. Seeds of annual tropical and subtropical grasses are easier to produce and to hande.

In some countries seed certification schemes are in operation. To be certified, a seed crop should be grown from seed of an approved source, of known variety or cultivar, in clean fields not contaminated with other varieties of the same species and, in the case of a cross-fertilized grass, isolated by a distance, usually of 100–200 metres, from other varieties.

Introduction and breeding

The majority of useful tropical pasture and fodder grasses originated in Africa, a smaller number from tropical and subtropical South and

Central America, a few from India and southern Asia and practically none from Australia. Early development of pasture grasses can perhaps be linked with the large number of grazing ungulates numerous in Africa, less numerous in America and Asia and absent in Australia. In places of their origin, during a number of centuries, the grasses, which are now under cultivation, have developed various forms adapted to specific conditions and the number of such species is particularly large in East Africa. This concentration of valuable grasses with their innumerable varieties and forms has been noted by Hartley & Williams (1956), who, following Vavilov's concept of centres of concentration of cultivated plants and their varieties, suggested that a centre of grass variability, second in importance to the Mediterranean centre with its numerous cultivated temperate grasses, exists in East Africa. The third centre, suggested by Hartley and Williams, is in the area of southern Brazil–Uruguay–northern Argentine where numerous species of *Paspalum* occur. Still another centre can perhaps be suggested in central America–southern Mexico for *Euchlaena mexicana*, species of *Tripsacum* and for maize which is much used for fodder in the tropics and elsewhere.

From these centres of distribution and variability fodder and pasture plants were spread to other tropical, subtropical and even warm temperate regions introduced for testing and use in cultivation, or, at least at the early stages of introduction, often also accidentally. Some grasses were introduced more than a century ago, their origin was little known and certain species were in the past considered to be indigenous in countries of their adoption. So were e.g. *Melinis minutiflora*, *Hyparrhenia rufa* or *Panicum maximum* considered in America, both southern and northern, where they were introduced a long time ago, fully naturalized and even penetrated into semi-natural grasslands. *Melinis minutiflora* was even first described from American (Brazilian) specimens. It was only relatively recently that the origin of these species was traced (Parsons, 1972). Only relatively scanty material penetrated in the past into the countries of adoption and although it has been more recently supplemented by numerous fresh introductions the variability of these species still remains much greater in the areas of their origin. My studies of variability in *Melinis minutiflora* has shown that the types cultivated in distant countries are remarkably uniform, whereas the material obtained from Africa, and especially in East Africa, where this grass is indigenous, showed a considerable variability. The most profitable source of material for the selection of superior varieties and for the improvement of the existing ones can be found in the areas and countries where the species originated and where the gene pool is much wider than in the areas where occasional material introduced in the past gave rise to locally developed forms. This is almost equally true for both cross-pollinating and apomictic species and varieties. This has been recognized for a number of years and numerous valuable forms of grasses and legumes already known in cultivation as well as of new, not yet tested species, were collected by expeditions and single breeders visiting the countries where the gene sets are the widest. A considerable number of seed and root samples were introduced from East Africa and Africa in general to America and Asia, and particularly rich material reached Australia as a result of work by J. F. Miles and those who followed his pioneering efforts. The exchange of samples between research stations and of various countries has also been developed.

When collecting seeds of cross-pollinating species large samples taken from single communities are desirable in order not to miss any important genes, whereas small samples of the apomicts are adequate but a large number of samples of apomictic grasses collected throughout the country is advantageous. The best known and well performing cultivars already grown in various countries should also be acquired and tested.

The objectives for breeding tropical grasses are essentially similar to those for temperate species: high herbage production, high seed yields and high herbage quality, seedling vigour and the ease of establishment, and resistance to diseases and parasites. High herbage production is perhaps most difficult to achieve and to evaluate because, apart from the potential of the grass itself, it depends on a number of environmental factors, in the first instance soil fertility, moisture regime and the pattern of utilization. The problem of extending herbage production into the dry season is common with the same problem for semi-arid areas outside the tropics. Breeding for frost and cold tolerance is hardly applicable to true tropics but is needed for subtropical areas and for high altitudes in the tropics.

Breeding for high seed production is particularly important for tropical grasses, except, of course, for those which can be easily and cheaply propagated vegetatively. One of the main obstacles for obtaining high seed yields in the tropics is that fertile shoots appear gradually during a long period of time and the flowering and seed ripening is extended for a much longer period than in temperate grasses, and breeding for obtaining a well expressed peak of flowering can be important. Other aspects of breeding for seed are: increasing the size of flowering heads, increasing the number of fertile shoots per plant or per unit area (although this may make the herbage more stemmy and decrease its quality), and resistance to seed diseases. The development of cultivars with good seed retention, so that seed or spikelets do not shed easily, would be of great value, but there is only a small chance of finding mutants of this nature. This chance can perhaps be increased by radioactive irradiation.

The technique of breeding cross-fertilized tropical grasses is essentially the same as applied to temperate species. Simple breeding based on single plant mass selection often implies an examination of hundreds or even thousands of single plants, selection of best performing ones, testing their combining ability, i.e. the ability to transfer their valuable characters for which they had been selected, to the progenies, and producing the pedigree, or breeder's seed from a mixture of plants with good combining ability. The new cultivar is then to be compared with the original population and with other cultivars. This all requires time and a considerable space. One of the main difficulties in breeding pasture grasses is that the selection is based on the performance of single spaced plants which may behave differently in close swards. For creeping grasses (*Chloris gayana, Cynodon nlemfuensis, Paspalum notatum* and some others) this problem does not exist because a single plant represents at the same time a piece of sward. To grow single plants of creeping grasses requires, however, more space than the non-creepers need.

Other methods of breeding include crossing plants with different useful characters in order to combine them in the same plants, and this includes interspecific crosses between elephant grass and pearl millet, maize and teosinte and between various types of sorghum some of which proved to be most useful.

Breeding apomictic grasses is a relatively new venture as grass breeders had been confronted with apomixis only when they began breeding cultivars of tropical species. Breeding apomictic grasses is simple and quick. A number of introductions are compared, the best ones selected and their seed can be immediately bulked for variety trials. The number of introductions for the original comparison should be, however, large. Another method is finding sexual, usually diploid plants and treating them as cross-pollinating species, or selecting the best plants and by doubling their chromosome numbers trying to revert the plants to apomixis thus fixing their superior qualities.

Application of strong agents such as colchicin or radioactive irradiation can hardly produce quick practical results, except perhaps in attempts at obtaining useful abnormalities, e.g. firm retention of seed on the panicles or spikes.

The majority of tropical grass varieties now in cultivation are natural, ready-made cultivars picked up in natural grasslands or found amongst the introduced types and selected for their outstanding performance. However, some bred cultivars exist and their number is rapidly increasing. *The International Code of Nomenclature of Cultivated Plants* (the second edition appeared in 1969) requires that the names of cultivars should follow certain rules of the Code and that they should be registered. Registration can be done by publication in agricultural or botanical periodicals as is the practice in the USA where they are normally registered and described in *Crop Science*. In Australia, which, together with India, is now a leading country in breeding tropical grasses and herbage legumes, a Herbage Plant Registration Authority was established by CSIRO (Commonwealth Scientific and Industrial Research Organization) in 1965. Origin, description and agricultural characteristics are given for 46 cultivars of tropical grasses and 21 cultivars of tropical legumes (apart from the cultivars of temperate and subtropical species) in the *Register of Australian Herbage Plant Cultivars* (Barnard, 1972). Brief descriptions of cultivars grown or developed in Kenya can be found in the *East African Agricultural and Forestry Journal* (Bogdan, 1965).

The More Important Species

Acroceras Stapf

About fifteen, mostly creeping species which grow in swampy situations of tropical Africa and of Indo-Malayan region.

Acroceras macrum Stapf. Nile grass $2n = 36$

Creeping perennial with stolons and rhizomes. Stems sparingly branched, slender, up to 70 cm high. Leaves 5-20 cm long and 1-8 mm wide. Panicle narrow, scanty, up to 20 cm long, with one to five erect or suberect branches or reduced to a single slender spike. Spikelets oblong, 4-5 mm long; the upper, fertile floret is 4-4.5 mm long, smooth, whitish.

Occurs in tropical East Africa from Ethiopia to South Africa, also in Angola and South West Africa in seasonally flooded grasslands, on swamp edges and stream banks in the areas with over 800 mm of annual rainfall. It forms dense cover, sometimes reaching some 60 cm in height and is used for grazing and haymaking. *Acroceras macrum* has also been used for the improvement of natural moist pastures by planting splits or rhizome cuttings either into the existing natural grass or into ploughed grassland after an arable crop. It was persistent although not very productive in South Africa and Kenya (Gosnell, 1963) and had also a reasonable success in Rhodesia, Swaziland and Surinam. Attempts to grow *A. macrum* on dry land were not encouraging because the established pastures were not sufficiently productive. When cut for hay it produces 5-8 t hay/ha (Whyte *et al.*, 1959), and comparable yields, about 7 t DM/ha, were obtained in a trial in Rhodesia (Rodel & Boultwood, 1971) in which *A. macrum* was one of the least yielding species and suffered from a severe drought with only 20 per cent recovery in the following season. CP content in the herbage was 8.7 per cent, CF content 30.7 per cent and NFE 44.7 per cent (Dougall & Bogdan, 1965); a high content of EE, almost 6 per cent, was noted as unusually high for a non-aromatic grass. The content of P was low, 0.11 per cent, and of Ca adequate, 0.45 per cent. Rodel & Boultwood gave, however a very high content of CP 22 per cent in the whole plant and Dirven (1962) reports 21.3 per cent CP in the leaves and 7.9 per cent in the stems and CP digestibility about equal for the leaves and the stems, 73 per cent and 74 per cent; CF content is given as 30.0 per cent in leaves and 38.5 per cent in stems.

Agrostis L.

Spikelets small, with one floret. Essentially a temperate genus with 150–200 species found mainly in Europe or Asia. Several species occur in the tropics but only at high altitudes. Some species can be numerous in montane grasslands but never dominate in the sward.

Agrostis producta Pilger $2n = 28$

Tufted perennial 30–80 cm high with stem bases coated with fibrous remains of old leaf sheaths. Leaves up to 25 cm long and 2 mm wide, flat or involute. Panicle 5–20 cm long. Spikelets 3.5–4.5 mm long. Floret 2–2.5 mm long; lemma pilose, terminating in five acute points; a geniculate awn 4–7 mm long arises from the back of the lemma near its base.

Occurs in East Africa, Sudan and Malawi, at 2,400–3,800 m alt, in grasslands. It is perhaps the most common species of *Agrostis* in East African mountains and highlands and can constitute a noticeable proportion of the sward.

Agrostis schimperana Steud. (*A. alba* L. var. *schimperana* (Steud.) Eng.)
$2n = 42$

Perennial forming small tufts which can also develop stolons. Stems slender, 30–120 cm high. Leaves flat, 6–15 cm long and 3–5 mm wide. Panicle linear to narrow oblong, 6–20 cm long, dense or loose. Spikelets 2–2.5 mm long; glumes as long as the spikelet. Floret 1.5 mm long, glabrous and awnless.

Occurs in African highlands and mountains, from 1,400 to 4,000 m alt. in Ethiopia, East Africa and southwards to Zambia, alongside streams or in wet places in highland grassland. The plants are slender and usually leafy, and are well grazed by stock.

Tried under cultivation in Kenya, at 1,800 m alt, in a slightly swampy stream valley, *A. schimperana* produced a good crop of palatable herbage; the competitive ability was, however, poor and in the third year of growth it was replaced by other grasses and sedges. An analysis at the early flowering stage has shown an outstanding quality of herbage which contained 17.9 per cent CP, 2.1 per cent EE, 28.2 per cent CF, 41.1 per cent NFE, 0.22 per cent P and 0.34 per cent Ca (Dougall & Bogdan, 1958). Seed was well produced and easy to harvest and to sow, but the seedlings were easily overgrown by weeds and needed a clean seedbed.

Alloteropsis C. Presl.

A small genus of 8–10 species spread in tropical and South Africa, tropical Asia, Indonesia and Australia. One of the easily distinguishable characters by which it differs from most genera of the Paniceae tribe is the presence of rather short, fine and straight awn on the lemma. *A. semialata* described below is widely spread and so is annual *A. cimicina* (L.) Stapf; the other species occur in limited areas. Some species have a faint scent of coumarin.

Alloteropsis semialata (R.Br.) Hitchc. Tongolonakanga (Madagascar)

$2n = 18, 54$

Tufted perennial 20–90 cm high with shoot bases coated with densely hairy scales. Leaves glabrous or hairy, 2–3 mm wide. Panicle digitate with two to six racemes. Spikelets in distant pairs, 3.5–6.5 mm long. Lower glume 2 mm long, shorter than the upper glume. Lower floret usually male, upper floret bisexual, fertile, as long as the spikelet, with long hairs on the lemma margins and with a short straight awn.

Occurs naturally in tropical Africa, Asia and Australia, in open grassland or in woodland and can be locally abundant. In Madagascar it is regarded as an accidentally introduced species now common in plateau grasslands (Cabanis, *et al.*, 1970). In Australia it is one of the main grasses in thin, lowland woodland, and can withstand a certain degree of waterlogging. In New Guinea *A. semialata* is considered to be a good grazing grass (Henty, 1969) and an analysis in Kenya showed a good herbage quality: cut at the early flowering stage, when the grass was leafy, the herbage contained 15.6 per cent CP, 1.1 per cent EE, only 14.7 per cent CF, 59.4 per cent carbohydrates (NFE), 0.14 per cent Ca and 0.30 per cent P (Dougall & Bogdan, 1958). It is an early grass and in Kenya fresh herbage is produced at the very beginning of seasonal rains when most of the other grasses are still dormant.

Andropogon L.

Perennials with non-aromatic leaves. Racemes of the panicle paired, digitate or very occasionally single, and in some species gathered into complex spathate inflorescences. Racemes articulate, with thin or swollen joints. Spikelets paired, one sessile the other on a pedicel which is rather similar to the joint. Sessile spikelets all alike in shape and bisexual or very rarely one at the bottom of the raceme can be male or neuter. Lower glume of the sessile spikelet with two keels. Florets two: the lower male or neutral, or reduced to a lemma; upper bisexual, with a glabrous or scaberulous, geniculate and spirally twisted awn. Pedicelled spikelet male or neutral and can be reduced to glumes.

A large and heterogenous pantropical genus. In the past it was more widely understood but later reduced to a more narrow genus when *Sorghum*, *Hyparrhenia* and a number of other taxa were excluded. Its species often occur in grasslands where they can be grazed but are mostly of medium quality and, with a few exceptions, are of not much significance for the animal industry.

Andropogon gayanus Kunth. Gamba grass; Sadabahar (India) (Fig. 6)
Tufted perennial. Stems usually 1–2 m high but can reach up to 3 m. Leaves glabrous or hairy, up to 45 cm long and 5–15 mm wide, with a strong white midrib. Inflorescence a large spathate panicle with up to six groups of primary branches, 2 to 18 in a group; final branches filiform, 5–8 cm long, terminating in a pair of racemes. Spathes supporting primary branches have well-developed blades; spatheoles supporting the rays have somewhat inflated sheaths and reduced or absent blades. Racemes 4–8 cm long, hairy at least on one side, each with ten to fourteen joints which are 4–5 mm long, inflated, usually hairy on sides, each bearing a pair of spikelets. The sessile spikelet of the pair 7–9 mm long, bisexual, with a kneed and spirally twisted awn 1–4 cm long. Pedicelled spikelet male, its pedicel similar to the raceme joint.

There are three generally recognized varieties:
1. Var. *gayanus*: joints and pedicels ciliate on one margin only; awns 1–2 cm long.

The other two varieties have joints and pedicals ciliate on both margins and the awns are 2–4 cm long;
2. Var. *squamulatus* (Hochst.) Stapf: a moderately vigorous plant not exceeding 150 cm in height; pedicelled spikelet scaberulous and puberulous.
3. Var. *bisquamulatus* (Hochst.) Hack.: large vigorous plant often exceeding 2 m in height; pedicelled spikelet hairy to villous.

There is also var. *tridentatus* Hack. which occurs in semidesert parts of the Sahel zone of West Africa (Toutain, 1973).

Var. *squamulatus* occurs naturally throughout tropical Africa, except in very dry or very humid areas, from sea level to 2,300 m, or occasionally 2,600 m alt, and is more common in western than in easten Africa. Var. *bisquamulatus* is distributed in West Africa only, up to 2,000 m alt, but is more common below 1,000 m and both varieties often dominate over other grasses in West African savannas. Var. *gayanus* occurs in West Africa in seasonal swamps where it often forms almost pure stands.

Environment. The three varieties occur naturally in the areas where the average minimum day temperature of the winter month of January (or July in the southern hemisphere) does not drop below 4.4°C (Bowden, 1964); there are however reports that *A. gayanus* can tolerate light frosts. Var. *squamulatus* and *bisquamulatus* grow in the areas within the annual rainfall limits of 400 and 1,500 mm and are sufficiently drought resistant to withstand 2 to 9 months of drought and favour the areas where the dry season lasts 3 to 5 months. The plants can stay green long

Fig. 6   *Andropogon gayanus.*

into the dry season and even produce some growth and, in the areas with less severe droughts, they can remain green throughout the year. Var. *gayanus* tolerates deep seasonal flooding with the water level reaching up to 2 m above the ground, whereas the two other varieties do not grow on seasonally flooded land; they show however no preference to any particular soil except that they avoid heavy clays. *Andropogon gayanus* is resistant to grass fires and develops new leaves and shoots a few days after burning.

Introduction. *Andropogon gayanus*, especially var. *bisquamulatus*, has been introduced into cultivation, mainly in West Africa, and particularly in Nigeria, where considerable experimental work has been done. *Andropogon gayanus* is valued for its productivity, drought resistance, reasonable dry season production and excellent palatability. Although it can be relatively easily established from seed, seed yields so far obtained in practice are not sufficiently high for the establishment of more than very few hectares from seed harvested from 1 hectare, the factor which inhibits a rapid spread of farm-scale cultivation. So far the main practical value of *A. gayanus* is in its use in natural stands. Its potential value as a cultivated grass is however high and can materialize when the technique of seed harvesting and cleaning can be mastered. Apart from Nigeria it is grown on experimental, and, to a lesser extent, on a field scale in practically all West African countries and had been tried in Uganda and Rhodesia. Outside Africa *A. gayanus* is grown in Australia (mainly in the north), in tropical India, in central and northeastern Brazil as a valuable hay or green-fodder crop (Whyte *et al.*, 1959) and is under trial in Jamaica.

Establishment. Vegetative propagation is possible because the tufts of var. *bisquamulatus* are large and can reach up to 1 metre in diameter producing numerous tillers so that splits can be spaced 50×50 cm. Nevertheless seed sowing is usually recommended and a good seedbed for satisfactory establishment should be prepared. Seed is broadcast or sown in rows and is covered by 1–3 cm of soil, the optimum depth for germination and emergence. Haggar (1969) reports that sowing in rows gives higher herbage yields than broadcasting. Establishment with companion annual crops has been satisfactory (Haggar, 1969) and in the first year of growth *A. gayanus* + maize or + soyabean gave higher yields of fodder than *A. gayanus* sown with *Mucuna* (*Stizolobium*) sp. or with *Pennisetum pedicellatum*. These companion crops did not affect herbage yields of *A. gayanus* in the following or subsequent years. Whyte *et al.* (1959) also report successful establishment with sesame and millet as cover crops. In laboratory and in small observational plots naked caryopses germinate better than those enclosed in hard glumes but storage of naked caryopses is risky and spikelets are usually sown. Sowing rates are uncertain because of the varying quality of seed with its purity ranging from under 1 per cent to 60 per cent; germinability of well-developed caryopses is usually good and is mostly 60–80 per cent. It is assumed (Bowden, 1963) that 2–2.5 kg of pure germinating seed (PGS)/ha can be adequate, but as seed is mostly of low and often of unknown purity, up to 45 kg seed/ha is often recommended although Whyte *et al.* (1959) write that 5 kg/ha is sown in Brazil. In laboratory trials (Bowden, 1964) seed imbibition takes 1 day, the rootlet appears 2–6 days after seed wetting or sowing and the coleoptile emerges 1 day later than the rootlet. Field emergence occurs usually 5 to 10 days after sowing and only one seminal root develops and penetrates to a depth of 10–20 cm in 10 to 20 days; adventitious roots appear about 10 days later. Up to about 120–140 kg of superphosphate per ha is recommended for application at sowing and dressing with fertilizer N is done when the grass reaches some 20 cm in height.

Management, fertilizing. In natural stands *A. gayanus* is usually burnt towards the end of the dry season and then grazed, continuously or rotationally. The effect of the frequency of cutting on grass yields studied in Nigeria (Rains & Foster, 1956–7) in sown swards, showed, as expected, that frequent cutting, twelve times a year, when the grass was 60 cm tall, produced lower yields of DM, just under 10 t/ha, than when it was cut seven times, at a height of 150 cm and yielded about 14 t/ha. Average yearly CP yields during the 3 years of trial were however equal, 384–396 kg/ha, for both treatments. *Andropogon gayanus* responds well to fertilizer N, but P is usually effective only when applied with substantial amounts of N. In a trial in Nigeria (V. A. Oyenuga, from Bowden, 1963) the best responses to N applied alone were at 250 kg ammonium sulphate/ha which increased the yields of DM from 1.46 t to 3.83 t/ha and further 250 kg of the fertilizer increased the yields only slightly, to 4.35 t. However, with 250 kg of superphosphate as a basic dressing, the best response was obtained from 750 kg ammonium sulphate/ha which increased the yields to 6.38 t/ha, but P alone produced no effect. Similar results were obtained by Haggar (1966). Normal yield increases from applied N were reported to be of the order of 20–25 kg DM/ha and 2 kg of CP from each kg of applied N up to a certain limit of N fertilizer used. Applied N also produced good effect on natural stands of *A. gayanus* and Adegbola *et al.* (1968) trebled the yields of herbage by applying 100 kg N/ha and double or treble yields were also cited by Bowden (1963) from the application of 20 t f.y.m/ha. Occasional responses were also obtained from applied S.

Association with legumes. *Centrosema pubescens* and *Stylosanthes guianensis* were the best companion legumes in a Nigerian trial (Adegbola & Onayinka, 1966) and *Clitoria ternatea* gave good results in Australia.

Herbage yields. *Andropogon gayanus* is one of high-yielding grasses in West Africa although it can be outyielded by *Pennisetum purpureum*, *Panicum maximum* or *Melinis minutiflora*. Reasonably high yields were reported by Haggar (1966), who obtained 2.43 t DM/ha from unfertilized grass and 8.57 t from grass given 112 kg N + 30 kg P_2O_5/ha. Yields of about 4 t DM were also reported from Australia and 7.1–7.8 t hay/ha from Cameroon by Barrault (1973). Fifty-seven t fresh fodder were recorded in India and perhaps the highest yield of 76 t fresh forage/ha were recorded in Mali by Derbal *et al.* (1959).

Conservation. Hay is sometimes prepared but its digestibility is usually lower than that of fresh grass. Ensiling has also been tried with varying but mostly poor results, partly because of the very slow fermentation.

Chemical composition, nutritive value. *Andropogon gayanus* is a grass of medium nutritive value. V. A. Oyenuga (Bowden, 1963) gives CP content at the early flowering stage, at which it is often utilized, as 6.1 per cent and CF content as 33.7 per cent. In this trial CP content decreased from 10.1 per cent 4 weeks after the previous cut, to 4.8 per cent at the full flowering stage; the content of CF varied relatively little, from 30 to 34 per cent, Sen & Mabey (1966) gave perhaps the highest CP recorded, which was 12.9 per cent in herbage cut every 8 weeks; it remained high in the 12-week-old herbage and decreased to 8.4 per cent and 5.4 per cent in 4- and 6-month-old herbage, respectively. Butterworth (1967), in his review of grass digestibility in the tropics, gives CP content from 3.8 to 6.5 per cent, that of DCP from 1.5 to 3.4 per cent and of CF from 30 to 36 per cent. CP digestibility in trials with cattle was 39–47 per cent and with sheep 30–52 per cent. In silage CP content ranged from 5.5 to 7.1 per cent; its digestibility was determined as 21 to 44 per cent and DCP content was 1.2–2.7 per cent, lower than in fresh herbage. According to Sen and Mabey (1966), P content ranges from 0.14 to 0.21 per cent in the herbage up to 4 months old and is nearly adequate for animal requirements but Ca content is on the low side.

Animal productivity. Little information is available about the productivity of animals grazed on pure sown *A. gayanus* or fed with it. In natural grasslands dominated by *A. gayanus* average liveweight gains of steers were small, 140–170 g per animal per day; 112 kg fertilizer N/ha increased the liveweight gains to 300 g, but 50 kg N produced little effect (Adegbola *et al.*, 1968).

Flowering, reproduction. In the first year of growth flowering begins 10–15 weeks after seedling emergence, usually at the end of the rainy season, and the old plants established in previous years flower at about the same time. Flowering usually begins about 3 hours after sunrise (Bowden, 1964) although some flowering can be observed until the afternoon; direct sunlight or heat encourage the opening of florets. The sessile spikelet flowers first and the pedicelled spikelet of the same pair on the following day. Diploids with $2n = 20$ chromosomes and tetraploids ($2n = 40$) have been found in var. *bisquamulatus* in southern Nigeria but only diploids were so far encountered in northern Nigeria. Only tetraploids were recorded for var. *squamulatus*. Irregular chromosome numbers – $2n = 35, 42, 43$ and 44 – were also reported.

Seed production. Haggar (1966) found in Nigeria that the majority of panicles are produced by the stems formed before the beginning of the wet season. Grazing may not interfere with the development of panicles but only if it is completed early, in June, some 5 months before seed ripening. In a trial in Nigeria (Haggar, 1966) stands of *A. gayanus* grazed only in June, produced 47 panicles per square yard when fertilized with N and 18 panicles in unfertilized plots; when grazed in June and July the number of panicles was greatly reduced. N was effective when applied at 56–112 kg/ha but smaller dressings produced no effect. Bowden (1964) observed in Nigeria that the average number of fertile stems per plant is about 30, of racemes per stem 20 and of spikelets per raceme pair 33; the number of spikelets per plant would then be about 20,000. Some 60 per cent of mature spikelets contain caryopses, the number of which per plant was then calculated as being about 12,000. A conservative estimate of 4 plants per m^2 gives some 480 million caryopses per harvest; the actual numbers of caryopses in the harvested seed are considerably smaller because a large proportion of spikelets containing caryopses is lost during seed ripening and harvesting. Actual yields of uncleaned seed are of the order of 20–100 kg/ha. Haggar (1966) harvested 25.1 kg unthreshed seed/ha from unfertilized plots and 38.7, 57.7 and 74.9 kg/ha when fertilized with 56, 112 and 224 kg N/ha, respectively. P produced little effect and only at 224 kg N/ha interaction with P was observed and the maximum seed yield obtained was 86.5 kg/ha. In India (Mishra & Chatterjee, 1968), seed yields reached 90 kg/ha when the grass was fertilized with 27.8 kg N/ha and 69 kg seed were obtained from unfertilized plots. P again produced little or no effect and cutting the herbage early in the season did not affect seed yields. The yields recorded for the Indian trial were for uncleaned seed containing 5–10 per cent caryopses. In Brazil, three harvests per year have been taken with up to 30 kg seed/ha resulting from each cut (Whyte *et al.*, 1959). Uncleaned seed as harvested usually contains a large amount of chaff, including pieces of stems, leaves, racemes, awns, etc., and cleaning, to obtain easily-flowing seeds which can be drilled, is difficult. Seed of *A. gayanus* germinate well in the first year or two, to about 50–80 per cent; under open storage germinability declines in the third year to 30–70 per cent and to nil in 4 to 6-year-old seed.

Andropogon pseudapricus Stapf (*A. apricus* Trin. var. *africanus* Hack.)
Annual or short-lived perennial tufted grass 60–150 cm high. Leaves 12–30 cm long and 2–4 mm wide. Inflorescence a spathate panicle of five to seven tiers with up to five rays each. Spathes are transitional in shape and size from leaves to spatheoles which are 10–15 cm long. Peduncles filiform, about as long as the spatheoles and bear two racemes 2.5–6 cm long. Joints and pedicels obconical. Sessile spikelets 5–6 mm long with an awn 5–7 cm long and with a bristle on the lower glume.

Occurs in western, central and south-eastern Africa. Common and can be co-dominant in grasslands of Sahelo–Sudanese region of Mali, Senegal and Niger, under an annual rainfall of 500–700 mm. Late-flowering grass of medium quality. CP content in the rainy season has been determined as 7.4 per cent (Boudet, 1970). In Mali, when sown in rows, it yielded 15–18 t green fodder/ha (Derbal *et al.*, 1959).

Aristida L.

Densely tufted perennials or annuals. Leaves linear, narrow, often convolute. Panicles from narrow, dense and spikelike to wide and loose. Spikelet with one floret; glumes long, often exceeding the lemma which is cylindrical, spindle-shaped or laterally compressed, with convolute margins and a three-branched awn. Awn with or without a column. A large genus with about 260 species, common in the tropics and subtropics throughout the world, mostly in arid or semi-arid areas when they can be pioneers and occupy denuded pastoral land. A number of species are also indicators of overgrazing and pastoral digression. Herbage quality can be satisfactory in young plants, but during the flowering, especially at its later stages and early fruiting, the plants are often avoided by grazing stock because of hard and prickly florets provided with sharp callus which can be irritating and even dangerous to the animals. Leaves and stems can also be unpalatable at the advanced stages of growth and are eaten only if more palatable grasses are not available.

Aristida adscensionis L. $2n = 22$

Annual 10–100 cm high with thin stiff stems forming erect or slightly spreading tufts. Leaves up to 20 cm long and 3 mm wide. Panicles up to 30 cm long, lax or more often dense and narrow with erect branches. Glumes unequal. Lemma 5–13 mm long, equal to or exceeding the longer glume, narrow, tightly clasping the palea and the seed at maturity, with three subequal to very unequal awns; the central awn 7–25 mm long, longer than the side awns. The callus is very sharp.

A very variable species, the extreme forms of which are known as *A. submucronata* Schumach., *A. curvata* (Nees) Trin. & Rupr. and under some other names. *Aristida adscensionis* is common throughout the dry tropics as a pioneer of bare soil, or in mixed swards with other annual or perennial grasses. Its content in the herbage increases under heavy grazing or in the years that follow droughts. The nutritive value of *A. adscensionis* is a controversial issue; in some areas it is regarded as unpalatable, in others poor, mediocre or fair, and in Brazil its hay is described as tender and nutritious. It can be of fair value before flowering or at the early flowering stage, when the CP content can be as high as 9 per cent (Dougall & Bogdan, 1965), although a CF content of 37 per cent was high even at this stage. At the early stages of growth *A. adscensionis* can be well grazed by cattle, but the animals avoid eating it at the late flowering and seeding stages because of the sharp seeds and stiff awns. The sharp seeds can penetrate into sheep's wool and through the skin, and are deposited on the inner side of the skin or even enter the flesh, as was observed in Kenya, causing considerable discomfort to the animals which lose appetite and weight.

Aristida browniana Henr. (*A. stipoides* B.Br.; *A. muelleri* Henr.)

Tufted perennial with easily splitting tufts. Stems 30–80 cm high, erect, rigid. Leaves, narrow, convolute. Panicle narrow, up to 20 cm long. Glumes narrow-lanceolate, the lower 8–13 mm long, the upper 15–24 mm long. Lemma 8–9 mm long. Awns 35–65 mm long on a twisted column which is 20–40 mm long.

Widely spread in arid and semi-arid areas of the northern and western parts of Australia where it can often be abundant. The grass is stemmy and tough and is grazed mainly when young.

Aristida contorta F. Muell. Bunched kerosene grass (Lazarides, 1970); Mulga grass (Siebert *et al.*, 1968)

An annual or short-lived perennial with slender stems forming compact tufts up to 30 cm high. The narrow leaves tend to curl at maturity. Panicles erect or bent. Floret cylindrical, 6–7 mm long, and terminates in a spirally twisted column bearing three subequal awns 4–7 cm long.

Occurs in dry areas of Australia where it is common and provides useful grazing at all stages of growth, except at the late flowering and early ripening stages, but is grazed again after the florets with their sharp callus, harmful to grazing animals, have fallen to the ground. *Aristida contorta*-dominated pastures can even be used for stock fattening (Lazarides, 1970). Siebert *et al.* (1968) found 11.4 per cent CP in fresh green herbage, 7.0–7.6 per cent at mature stages and 4.8 per cent in the fully ripe material. A relatively low level of CF, about 27 per cent, determined in fresh green herbage was maintained at the later stages of growth.

Aristida jerichoensis Domin ex Henr. (*A. ingrata* Domin var. *jerichoensis* Domin)

Annual resembling *A. adscensionis* but differing from the latter mainly by the shorter lemma ('seed') which is shorter than the glumes. Common in northern and western Australia and also in Queensland and New South Wales, in semi-arid areas, mainly in open, and often overgrazed grasslands. The nutritive value and the importance for grazing are probably similar to those of *A. adscensionis*.

Aristida mutabilis Trin. & Rupr.
Annual 5–70 cm high. Leaves linear, up to 15 cm long, flat or convolute. Panicle 5–20 cm long, loose or contracted. Glumes subequal, 4–7 mm long. Lemma spindle-shaped, 4–5 mm long. Awns slender, 10–30 mm long, on a twisted column 2–5 mm long articulated just below the branching of awns.

Occurs in tropical Africa, mainly north of the equator, and from Mauritania to Somalia, Arabia and India. Common in arid and semi-arid areas, usually on light-textured soils, in bushland or in open areas, in situations where perennial grasses cannot grow either because of the lack of adequate rains or because the perennials have been grazed out. *Aristida mutabilis* is one of the commonest pioneer grasses on denuded land under arid conditions. In dry years it uses the little rain that falls and remains alive but dwarfed and even produces a few seeds per plant. A sample of herbage collected at the flowering stage in one of the better years contained 8.7 per cent CP, 1.5 per cent EE, 30.7 per cent CF, 44.7 per cent NFE, 0.45 per cent Ca, but only 0.11 per cent of P, an amount inadequate for cattle requirements (Dougall & Bogdan, 1965).

Aristida pruinosa Domin
Tufted perennial with robust stems up to 100 cm high. Leaves rigid, long, about 3 mm wide. Panicle erect, up to 40 cm long. Spikelets 10–16 mm long on long erect pedicels. The lower glume 10–12 mm and the upper 15–16 mm long. Awns 2–3 cm long and without a column.

Common and often dominant grass in north-western Australia, in tropical and subtropical woodland in the grass understorey and also in open grassland. A grass of a relatively low palatability but the leaves are well eaten by the animals when the plants are young.

Astrebla F. Muell. ex Benth. Mitchell grasses

Tufted perennials with the inflorescence consisting of one to three spikes. Spikelets with two to four bisexual florets and one to three poorly developed empty florets in the upper part of the spikelet. The densely hairy lemma has three long bristles of which the central one can become a stout awn.

The four species of this genus occur in dry grasslands of Australia where they are of considerable importance for sheep grazing. They do not normally grow in pure stands but are mixed with other grasses, perennial or annual; the annuals can be abundant in favourable years with good rains.

Jozwick (1970) has shown that the four species he studied (*A. lappacea, A. pectinata, A. elymoides* Bailey & F. Muell ex Bailey and *A. squarrosa* C. E. Hubbard) are flexible in their relation to the environment. Their life cycle can be completed under a variety of conditions and they are adapted to regular and also occasional droughts and can exist under a variety of temperatures although the optimum growth was observed under 30/25°C of day/night temperatures and the greatest number of tillers under 28/23°C. Photoperiods have little effect on the growth and development of the four species and on the initiation of floral parts but the number of inflorescences per plant increases with the decrease of photoperiods.

When grazing Mitchell grasses the sheep do not pick up single tillers but graze the whole plant from the top down (Weston, 1962). It was also observed that cutting close to the ground level can kill the grass and for maintaining its dominance the grass should not be grazed too low.

Astrebla lappacea (Lindl.) Domin (*Danthonia lappacea* Lindl.). Wheat Mitchell grass $2n = 40$

Stems geniculately ascending, usually branched, 30–80 cm high. Leaves 4–5 mm wide. Spikes one or two, 5–30 cm long, with the spikelets distant in the lower part and dense in the middle and upper parts. Spikelets 7–13 mm long. Lemma of the lowermost floret 8–13 mm long, two-lobed, with a straight awn 4–14 mm long between the lobes.

Occurs mainly in the northern part of Western Australia in dry pastures. Weston (1969) observed that in February and March sheep eat shoots and stems of *A. lappacea*, whereas later on, from April to July, the animals pick up mainly the flowering heads. The material they ate contained 9.5–10.8 per cent CP in the shoots, 13.3 per cent in the leaves and 8–10 per cent in the flowering heads. P content in the stems ranged from 0.14 to 0.20 per cent in February and it was about 0.24 per cent in the leaves. Flowering heads contained 0.27 per cent P at the early flowering stage but its content was reduced at the later stages of flowering.

Astrebla pectinata (Lindl.) F. Muell. ex Benth. (*Danthonia pectinata* Lindl.). Barley Mitchell grass $2n = 40$

Similar to *A. lappacea* but the tufts are more compact and the stems taller than in *A. lappacea*, and the leaves slightly wider; the spikelets are longer (10–17 mm) and the lemma terminates in three long bristles. Occurs all over Australia, except Victoria, in dry grassland. The grass is eaten at all stages of growth but is not very palatable to the animals (Lazarides, 1970). In the central arid zone of Australia, partly dry herbage contained 6.8 per cent CP and completely dry herbage 4.5 per cent. The content of CF was unexpectedly low and ranged from 26.1 to 28.5 per cent, whereas very high content was recorded for NFE: 56.8–58.3 per cent (Siebert *et al.*, 1968).

Avena L.

Some 70 species of temperate areas; one species is widely cultivated.

Avena sativa L. Oat, Avoine; Avena

Erect annual up to over 1 m high. Commonly grown in temperate areas for grain or fodder, often in mixtures with vetch (*Vicia sativa*). Oats are also sometimes grown in the tropics in the high altitude areas of Colombia, Brazil, Kenya and some other countries. In Colombia, e.g., it is used in the upper temperate zone above 1,600 m alt, and yields 10–15 t fresh forage/ha (Crowder *et al.*, 1970). Higher yields were recorded in Madagascar where oats have been tried, as a catch crop after rice has been harvested, and if sown early and on fertile soil yielded up to 52 t fresh herbage/ha (Granier & Razafindratsita, 1970). In northern India oats can be grown in mixtures with berseem (*Trifolium alexandrinum*).

Axonopus Beauv.

Spikelets with one glume, lanceolate or oblong, awnless, dorsally compressed, almost sessile and arranged in two rows on a slender rhachis of the spike. The spikes are in a subdigitate or elongated panicle. Florets two, the lower empty and reduced to a lemma, the upper bisexual. Perennials or annuals; 35 species indigenous in tropical America and the West Indies.

Axonopus affinis Chase (*A. compressus* (Chase) Henderson). Carpet grass; Narrowleaf carpet grass; Mat grass; Zacate amargo

Creeping perennial spreading by stolons and forming dense low growth similar to *A. compressus* from which it differs by narrower leaves (usually 2–6 mm wide), slightly smaller spikelets (2–2.2 mm long) and by the upper glume and the lower lemma equal or only slightly exceeding the rest of the spikelet. It is an octoploid ($2n = 80$) and Gledhill (1966) remarks that it is more vigorous than *A. compressus* which is a tetraploid; $2n = 54$ have also been recorded.

In its native American tropics *A. affinis* is less widely spread than *A. compressus*. It is more frost tolerant than *A. compressus* and outside the areas of its origin has spread mainly to the cooler areas of the USA and South Australia but can also be found in the tropics of the Old World. *Axonopus affinis* has been cultivated for a number of years in the USA as 'carpet grass', mainly on poor soil and is reputed to be slightly less palatable to the animals than *A. compressus*. Similarly with *A. compressus A. affinis* is also grown for soil conservation and as a lawn grass, is a good seeder, seed can be easily harvested, and combine-harvesting was practised in the USA (Wheeler, 1950). Seed is small and there are about 2.5–3 million seeds per kg. Seed is sown at 5–15 kg/ha and satisfactory stands can be established with a minimum or no seed bed preparation. In mixed stands with other low-growing grasses *A.*

affinis can compete with *Pennisetum clandestinum* or *Paspalum notatum* on poor unfertilized or sparingly fertilized soil but is suppressed by the companion grasses on good or well fertilized soil. It should be noted that *A. compressus* and *A. affinis* cannot be easily distinguished by their general appearance because the leaf width can vary and wrong naming cannot be excluded. Moreover, the existence of mixed types, possibly of hybrid nature, can be confusing.

Axonopus canescens (Nees) Pilger

Tufted perennial resembling *A. scoparius* but less vigorous, smaller in all parts and non-stoloniferous. In Venezuela and other parts of South America it can be a dominant or co-dominant grass in savanna vegetation.

Axonopus compressus (Swartz) Beauv. Carpet grass; Broadleaf carpet grass; Zacate amargo (Fig. 7)

Perennial spreading by rhizomes or stolons. Stems slender, compressed, one-to-three-noded, 15–60 cm high. Leaves broadly linear or lanceolate, 4–15 cm long and 4–10 mm wide, flat or folded. Spikes usually two to four, slender, dense, 3–10 cm long. Spikelets 2.2–2.8 mm long. Upper glume and lower lemma markedly exceed the rest of the spikelet.

Gledhill (1966) in his *Axonopus* studies, mainly on species which occur in Africa, considers *A. compressus* to be the central and most widely distributed species in a complex or a group of allied taxa. This complex includes *A. compressus*, *A. affinis*, *A. arenosus* D. Gledhill, *A. flexuosus* (Peter) Hubbard ex Troupin, and *A. brevipendunculatus* D. Gledhill, and perhaps other species not studied by Gledhill. Similarly with other species of *Axonopus* they originate from the Central and South American tropics but had spread outside the area of their origin: *A. compressus* to a number of other tropical or subtropical countries, *A. affinis* to USA, Australia and elsewhere as a predominantly cultivated grass, and the three other species mainly to Africa where they occupy habitats suitable for their ecology: *A. arenosus* with saline soil and *A. flexuosus* – places subjected to flooding. Similarly with most of the other species of *Axonopus*, *A. compressus* is a cross-pollinating, heterogamous plant and can hybridize with other species of the complex, the hybrids being mostly sterile as has been shown by Gledhill. Being unable to reproduce sexually, the hybrids, nevertheless, survive as sterile clones and can be numerous in the areas where they had originated. *Axonopus brevipedunculatus* is an apomict, the only apomictic species of *Axonopus* so far known; nevertheless, as a male parent, it hybridises easily with *A. compressus*. *Axonopus compressus* is the most widely spread species of the complex, both in the area of its origin and elsewhere and is represented by a number of forms, the differences between these forms can be accentuated by different environmental conditions of the new habitats in the countries of their adoption.

Axonopus compressus occurs naturally in Mexico, Central America, tropical South America and the Caribbean islands (West Indies). During the last 30–150 years it has spread, mainly as accidental introductions, into the extreme south-eastern parts of the USA, Africa, south-eastern Asia, including India, the Philippines, Australia and the Pacific islands. This species has naturalized in most of these countries where it often forms almost pure stands and is used for grazing. It thrives on fertile sandy loams but is more common on poor sandy soils where it can compete with other grasses requiring better soil. It often also invades run-down old sown unfertilized pastures and Roberts (1970) reports from Fiji invasions of old *Brachiaria mutica* pastures by *A. compressus*.

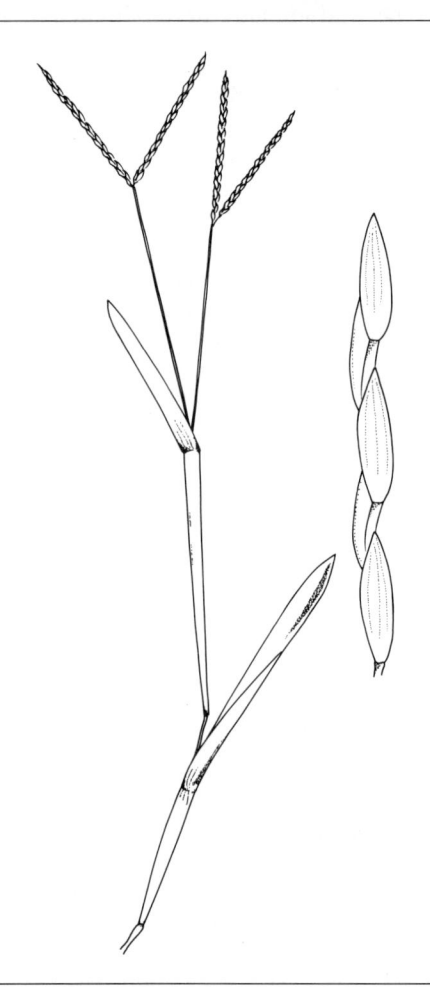

Fig. 7 *Axonopus compressus*.

Although *A. compressus* grows mainly on wet soil it cannot withstand waterlogging or flooding.

Herbage yields are moderate and Dirven (1971) reports from Surinam that 5 m² yielded 10.64 kg green material which corresponds to over 10 t/ha.

The quality of *A. compressus* herbage is reputed to be comparatively low although in young plants the content of CP can be high and, in Dirven's trials (1971), 3-week-old plants given 100 kg N/ha plus P and K fertilizers, contained 22.3 per cent CP and 28.9 per cent CF which changed to 17.8 and 29.5 per cent, respectively, when the plants reached an age of 6 weeks; the content of NFE was however low in the young plants. *Axonopus compressus* is a good seeder and is grown to a limited extent for grazing and also for soil conservation or as a lawn grass. It is a tetraploid with 40 2*n* chromosomes, although 50 and 60 2*n* chromosomes have also been encountered.

Axonopus micay H. García-Barriga. Pasto Micay $\qquad$ 2*n* = 40

A species allied and rather similar to *A. scoparius* from which it differs in the non-tufted, creeping, rhizomatous habit; it is also less vigorous and is 30–60 cm, and occasionally 100 cm, tall. The leaves are 10–22 mm wide, narrower than those of *A. scoparius*, and the panicles are smaller and consist of 4 to 12 spikes. Occurs in Colombia and Venezuela, where it can form swards readily grazed by cattle and is also grown for grazing or fodder. Vegetative propagation is easy but the plant can be severely damaged by a fungus disease, possibly *Xanthomonas axonopeois* (García-Barriga, 1960). Two samples of *A. micay* herbage analysed in Colombia contained 4.73 and 5.68 per cent of CP and 34.3 and 38.0 per cent CF, respectively. The contents of P was 0.16 per cent and of Ca 0.28–0.30 per cent, adequate for cattle requirements (Blasco & Bohórquez, 1968).

Axonopus purpusii (Mez) Chase. Guaratara (León & Sgaravatti)
$\qquad$ 2*n* = 20

Tufted or sometimes stoloniferous perennial resembling *A. compressus* but taller and more vigorous. The easiest way to distinguish it from *A. compressus* or *A. affinis* is by its densely hairy spikelets, the hairs being usually brown or rusty in colour. Occurs naturally in Mexico, Central America and tropical South America, reaching Argentina in the south. It grows in the continental type of South American savanna (Black, 1963) where it can be a co-dominant grass readily grazed by cattle.

Axonopus scoparius (Flügge) Hitchc. Capim Columbia; Capim imperial; Micay $\qquad$ 2*n* = 20

Perennial 0.6–2 m in height, forming tufts and also spreading by long stolons. Leaves glabrous, or hairy on the upper surface, 10–60 cm long and 15–35 mm wide. Inflorescences terminal and lateral, long, with numerous, 10 to 100 (mostly 15–25), spikes 10–25 cm long. Spikelets slightly hairy, 2.5–3 cm long.

Occurs naturally in tropical South America and is often abundant in savannas. This vigorous grass has succulent stems and is readily eaten by cattle at practically all stages of growth. It is cultivated in Colombia for soilage or silage and is planted by rooting stem-cuttings or pieces of stolons in 50 or 100 cm rows. Crowder *et al.* (1970) report from Colombia that yields of DM are around 10–14 t/ha/year. In Brazil a yield of 21.9 t dry fodder/ha during 13 months of trial was obtained by Zúñiga *et al.* (1967) from unfertilized plots and some 10 per cent higher when fertilized with NPK. In another trial in Brazil (Pereira *et al.*, 1966) *A. scoparius* yielded only 5 t fresh material, unfertilized and only slightly more when fertilized, but irrigation in the dry season and fertilizing increased the dry-season production from under 1 t/ha to about 8 t. Although the grass is palatable, the plants analysed by Blasco & Bohórquez (1968) contained only 3.9–5.8 per cent CP and Butterworth (1967) quotes only a slightly higher figure of 6.2 per cent. The low CP content given by Blasco & Bohórquez can be explained by an advanced stage of growth of plants collected just before the main rains and higher content of CP can be expected in younger material. CF content has been given as 30–35 per cent. The contents of P was low (0.05–0.13 per cent) and of Ca just adequate (0.32 per cent). The digestibility in Costa Rica (from Butterworth, 1967) in trials with cattle was 51.5 per cent for DM, 48.6 per cent for CP, 75.7 per cent for CF and 58.4 per cent for NFE. A bacterial disease caused by *Xanthomonas axonopeois* can destroy the plants and planting healthy cuttings is essential. Clones resistant to *Xanthomonas* have been selected and vegetative material for planting distributed as clones 60 and 72 (Crowder *et al.*, 1970).

Beckeropsis Fig. & De Not.

Five to six species, mostly tall perennials, of tropical Africa.

Beckeropsis uniseta (Nees) Robyns (*Pennisetum unisetum* Benth.). Natal grass; Beck grass $2n = 18$

Tufted perennial 1–2 m high. Stems numerous, stout, erect, much branched in the upper part. Leaves linear-lanceolate, much narrowed and almost petiole-like at the base, 15–60 cm long and 6–25 mm wide. Panicle large, with the branches supported by small leaves, and numerous spikes which are 2–4 cm long and borne on long fine peduncles. Spikelets 2.5–3 mm long; glumes minute, lower floret empty, upper floret bisexual. Grain 2 mm long.

Occurs naturally throughout tropical Africa reaching Natal in the south, in areas with an annual rainfall ranging from 800 to 1,500 mm. Common in thin woodland or in scattered-tree savanna grasslands but disappears gradually after all the trees had been removed. A good grazing grass, leafy when kept low; the leafy stage can last for a considerable period because the stems appear rather late in the season, but it is stemmy and almost unpalatable to cattle when the stems develop. *Beckeropsis uniseta* was extensively tried at Kitale in Kenya as a ley grass for grazing or hay, but it proved to be short-lived, lasted 2 or 3 years only, and the interest in this species was lost. However, in Rhodesia, it gave encouraging results, and yielded, when well fertilized, 11.45 t DM/ha/year during the 4 years of trial, a medium yield for 30 grasses under trial, but it suffered from a drought and only 46 per cent recovery was observed in the year that followed the drought. The average CP content in the herbage was high: 16.1 per cent (Rodel & Boultwood, 1971). *Beckeropsis uniseta* is a very good seed producer and can be easily propagated from seeds; 10 kg seed/ha is a recommended minimum sowing rate.

Bothriochloa Kuntze

Tufted or stoloniferous perennials. Racemes articulate with fine joints arranged in digitate or sometimes elongated panicles which are usually aromatic. Spikelets dorsally compressed, in pairs, one sessile the other on a pedicel; sessile spikelet with two florets, the lower male or neutral, the upper bisexual, with a fine geniculate and spirally twisted awn; pedicelled spikelet male or neutral. A relatively small genus of about 20 species which occur in tropical, subtropical and warm temperate areas of both hemispheres. Some species are valuable grazing grasses. This genus is closely allied to *Dichanthium* and *Capillipedium* and *B. intermedia* is a central species not only in *Bothriochloa* but in all the three genera. It crosses easily under natural environments with *B. ischaemum* (L) Keng., *B. ewartiana* (Domin) C. E. Hubbard, *C. parviflorum* (R.Br.) Stapf and *D. annulatum*, the species genetically isolated from each other except through *B. intermedia* (De Wet & Harlan, 1970b). Natural hybrids resulted from these crosses are usually taxonomically recognized as species. These genetical links lead to a suggestion of bulking the three genera and consider their species as those of *Bothriochloa*.

Bothriochloa barbinodis (Lag.) Herter (*Andropogon barbinodis* Lag.)

Differs from *B. saccharoides* by larger leaves, larger sessile spikelets which are 5–6.5 mm long and longer, 15–28 mm, awns. Distributed in the same areas and has about the same grazing value. The herbage can be damaged by Rhodes grass scale (*Antonina graminis*) but a parasite, *Neodusmetia sangwani*, which controls the scale to a certain extent, can reduce the damage and increase grass yields (Schuster & Boling, 1971). Chemical analyses at the flowering and fruiting stages showed the contents of CP from 7.1 to 11.7 per cent, EE 1.2 to 1.7 per cent, CF 28.0–35.5 per cent and NFE 46.2–52.7 per cent (Vonesch & Riverós, 1967–8).

Bothriochloa insculpta (A. Rich.) A. Camus (*Andropogon insculptus* A. Rich.; *Amphilophis insculpta* (A. Rich.) Stapf). Sweet pitted grass

Stoloniferous perennial. Stems slender, 30–120 cm high, erect or geniculately ascending. Leaves up to 25 cm long and 3–5 mm wide, pale green or bluish, usually glabrous. Panicle subdigitate, 7–10 cm long, its primary axis slender, 1–3 cm long. Racemes 5–12, usually over 5 cm long, olive green in colour, slightly silvery villose. Joints and pedicels similar, slender, 3 mm long. Spikelets in pairs, the sessile 4–5.5 mm long with three pits (concave glands) and a spirally twisted awn about 2 cm long. Pedicelled spikelet male or neutral, 3.5–5 mm long, with one pit.

Var. *insculpta* is a usual form but var. *vegetior* (Hack.) C. E. Hubbard occurs only occasionally; it is more vigorous than the main form, with a longer panicle which has a well-developed long axis and more numerous racemes. Possibly a polyploid or an interspecific hybrid.

Occurs in tropical Africa, Arabia and South-East Asia, in moderately dry areas, on various soils, including black, slightly waterlogged clays, in natural grasslands and also as a pioneer grass which can invade abandoned arable land, usually during the second stage of plant succession, after the weed stage, and can then form almost pure stands. *Bothriochloa insculpta* can withstand heavy grazing and its proportion in natural grasslands tends to increase with the increase in grazing density. The panicles are strongly aromatic but the leaves are only slightly scented and the grass is well liked by stock; it retains its palatability well into the dry season and Stewart & Stewart (1971) report that *B. insculpta* is one of the grasses well grazed by wild ungulates at the end of the dry season. At the early flowering stage the herbage analysed in Kenya contained 7.95 per cent CP, surprisingly little EE for an aromatic grass (1.45 per cent), 32.6 per cent CF and 49.8 per cent NFE; mineral contents was low: 0.32 per cent Ca and 0.12 per cent P (Dougall & Bogdan, 1960).

In Kenya *B. insculpta* has been tried with a moderate to good success for reseeding denuded pastoral land in moderately dry areas with an annual rainfall of 700–800 mm. It was established without any soil cultivation and responded little to a tilled seedbed or to covering with cut branches as a protective measure against early grazing; seed pelleting produced a negative effect (Bogdan & Pratt, 1967). It has also been tried in normal cultivation but because of insufficient productivity met with only a moderate success. Seed is in the form of a pair of spikelets with attached joint and pedicel and there are about 1,200,000 such 'seeds' per kg.

An apomict with 60 $2n$ chromosomes.

Bothriochloa laguroides (DC.) Herter (*Andropogon laguroides* DC.; *A. saccharoides* Sw. ssp. *laguroides* (DC.) Hack.) Cola de zorro (Argentina)

Allied to *B. saccharoides*, from which it differs by more compact panicles, glabrous nodes of the stem and slightly smaller spikelets. The whole plant is smaller and less robust. Distributed in about the same areas.

Bothriochloa laguroides has an outstanding ability to establish itself from seed and appears in large quantities during the first stages of fallow, and although the quality of herbage is rather mediocre it is regarded as a useful grass. *Bothriochloa laguroides* is drought resistant, can grow on soil of various degrees of fertility and is considered to be suitable for reseeding worn-out pastures in the areas of non-intensive animal husbandry. Seed is well produced but fluffy and difficult to handle (Burkart, 1969). At the flowering and fruiting stages the plants contained 11.8–14.7 per cent CP, 1.1–1.7 per cent EE, 28.9–34.0 per cent CF and 45.0–47.6 per cent NFE (Vonesch & Riverós, 1967–8).

Bothriochloa pertusa (L.) A. Camus (*Andropogon pertusus* (L.) Willd.; *Amphilophis pertusa* (L.) Stapf). Sweet pitted grass; Hurricane grass.

Stoloniferous or tufted perennial similar to *B. insculpta* but usually smaller, with slightly smaller panicles and shorter spikelets, with one pit on the sessile spikelet and none on the pedicelled spikelet. It seems that the presence or the absence of pits may depend on plant vigour. The number of chromosomes is $2n = 40$ or 60, i.e. the plants are tetraploids or hexaploids. The grass, at least some of its forms, is less leafy than *B. insculpta* although good leafy types have also been encountered.

Bothriochloa pertusa is somewhat more widely spread in the Old World tropics than *B. insculpta*, and has been introduced to the West Indies where it has naturalized. Occurs in dry areas with an annual rainfall ranging from 500 to 900 mm, mainly on well-drained soils. It is valued as a pasture grass in India (Rao, 1970) and Burma (Rhind, 1945). In the West Indies it is also considered to be a good grazing grass (Whyte et al., 1959), and Pérez Infante (1970) in Cuba obtained average annual yield of 15 t DM/ha of which 40 per cent was produced in the dry season under sprinkler irrigation. In some areas, e.g. US Virgin Islands, some forms of *B. pertusa* are regarded as pasture weeds difficult to eradicate and to replace by better grasses (Oakes, 1968b). This species is reported to be an apomict but in India Gupta (1969–70) observed both apomictic and sexual reproduction; sexual reproduction being predominant.

Bothriochloa radicans (Lehm.) A. Camus (*Amphilophis radicans* (Lehm.) Stapf

Tufted perennial. Stems 50–70 cm high, but sometimes taller, erect or ascending and then rooting from the lower nodes. Leaves up to 10 cm long and 3–4 mm wide, green, glabrous or sparingly hairy. Panicles similar to that of *B. insculpta* but smaller and more compact, with shorter racemes, only slightly scented. Sessile spikelet unpitted or occasionally with a single pit; awn 5–9 mm long.

Distributed in tropical, mainly eastern Africa, from Sudan and Somalia to South Africa, up to 1,800 m alt, on various soils but mainly on dark volcanic or black, heavy, somewhat waterlogged soils, in grasslands or in thin bush. The tufts have sterile shoots covered with short leaves which makes the grass leafy. It is well liked by cattle and has apparently a high nutritive value. Analysed in Kenya at the early flowering stage it contained 10.25 per cent CP, 3.69 per cent EE, 40.1 per cent NFE, a relatively low content of CF, 28.7 per cent, and adequate amounts of Ca and P, 0.45 and 0.27 per cent respectively (Dougall & Bogdan, 1958). An apomict with 40 $2n$ chromosomes.

Bothriochloa saccharoides (Sw.) Rydberg (*Andropogon saccharoides* Sw.). Capim bóbó (Brazil)

Tufted perennial 60–150 cm high. Stems robust, with hairy nodes. Leaves linear, glabrous or sparingly pilose. Inflorescence a compact feathery, whitish, ovate panicle 7–11 cm long. Principal rhachis of the panicle 4–12 cm long, with 3–4 branches. Racemes (spikes) 15–35, fragile, 3–6 cm long, with 5–10 pairs of spikelets. Joints and pedicels with numerous long hairs on margins. Sessile spikelet bisexual, lanceolate, 3–5 mm long; glume without a pit; lemma of the upper floret with an awn 10–15 mm long. Pedicelled spikelet empty, 2–4 mm long.

Occurs in warm American countries from the southern USA to southern Brazil and Argentina, predominantly in semi-arid areas and produces satisfactory grazing for cattle, mainly late in the season. DCP content is low and this grass gives mediocre hay (Burkart, 1969). Early authors (Pio Corrêa, 1926) report however that *B. saccharoides* supplies fodder of good quality with 9.16 per cent of CP according to an analysis, and can even serve for cattle fattening; these data may however refer to *B. laguroides*. In North America, CP content was low in winter, about 2 per cent of the oven-dry material; it reached 9 per cent in summer but was reduced again to 4 per cent in August (Willard & Schuster, 1973). CF content was high, 30 per cent in late spring and early summer and 38–40 per cent in other months of the year. NFE content was medium, 46–50 per cent throughout the year except in spring when it was much lower, about 42 per cent. $2n$ chromosome number can be 60 or 120 and Argentinian plants have only 60 somatic chromosomes.

Bouteloua Lag. Grama grass

Inflorescence of one to several short spikes on an elongated axis. Spikes with one to several spikelets. Spikelets with two glumes and one bisexual fertile floret; there are one to a few sterile rudimental florets above the fertile one. Fertile floret with three points or very short awns; rudimental florets with longer awns.

An American genus with about 40 species distributed from Canada to Argentine but mainly in south-western and mid-western USA where they often form an essential part of the sward. A few species occur in the drier parts of Mexico, Central America, West Indies and South America. Most of the species are good grazing grasses.

Bouteloua curtipendula (Michx.) Torrey. Side-oats grama
Perennial 40–100 cm high, forming slowly spreading tufts by means of short scaly rhizomes. Leaves 15–25 cm long and 3–6 mm wide. Panicle 10–20 cm long with 30–50 spikes, spreading or pendent, 1–2 cm long, with five to eight spikelets. Rudimental floret one, with a short central awn 4 mm long and still shorter side awns.

Very common in mid-western USA and in Mexico and less common in Central America, Bolivia, Peru, Paraguay, Uruguay and northern Argentina. Occurs in arid and semi-arid areas in grass plains, woodlands and on rocky slopes. Produces good grazing in its natural habitats and also used in the USA for improving worn-out natural pastures or for establishing grass cover in thinned or cut pine forests and superior types regarded as cultivars, have been selected. Established from seed which is well produced, can be easily harvested (Whyte *et al.*, 1959) and sown at 15–25 kg/ha; there are about 400,000 seeds/kg. *Bouteloua curtipendula* has been much studied in the USA and found to be variable and of apomictic reproduction. Plants of various ploidy from tetraploids, with the chromosome number $2n = 28$, to pentaploids, hexaploids, octoploids, and decaploids were found as well as those with irregular chromosome numbers from $2n = 74$ to $2n = 101$.

Bouteloua heterostega (Trin.) Griff.
Tufted, usually semiprostrate or prostrate perennial, sometimes stoloniferous. Spikes 4–10 per panicle, 1.5–3 cm long with four to seven spikelets. Spikelets 12 mm long (including awns), with one rudimental floret provided with three awns 6–8 mm long.

Occurs in West Indies, mainly in Cuba, and is a good grazing grass. A closely allied *B. americana* (L.) Scribn. is also common in the West Indies.

Bouteloua megapotamica (Spreng.) Kuntze
Tufted and also stoloniferous low-growing perennial 10–30 cm high; leaves rigid, up to 10 cm long. Panicle with a few (two to five) spikes. Rudimental florets three or four with three long awns each.

Occurs in Argentina, Chile, Uruguay, Brazil, West Indies.

Bouteloua repens (H.B.K.) Scribn. & Merr.
Leafy stoloniferous perennial with erect or ascending stems 50–60 cm high. Leaves about 10 cm long and 4 mm wide. Inflorescence 10–15 cm long with about 12 rather lax spikes which are 2.5 cm long. Spikelets with purple glumes. Rudimental floret one with three awns. Common in the drier areas of Central America.

Brachiaria Griseb. Signal or Palisade grasses

Perennial or annual, tufted or creeping grasses. The panicle consists of a few (sometimes one) to several distant (not crowded) racemes, with sessile or subsessile spikelets arranged in two rows on a usually flattened rhachis. Of the two florets of the spikelet the lower is male, with soft lemma and palea, and the upper fertile, bisexual or sometimes female, flat on one side and convex on the other. The caryopsis is enclosed in hard, crustaceous lemma and palea.

Brachiaria brizantha (Hochst. ex A. Rich.) Stapf. Palisade grass; Signal Grass. (Fig. 8)

Tufted perennial with erect or suberect stems up to 1 m high. Leaves glabrous or hairy, linear or linear-lanceolate, 5–40 cm long and 6–15 mm wide. Panicle of two to eight racemes which are 5–15 cm long and have narrow rhachis. Spikelets 4–6 mm long, glabrous or with a few hairs near the tip, in two rows which look however as one row. Lower glume broadly ovate and clasps the spikelet for half of its length. Fertile floret 4–5 mm long with a short blunt point.

Occurs throughout tropical Africa under an annual rainfall over 800 mm, mainly in grasslands with scattered bush. *Brachiaria brizantha* has been cultivated experimentally in East and West Africa, Madagascar, Sri Lanka, Australia, Fiji, Surinam and perhaps in other tropical countries, mostly with moderate success. It is propagated by seed which is however only sparingly formed as the percentage of empty spikelets can be very high. Vegetative propagation of this non-creeping grass is impracticable. In the countries of its origin *B. brizantha* varies considerably and good forms can be selected; it is however in general inferior to other cultivated species of *Brachiaria*. In Sri Lanka (Fernando, 1961) a mixture with *Alysicarpus vaginalis* gave liveweight gains in cattle of 464 kg/ha in 260 days and mixtures with *Centrosema pubescens* or *Pueraria phaseoloides* gave 632–648 kg; mixtures with *Stylosanthes guianensis* performed well in Malaysia. These and other data in old publications, should however be used with caution, as in the past other species of *Brachiaria* were sometimes grown under the name of *B. brizantha*. CP content of the herbage was reported to be 10.7 per cent in the DM (Edwards & Bogdan, 1951), but Butterworth (1967) gives lower figures: 8.7 per cent for so called 'standing hay' and 6.1 per cent and 4.7 per cent for herbage 50- and 80-day-old, respectively, with CP digestibility of 36 and 40 per cent. *Brachiaria brizantha* is an apomict and tetraploids ($2n = 36$) and hexaploids ($2n = 54$) have been found.

Brachiaria decumbens Stapf. Surinam grass (Jamaica) (Fig. 9) $2n = 36$

Perennial with the flowering stems 30–60, sometimes up to 100 cm high, ascending from prostrate, often long, many-noded bases rooting at the nodes. Leaves hairy, lanceolate, 4–14 cm long and 8–12 mm wide.

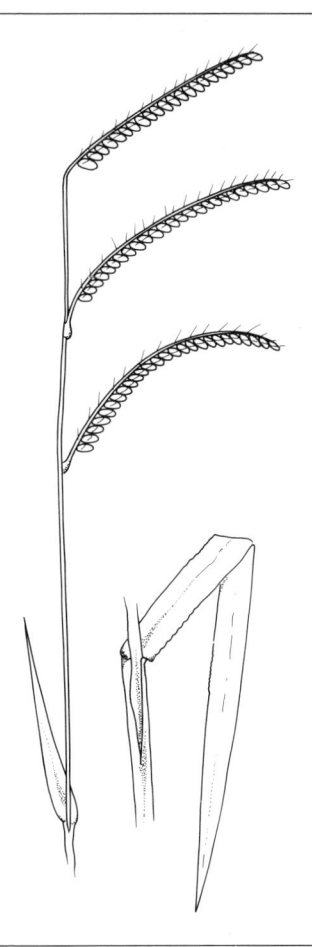

Fig. 8 *Brachiaria brizantha.*

Panicle with two to five spreading or suberect racemes; racemes 2–5 cm long, with flat rhachis. Spikelets hairy, 4–5 mm long; lower glume one third to one half the length of the spikelet. Fertile floret 3.5–4 mm long.

Occurs naturally in tropical East Africa, at altitudes over 800 m, under a moderately humid climate, in open grassland or in grassland with scattered bush, on fertile soil. The grass forms low leafy herbage and is well liked by stock; it can withstand a considerable grazing pressure and in Uganda heavy stocking converted natural grassland with the dominance of unpalatable *Cymbopogon afronardus* into a mixed pasture dominated by *B. decumbens* and the application of fertilizer N increased its content in the sward still further (Harrington & Thornton, 1969).

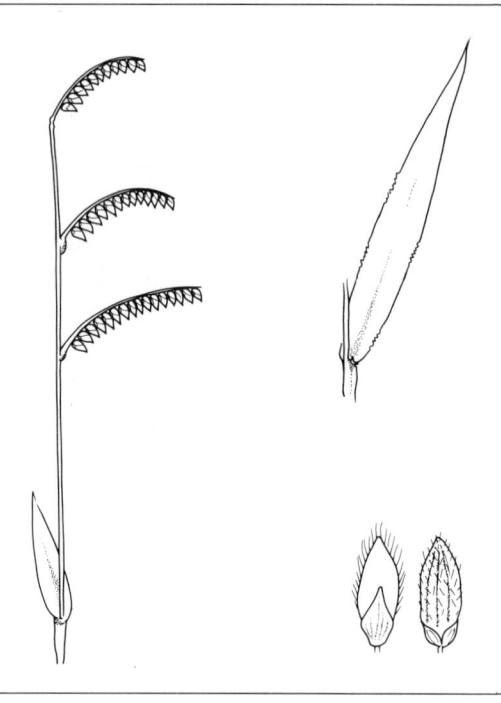

Fig. 9 *Brachiaria decumbens.*

Brachiaria decumbens was tried on a small scale in Kenya with a moderate success but better results were obtained in some other countries where it was introduced from East Africa, and where yields of DM obtained in experimental plots ranged from 8 to 15 t/ha, and sometimes were much higher. In Sarawak, *B. decumbens* gave the highest yields of the four grasses under trial and produced 9.9 t DM/ha without fertilizers; given 112 kg N/ha it yielded 14.0 t, and 19.7 t at a double rate of N; still higher rates produced no further yield increases. CP yields were 457, 736 and 1,160 kg/ha/year, respectively (Ng, 1972). In Colombia, yields of 14.6 t DM/ha were obtained on fertile soil with no applied fertilizer, and fertilizer N increased them to 20 t when the grass was cut every 6 weeks (Crowder *et al.*, 1970); slightly lower yields were obtained in Brazil. Good results were also obtained in Surinam, Jamaica and Australia where very high yields, over 36 t DM/ha, were reported (Grof & Harding, 1970). In Australia *Desmodium heterophyllum* was found to combine well with *B. decumbens* in a grass/legume mixture.

Satisfactory to good quality of herbage has been reported from nearly all countries experimenting with *B. decumbens* and CP content ranged from 6.1 to 10.1 per cent, depending on the rates of fertilizer N, in Sarawak, 8.2 to 13.1 per cent (2.6–9.7 per cent DCP) in Trinidad (from

Butterworth, 1967) but low content of CP, 6.0 per cent, and high content of CF, 37 per cent, were reported by Blasco & Bohórquez (1968) from Colombia. In the last decade *B. decumbens* has become particularly popular in the Caribbean area, especially in Jamaica, as a grass suitable for the replacement of Pangola grass (*Digitaria decumbens*) badly affected by stunting virus.

Brachiaria decumbens is readily grazed and liveweight gains of grazing steers were reported to be of the order of 0.6 kg/ha/day/animal (Crowder *et al.*, 1970). It was observed, however, that continuous grazing of *B. decumbens* herbage for a long period of time can sometimes result in a form of scouring adversely affecting animal productivity.

Brachiaria decumbens can flower profusely but seed is only scarcely formed; moreover poor germination, attributed mainly to impermeability of hard floral scales firmly enclosing the caryopsis, was observed. Dipping in concentrated sulphuric acid for 10–15 min. increased seed germination from 1 to 30 per cent (Grof, 1968). Because of the scarcity of seed supply the grass is usually propagated vegetatively by cuttings or tuft splits which take well under wet weather although poor survival of planted sprigs was observed in Colombia. Reproduction is apomictic. Cv. **Basilisk** has been recently selected and registered in Australia.

Brachiaria dura Stapf

A densely tufted perennial up to 70 cm high with the stems geniculately ascending at the base. The linear flat or convolute leaves are up to 30 cm long. The single raceme (rarely two) is erect or arching, 3–4 cm long. Spikelets glabrous, 4 mm long.

Occurs naturally in central Africa. In Zambia it grows at about 1,000 m alt under an annual rainfall of 800–900 mm, often on old fallow land on sandy soil of low fertility, where it is common and can be abundant. The roots are covered with a dense growth of hairs, and exudate a glutinous substance cementing sand grains on to the surface of the roots, which is considered to be an adaptive feature on sands. The plants are highly palatable to grazing animals and contain about 7.5 per cent CP and over 40 per cent of CF; CP content decreases to 6.4 per cent at flowering, but is higher than in other grasses of the same plant communities. Attempts to establish *B. dura* from seed failed, but it can be planted from tuft splits (Verboom, 1966).

Brachiaria humidicola (Rendle) Schweickt. Creeping signal grass; Coronivia grass (Fiji) $2n = 72$

In cultivation this grass is better known as *B. dictyoneura*, and under this name a productive type was introduced from Rhodesia to Kenya and from Kenya, and perhaps also direct from Rhodesia, to other tropical countries. True *B. dictyoneura* (Fig. & de Not.) Stapf differs from *B. humidicola* in being a tufted perennial, whereas the latter is strongly stoloniferous; there are also differences in minor details and in the number of chromosomes, which are $2n = 42$ in *B. dictyoneura* and $2n = 72$ in *B. humidicola*. Reports on '*B. dictyoneura*' from Fiji and other countries are considered here as referring to *B. humidicola*.

Perennial, with flowering stems up to 50 cm high and numerous stolons forming a dense cover. Leaves lanceolate, usually glabrous and somewhat crisp. Panicle of three to five racemes 2–5 cm long. Spikelets about 5 mm long, slightly hairy. The lower glume is as long as the spikelet. Fertile floret is 4 mm long.

Brachiaria humidicola is indigenous to East and South-East tropical Africa where it occurs in relatively moist areas and situations, and has been introduced to Australia and Fiji. In the year of establishment, and often also in the following year, the flowering heads are numerous, but seed is very sparsely formed and the plant is easily propagated by cuttings or rooting tuft-splits; quicker establishment is achieved if the splits contain one or a few tillers than if large clumps are planted. Very few flowering heads appear in well-established swards.

In Kenya *B. humidicola* (as *B. dictyoneura*) was experimentally grown and grazed in the late 1940s, and good yields of herbage were obtained, but at present it is used, with excellent results, only for sports ground cover or for soil conservation, but experiments with *B. humidicola* as a pasture plant are still in progress in Rhodesia. *Brachiaria humidicola* is considered a promising introduction in Australia, and in Fiji it is one of a few most widely used grasses in moist areas where it produces high yields of herbage, and shows a good response to fertilizer N. In a fertilizer trial it yielded 10.8 t DM/ha, unfertilized, and 33.7 t when given 450 kg N/ha (Roberts, 1970). In another Fiji report, exceptionally high gains of steer liveweight were recorded. The rapid formation of dense cover hinders the establishment of grass/legume mixtures, and when the legumes are sown before the grass is planted they can suppress grass establishment. Nevertheless some success was achieved with sod seeding of *Macroptilium atropurpureum* (Siratro) and *Centrosema pubescens* into closely mown stands of *B. humidicola*. In Rhodesia admixture of *Trifolium repens* or *Lotononis bainesii* increased the yields of herbage.

Brachiaria miliiformis (Presl) Chase $\qquad 2n = 54$–56 and 72

A glabrous annual 30–50 cm high. The panicle consists of three to four racemes which are 3–4 cm long. The spikelet is 4–5 mm long and the fertile floret 3 mm long.

This annual occurs naturally in India, Burma, Sri Lanka, Coco Islands, Malaysia and Western Australia, and is reputed to be a good fodder grass. In a comparison with *B. brizantha* in Sri Lanka (Santhirasegaram & Ferdinandez, 1967), the two grasses produced equal yields when unfertilized, but when given 125 kg N/ha *B. miliiformis* outyielded *B. brizantha* by 50 per cent, and by 400 per cent with 350 kg of fertilizer N. The reduction of daylight by 25 per cent reduced the herbage yields.

Brachiaria mutica (Forsk.) Stapf (*Panicum muticum* Forsk., *P. purpurascens* Raddi, *P. barbinode* Trin.) Pará grass; Mauritius grass, Malohillo (León & Sgaravatti); Angola grass, Capim angola (Brazil); Egipto (Mexico, Parsons); Amirable (Parsons); Penhalonga grass (Madagascar) (Fig. 10) $2n = 36$

Perennial with flowering stems 1–2 m high ascending from long, many-noded prostrate shoots freely rooting at the nodes and forming dense cover. Leaves glabrous, or occasionally slightly hairy, linear to lanceolate, 10–30 cm long and 8–20 mm wide. Panicle of numerous (10–20) single, or sometimes paired or grouped and occasionally branched racemes 2.5–15 cm long. Spikelets 3–4 mm long, glabrous, in two rows

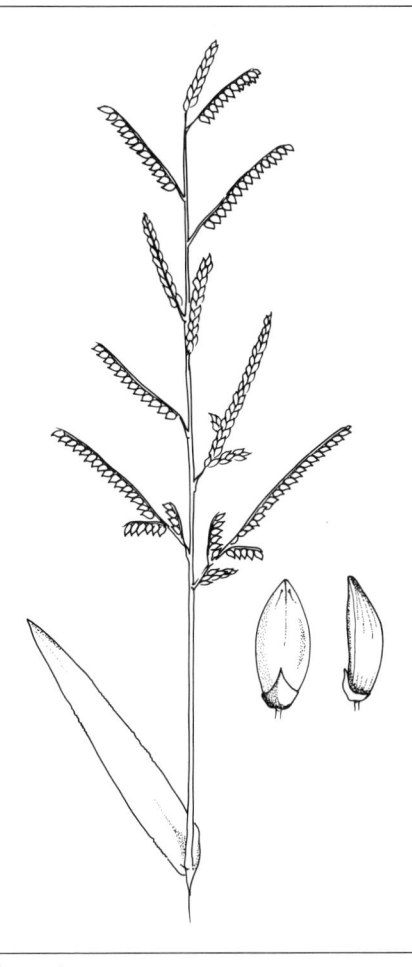

Fig. 10 *Brachiaria mutica.*

and, if one spikelet is sessils and the other on a pedicel, seemingly in four rows. Lower glume about one third to one half the length of the spikelet. Fertile floret about 3 mm long, pale yellow when ripe.
In Vol. 9 of *Flora of Tropical Africa* Stapf states that *B. mutica* 'is apparently a native of South America and West Africa', but Parsons (1972) maintains that it was introduced, probably accidentally, to America from Africa a long time ago and was established in Brazil in the 1820s if not earlier. It spread then to other parts of America where it naturalized. *Brachiaria mutica* is one of a few tropical pasture grasses which are cultivated on a large farm scale, and in some countries, especially in the South and Central American tropics, it has become a grass of considerable economic importance. Large-scale cultivation has also been reported from Australia, Fiji, the Philippines, Puerto Rico, Cuba, humid West Africa and Rhodesia. This wide spread of *B. mutica* can be explained by the ease of vegetative propagation, the competitive vigour of the grass and high yields and good quality of herbage.

Brachiaria mutica forms colonies on stream banks and in seasonally flooded valleys and lowlands and can withstand waterlogging and long-term flooding but it cannot grow on dry land in arid or semi-arid areas. It is therefore suitable for cultivation either in the humid tropics and subtropics or on moist or irrigated soil in less humid areas.

Establishment. Establishment by cuttings or pieces of creeping shoots is the usual practice, and planting 1–2 metres apart can give good results; the weeds that appear are later suppressed because of the high competitive vigour of the grass. In Taiwan (Wang *et al.*, 1969) newly established stands contained only 27–35 per cent of *B. mutica* in the first year, 91–93 per cent in the second year and 100 per cent in the third year. In Fiji (Roberts, 1970), the only weed not suppressed by *B. mutica* was *Mimosa pudica*, which can constitute up to 50 per cent of the herbage.

Mixtures with legumes. The competitive vigour of *B. mutica* interferes with the co-existence of legumes and their establishment in mixtures often fails, although *B. mutica/Centrosema pubescens* mixtures were established in Fiji and Queensland and mixtures with *Stizolobium deeringianum, Cajanus cajan, Lablab purpureus* and *Pueraria phaseoloides* in Colombia (Lotero *et al.*, 1960); in Colombia, however, only the last named legume, although slow in establishment, persisted throughout the five cuts taken in the trial. The technique of establishing *B. mutica*/legume mixtures in northern Queensland (Seton, 1962) was to drill the legume on ridges and to plant the grass between the ridges 2–3 months later. Mixtures with kudzu (*P. phaseoloides* and other legumes usually give lower yields but better herbage quality than pure grass.

Management, fertilizing. Low cutting or grazing *B. mutica* swards is favoured and cutting 1–7 cm above the ground level can result in some

20 per cent higher yields of herbage than cutting at 15-20 cm. Fertilizing, especially with N, is needed for high herbage production and in India (Rai et al., 1966) 100 kg N/ha increased the yields of DM by 49.2 per cent, of TDN by 79.2 per cent and of CP by 233 per cent, increasing the stocking capacity of the pasture and raising the animal production. Gains of 21-47 kg DM per kg of applied N have been recorded; responses to potassium have also been observed. Much higher rates of N are sometimes used and the management of *B. mutica* swards in central Venezuela during the dry season includes irrigation every 4 or 7 days, the application of 200 to 400 kg N/ha annually and cutting every 8 to 10 weeks (Novoa & Rodriguez-Carrasquel, 1972).

Herbage yields and quality. Herbage yields vary enormously depending on the conditions of growth and range from 9 to 135 t fresh herbage or from 3 to 39 t DM/ha, the highest mentioned yield of DM being reported from Puerto Rico (Vásquez, 1965) for irrigated *B. mutica* fertilized with 450 kg N/ha, but usually yields of 5 to 12 t DM/ha can be expected. In Surinam DM yields were determined as 16 t/ha/year and CP yields as 1,872 kg (Dirven, 1963).

Chemical compositions of herbage vary and various authors report CP contents ranging from 2.8 to 16.1 per cent, CF from 28 to 34 per cent, NFE from 41 to 57 per cent and EE from 0.9 to 3.9 per cent; high contents of P which can reach 0.80 per cent, but are usually lower, have been reported. Dirven (1962) gives CP contents in the leaves ranging from 10.5 to 14.0 per cent and from 3.4 to 5.9 per cent in the stems; and in another Surinam trial the content of CP decreased with the increased intervals between cuttings, a usual phenomenon. Herbage digestibility varies within very wide limits and in Butterworth (1967), who quotes the data obtained by several authors, it ranged from 39 to 63 per cent for DM, 17 to 81 per cent for CP, 40 to 78 per cent for CF, and 21 to 84 per cent for EE. TDN content ranged from 41 to 71 per cent and DCP from 0.5 to 11.8 per cent. Chapman et al. (1960) give 63 per cent digestibility for DM, 64 per cent for CP, 50 per cent for CF and 69 per cent for NFE and these figures can perhaps be accepted as average, except for CF, for which an average digestibility can be higher. Combellas and Gonzáles (1973) give, however, higher figures for CP digestibility, 68-75 per cent at its contents of 9.2-14.7 per cent in the consumed herbage. It is worth noting that Dirven (1962) determined CP digestibility in the leaves as being 54-62 per cent and 72-77 per cent in the stems in spite of the fact that the leaves contained twice as much CP as the stems.

Conservation. Ensiling *B. mutica* herbage gave relatively poor results. Attempts to ensile the herbage with 1 kg molasses added to 1,000 kg ensiled material resulted in a high pH of silage which contained large amounts of butyric acid and considerable losses of DM. A better quality silage was obtained with 7-8 kg molasses but its uptake by the animals was relatively low and so were milk yields from the cows fed on the sillage (see in Catchpoole & Henzell, 1971). *Brachiaria mutica* herbage dries slowly when cut, is hardly suitable for haymaking and is used mainly for grazing.

Animal production. Milk and beef production from fresh herbage of *B. mutica* can be reasonable to high, especially if the grass is well fertilized. In Fiji, 1 ha of unfertilized pasture supported 3.2 cows, pasture fertilized with 192 kg N/ha supported 4.2 cows and fertilized with 384 kg N 5 cows; milk yields per day per ha were 11.2, 22.0 and 29.6 kg, respectively (Roberts, 1970). In South America, steers which graze unfertilized grass can gain 0.60 kg liveweight per day and up to 0.80 kg when grazing on pastures fertilized with N. In Florida, liveweight gains of steers reached about 900 kg/ha, while much lower, but still reasonable gains, 309–356 kg/ha, were reported from Mexico (Carrera & Ferrer, 1963).

Flowering, seed production. In Taiwan (Wang, 1961), *B. mutica* flowered under a wide range of natural day lengths, from 10 hours 19 min. to 13 hours 42 min., or under photoperiods about 2 hours shorter or longer than the average day length, and the shorter photoperiods hastened flowering. Flowering usually occurs in early morning hours. In the Northern Territory of Australia, *B. mutica*, established vegetatively, apparently from a clonal material, did not produce viable seed and to obtain germinable seeds the plants had to be grown from seed (Wesley-Smith, 1973), which suggests cross-pollination rather than apomixis, the latter being a mode of reproduction found in a number of perennial species of *Brachiaria*. Grof (1969a) had, however, indicated earlier that *B. mutica* can perhaps produce seed only under the conditions of the humid tropics. Seed yields are generally low but the percentage of seeds containing caryopses is normally greater than in other perennial species of *Brachiaria*. Moderate rates of fertilizer N can increase seed yields, and in Grof's trials (1969a) unfertilized *B. mutica* yielded 13.3 kg seed/ha, 25.1 kg when given 56 kg N/ha and 30.8 kg at a double rate of N. These yields were obtained at the correct date of harvesting, soon after the anthesis has ended; a delay of harvesting by 1 week reduced the yields to 7.0 kg, unfertilized, and 13.5 kg under the single rate of fertilizer N. Moreover, seed harvested early had 75–77 per cent purity and 51–57 per cent germination, whereas those harvested 1 week later showed 76–83 per cent purity but only 31–40 per cent germination. There seems to be no post-harvest dormancy and seed can be sown almost immediately after it has been harvested.

Brachiaria plantaginea (Link) Hitchc. (*Panicum plantagineum* Link). Marmalade grass.

Annual with the stems up to 1 m long, often branched, usually decumbent and rooting at the lower nodes. Leaves glabrous, linear-

lanceolate, 5–25 cm long and 8–15 mm wide. Panicle of 3–11 racemes, the lowest raceme 6–10 cm long. Spikelets 4–5 mm long, glabrous. Lower glume clasps the spikelet and is somewhat distant from the rest of the spikelet.
Occurs naturally in western tropical Africa extending to Zaire and the Cameroons. Introduced, perhaps accidentally, to North and South America and can be found in south-western USA, Mexico and further south to Bolivia and Brazil where it has been naturalized. *Brachiaria plantaginea* is a popular grass in Brazil where it is cultivated to a small extent for feeding green and produces a large bulk of high-quality herbage. Seed is abundantly formed and yields of 670 kg/ha have been reported (Whyte *et al.*, 1959).

Brachiaria radicans Napper. Tanner grass $2n = 36$

Perennial. Stems up to 120 cm long, usually ascending from a long creeping base. Leaves lanceolate, cordate at the base, 7–16 cm long and 7–14 mm wide. Panicle of six to nine solitary racemes, the lowest 4–6 cm long. Spikelets subsessile, ovate, 3.5 mm long, in two clear rows.

This species is sometimes confused with *B. rugulosa* Stapf which, however, differs clearly by the presence of long hairs (cilia) on raceme margins and by some less definite characters. It can also be confused with *B. mutica* as the difference between the two species lies mainly in the inflorescence which has more numerous and longer racemes in *B. mutica* than in *B. radicans* and in *B. mutica* they can be paired or branched. Renvoize in a private communication (1974) states that 'in many ways such differences are scarcely sufficient to justify specific recognition.... Nevertheless, it is possible to distinguish plants which have been named *B. radicans* from *B. mutica* and for this reason I consider that *B. radicans* is worthy of recognition'.

In a wild state *B. radicans* occurs apparently throughout tropical Africa and has been recorded from Nigeria, Cameroons, Zaire, Rwanda, Ethiopia, Sudan, Uganda and Tanzania (Renvoize, 1974), on swampy ground and at the margins of lakes and streams where it can form extensive colonies of low leafy herbage under tall stems, especially when grazed. It has also been introduced into cultivation, the cultivated form originating from South East Africa: Joe Tanner, a Rhodesian farmer, brought *B. radicans* from a farm near Durban to the Marandella Grassland Research Station where it was grown as a pasture grass and later introduced to some other African countries and to French Guyana and Brazil in South America (Renvoize, 1974), and was grown in these countries and in Puerto Rico with reasonable success. In Puerto Rico (Sotomayor-Ríos *et al.*, 1973) it was comparable with six other grasses under trial. When well fertilized with NPK, irrigated whenever necessary and cut at the optimum stages of growth it yielded over 34 t dry fodder/ha/year; the herbage contained 7.8 per cent CP and CP yields up to 2.5 t/ha were recorded. In Brazil it was found, however, to be toxic to cattle and buffaloes. In an experiment with cows and heifers in 1970, symptoms of poisoning appeared after 8–37 days of grazing, but the animals recovered when removed from the Tanner grass pasture (Andrade *et al.*, 1971a). An analysis of Tanner grass herbage revealed that the grass contained 0.55–0.90 per cent nitrate (KNO_3 equivalent), the content of nitrates being much higher than in *B. decumbens*, *B. ruziziensis* or *B. brizantha* in which their contents ranged from 0.025 to 0.058 per cent and the toxic effects of Tanner grass were ascribed to high contents of nitrates (Andrade *et al.*, 1971b). In another trial in Brazil, buffaloes avoided Tanner grass, but when forced to graze it some animals died and the other developed symptoms of poisoning (Oschita *et al.*, 1972).

Brachiaria ruziziensis Germain & Evrard. Congo Signal grass; Congo grass; Ruzi grass; Kennedy Ruzi (Australia)

Perennial with flowering stems up to 1 m high, arising from many-noded creeping shoots which form dense and leafy cover. Leaves hairy, lanceolate, 10–25 cm long and 10–15 mm wide. Panicle of three to six racemes which have a flattened rhachis 3–5 mm wide; the lower racemes are 6–10 cm long. Spikelets hairy, 5 mm long. Lower glume 3 mm long, arising 0.5–1 mm below the rest of the spikelet and clasping the spikelet. Fertile floret 4 mm long.

Brachiaria ruziziensis is closely related to *B. decumbens* but is usually larger and differs in having the lower glume distant from the rest of the spikelet. It has also been confused in the past with *B. brizantha*. This species is indigenous to East–Central Africa where it occurs under humid, but not waterlogged, conditions and has been recorded in Zaire and western Kenya. It was first introduced into cultivation in the Congo (now Zaire) where, together with *Setaria anceps*, it forms the basis of sown pastures; a mixture with *Stylosanthes guianensis* maintained a high productivity of pasture in the dry season. A productive and palatable mixture with *Desmodium intortum* was successfully grown in Rwanda. *Brachiaria ruziziensis* has also been cultivated in Kenya, mainly on an experimental scale, with encouraging results. It was introduced to Queensland in Australia, the Philippines, Surinam and, on a small scale, to some other tropical countries. In the humid tropics of Queensland, the five introductions tested yielded 2–2.5 t DM/ha, the yields being lower than those of the other grasses under trial (Grof & Harding, 1970). Very high yields of fresh herbage, 133–145 t/ha, and unusually high liveweight gains of beef cattle, up to 1,300 kg/ha/year, were reported from Surinam (Hunkar, 1969), but milk yields of 5.8 l per day were only moderate. Butterworth (1967) gives in his review high CP contents, ranging from 9.9 to 13.9 per cent, and DCP 6.9–9.6 per cent. The proportion of nitrate N (which can produce an adverse effect on the animals) in the total N can be high in *B. ruziziensis* herbage. The grass is propagated either by seed or vegetatively by rooting cuttings.

In trials under controlled environment (Deinum & Dirven, 1967) in which three levels of day/night temperatures (26°/20°C, 30°/23°C and 34°/25°C) and three levels of light intensity were applied, low temperature combined with high light intensity encouraged tillering. High light intensity gave the highest yields of DM: under high temperatures for the first cut and under lower temperatures for the second cut. Low temperatures also resulted in high content of N and of CP in the herbage and in a low content of CF.

Flowering is often abundant, but yields of viable seed are relatively low, up to 100 kg/ha (Toutain, 1973) and so is germination of untreated seed, but in trials in Australia (McLean & Grof, 1968) sulphuric acid treatment increased the germination from 17 to 40 per cent. Reproduction is apomictic. Cv. **Kennedy** has been selected in Australia.

Bromus L.

Some 50 species of temperate regions and a few species also occur at high elevations in the tropics. Annuals or perennials and a few perennials, including *B. inermis* Leyss, are good grazing and fodder plants.

Bromus unioloides Kunth (*B. catharticus* Vahl, *B. willdenowii* Kunth).
Rescue grass; Prairie grass; Criolla; Cebadilla; Nakuru grass.

This species was also known in East Africa under the wrong name of *B. marginatus*. Ravens (1960) regards the cultivated rescue grass as *B. willdenowii* and the South American wild types as *B. unioloides*; this treatment is, however, not generally accepted (see e.g. in Clayton, 1970), and *B. unioloides* is the commonly accepted botanical name for rescue grass. $2n = 28, 42$ or 56

Tufted short-lived perennial up to 100 cm high, with loose pendent panicles, and large laterally compressed spikelets 15–40 mm long, with 6–12 florets. Large florets terminate in a short (3 mm long) awn.

This grass originates from temperate and subtropical areas of South America, and is now cultivated in a number of warm-temperate and subtropical countries: southern Brazil, Argentina, Uruguay, southern Australia and south-eastern states of the USA, and also at high altitudes in the tropics, e.g. Colombia and Kenya. In Kenya it is, however, losing its popularity largely because it seldom lasts more than two years. The type grown in the Nakuru district of Kenya was introduced to some countries in which it is known as Nakura grass or cv. Nakuru.

Bromus unioloides is grown from seed sown at 10–30 kg/ha; the young plants are strong and establishment presents no difficulties. The grass is used mostly for hay and yields of 4–5 t hay/ha from the first cut were recorded from South America, and annual yields of 8 t/ha from Queensland. The herbage is usually of satisfactory quality and contents of 4.1–4.3 per cent DCP were recorded (Reichert & Trelles, 1923). It can be grown in mixtures with lucerne and in temperate areas also with sainfoin or red clover. Seed production is good. Seeds are large and there are 135,000 seeds per kg (Whyte *et al.*, 1959). The better-known cultivars are **Lamont**, **Nakuru**, **Priebe** and **Chapel hill**, and Whyte *et al.* also mention cv. **Pergammo** and **Angel gallardo**.

Cenchrus L.

Tufted or occasionally rhizomatous perennials, or annuals. Panicles dense, spikelike. Single spikelets, or more often groups of two to seven spikelets, are surrounded by bristles, usually hard, flattened and united in the lower part forming an involucre or a cup. Spikelets lanceolate or ovate, pointed, with two florets of which the upper is bisexual. The 25 species of the genus are distributed in warm, predominantly tropical and subtropical areas of the world, mostly in arid and semi-arid regions. *Cenchrus ciliaris* and *C. setigerus* are excellent grazing perennials, but species with hard and retrorsely scabrid involucres, and especially *C. tribuloides* in North America, are undesirable in pastures because the involucres penetrate into sheep's wool creating a problem for wool producers.

Cenchrus biflorus Roxb. (*C. barbatus* Schum.)

Annual 10–100 cm high. Leaves 3–30 cm long and 2–7 mm wide, flat, green or bluish-green. Spikes dense, cylindrical, 3–15 cm long and 10–12 mm wide, pale, rarely purplish. Involucres with numerous bristles connate at the base, the outer retrorsely scabrid, at length horizontally spreading, the inner 4–7 mm long, rigid, flattened towards the base. Spikelets in groups of one to three, 2–2.5 mm long. Lower floret barren, upper bisexual, fertile.

Occurs mainly in the northern parts of tropical Africa, and also in Mozambique, and on Indian plains, in dry areas and habitats. In the Sahel Zone of Africa (south of the Sahara) Bartha (1970) observed that *C. biflorus* in the early stages of growth is liked by all grazing animals and is grazed well again when the prickly spikelets have shed. During the rainy season it can be cut a few times, persists almost throughout the dry season, and is of importance as a reliable source of fodder. When ensiled, the fermentation softens the hard and prickly bristles and the silage can then be eaten by the animals. According to his analysis the herbage contained 10.0 per cent CP, 1.6 per cent EE, 42.8 per cent NFE, 34.6 per cent CF, 0.28 per cent Ca and 0.35 per cent P.

Cenchrus ciliaris L. (*Pennisetum ciliare* (L.) Link, *P. cenchroides* Rich.). Buffel grass; African foxtail; Anjan; Dhaman; Kolukattai. (Fig. 11a)

Tufted or rhizomatous perennial; rhizomes, if present, are short, stout and not numerous. Stems erect of ascending, up to 140 cm high, stout or slender. Leaves glabrous or hairy, 7–30 cm long and 2–5 mm wide, green or bluish, flat. Panicle in the form of a single cylindrical spike (ear), erect or nodding, 3–15 cm long and 8–20 mm wide, pale or purplish, bearing numerous clusters of spikelets, each cluster being surrounded by an involucre of bristles. The outer bristles slender, shorter or slightly longer than the spikelets, the inner bristles longer than the spikelets, about 12 mm long, ciliate, often flexuose, slightly flattened at the base; one bristle is longer and more stout than the others. Spikelets solitary or in clusters of two to three, lanceolate, 3.5–5 mm long; the lower floret male or empty, the upper fertile. Caryopsis oblong, dorsally compressed, 2–2.3 mm long.

Distributed in tropical and subtropical Africa, North Africa, Madagascar, Canary Islands, Arabia, tropical and subtropical India and Pakistan, mostly in dry areas, from sea level to 2,000 m alt.

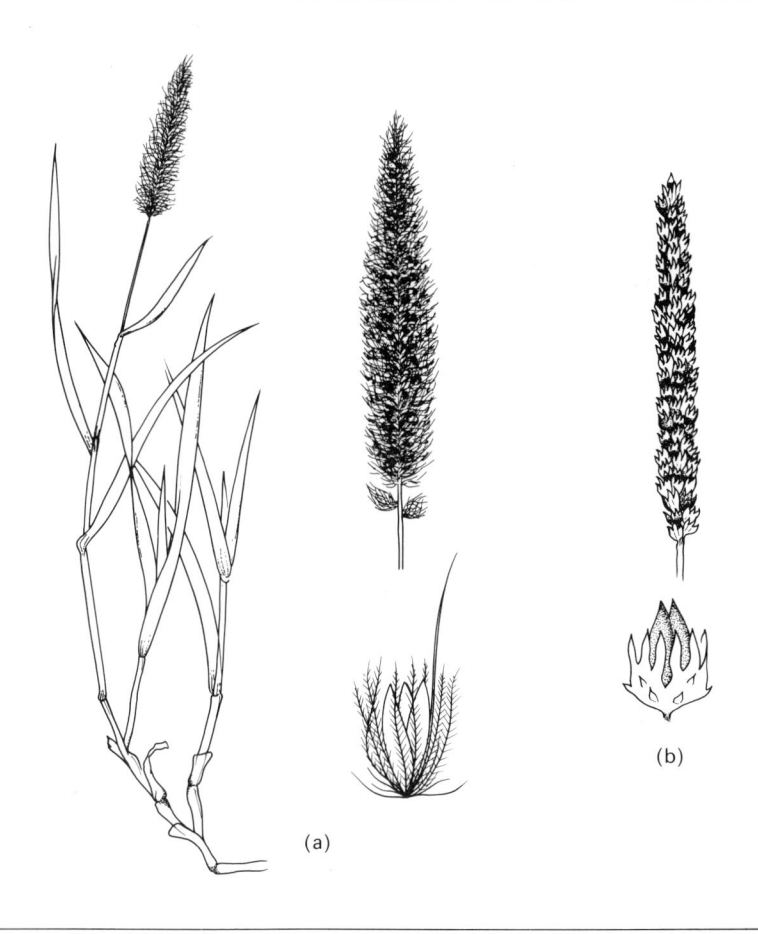

Fig. 11 (a) *Cenchrus ciliaris*, (b) *C. setigerus*.

Environment. *Cenchrus ciliaris* is a drought-resistant grass and can grow under an annual rainfall from 270 to 300 mm upwards in the areas of Kenya with two rainy seasons a year and from 400 mm in the areas with a single wet season and under similar or slightly higher rainfall limits elsewhere. The drought resistance of *C. ciliaris* can be partly explained by the very deep root system. It thrives at high air temperatures, the optimum temperature for photosynthesis being about 35°C. *Cenchrus ciliaris* grows best on loams and can do well on loose sandy soils or on alluvial silt but avoids heavy clays or soils deficient in calcium. In Australia it withstood 5-day flooding without any loss of plants and 20-day flooding with losses of 15–70 per cent plants depending on the cultivar, but only if the grass had not been recently cut. (Anderson, 1970). Field germination of seed and seedling emergence were greatly reduced by flooding for 10 and 40 days, the reduction also depending on the cultivar: in cv. American and Boorara the emergence under short-term flooding was reduced to 35 per cent of that without flooding and to 4–6 per cent under long-term flooding, in cv. Biloela, Nunbank and Tarewinnabar to 8–16 per cent and 0.5–7 per cent, respectively, and practically to nil in cv. Molopo and Gayndah (Anderson, 1972).

Introduction. *Cenchrus ciliaris* is cultivated in the countries of its origin and has also been introduced to America and Australia, the first introductions to Australia in the nineteenth century being apparently accidental. The grass is valued mainly for its remarkable drought resistance and for good herbage quality and palatability observed in dry tropical and subtropical areas of Africa, India, Australia and in southern USA; in more moist areas where more productive grasses can be grown, *C. ciliaris* is not particularly popular. Because of its relatively easy establishment and persistence this grass is much used for reseeding denuded arid pastoral land or for improving worn-out pastures in Australia, India, Pakistan and East Africa.

Establishment. *Cenchrus ciliaris* is established almost exclusively from seed although it is also recommended in India to raise young plants in the nursery and transplant them later to the field. Direct vegetative propagation from tuft splits or rhizomes is also advocated in India and Bolivia (Rossiter & Delgadillo, 1971). In Australia, Coaldrake & Russell (1969) successfully established *C. ciliaris* by seed broadcasting into burnt brigalow grass and without any soil tillage, but in trials in Kenya and elsewhere seedbed preparation was found essential for achieving reasonable establishment in denuded pastoral land. In trials in Tanzania, rolling after sowing improved the establishment and so did cattle trampling, although to a lesser extent. Seed of *C. ciliaris* germinate best from a depth of 1–2 cm. Various seed rates have been tried and suggested, and 3 to 12 kg/ha are the rates usually recommended;

Rossiter & Delgadillo (1971) advise 6–8 kg/ha for drilling in rows and 12 kg for broadcast sowing. Sowing whole clusters of spikelets (as harvested) is usually preferred to sowing naked caryopses and in Tanzania 5-fold better seedling emergence was observed from clusters than from naked caryopses (Brzostowski, 1961). Under laboratory conditions naked caryopses germinate well, often better than those in involucres, but in the field the clusters almost invariably give better results. Lahiri & Kharabanda (1962–3) in India found water-soluble germination inhibitors in the glumes and this was also reported from Rhodesia. The presence of inhibitors can be regarded as an adaptation in that seeds are prevented from germination under the effect of occasional off-season rains and germinate only if the soil has been moist for a sufficient time to allow for leaching the inhibitors and for the seedlings to survive. In Australia (Bryant, 1961) naked caryopses germinated in the soil if they had been freshly de-husked and mixed with superphosphate immediately before sowing; even a 12-hour delay in sowing the superphosphate-mixed seed reduced the germination. A certain proportion of seeds can remain in the soil for a few years without losing their ability to germinate. When oversowing natural grassland seed pelleting with f.y.m. and lime improved the establishment in Australia and in Tanzania but produced no effect in Kenya. Insecticides are sometimes needed for preventing harvester ants from carrying the seeds away and seed dressing with aldrin in Kenya and with lindane or aldrin (with or without pelleting) in Australia (Champ *et al.*, 1961) resulted in improved establishment. Moderate rates of superphosphate at establishment are recommended, and Chakravarty & Verma (1972) have shown that hand weeding, if it can be afforded, can considerably help the establishment. In a trial in India weeding once a year resulted in an 86 per cent increase in herbage yields in the next, second year of growth but two weedings gave only slight further yield increases.

Management, fertilizing. Cenchrus ciliaris is predominantly used for grazing, but in India the herbage is often cut and fed green. It is also reported from India and Bolivia that *C. ciliaris* makes good hay or silage. In some parts of Australia it is also used for off-season grazing. The establishment is slow and the grass can be utilized some 9–12 months after sowing. Close cutting or grazing, to 5–10 cm from the ground level, is reported to give higher yields and better herbage quality than cutting at higher levels and the basal shoots, stimulated by close cutting, grew vigorous and productive (Khan, 1970). On the other hand Pandeya & Jayan (1970), in India, showed that frequent close clipping reduced the yields of the next cut, and it has also been noted that close cutting reduces the content of carbohydrates in the plant and the volume of roots. Green shoots often sprout during the dry season from old, seemingly dry stems, as was observed in Tanzania and elsewhere; these shoots contributed to the volume of dry-season grazing and it is not advisable to top the stems after the wet season grazing. *Cenchrus ciliaris* is very persistent and in trials in Tanzania, where it was grown in mixtures with *Chloris gayana*, it lasted much longer than the companion grass and its proportion in the herbage increased under grazing (Brzostowski & Owen, 1964). *Cenchrus ciliaris* usually responds well to fertilizer P and in northern Australia P applied to a *C. ciliaris/ Stylosanthes humilis* mixtures benefited more the grass than the legume. The reaction to N varies: in Tanzania, where *C. ciliaris* was grown under low rainfall and on reasonably good soil, responses to fertilizer N were obtained only in exceptionally wet seasons, whereas in Australia, especially in Queensland, and also in South Africa, good responses to N, and especially to PN, were frequently observed.

Diseases. The only serious diseases of *C. ciliaris* are ergot and smut which attack the spikelets, especially during the rains, and can damage or even destroy seed crops. Some types of *C. ciliaris* are however almost free from spikelet diseases, whereas some others are much affected and selection of resistant cultivars is perhaps the only practical answer for controlling fungus infections.

Association with legumes. A grass of dry areas, *C. ciliaris* has been tried mainly in mixtures with relatively drought resistant perennial or annual legumes. In northern Australia, *C. ciliaris* is often grown on a farm scale in mixture with *Stylosanthes humilis*. Both components are sown broadcast but the best results were obtained from drilling *C. ciliaris* in rows and growing the legume between the rows. If the main emphasis is on *S. humilis* then *C. ciliaris* is grown in wider rows, 180 cm (6 ft) apart, otherwise in 90 cm rows. All the mixtures with *S. humilis* yielded considerably more herbage than pure grass and heavy grazing or frequent cutting increased the content of the legume in the sward. Good results were obtained in Queensland and in Tanzania from *Macroptilium atropurpureum* mixtures; in Queensland liveweight gains of cattle grazed on such a mixture were equal to those obtained from *C. ciliaris* given 168 kg/N/ha (t'Mannetje, 1972). Of other perennial legumes *Stylosanthes guianensis* in Tanzania and *Medicago sativa* in Australia showed good promise. In Bolivia, mixtures with *S. humilis* are recommended for the driest parts of country, with 600–700 mm of annual rainfall, with *Glycine wightii* plus *M. atropurpureum* for the areas with 700–800 mm of rain and with *M. atropurpureum* for still wetter areas. In Malawi, a reasonable success was obtained from *Desmodium intortum* mixtures. In India, annual *Cyamopsis tetragonoloba*, *Vigna aconitifolia* and *Lablab purpureus* were compatible with *C. ciliaris* when grown in alternate grass–legume rows. Mixtures with *Centrosema pubescens* were also tried. *Cenchrus ciliaris* was also grown, mainly experimentally, with *Chloris gayana* in Tanzania, and with *Lasiurus hirsutus* and *Dichanthium annulatum* in India, the two grasses, especially *D. annulatum*, showing lesser competitive vigour than *C. ciliaris*.

Herbage yields. Cenchrus ciliaris is not a heavy producer and herbage yields usually range between 2 and 8 t DM/ha or, according to Toutain (1973), 10–20 t green material or 3–6 t hay/ha/year. In Malawi, it yielded 7–8 t DM/ha when fertilized with N and 4–5 t when given no fertilizer N. In Tanzania (Walker, 1969) DM yields ranged from 3.58 to 5.60 t/ha, depending on the amount of applied N and grass receiving no N yielded 1.90 t; in old trials in northern Nigeria 2–3 tons DM/ha/year were obtained. In the dry areas of Pakistan *C. ciliaris* is reputed to be one of the best yielding grasses and in India Narayanan & Dabadghao (1972) report that 9.11 t green herbage/ha/year was obtained in the areas receiving less than 300 mm annual rainfall and they estimated that 22.4–28.0 t can be obtained under an annual rainfall ranging from 380 to 780 mm.

Chemical composition, nutritive value. Cenchrus ciliaris is a grass of high nutritive value and although CP content can range from 3 to 16 per cent it seldom drops below 6–7 per cent in the herbage cut green and about 3 per cent was recorded for completely dry plants harvested in the middle of the dry season. Similarly with other tropical grasses the content of CP in *C. ciliaris* is high during the early growth and decreases as the season advances and Milford (1960) reports decreases from 13.8 per cent to 7.2 per cent and from 10.8 to 7.1 per cent in the two varieties he studied. The content of NFE is around 45–50 per cent and that of CF ranges mostly from 29 to 40 per cent; Bredon & Horrell (1961) traced the increase of CF content from 30 per cent in the early growth up to over 40 per cent in the dry season. The content of EE ranges from 1.0 to 2.6 per cent. P content is usually adequate for the animal requirements and is reported to range from 0.30 to 0.65 per cent in India but was below 0.15 per cent in East African plants. DM digestibility according to Milford (1960) varies but is mostly 50–60 per cent and so is CP digestibility (50–61 per cent) although it can be as low as 30 per cent and even lower or as high as 74 per cent. CF digestibility of 35 to 59 per cent is given by Narayanan & Dabadghao (1972), and from 32 to 76 per cent by Butterworth (1967), who also quotes the digestibility of NFE as ranging from 43 to 73 per cent. Cell sap of *C. ciliaria* herbage has pH values around 5.4 and is more acid than in the majority of grasses; low pH of fresh herbage has also been recorded for species of two other genera allied to *Cenchrus*: *Setaria* and *Pennisetum* (Dougall & Birch, 1966). The content of water-soluble oxalate in *C. cilliaris* was determined in Australia (Jones & Ford, 1972) to be 1.4–1.8 per cent, the amounts which are not likely to cause any adverse effect on grazing animals. *Cenchrus ciliaris* is well grazed by all kind of stock and retains its palatability at the later stages of growth better than the majority of other tropical grasses, and is grazed, to some extent, even when 'dead dry'.

Animal productivity. Narayanan & Dabadghao (1972) consider about 1 ha of sown *C. ciliaris* for one head of cattle or six head of sheep as usual stocking rates in India but they can fluctuate depending mainly on the rainfall and soil fertility. In Kenya, gains of 50 to 80 kg beef per ha were obtained when the animals were grazed on *C. ciliaris*, which was more productive than the other grasses in the trial, and fed hay during the dry weather (Pereira *et al.*, 1961). In trials in Paraguay a pasture oversown with *C. ciliaris* supported 2.6 yearling steers/ha; daily liveweight gains of the animals were 0.49 kg/head or 1.3 kg/ha (Simpson & Fretes, 1972), a good performance for a dry-area grass.

Flowering, reproduction. The formation of floral parts of *C. ciliaris* depends to some extent on photoperiods and in Texas, USA (Evers *et al.*, 1969), the 12-hour photoperiod was more favourable compared to the 10- and 14-hour photoperiods: it induced earliest flowering giving the highest number of ears per plant and seeds per ear. The effects of photoperiods were, however, not uniform and differed in the eight lines used in the trial.

Cenchrus ciliaris is an apomictic grass (Fisher *et al.*, 1954) and had been considered an obligatory apomict until a sexual plant was found in the USA. As in a number of other apomicts, apomixis in *C. ciliaris* is linked with pseudogamie (Snyder *et al.*, 1955) under which pollination is needed not for the formation of the embryo but for the initiation of endosperm, otherwise no viable seed can be developed. Chromosome numbers vary in different lines and $2n = 32, 34, 36, 40, 44, 52$ and 54 have been found. Ramaswamy *et al.* (1969) have found some correlation between the chromosome number and plant morphology and vigour: tetraploids with $2n = 36$ were more variable than plants with $2n = 44$, but the latter showed more vigour; these correlations were however not always clear.

Variability, cultivars. *Cenchrus ciliaris* is a very variable species and in a study of 65 introductions by Chakravarty *et al.* (1970) stem height in spaced plants ranged from 45 to 140 cm, the number of tillers per plant from 6 to 98, leaf length from 15 to 38 cm and width from 2.1 to 5.3 mm, leaf area per plant from 2.6 to 12.4 cm^2, ear length from 4.8 to 10.7 cm, the average number of spikelets/ear from 35 to 250 and the weight of 1,000 clusters from 1.20 to 5.87 g. The presence and the intensity of antocyanin colouration of stems, ears and stigmas also varied and there seems to be an almost unlimited number of combinations of the agronomically important characters for selection of cultivars suitable for various climatic and soil conditions. Amongst the local and introduced material numerous types were selected as ready-made cultivars. Those grown and approved in Australia and described by Barnard (1972) belong to two major types:
1. Tall, vigorous, stoloniferous cultivars often with bluish leaves and hard stems which are regarded as suitable for cattle grazing.
2. Low-growing, less vigorous, non-stoloniferous cultivars with finer and softer stems, more suitable for sheep.

To group 1 belong: **Biloela**, selected at the Biloela Research Station, Queensland, from Tanzania introductions: the leaves are bluish; being a good seed producer this cultivar is well known outside Australia. Cv. **Molopo** originating from South Africa is more rhizomatous, later flowering and with somewhat poorer seed productivity than Biloela; it is also more cold-tolerant and suits cooler areas and for cool season grazing. Cv. **Boorara** has slightly finer stems and is more leafy than Biloela. Cv. **Lawes**, of South African origin, differs from Biloela by its wider leaves. Cv **Nunbank** is of Uganda origin and is rather similar to cv. Lawes. Cv. **Tarewinnbar** of Kenya origin differs from Biloela in having green leaves without any bluish tint, more robust stems and purplish ears. Cv. **Higgins grass** developed in Texas, USA, is reputed to combine good persistence with high yields of forage and seed.

To group 2 belong: Cv. **Gayndah**, which originates from Kenya; it produces numerous tillers and has been noted in northern Australia for its good compatibility with *Stylosanthes humilis*. Cv **West Australian** is an old type derived from the nineteenth century introductions; it has numerous tillers and narrow leaves; grown in north-east Australia with good success, it was outyielded in Queensland by a more robust cultivar of the Biloela type. Cv. **American** was introduced as commercial seed from USA; it is similar to Gayndah but is less dense and flowers earlier; its seedlings are said to be tolerant to soil acidity. Of the African cultivars, cv. **Chipinga** is a Rhodesian selection and is grown in Rhodesia and elsewhere. Cv. **Mbalambala** originates from south-eastern Kenya; it is of semi-prostrate habit, forms broad tufts with numerous tillers and is very palatable to cattle. Both cultivars are well known outside the countries of their origin and selection. Cv. **Higgins grass** is the first bred cultivar developed from a single sexual plant found in Texas, USA. Cv. **Pusa giant** developed in India is described as a hybrid between an Indian type of *C. ciliaris* and an American type of *Pennisetum ciliare* (from Narayanan & Dabadghao, 1972); as both names are synonyms and *C. ciliaris* is an apomict, the hybrid nature of Pusa Giant needs confirmation. Pusa Giant is highly productive and its herbage has a high content of CP; it is also tolerant to cold. *Cenchrus glaucus*, a species of uncertain status (Bor. 1960), can be provisionally considered as cv. **Blou buffel** of *C. ciliaris*. It is very leafy, with the main bulk of leaves high on the stem, and with a high leaf : stem ratio, 16 : 9; in India it yielded over 11 t fresh material/ha as a rainfed crop and over 56 t under irrigation (Narayanan & Dabadghao, 1972). Seed setting of Blou Buffel is good but germination is rather low for the species.

Seed production. *Cenchrus ciliaris* produces reasonable or sometimes high yields of good quality seed. Yields of 150–210 kg/ha were obtained in Tanzania (Brzostowski & Owen, 1966) without fertilizer N. In Queensland (Humphreys & Davidson, 1967) only 8 kg of seed/ha were harvested with no irrigation and no fertilizer N, but the application of

84, 168, 336 and 672 kg N/ha raised seed yields to 47, 168, 260 and 504 kg respectively, and split application of the larger rates gave still higher yields. Seed yields depend also on the cultivar. Seed is harvested either by hand, or mechanically in Australia and in Tanzania where a stripping machine was developed which gave better quality seed than hand harvesting. Seed production and harvesting of *C. ciliaris* is somewhat easier than of some other tropical grasses, such as *Panicum maximum*, because the seed is shed less readily. Most of the data available indicate that post-harvest maturation is needed and that it lasts for a few to 18 months and only the seed stored for at least 6 months, and, in some cases, up to 18 months, is recommended for sowing. However, in Tanzania, 90 per cent germination was observed in 1-month-old seed; the highest germination, 92.5 per cent, was reached after 18 months of storage and then declined with further storage, although $3\frac{1}{2}$-year-old seed still germinated to 80 per cent and 5-year-old to about 60 per cent; 8-year-old seed germinated to 4 per cent. Seed of *C. ciliaris* germinate fast, most of the seeds within 24 hours (Brzostowski & Owen, 1966). Seed is fluffy and therefore difficult to drill or even to sow by hand, and it should be mixed with fine soil or hammermilled to beat off the bristles. Spikelet clusters differ in size and weight and in India (Chakravarty *et al.*, 1970) the weight of 1,000 clusters ranged from 1.2 to 5.9 g. Bogdan (1966a) gives the weight of 1,000 clusters containing one caryopsis each as 2–3 g or 350,000–450,000 clusters per kg, but Whyte *et al.*, (1959) cite much smaller numbers per kg: 90,000–200,000. Seeds of the same plant can be large or small but no intermediate-sized seeds were found (Lahiri & Kharabanda, 1961). The larger seeds germinate better and produce larger seedlings.

Cenchrus pennisetiformis Hochst. & Steud. ex Steud. $2n = 42, 54$

Perennial but sometimes behaving as an annual. Stems up to 50 cm high. Leaves 3–15 cm long and 1–4 mm wide. Spikes cylindrical, 2–7 cm long, pale to purplish. Bristles of the involucres numerous, the outer slender, filiform, up to 7 mm long, the inner up to 12 cm long connate at the base into a cup 1–3 mm long, the free parts densely ciliate around the spikelets. Spikelets in clusters of two to five, 4–5 mm long. Lower floret male or barren, upper bisexual, fertile.

Occurs in northern tropical Africa: Sudan, Eritrea, Socotra, Egypt and tropical Arabia reaching the Mediterranean area in the north and India in the east. A valuable grass in arid and semi-arid areas which remains green for a considerable part of the dry season. Suspected to be a product of natural hybridization between *C. ciliaris* and *C. setigerus* as it grows usually in the areas where these two species occur, and its nutritive value approaches that of *C. ciliaris*.

Cenchrus prieurii (Kunth) Maire (*Pennisetum prieurii* Kunth) $2n = 34$

Annual up to 80 cm high. Leaves 10–30 cm long and 5–10 mm wide.

Spikes 8–12 cm long and 2–4 cm wide, pale to purplish. Involucre of numerous bristles connate at the base only. Outer bristles up to 10 mm long, inner 10–20 mm long densely ciliate in the lower part; one bristle is longer than the rest and up to 30 mm long. Spikelets in pairs, ovate, acute, 4–5 mm long. Lower floret barren, upper bisexual, fertile. This species can be distinguished in the field by its wide and somewhat silky spikes.

Occurs in the Sahel zone of north-western tropical Africa, and also in Sudan, Ethiopia and northern Kenya, in semi-arid and arid regions where it can be numerous. An excellent grazing grass much liked by animals. The plants analysed by Bartha (1970) contained 9.1 per cent CP, 1.8 per cent EE, 42.8 per cent NFE, 37.1 per cent CF, 0.23 per cent Ca and 0.15 per cent P.

Cenchrus setigerus Vahl. Anjan grass; Black kolukattai; Birdwood grass (Fig. 11b)

Tufted perennial without rhizomes, rather similar to smaller, non-rhizomatous types of *C. ciliaris* but of more erect habit. Morphologically it differs from the latter mainly by the structure of the involucre surrounding the spikelets: its inner bristles are short, thick and fused together up to one third of their length; the outer bristles are finer but also short and fused with the lower, solid part of the involucre. Two or three spikelets, or occasionally a single spikelet, are firmly enclosed in the hard involucre. The colour of the involucre is dark-purple or dark violet in India, hence the name black kolukattai, but in East African plants the involucre is usually straw-coloured.

Cenchrus setigerus occurs in tropical East Africa, north-east Africa, Arabia, southern Pakistan and in India. It is as drought resistant and heat tolerant as *C. ciliaris* and even reported to grow under an annual rainfall of 200 mm (J. N. Whittet, from Barnard, 1969). Soil requirements are also similar to those of *C. ciliaris* except that in East Africa *C. setigerus* also occurs on black heavy clays with impeded drainage. *Cenchrus setigerus* has been introduced to dry parts of the southern United States, and to northern Australia and Queensland and has shown good performance in dry areas both in sown pastures and as a grass for the improvement of worn out or denuded natural pastures. Similarly with *C. ciliaris*, it requires a proper seedbed for good development and germinates best from a depth of 1–2 cm; the seedlings are more tolerant to soil salinity than those of *C. ciliaris*. The pattern of growth is rather similar to that of *C. ciliaris* and the same management applies. In the Northern Territory of Australia it did well in mixtures with *Stylosanthes humilis*, and in dry areas of India sowing annual *Vigna radiata*, *V. aconitifolia* and/or *Cyamopsis tetragonoloba* between the widely spaced rows of the grass gave higher DM yields than the grass alone. Dabadghao & Marwaha (1962) in India estimated the relative palatability of *C. setigerus* to be 32–56 per cent of that of *Dichanthium annulatum*, one of the most palatable Indian grasses, and only at the ripe stage it was equal to or exceeded the palatability of *D. annulatum*; in this trial *C. setigerus* was less palatable than *C. ciliaris*.

In contrast with *C. ciliaris*, 'seeds' of *C. setigerus* are not fluffy and therefore easier to harvest and to drill. The existence of seed dormancy has not so far been proved: some authors observed satisfactory immediate post-harvest germination, whereas others, e.g. J. N. Whittet (from Barnard, 1969), showed that germination can improve with storage. Germination is normally considered to be 10–20 per cent and Mukherjee & Chatterji (1970) report that hulled seeds germinated 15–65 per cent and unhulled seeds 2–10 per cent.

This species is relatively uniform and its variability is not great. Similarly with *C. cilaris*, *C. setigerus* is an apomict (Fisher *et al.*, 1954) and being pseudogamous it requires pollination for the development of the endosperm. Two levels of ploidy were observed: tetraploids with $2n = 36$ and hexaploids with $2n = 54$, but irregular chromosome numbers ($2n = 34, 37$) were also found. Ramaswamy *et al.*, (1969) found some connection between the chromosome number and plant morphology and development: the hexaploids were more vigorous, taller, had more abundant tillers and denser spikes than the tetraploids.

Chloris Sw.

Tufted, rhizomatous or stoloniferous perennials, or annuals. Inflorescence a digitate panicle, occasionally with an elongated axis (*C. roxburghiana*), with a few to numerous spikes. Spikelets with two to six, but usually two to four, florets; lower floret bisexual, the others sterile, or sometimes male or occasionally bisexual, gradually reduced in size from the lower floret upwards. Lemma of the first to the third florets usually with a straight awn. The genus includes about 40 species distributed in tropical and warm regions throughout the world, in moderately wet to semi-arid areas. Palatable grasses mostly of good grazing value and *C. gayana* is an important cultivated pasture species.

Chloris acicularis Lindl. (*Enteropogon acicularis* (Lindl.) Lazarides). Curly windmill grass

Tufted perennial 50–100 cm high with erect or ascending stems. Leaves mostly basal, up to 20 cm long and 4 mm wide, usually curled at maturity. Inflorescence digitate, with 3 to 14 rigid racemes spreading in various directions. Spikelets about 8 mm long, with two to three florets; lower floret bisexual, upper florets empty. Lemmas with straight awns 10–15 mm long.

Widespread in dry parts of Queensland and other semi-arid areas of Australia. A useful grazing grass in dry grasslands (Tothill & Hacker, 1973).

Chloris gayana Kunth. Rhodes grass; Pasto rodes (Fig. 12)

Stoloniferous creeping or occasionally tufted perennial, with erect or geniculately ascending stems 0.5–2 m high. Leaves glabrous, 15–50 cm long and 2–20 mm wide; on the stolons the leaves are shorter and arise in groups of two to four from each node. Panicle digitate or subdigitate of 3 to 20 dense, spikelike racemes 4–15 cm long. Spikelets have three to four florets; the lowest, fertile floret 2.5–3.5 mm long, with an awn 1–10 mm long. The second floret is slightly shorter, with a shorter awn of awnless, male or sometimes fertile. The top florets are reduced to minute scales. Caryopsis spindle shaped, about 2 mm long.

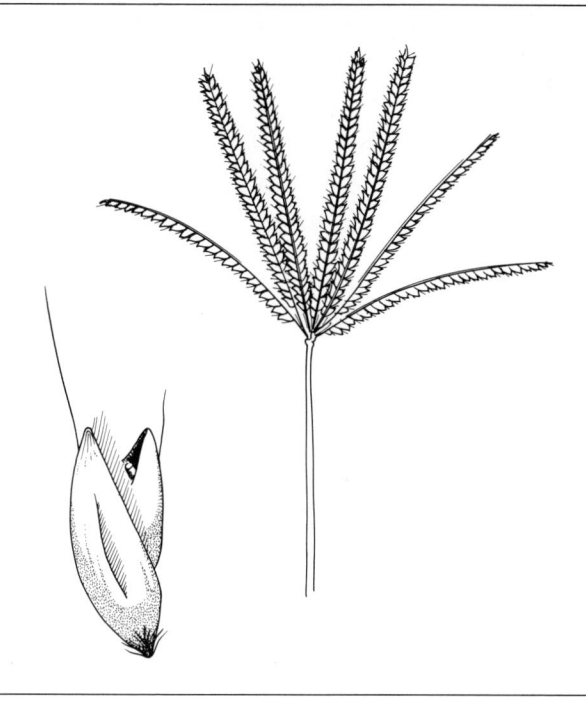

Fig. 12   *Chloris gayana.*

Occurs naturally in most tropical and subtropical areas of Africa, in open grassland or in grassland with scattered bush or trees, lake margins or seasonally waterlogged plains, up to 2,000 m altitude, rarely higher. Often a pioneer grass in fallows or abandoned cultivation, following the weed stage in plant succession.

Environment. *Chloris gayana* thrives under a wide range of temperature, and 35°C was the optimum temperature for photosynthesis in trials by Murata *et al.* (1965). Linear increases in respiration intensity were

Fig. 13 *Chloris gayana.*

observed in response to temperature increases from 10°–50°C. On the other hand *C. gayana* is tolerant to cold, and in Queensland, at 27°S., survived to 97 per cent in a winter during which the temperature dropped to −9°C (R. M. Jones, 1969). Most successful seed germination was observed at alternating or constant temperatures in the region of 20°–35°C, although in Russia adequate field emergence was obtained at an air temperature of 10°–12°C. *Chloris gayana* grows most successfully under daylengths from 10 to 14 h; under photoperiods shorter than 10 h herbage yields are considerably reduced, and yield reduction can be even greater at photoperiods longer than 14 h (Wang, 1961). Although *C. gayana* thrives under ample rainfall and responds well to irrigation in dry areas, it is relatively drought resistant and can grow under an annual rainfall of about 600 mm, but not much below this level. It grows under various soil conditions except on very heavy clay, or soils of high acidity, but tolerates highly alkaline soils. It develops particularly well on soils of loose texture such as volcanic ash. Salt tolerance is high and *C. gayana* can grow on soils rich in Na in the form of $NaHCO_3$ or Na_2SO_4, but is less tolerant to NaCl or $NaNO_3$, and good establishment on saline-sodic soils was observed at Tucuman in Argentina (Díaz & Lagomarsino,

1969). It can accumulate large amounts of Na in the leaves without any harm to the plant, but cannot tolerate $MgCl_2$ and relatively high concentrations of Mg in the leaves can be toxic to the plant. It is also tolerant to relatively high contents of Li in the soil. *Chloris gayana* can withstand flooding for up to 10–15 days but suffers from longer periods of flooding.

Introduction. *Chloris gayana* was first introduced into cultivation by Cecil Rhodes in South Africa in 1895, and taken to USA a few years later. It is now cultivated on an experimental, and often also on a farm scale in practically all tropical countries, and in a number of subtropical or even warm temperate countries; e.g. in Japan, southern Russia (Caucasus) and USA. Seed is commercially produced in Australia, USA and some other countries including Kenya where in 1970 350 ha of *C. gayana* was grown for seed. The popularity of *C. gayana* is due to its good seed production, the ease of establishment and its creeping habit. Herbage yields are reasonable and often high, but they seldom reach the yields of other particularly productive grasses, and the quality of herbage is not particularly high except in the first month or two of the growing season. Productive life usually lasts about 3 years unless heavily fertilized, and then it extends to 5 or more years. *Chloris gayana* appears to be most suitable for those farms where ease of establishment is more important than particularly high production.

Establishment. *Chloris gayana* can be established from pieces of rooting stolons, but propagation by seed is the normal practice. The seedbed should be well prepared and the seed is broadcast with subsequent rolling, or covered lightly by dragging light chains or a bunch of tree-branches stripped of their leaves over the land; harrowing is not recommended as it can bury the seed too deep in the soil. Seed can also be sown with a cereal drill (Kyneur & Tow, 1958) from which the delivery tubes are removed, and the seed diluted with sawdust is allowed to flow freely onto the soil surface, and covered by light chains attached to the drill. Seed of *C. gayana* requires shallow planting, and does not germinate below a depth of 2.5 cm or below 1.5–2 cm in easily compacted soil. *Chloris gayana* can be established under maize without any serious reduction of maize yields; seed is sown on the surface after the last weeding when the maize plants reach 60–90 cm in height. It can be established under wheat but not under oats. Attempts to establish *C. gayana* on heavy clay usually result in failure (Leslie, 1965); seed germinates well but only a small proportion of seedlings can reach the surface. In trials on reseeding denuded pastoral land (Bogdan & Pratt, 1967) *C. gayana* was the easiest grass to establish on bare uncultivated soil, and did not respond to presowing cultivation or to seed pelleting, but showed a good response to cut-branches cover which prevented early grazing, also to seed dressing with insecticides. Insecticide dressing also gave good results in Tanzania. In Queensland reasonable establishment in brigalow grassland was achieved by scattering seed on burned grassland (Coaldrake & Russell, 1969). Mulching with hay improved establishment in Rhodesia. Seed rate recommendations vary from 5 to 20 kg seed/ha, but they are usually given for seed of unknown quality and are, therefore, of not much significance. A better recommendation is to sow 0.5–1 kg pure germinating seed per ha, then e.g. seed containing 10 per cent PGS (pure germinating seed) should be sown at 5–10 kg/ha. Boonman (1972a) has however found that sowing rates varying from 0.2 to 1.8 kg pure germinating seed per ha resulted in about equal yields of herbage and seed. Fertilizer P is normally applied to seedbed at some 50 kg P_2O_5/ha, whereas fertilizer N can be given later. The seedlings are small and weak, and weed competition is usually the main difficulty at establishment, but early cutting or grazing can reduce this competition. Satisfactory, although slow establishment can be achieved even if the seedlings are sparse or unevenly distributed, as the creeping habit of the grass helps to fill the gaps between the plants later in the season.

Managing, fertilizing. Grazing or cutting for hay can begin when the grass reaches some 50 cm in height, but further use can vary, although in general rotational or deferred grazing can be recommended, under which three to five grazings per season can be made. Longer intervals between cuttings or grazings result in higher yields but decrease CP content of the herbage thus reducing its nutritive value. Good responses by *C. gayana* to fertilizer N were noted, and linear responses to up to 300 kg of applied N/ha were observed (Henzell, 1963). These responses depend on the rainfall: in dry years *C. gayana* responds only to limited rates of fertilizer N, and to much higher rates in wet years and applied N can increase the yields of CP to a greater extent than those of DM. Split application is advisable if N rates are sufficiently high. Interaction with P fertilizer is usually observed, and high yields are obtained only when N is given after a basic application of P. The usual practice in East Africa is to apply about 50 kg P_2O_5/ha at establishment, and to give N much later in the season; P is then given every year, or every second year, early in the season, and N after each cut or grazing except the last one in the season. Fertilizer N increases the proportion of leaf in the herbage, but when applied after flowering it can increase the proportion of stem. N absorption and transfer to the tops can be noticed 5–7 days after application (Barnes, 1960). *Chloris gayana* is often grown under irrigation in arid and semi-arid countries, mainly outside the tropics, in Texas, California and Mississipi in the USA, and in Israel. In trials in Kenya irrigated *C. gayana* produced about 23 t DM/ha for a 2-year period.

Association with legumes and other grasses. *Chloris gayana* has been successfully grown in mixtures with *Centrosema pubescens, Clitoria ternatea, Desmodium leiocarpum, D. uncinatum, Glycine wightii, Macroptilium atropurpureum, Medicago sativa, Stylosanthes guianensis, S. humilis* and *Trifolium repens*, and tried with some other pasture and fodder legumes. Total grass+legume yields are usually higher than those of *C. gayana* grown alone; however only seldom does the companion legume increase the yields of grass itself but often reduces them, especially in the establishment year, as was observed for *C. gayana/T. repens* mixtures in Kenya and for mixtures of *C. gayana* with *S. guianensis* or *Cajanus cajan* in Zambia. CP contents of total herbage are invariably higher than those of pure *C. gayana* stands, and the legumes can increase CP yields up to 200 per cent or even higher. Under heavy or frequent grazing, the legumes, except *T. repens*, are usually suppressed by *C. gayana*, but the reverse occurs under light or infrequent grazing. *Chloris gayana* can be grown in mixtures with perennial tufted grasses, and good results were obtained from mixtures with *Cenchrus ciliaris* in Tanzania and with *Setaria anceps* in Kenya. *Chloris gayana* is however less persistent than the majority of tufted species and in the fourth or fifth years the tufted grasses may still grow strong, whereas *C. gayana* may begin dying out, leaving gaps between the companion grasses.

Pests and diseases. Of the several fungus and virus diseases recorded on *C. gayana* only two inflict major damage: a *Helminthosporium* disease causing die-back of leaves and shoot bases, apparently inflicted by a species of *Cochliobolus*, and *Fusarium gramineum* attacking spikelets and causing considerable seed losses in wet years. The two diseases appear to be genetically controlled and resistant types exist. In Rhodesia and in Australia streak virus has been observed on the leaves, but its economic importance is not great. Of the pests, Rhodes grass scale (*Antonina graminis*) which parasitizes more on other grasses than on *C. gayana*, and a grain-damaging thrips (*Chirothrips mexicana*) can inflict considerable damage but mainly outside the tropics.

Herbage yields. Herbage yields range from 1.5 t to 25 t DM/ha depending on soil fertility and fertilizers applied, frequency of cutting or grazing, rainfall and other factors. Yields likely to be obtained from well-managed *C. gayana* grown under farm conditions are of the order of 5–8 t DM/ha. Applied N is one of the major factors influencing herbage yields, and in South Africa 434 kg N/ha increased hay yields from 1.55 to 6.01 t/ha (Haylett, 1970). In Queensland (R. J. Jones, 1970), the total yield of DM for the 4 years of a trial was 25 t/ha without applied N, and almost 50 t when 335 kg N/ha were given, averaging over 6 t DM per year and double this amount, respectively. Maximum yields are usually obtained in the second year of growth, and in Zambia (Van Rensburg, 1969) average yield of 15 cultivars under trial reached 3,250 kg DM/ha in the year of establishment, and 7,719 kg in the following year.

However, if establishment is particularly good and the season long, the first-year yields can be higher than in any of subsequent years, as was observed in Fiji (Roberts, 1970) where fresh material yields were 119 t/ha in the year of establishment, 95 t in the second year and 44 t in the third year, and also in Thailand where the grass well fertilized with NPK and irrigated yielded 20.1 t DM/ha in the first year and 16.0 t in the second year (Holm, 1972). Yields of DM and especially of CP are normally higher for *C. gayana*/legume mixtures than for pure *C. gayana*.

Conservation. *Chloris gayana* produces uniform herbage in even stands, and is one of the most satisfactory tropical grasses for use as hay. As early as 1930 D. C. Edwards in Kenya recognized this quality, and recommended it for haymaking. *Chloris gayana* is less suitable for ensiling, and although satisfactory results from its ensilage were reported from Nigeria and South Africa (from Bogdan, 1969), reports from other countries are less encouraging. The ensiled material does not settle well, and the content of sugars is not sufficiently high for lactic acid fermentation, even with added glucose. Losses of DM were up to 20 per cent and of N 3–9 per cent, which are not considered to be excessive. Wilting of the herbage produces a beneficial effect on ensilage provided that the water content does not decrease much below 40 per cent (Catchpoole & Henzell, 1971).

Chemical composition, nutritive value, digestibility. CP content of *C. gayana* herbage usually ranges between 4 and 13 per cent, although in young leaves it can be as high as 16–17 per cent and as low as 3 per cent, and depends mainly on the age of plant, season and applied N. In Uganda (Bredon & Horrell, 1961) CP content of 16 per cent early in the season was reduced to 11, 6.5 and 5.5 per cent after each monthly interval, and 4.5 per cent, or below the subsistence level throughout the dry season. In Australia (Henzell, 1963) unfertilized grass contained about 6.3 per cent CP (or 0.96–1.10 per cent N), but increased progressively with the increase of rates of fertilizer N, reaching 9.5–9.8 per cent when 450 kg N/ha were applied. However, in some countries N produced little or no effect on the content of CP in the herbage. The content of CF usually ranges between 30 and 40 per cent but can be 25 per cent in the early cut hay and reach over 45 per cent in the late cut hay. The content of NFE usually ranges from 40 to 50 per cent and of EE from 1–2.5 per cent. The carotene content is usually sufficiently high for the normal requirements of cattle. The contents of P and Ca are about equal to those found in the majority of tropical grasses, but the contents of K and Mg are usually lower, Na content is high and exceeds three to five-fold its content in other tropical grass spceies. Palatability of reasonably young grass is good, but it falls rapidly with plant age and is, in general, lower than in the majority of other important cultivated tropical grasses, although Milford & Minson (1966) report that in their trials the intake of *C. gayana* herbage by cattle depended little on the season in contrast with five other species under trial. Digestibility of total herbage is relatively low, 40–60 per cent according to Milford (1960), while digestibility of CP varies considerably and can be as high as 70 per cent at 13.8 per cent CP content, and only 4.7 per cent when the content of CP decreased to 5.4 per cent, resulting in a negative N balance of the grazing animals. Digestibilities of NFE and CF vary to a lesser extent and according to Butterworth (1967) range from 30 to 79 per cent and from 46 to 84 per cent, respectively. Gross energy of *C. gayana* hay was determined as 4.02–4.19 kcal/g and digestible energy as 2.08–2.51 kcal (from Bogdan, 1969).

Animal production. Stocking rates for cultivated *C. gayana* range from 0.25 to 1.50 ha per head of cattle depending on herbage productivity. The productivity of animals grazed on *C. gayana* can be high in the first year of utilization, and 366 kg liveweight increase per ha per year were recorded in Kenya, but liveweight gains considerably decreased in the following 2 years, and were about equal to those obtained from unimproved natural grassland. In Rhodesia, the application of 100 kg N/ha doubled the stocking rates and increased liveweight gains to over 600 kg/ha, but such spectacular gains from fertilizer N are well above the usual data. *Chloris gayana* is capable of producing high liveweight gains mainly in its first or second month of growth, when it can be more productive and nourishing than a number of other tropical grasses, but in the following months the gains decrease rapidly. In general *C. gayana* produces less beef per ha than other good tropical grasses even under equal herbage yields. Grass/legume mixtures give higher animal production than pure *C. gayana* as was observed in Australia, Uganda, Madagascar and a number of other countries. Hamilton *et al.* (1970) report that cows grazed on *C. gayana* gave 6–7 kg milk/day or about 70 per cent of their capacity at the optimum regime of feeding.

Flowering, reproduction, genetics. Flowering of *C. gayana* in the tropics is apparently not influenced by day length: in Taiwan (Wang, 1961) changes of day length from 10 h 20 min to 13 h 40 min did not affect flowering, but under photoperiods of 7–10 h it was delayed up to 15 days, of 14–17 h up to 125 days. Flowering begins some 15 days after the panicle emerges at the middle of the raceme, and proceeds towards both ends, a few to several spikelets flowering in the same day. In each spikelet the lower, fertile floret flowers first and the second floret 3–4 days later, about 75 per cent of the first florets of each raceme complete their flowering in 3 days. Under equatorial conditions of Kenya flowering can occur between 12.30 p.m. and 5.30 p.m. depending on the weather, the spikelets and florets opening only under direct sunlight. All florets ready for flowering open simultaneously on all plants in the field, and the elongation of filaments and the release of pollen take 6–12 min.

to complete. The caryopses ripen some 23–25 days after the anthesis (Bogdan, 1959). It was assumed in the past that *C. gayana* was an apomict, but field and greenhouse trials supplemented with histological investigations (Bogdan, 1961b; R. J. Jones & Pritchard, 1971) have shown the grass to be a typical cross-pollinating plant, with self compatability of 1 to 4 per cent. Studies of the breeding behaviour of *C. gayana* were facilitated by the discovery of anthocyanin-free plants easily recognizable by the pale-yellow coloration of the panicles, the absence of pigment being dependent on a single recessive gene. If a mother plant recessive for the absence of anthocyanin is crossed with an anthocyanin-containing (normal) plant, the crossing is immediately detected in the first generation, even at the seedling stage, by the purple colouration of the seedlings (Bogdan, 1963b). There are diploids and tetraploids with $2n = 20$ and $2n = 40$, respectively, (triploids with $2n = 30$ were also found) but no connection between the ploidy and plant vigour was established. *Chloria gayana* is an extremely variable species, both in morphology and in agriculturally important characteristics, and even apparently uniform populations contain various types which can be detected only in single spaced plants; a certain type prevails however in each distinct population. Most of the existing cultivars represent wild populations brought into cultivation and found superior to other wild populations. Although the cultivars can change as a result of cultural practices, true breeding, i.e. the improvement of existing populations by e.g. single plant selection requires considerable field work, because each single plant being a creeper has to be grown in a separate plot. The only tropical cultivar of *C. gayana* developed by the improvement of the original population, and not simply selected by comparison with other existing natural populations, seems to be cv. Pokot, developed in Kenya, although frost-resistant types developed in Texas, USA, could also have resulted from a selection of resistant plants.

Cultivars. There is a large number of cultivars with local names or the introduction numbers only, but only a few better known ones are given below: these have been developed mainly in Australia, Kenya and Rhodesia. **Pioneer**. Diploid. Registered under this name in Australia in 1973, being previously known as **Commercial**, a cultivar of unknown, probably South African origin. An early type which requires fertile soil and relatively high rainfall. **Katambora**. Diploid. Originates from Katambora, Zambia, where it grows on the banks of the Zambezi River. Stolons thin and long. Valued for its fast growth and drought resistance. A late-maturing cultivar frost-susceptible in Queensland. **Samford**. Tetraploid. Developed in Australia, at Samford, where it was introduced from Kenya via Sierre Leone. In Australia it gives high herbage yields and its palatability is outstanding. **Callide**. Tetraploid. Apparently identical with cv **Mpwapwa** and **Giant** and perhaps with cv.

Kongwa grown in Tanzania. Introduced to Australia and Kenya from Tanzania. A coarse type with thick stolons and stems, broad leaves, long racemes, long awns and long hairs on spikelets. Seed is more fluffy than in other cultivars. Drought resistant and palatable. **Masaba**, Tetraploid. Previously known in Kenya as **Endebess**, a village near Mt Elgon, but later renamed **Masaba**, which is the local name for Mt Elgon, where it was originally found. Productive and leafy, but susceptible to spikelet disease and its seed production is erratic, although in a 4-year trial (Boonman, 1973) it was one of the best seeding types. **Mbarara**. Tetraploid. Selected in Kenya from material introduced from Mbarara in Uganda. A fine-stem type with long thins stolons. An outstanding seed producer. **Nzoia**. Diploid. Although the name originates from the River Nzoia in western Kenya, it is probably of South African origin. A leafy type. Susceptible to a *Helminthosporium* disease which reduces its persistence. It often has two seeds per spikelet and seed production is good. The *C. gayana* grown in Israel belongs to the same or a similar type. **Pokot** (Suk). Tetraploid. Originates from Pokot district of Kenya. It is remarkable in the vigour of herbage and of stolons which can develop even in well-established dense stands. Very leafy, but seed production is only moderate. A bred cultivar based on single plant selection.

Seed production. *Chloris gayana* is a cross-pollinating plant and different cultivars, especially of the same ploidy, should be grown for seed at some distance apart; 200 m should be sufficient. For seed production *C. gayana* can be established in the usual way, and Boonman (1972a) has shown that neither sowing rates nor the width of inter-row space have much effect on seed yields. In Kenya the number of fertile tillers in well managed stands ranged from 110 to 260/m^2 in the year of establishment, and from 160 to 320/m^2 in the following year, and cultivars with larger numbers of fertile tillers gave higher seed yields. In the year of establishment the panicles begin to appear some 4–5 months after sowing, their number reaching a maximum 2–3 weeks later. The application of fertilizer N is necessary for good seed production, and in Kenya the optimum rate of N was 100 kg/ha, although this is not necessarily adequate in other countries. Harvesting is done either by hand with sickles, or by reaper-binder, and threshing by sticks or by stationary combining. Direct combining used in the past in Kenya results in heavy losses of seed caught in large masses of stems and green leaves. It is essential to keep the seeding swards clean of weeds because *C. gayana* seed is more difficult to clean than that of almost any other tropical grass, and in Kenya seeds of *Digitaria velutina*, *Tagetes minuta* and *Ageratum conyzoides* were particularly difficult to separate. In the tropics *C. gayana* often produces two crops of seed per year, and seed yields range very widely from 65 to 650 kg/ha/year. Particularly high figures were reported from Tanzania; in later years, however, when the seed was better cleaned, yields did not exceed some 300 kg/ha. Seed yield figures are however of no great importance if the percentage of caryopses is not indicated, and yields of pure germinating seeds (PGS) or, in other words, pure viable seeds (PVS), which really matters, are not given. In Kenya, yields of PGS are usually of the order of 40–50 kg per harvest, or 80–100 kg per year (Boonman, 1972a); yields can however be higher in other countries. These yields present only a fraction of potential yields, which with >500 seeds per panicle, and 2 million panicles per ha can produce 1,000 million seeds or 500 kg of PGS per harvest. Seed of *C. gayana* as sown is in the form of a spikelet (but without glumes which remain attached to the raceme axis after threshing) containing one caryopsis. There are about 2 million such seeds per kg. Seed can be sown almost immediately after harvest, but germination improves in the first few months of storage, reaching its maximum in 6–12 months after harvesting, remains high for about 4 years and then declines to almost nil in the sixth or seventh year.

Basic information from Bogdan (1969).

Chloris polydactyla (L.) Swartz. Pasto borla

Tufted perennial 80–150 cm in height. Leaves about 30 cm long and 7–12 mm wide. Panicle digitate, with 9–25 spikelike, flexuous or pendent racemes 8–15 cm long. Spikelet usually with three florets; the lower floret fertile, 2–3 mm long, long, densely ciliate on margins and with an awn 2–4 mm long; the second floret much reduced in size and the third floret rudimental. Glumes unequal, the upper as long as the spikelet. Caryopsis 1.3 mm long. It resembles *C. gayana* but has longer and flexuous racemes, which are straight or bent in *C. gayana*, and smaller spikelets; it also has a tufted habit; Parodi, however, reports a stoloniferous form (f. *stolonifera* Parodi) with the stolons similar to those of *C. gayana*.

Occurs in Florida, USA, and extends southwards to South America, reaching about 25°S. in Argentina; also in the West Indies. Common in grasslands in some parts of Brazil and Argentina where it can constitute a considerable portion of herbage and provide grazing during a long period in summer. *Chloris polydactyla* is well liked by cattle but herbage quality is relatively low and at the stage of flowering CP content was 5.5 per cent, DCP 2.0 per cent, CF 32.5 per cent, EE 2.3 per cent and NFE 49.7 per cent (Reichert & Trelles, 1923). Seed is abundantly formed and germinates well.

Chloris roxburghiana Schult. (*C. myriostachya* Roxb.) $2n = 20$

Tufted perennial 50–150 cm high with fan-shaped arranged basal leaf-sheaths and shoots. Leaf blades flat or folded, 10–20 cm long and 2–10 mm wide. Panicle a compact, feathery, oblong head usually 10–15 cm long; the numerous short racemes are densely beset with spikelets each containing three to four florets; the lower floret is 1.5–2 mm long with a fine awn 8–17 mm long; the second to fourth florets are reduced to awned scales. Occurs naturally in tropical East Africa from Sudan to Angola and also in southern India, up to 1,500 m alt, in grassland or in open bush, often as a pioneer grass of abandoned cultivation or cleared woodland where it can form almost pure stands, and is a relatively short-lived perennial. The herbage is of good grazing value containing up to 16 per cent CP and about 30 per cent CF in the DM at the flowering stage (Dougall and Bogdan, 1960) and is readily grazed. *Chloris roxburghiana* has been tried in Kenya for reseeding denuded grassland by Jordan (1957) with good success and with moderate success by Pratt (Bogdan & Pratt, 1967). Each panicle contains about 1,000 seeds, and harvesting is easy; it is however difficult to separate the caryopses from the entangled awns. Naked caryopses are sown; they are small: 6.5 to 13 million per kg, depending on the variety.

Chloris virgata Sw. (*C. virgata* Sw. var. *elegans* (Kunth) Stapf, *C. elegans* Kunth) $2n = 20, 40, 26$ or 14

Annual with erect or ascending stems up to 1 m high, occasionally rooting from the lower nodes. Leaves 10–30 cm long and 2–6 mm wide. Panicle digitate with 4–12 suberect dense spikes 2–10 cm long, feathery at maturity. Spikelets with two to three florets; glumes unequal, the lower 1.5–2.5 mm and the upper 2.5–4.5 mm long. Lemma of the first (lower) floret 2.5–4 mm long, laterally compressed, with a gibbous keel, a tuft of hairs 1.5–4 mm long at the apex and an awn 5–15 mm long. Lemma of the second floret glabrous, oblong, 2.0–2.5 mm long and with a shorter awn. The third floret reduced to a clavate scale. Usually only the first floret contains a caryopsis.

Distributed throughout the world's tropics, mainly in semi-arid areas where, in favourable years with a relatively high rainfall, it can be abundant in grazing land with loose cover of perennial grasses. True arid areas are avoided and alluvial soils preferred. *Chloris virgata* is leafy, the stems are soft and it is readily eaten by stock. Analysed in Kenya in a favourable year, at the full-flowering stage, the whole plant contained 12.9 per cent CP, 1.8 per cent EE, 31.1 per cent CF, 40.4 per cent NFE, 0.19 per cent P and 0.55 per cent Ca, showing a good nutritive value (Dougall & Bogdan, 1965). *Chloris virgata* has been used with some success in an initial stage of reseeding denuded pastoral land at medium altitudes (about 1,500 m) of Kenya. Seed is well produced; it is in the form of a spikelet with one caryopsis and there are about 2 million such seeds/kg (Bogdan & Pratt, 1967), about the same number as for *C. gayana* seed.

Chrysopogon Trin.

Inflorescence a panicle with whorled filiform branches terminating in a raceme reduced to three spikelets arising from the same point, of which the middle, sessile, spikelet is fertile and contains one bisexual floret provided with a kneed awn spirally twisted below the knee. The side spikelets are on pedicels (which are often covered with rufus hairs) each having a male or empty floret and can be provided with one or two bristles. In some species the terminal raceme can have a pair of spikelets below the top three. Mostly perennials some of which are good pasture grasses mainly in semi-arid or arid areas.

Chrysopogon acicularis (Retz.) Trin. $2n = 20$

Creeping stoloniferous perennial with the stems 20–60 cm high. The stolons have short internodes and are densely covered with leaves which are 2–10 cm long and 2–6 mm wide. Panicle dense, contracted, purplish. Spikelets 3–5 mm long, glabrous or slightly hairy; sessile spikelet with an awn 4–6 mm long; bristles short or absent.

Occurs naturally from India and Sri Lanka to southern Asia (Malaysia, Thailand), Australia and Polynesia, under an annual rainfall of 1,250–2,000 mm. Moist soils, usually sandy loams with pH 5.1–6.1, are preferred (Dabadghao & Shankarnarayan, 1973) and the plants can hardly survive in dry situations. In Malaya *C. acicularis* often occurs at roadsides, near villages, etc., where it is heavily grazed and also provides grazing in grasslands developed under grazing or cutting which follow forest clearing (Verboom, 1968). This grass is resistant to grazing, is palatable and considered to be one of the best grazing grasses at most stages of growth; however, at late flowering or seed ripening the seeds, with their sharp callus, can penetrate the skin of grazing animals and cause irritation (Bor, 1960).

Chrysopogon aucheri Stapf var. *quinqueplumis* Stapf. Daremo (Somalia)
$2n = 40$

Tufted perennial 25–80 cm high with numerous slender stems. Leaves up to 15 cm long and 2–3 mm wide. Panicle 5–10 cm long. Sessile (fertile) spikelet 6 mm long with an awn 2–4 cm long and a straight plumose bristle. Side spikelets with two plumose bristles each; the bristles are 10–12 mm long.

Occurs in northern Kenya, Somalia, Ethiopia, Arabia and further north in Iran, in dry lowland, but can be found at altitudes up to 1,500 m. In northern Kenya it grows under semi-desert conditions and often covers extensive areas of sandy or rocky soil; the plants do not form dense stands but grow scattered, leaving a considerable space between the tufts, which, in favourable years with relatively high rainfall, are occupied by numerous annual grasses. It is one of the most drought-resistant perennial tropical grasses. *Chrysopogon aucheri* var. *quinqueplumis* is leafy, readily grazed by all kinds of stock and is much valued by pastoralists. This grass has been recommended for reseeding denuded pastoral land in the direst parts of Kenya; the seed is however difficult to produce and although the caryopses can be well formed the spikelets shed early. Nevertheless attempts of seed bulking have been made

(Bogdan & Pratt, 1967). Seed is in the form of two-spikelet joints, only one of the spikelets containing a caryopsis, and there are about 450,000 such 'seeds' per kg.

Var. *pulvinatus* Stapf differs mainly in its growth habit: it forms dense cushions with numerous stems and short, up to 3 cm, leaves, the fine stems terminating in scanty panicles exerted from the cushion. It occurs, together with var. *quinqueplumis*, in Somalia where it is considered to be one of the best grazing grasses.

Chrysopogon fallax S. T. Blake

Densely tufted perennial 30–200 cm in height. Leaves 4–6 mm wide. Panicle 7–20 cm long. Sessile (fertile) spikelet 9–11 mm long, with an awn 20–40 mm long; each pedicelled spikelets is provided with a straight 10 mm-long bristle. Occurs mainly in northern Australia where it is a frequent and often co-dominant species usually associated with *Themeda australis* and *Sorghum plumosum*. It is considered a satisfactory, but not particularly valuable, grass. When burnt its reproduction by seed is much reduced.

Chrysopogon fulvus (Spreng.) Chiov. (*C. montanus* Trin; *Pollinia fulva* Spreng.). Goria; Guria (India)

Tufted perennial up to 50 cm high. Leaves 2–15 cm long and 2–3 mm wide. Panicle 4–12 cm long. Spikelets 3–7 mm long; sessile spikelet with an awn 2–3 cm long.

Chrysopogon fulvus is far from being uniform and robust forms which can be over 100 cm in height and have thick stems and long leaves and panicles also occur and the whole complex of forms is perhaps not specifically uniform, especially taking in account that three different ploidies are involved.

A variable grass which can be found growing naturally from South East Africa, throughout India and Pakistan and further east. It occurs under an annual rainfall of 250 to 1,250 mm where it prefers the drier habitats and grows on red soil both on rocky slopes and flat ground, but also found on black cotton soil. In India and in some areas of Pakistan *C. fulvus* often constitutes a considerable portion of grassland herbage, is one of the main species in the *Sehima/Dichanthium* and *Themeda/Arundinella* grasslands of India (Dabadghao & Shankarnarayan, 1973) and is considered to be a useful grass. It is relatively shallow rooted, the roots usually penetrated not deeper than 65–70 cm. Attempts have been made to oversow *C. fulvus* to poor natural pastures, e.g. those with *Heteropogon contortus*, and sowing 4 kg seed/ha resulted in satisfactory stands of the sown grass (Dabadghao & Shankarnarayan, 1970). Narayanan & Dabadghao (1972) state that *C. fulvus* (*C. montanus*) pastures can be established if the sown grass is not grazed during the establishment year but when well established in can produce 12.5 t fresh fodder/ha in five cuts. The grass is readily eaten by animals and is used almost exclusively for grazing. K. S. Sen, as quoted by Whyte (1964), gives the changes of CP content from the first to the third cut from 6.13 to 4.45 per cent, of CF from 36.8 to 31.6 per cent, NFE 46.0 to 51.6 per cent, CaO from 0.45 to 0.65 per cent and P_2O_5 from 0.39 to 0.20 per cent.

$2n$ chromosome numbers have been determined to be 20, 40 or 80 and this can imply that the grass can occur in a diploid, tetraploid or octaploid forms or that there are two taxons involved, each in the form of diploids and tetraploids. Brown & Emery (1957) suggest that *C. fulvus* (*C. montanus*) is a grass with sexual reproduction; this conclusion is based on the embryological evidence only and requires confirmation by progeny studies.

Other species of *Chrysopogon* of some importance for grazing are: *C. latifolius* S. T. Blake, which provides some grazing for sheep in Australia, *C. orientalis* (Desv.) A. Camus, common in India and *C. lancearius* (Hook.f.) Haines, also common in India where it often grows in bush or in shady places and provides good fodder.

Coix L.

Contains six closely allied species which occur in south-eastern Asia; they belong to tribe Maideae.

Coix lacrima-jobi L. Job's tears; Lagrimas de San Pedro; Adlay
$$2n = 10 \text{ or } 20$$

A robust, tufted glabrous annual 80 cm–2 m high. Leaves 20–50 cm long and 2.5–5 cm wide with rounded or subcordate base and a prominent midrib. Inflorescence a complex branched panicle which consists of a number of simple inflorescences arising from leaf axils of the main stem and its branches. Each simple inflorescence consists of a single female spikelet enclosed in a globose or near-globose, usually hard, cover from which only the stigmas protrude through a hole at the apex. A peduncle bearing male spikelets also protrudes from the same cover and bears up to 10–12 spikelets. The male spikelet is 7–8 mm long and has two florets. After flowering the male spikelets fall off together with the peduncle and the persistent bead-like cover contains one caryopsis. The cover is usually hard and shiny but in varieties used for human consumption it is soft and has thin walls.

Occurs in south-eastern Asia and, possibly as an early naturalized introduction, in tropical Africa. Grain varieties are grown in south-eastern and eastern Asia and in the Philippines (Purseglove, 1972), whereas those with hard cover and grown for beads or other ornamental uses, have now spread throughout tropical and subtropical areas, including America, and are cultivated on a small scale.

Coix lacrima-jobi usually grows in moderately humid areas, on fertile soil; in drier areas it can form colonies on stream banks. In their vegetative parts the plants resemble maize, except that they tiller freely and form tufts. They can be used for fodder and Derbal *et al.* (1959) report that in Mali *C. lacrima-jobi* produces 45–53 t fresh herbage/ha, that the herbage is palatable, especially when young, and provides good green fodder or can be made to silage. In Mali it is also recommended for mixtures grown for feeding pigs.

Cynodon L. C. Rich

Rhizomatous and stoloniferous or only stoloniferous perennials. Inflorescence a digitate or subdigitate panicle of a few to several one-sided narrow spikes (racemes) arranged in one to a few whorls. Spikelets in two rows, ovate or narrow ovate, laterally compressed, one-flowered. Glumes shorter than the spikelet. Caryopsis ellipsoid, laterally compressed.

Until recently only *C. dactylon* and *C. plectostachyus* were recognized as the main tropical species of importance. A recent revision (Clayton & Harlan, 1970) has shown, however, that true *C. dactylon* rarely occurs in the tropics and that the majority of tropical African plants previously thought to belong to *C. dactylon* are in fact either *C. nlemfuensis* or *C. aethiopicus*, both stoloniferous perennials without underground rhizomes, whereas rhizomatous *C. dactylon* occurs mainly in the subtropics and in warm temperate countries. It can be accepted with a high degree of probability that reports on the grazing of *C. dactylon* in tropical Africa, both in natural stands and in cultivation, actually refer to *C. nlemfuensis* and, to a lesser extent, to *C. aethiopicus*. The confusion in nomenclature was increased by mistakenly accepting the larger forms of *Cynodon* with a few whorls of racemes for *C. plectostachyus*. In view of the past nomenclature difficulties, the references to trials with '*C. plectostachyus*' and with tropical African '*C. dactylon*' are of uncertain value unless the plants under experiments have been identified by an experienced botanist, and collecting herbarium specimens from experimental plots is of particular importance for species of *Cynodon* grown in the tropics and subtropics.

Cynodon arcuatus J. S. Presl ex C. B. Presl $2n = 36$

A tetraploid characterized by its long, slender and flexuose racemes, low growth, not exceeding 20 cm, and broadly lanceolate leaves. A stoloniferous perennial without rhizomes which forms loose cover. Occurs in Madagascar and other Mascarene islands, South-East Africa, India, Indonesia, the Philippines and northern Australia. Not of much importance for grazing (Harlan *et al.*, 1970).

Cynodon barberi Rang. & Tad. $2n = 18$

A diploid somewhat similar to *C. arcuatus* but the racemes are straight and short. The upper glume sometimes exceeding the length of the rest of the spikelet. Occurs in India where it is an endemic. This is perhaps a relict species and if so it can be regarded as a link between *Leptochloa* and all other species of *Cynodon* (Harlan *et al.*, 1970).

Cynodon dactylon (L.) Pers. (*C. polevansii* Stent). Bermuda grass; Dhub (Doob); Hariali (India); Dog's tooth grass; Star grass; Chiendent; Zacate Bermuda; Pasto Bermuda; Pasto Argentina (Fig. 14).

Stoloniferous and rhizomatous perennial. Stems slender to stout up to 60 cm in height, stolons and rhizomes slender to robust. Leaves flat or folded, 3–12 cm long and 2–4 mm wide. Racemes (spikes) 1.5–8 cm long, three to six in a single whorl. Spikelets 2–2.5 mm long; the upper glume one-half to three-quarters the length of the spikelet.

Fig. 14 *Cynodon dactylon.*

Tropical, subtropical and warm temperate regions throughout the world, but rare and perhaps absent in tropical Africa although common in southern and northern Africa. Occurs in heavily grazed grasslands, roadsides, fallows and as a weed of arable land.

Cynodon dactylon is a variable species and several botanical varieties have been described; three or four of these are of importance in the tropics.

1. Var. *dactylon*, a tetraploid ($2n = 36$) originating probably from Turkey and Pakistan (Harlan, 1970) is the most widely distributed type; it is a common weed of arable land throughout the world in warm temperate areas, in subtropical areas and, to a lesser extent, in the tropics, e.g. in Mauritius. It can grow on alkaline and saline soils and tolerates droughts, but does not occur in particularly arid areas. In moderately dry areas of India and Pakistan and of the neighbouring countries var. *dactylon* (together with var. *aridus*) is of considerable importance for grazing and is highly valued as a pasture grass mainly because of the high quality of herbage: the herbage has a high percentage of dry matter and a high content of CP, which normally ranges from 8 to 15 per cent; yields are however not high. Var. *dactylon* can withstand a considerable grazing pressure and its proportion in the herbage and the ground cover increase under reasonably heavy grazing. It is utilized mainly in natural pastures, often in overgrazed land near the villages, and in India its cultivation has been recommended (Narayanan & Dabadghao, 1972). In the USA, var. *dactylon* has also been used as a grazing grass and a clonal variety, a vegetative progeny of a single superior plant, more leafy and with less abundant fertile stems than in ordinary local types, was found in Georgia and distributed as 'Tift Bermuda' (Wheeler, 1950), but replaced later by better cultivars of hybrid origin. As a weed of arable land *C. dactylon* is very common in the subtropics and in warm temperate areas of the USA and is one of the main noxious weeds, especially in the cotton-growing areas. Its rhizomes can be strong and stout and much branched, they spread rapidly and eradication is difficult and costly. A coarse and robust *seleucidus* race is common in the Mediterranean area and in southern Russia from where it has spread to the western hemisphere.

2. Var. *aridus* Harlan & de Wet, a diploid ($2n = 18$) which occurs in southern India extends westwards to South West Africa but is much less common than var. *dactylon*. It varies in habit and is small and unproductive in India but 'East African forms are large, robust and vigorous' (Harlan, 1970). This variety has also been found in Hawaii and in Arizona (USA), where seed was produced and the grass spread to other parts of the USA.

3. Var. *elegans* Rendle is a tetraploid which occurs in South Africa south of 13°S. latitude where it is common. It has much branched

stolons, rhizomes and stems; the latter are slender but usually wiry and the plants form dense low swards up to 30 cm in height and produce good grazing in South Africa. When kept low it makes good lawns and under the names of Cape Royal and Maadi River grasses is well known as a tropical turf grass. G. W. Burton suggests, however, that these forms may perhaps be hybrids between *C. dactylon* and *C. transvaalensis*, a South African species with very fine stems and leaves.

4. Var. *coursii* (A. Camus) Harlan & de Wet. A non-rhizomatous tetraploid which grows in the Madagascar highlands.

Hybrid cultivars of C. dactylon (Bermuda grass). In the southern United States *C. dactylon* var. *dactylon* has been a serious weed pest of arable crops since the eighteenth century and grazing utilization did not compensate for its negative effect on agriculture. The position changed when Burton (1947) in Georgia, USA, applied hybridization in developing types of Bermuda grass suitable for grazing and forage production and released, in 1943, cv. Coastal, a cross between a local var. *dactylon* and a South African type apparently belonging to var. *elegans*, both tetraploids. Cv. Coastal is almost seed sterile, produces relatively small numbers of unpalatable fertile stems and panicles and yields 1.5–4 times more herbage than the original local types. Its rhizomes can be eradicated relatively easily and it is less troublesome as a weed of cultivated land. Moreover, it is more resistant than the local plants to some common diseases, in the first instance to a *Helminthosporium* leaf disease common in the United States, and to root-knot nematodes. Cv. Coastal can be easily propagated by pieces of stolons and rhizomes and was soon multiplied and grown on a large scale. It sharply increased herbage production and revolutionized the livestock industry of the southern states (Harlan, 1970). Later, in the 1950s, other cultivars, particularly cv. Midland, more suitable for the areas north of Georgia, were developed by crossing Coastal Bermuda with various local types. In 1962 cv. Suwannee, an F_1 hybrid between an African introduction and cv. Tift Bermuda, suitable for the extreme south-east of the USA, was developed and released. It outyielded cv. Coastal and produced 4.23 t DM/ha when given 56 kg N/ha and 12.32 t at 224 kg N/ha compared with 3.58 t and 10.59 t/ha for Coastal Bermuda (Burton, 1962). The quality of herbage and feeding value of the hybrid cultivars were however little improved compared with the original local *Cynodon*. This was corrected by the development of a new hybrid, cv. Coastacross 1, released in 1967. Coastacross 1 is a cross between cv. Coastal and *C. nlemfuensis* var. *robustus*, a tetraploid introduction from Kenya. DM digestibility of the new cultivar was increased by over 12 per cent, from 53.5 per cent for Coastal Bermuda to 60.1 per cent for the new hybrid (Burton *et al.*, 1967) and the intake of herbage and liveweight gains, 717 g/day/animal for Coastacross 1 compared with 553 g for cv. Coastal, were reported to be higher than those recorded for previously released cultivars. The newly developed cultivars are now entering farm scale cultivation, but the bulk of information so far accumulated refers mainly to cv. Coastal. The hybrid cultivars, in the first instance Coastal Bermuda, have reached other countries, mainly in South and Central America and the West Indies and are cultivated, predominantly on an experimental scale, in Brazil, Venezuela, Colombia, Surinam, Mexico, Costa Rica, Puerto Rico, Jamaica, the Philippines and India.

Cynodon dactylon is essentially a tetraploid with the chromosome number $2n = 36$, although diploid varieties also exist. Amongst tetraploids $2n = 30$ has also been observed and Burton (1947) states that various chromosome fragments found during mitosis make the counting of chromosomes difficult. *Cynodon dactylon* is a cross-pollinated species and if grown as single-spaced plants from seed, taken even from one and the same plant, varies very considerably and hardly two plants that are alike can be found (Burton, 1947). The plants also differ in their agricultural characteristics: yields, leafiness, responses to fertilizers, etc. When developing superior varieties by hybridization, the plants grown from hybrid seed are evalued, the best F_1 are selected, multiplied vegetatively and used as clonal hybrid cultivars. The plants are highly self-incompatible, they produce no viable seed in the absence of foreign pollen and there is only a remote possibility of contamination of clonal cultivars with inferior plants.

Environment. Coastal Bermuda grass is sensitive to light intensity and in a trial by Burton *et al.* (1959) shading reduced the growth and yields, particularly when heavily fertilized: grass given 1,800 kg N/ha/year showed much more reduction in growth and yield than that given 224 kg N. In Burton's trial the high-N plots yielded significantly more than the low-N plots under full daylight and significantly less when the plants were grown under light reduced by 71 per cent, and particularly sharp reduction in weight was noted for underground plant parts – roots and rhizomes. High air temperatures, about 37°C, are required for the maximum photosynthetic activity as is typical for grass species with C_4 pathway of photosynthesis. On the other hand, Coastal Bermuda can tolerate frosts. It grows well but yields little under dry conditions and in dry areas can be dormant for as long as 6–7 months (Wheeler, 1950), depending on the length of the dry or cold seasons. Over 500 mm of annual rainfall are required for reasonable yields. Coastal Bermuda grows on a variety of well-drained soils ranging from sands to clays. The yields depend on the nutrients available, particularly on nitrogen, and low yields are usually observed on poor sandy soils, e.g. in Florida, although cv. Suvannee bred for the Florida conditions performed on sandy soils better than cv. Coastal. Bermuda grass can tolerate a considerable salinity of irrigation water and withstands flooding for a considerable period of time.

Establishment. Bermuda grass is established from pieces of rhizomes which are planted some 60–100 cm apart. Planting is done by machinery and tobacco-planting implements can be adapted; special machines for grass planting have also been developed. In countries where labour is not particularly expensive planting can be done by hand. P and K fertilizers are usually given before the 'roots' are planted and fertilizer N is sometimes added, although it is normally applied at later stages of establishment. Spraying the soil with pre-emergence herbicides such as 2,4-D or simazine immediately after planting controls the weeds and greatly facilitates the establishment. Grazing or cutting at the early stages of establishment is also recommended for weed control if the weeds are abundant. At the beginning of growth, the stolons usually spread first, underground rhizomes developing later. Utilization usually begins towards the end of the season or early in the following season.

Management, fertilizing. Frequency of defoliation has been studied, mainly by the use of clipping techniques. Total annual yields of forage increase as clipping intervals increase from 3 to 6 weeks but clipping at 8- and 12-week intervals do not usually increase annual yields. Herbage quality and digestibility decrease as the intervals between clippings increase. In several actual grazing trials, continuous grazing was as good as, if not better, than rotational grazing.

Of the fertilizers, N is the most needed nutrient and high responses to applied N have been obtained. The normal rates of N applied in the USA range from 100 to 400 kg/ha and higher rates, up to 1,350 kg/ha, were used in experimental work. Compared with high-rainfall areas, less N is usually applied in dry regions, such as Texas where yields of air-dry herbage increased from 4.9 t/ha for unfertilized stands to 11.4 t when 134 kg N/ha were given; a further 134 kg of N increased the yield by only 2.8 t (Holt & Lancaster, 1968). One kilogramme of applied N usually increases DM yields by 30–35 kg. P and K are normally given at about 100 kg P_2O_5a and 100 kg K_2O per ha, or at higher rates, especially for K when large amounts of N are used. On infertile sandy soils, a 4–1–2 ratio of $N-P_2O_5-K_2O$ is required for efficient fertilization by the three major plant nutrients. Of other nutrients S deficiency was reported and the use of N and P fertilizers containing S (ammonium sulphate, single superphosphate) are sometimes recommended. In trials with irrigated grass, very high yields of herbage were obtained if irrigation was combined with high fertilizer rates.

Legumes. Bermuda grass hybrids are often grown in mixtures with pasture legumes; the yields obtained from such mixtures are normally considerably higher than those of pure grass without applied nitrogen, and the herbage contains more protein, has a higher feeding value and can be more palatable. In southern USA *Trifolium incarnatum* L. has been used more than any other legume. *Vicia villosa* Roth., which contributes to late season grazing, and *Trifolium repens* are also frequent components of Bermuda grass mixtures. In Florida satisfactory results were obtained from a mixture with *Arachis glabrata* (Prine, 1964). It should be noted that Bermuda grass roots can produce an adverse effect on clover seed germination.

Herbage yields. Hybrid cultivars of Bermuda grass form taller and denser stands, they sometimes remain green longer into the dry season or winter and they give higher, often much higher, yields of herbage than the ordinary *C. dactylon.* Yields of DM vary very considerably ranging from 1 t/ha for unfertilized grass to 25 t when heavily fertilized and irrigated. The 25 t mentioned were obtained in Puerto Rico from irrigated grass receiving 1,350 kg N/ha (Brenes *et al.*, 1961) and only slightly lower yields were reported from Alabama in the USA. In trials with more moderate rates of N (150–400 kg N/ha) DM yields are usually within the limits of 4–10 tons/ha; such yields can be considered as normal high yields, and they are perhaps slightly higher than those of other pasture grasses grown under comparable conditions.

Conservation. Bermuda grass is utilized for cattle grazing although hay can also be prepared and fed to the animals, as well as dried and milled grass, sometimes in the pelleted form. Ensiling was met with little success: the grass is difficult to compress, pH of the silage is about 5.0 and fermentation is not of the lactic-acid type. Catchpoole & Henzell (1971) report in their review large amounts of ammonium N and low levels of organic acids, but Willis King in North Carolina obtained good performance from Coastal Bermuda silage properly made.

Chemical composition, nutritive value. Coastal Bermuda grass has been noted for its high content of DM which constitutes about 25 per cent, and sometimes even up to over 30 per cent, of fresh herbage. CP content ranges from 7 to over 18 per cent of the DM and is in general higher than in the majority of other tropical and subtropical grasses; it decreases with plant age but to a lesser extent than in other grasses, possibly because of high leaf:stem ratio. In spite of the high protein content the nutritive value of the herbage is not so high as it might be expected, although DM digestibility is reasonably high and has often been recorded to be about 50–60 per cent but can be as low as 45 per cent. CP digestibility, as reported from Trinidad (see in Butterworth, 1967), is usually high and can reach 77 per cent and is seldom below 50 per cent. Harlan (1970) states, however, that Coastal Bermuda is a grass of 'high production potential per land unit area and a modest to low potential for grazing animal'. The nutritive value seems to be inheritable and was improved in cv. Coastacross 1 in which DM digestibility was 12 per cent higher than in Coastal Bermuda. CF content is moderate, usually around 25–35 per cent, except when the herbage has not been cut or grazed for several months. The content of NFE usually ranges from 50 to 60 per cent. High content of HCN in the Coastal Bermuda herbage has been reported but no harmful effects on cattle were observed.

Animal production. Liveweight gains of cattle grazed on Coastal Bermuda, or fed with cut grass or hay, range mostly from 200 to 300 kg per ha per year or per grazing season when moderate rates of N and other fertilizers are applied, but can sometimes reach up to 900 kg at high N rates; such high production is rare but liveweight increases of 400–500 kg have often been achieved. Liveweight gains per unit area usually exceed those obtained from other tropical pasture grasses, although higher gains from *Digitaria decumbens* and occasionally from some other species were reported. Liveweight increases per animal are moderately high and hardly exceed those obtained from local unimproved *C. dactylon*; gains of 0.5–0.7 kg per animal per day during the grazing period can be considered as average. Good results were obtained from feeding pelleted, artificially dried grass: in Georgia, grazing Coastal Bermuda grass resulted in 0.62 kg daily liveweight increases, feeding artificially dried grass gave 0.48 kg/day and feeding the same dried grass but pelleted gave 0.94 kg (Hogan *et al.*, 1962). The presence of legumes in the sward increases liveweight gains; the increases can be considerable and make beef production more economic than on pure grass receiving fertilizer N. Hogg & Collins (1965) obtained 277 kg liveweight increase per year per ha from grass receiving 135 kg N/ha and 460 kg from a Bermuda grass/Louisiana white clover mixture, the cost of 1 kg liveweight increase being 19.2 and 7.2 cents, respectively.

It was observed in Florida (Kidder *et al.*, 1961) that a fungus found on dead leaves of Bermuda grass caused hyper-sensitivity to sunlight in cattle grazed on the aftermath.

Cynodon nlemfuensis and C. aethiopicus

Cynodon nlemfuensis Vanderyst (*C. dactylon* (L.) Pers. var. *sarmentosus* Parodi). Star grass

Stoloniferous perennial without underground rhizomes. Stolons stout, lying flat on the ground. Stems robust to slender, 30–70 cm high. Leaves flat, 5–16 cm long and 2–6 mm wide, often arching. Racemes (spikes) 4–10 cm long, usually green or pale but sometimes reddish. Glumes narrow-lanceolate in profile, the upper half to three-quarters of the length of the spikelet.

Var. *nlemfuensis*. Stems moderately robust to fine. Leaves 2–5 mm wide. Racemes 4–7 cm long, four to nine in a single whorl. Occurs in Ethiopia, Kenya, Uganda, Tanzania, Zaire, Zambia, Rhodesia and Angola.

Var. *robustus* Clayton & Harlan. Stems stout, leaves 4–6 mm wide.

Racemes 6–10 cm long, 6 to 13 in one or two whorls. Distributed in about the same areas as the main variety.

Both varieties occur from sea level to 2,300 m alt, in forest and bush clearings, in grasslands, as pioneer grasses on denuded land, roadsides, wasteland, old cultivations, cattle bomas (kraals), etc. Small forms of var. *nlemfuensis* can be mistaken for *C. dactylon* from which they differ mainly by the absence of underground rhizomes and it is hardly possible to distinguish them in those herbarium specimens in which the basal plant parts are absent.

Cynodon aethiopicus Clayton & Harlan. Star grass; Giant Star grass (Fig. 15)

As *C. nlemfuensis* but usually more vigorous and with more robust stems which can be taller. Leaves up to 25 cm long and up to 7 mm wide. Racemes 4–8 cm long, rather stiff and pigmented with purple, 6 to 17 in two to five whorls. Spikelets usually purple.

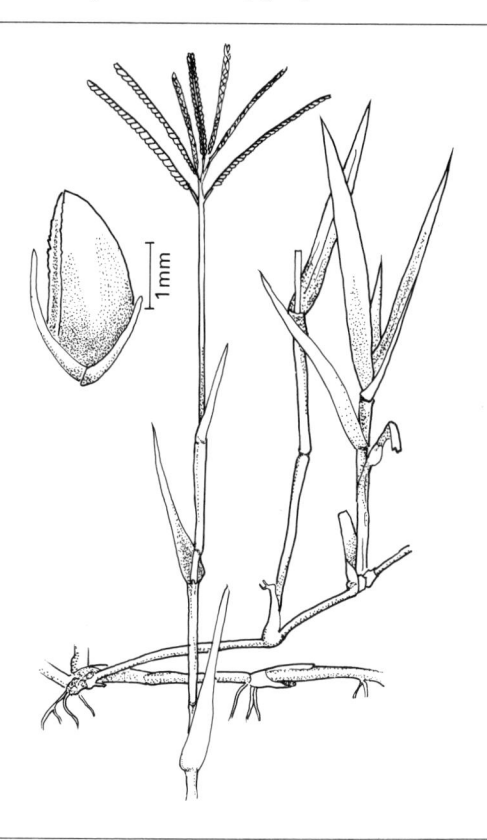

Fig. 15   *Cynodon aethiopicus.*

Occurs throughout East Africa from Ethiopia and eastern Sudan to Zambia and Rhodesia, at 800–2,300 m altitude, often at forest edges or on land from under cut forest and seems to be more moisture demanding and less drought resistant than *C. nlemfuensis*.

The extreme forms of the two species can be easily distinguished but there is a series of transitional forms and further taxonomic investigation is required in which a study of herbarium specimens should be supplemented by observation on living plants grown in plots.

Both species are utilized for grazing, and sometimes for hay, in the areas of their natural distribution, *C. nlemfuensis* being normally preferred to the more robust *C. aethiopicus*. There is no doubt that practically all reports on the use of **local** *C. dactylon* in tropical East Africa, and at least some reports from Rhodesia, Malawi and the neighbouring countries, can be referred to *C. nlemfuensis* and to a lesser extent to *C. aethiopicus*. There is also little doubt that the records of cultivation of Giant star grass or *C. plectostachyus* should actually be referred to *C. aethiopicus* or occasionally to *C. nlemfuensis*.

The information on the responses to grazing, yields, nutritive value, etc., that follows, refers to both species (*C. nlemfuensis* and *C. aethiopicus*) whenever quoted in the literature under correct names or as *C. dactylon* or *C. plectostachyus*.

In tropical East Africa *C. nlemfuensis* is particularly common on light-textured soil at the bottom of the Rift Valley where it often forms almost pure stands under relatively dry conditions. In particularly dry years it can be weakened by drought and invaded by annual grasses such as *Aristida adscensionis* or *A. kenyensis* but recovers in more moist years and retains its dominance, especially under reasonably heavy grazing. It is generally encouraged by heavy grazing pressure which in a Rhodesian trial increased its basal ground cover from 3.0 to 11.5 per cent (Rodel, 1970), but cutting for hay usually weakens the grass. Low grazing also produces a beneficial effect. In Tanzania *C. nlemfuensis* was successfully established from seed in bush in which the original herbaceous cover had been destroyed but the land not ploughed. It persisted well and, together with *Cenchrus ciliaris*, sharply increased the grazing capacity of land over that under the original annual grasses (Owen & Brzostowski, 1966).

Cultivation yields. Both species can be cultivated and are propagated by stolons or occasionally by seed. In Kenya it was successfully established at Muguga and Kitale. At Kitale it yielded 2,400 kg DM/ha on poor soil of western Kenya, but higher yields were obtained at Muguga, in the Virgin Islands and elsewhere. In Rhodesia high yields were obtained from *C. aethiopicus* (grown under the name *C. plectostachyus* but recently identified by Kew as *C. aethiopicus*) which yielded 14,450 kg DM/ha when given 450 kg N/ha + basic dressings of P and K (Rodel, 1970) and *C. nlemfuensis* yielded 12,270 kg/ha. In more recent trials in Rhodesia with *C. nlemfuensis*, Mills *et al.* (1973) obtained DM yields of the same order: 9.6–10.8 t/ha when given 225 kg N/ha and there were no yield decreases during the 3 years of growth; higher yields were obtained from the application of 675 kg N/ha but these declined from 17.2 t DM/ha in the first year to 12.7 t in the third year; still higher rates of N did not produce any further yield increases and *C. nlemfuensis* was less responsive to heavy rates of N than the three other grasses under trial. Under irrigation, heavy rates of N were more effective and the highest yield obtained reached 26.8 t DM/ha at 900 kg N. These high rates of N are hardly applicable in practice but the rates of up to 200–300 kg N/ha can be economical.

Both *C. nlemfuensis* and *C. aethiopicus* were occasionally grown, mainly in experimental plots, in mixtures with legumes; and *Stylosanthes guianensis*, *Centrosema pubescens*, cv. Louisiana of *Trifolium repens* and *Lotononis bainesii* produced satisfactory mixtures.

The diseases of both species, but chiefly of *C. nlemfuensis*, are mainly those affecting the spikelets and thus reducing seed formation and yields, the main diseases being loose smut and another spikelet disease caused by a species of *Fusarium*. As with *C. dactylon*, etiolated plants often occur but they do not affect yields to any appreciable extent.

Nutritive value, animal production. CP content in the herbage of both species is high and can reach over 20 per cent of the DM and is seldom below 8 per cent although as low a content as 5.2 per cent was found in so called 'standing hay', and the average CP content can be estimated as being between 10 and 15 per cent. Butterworth (1967) quotes for '*C. plectostachyus*' the contents of DCP ranging from 3.7 to 15.3 per cent but in mature hay it can be below 2 per cent. CF content was given from 23 to 40 per cent and its digestibility from 40 to 78 per cent; DM digestibility ranged from 42 to 64 per cent. In trials in Puerto Rico (Caro-Costas *et al.*, 1972) heavily fertilized grass containing 12–15 per cent CP yielded about 10.4 t TDN/ha. The grass is palatable and is readily eaten by cattle and selectively grazed by wild ungulates in Kenya (Stewart & Stewart, 1970).

Very high liveweight gains were obtained by Caro-Costas *et al.*, (1973) in Puerto Rico: steers grazed on heavily fertilized *C. nlemfuensis* gained 1,514 kg/ha/year as against 1,062 kg when grazed on *Digitaria decumbens*. Average daily gains per head were 0.60 kg and the pasture carried 7.4 steers/ha, with the average weight of 275 kg per steer. High animal production as also reported from Rhodesia where in trials with cows and calves the liveweight gains were of the order of 1,150–1,350 kg/ha for the summer season of 1970–1; these gains were obtained from grazing *C. aethiopicus* which received 335 kg N/ha. In these trials rotational grazing had no advantage over continuous grazing. Much smaller gains were reported for other trials in Rhodesia.

Occasional harmful effects of both species on the grazing animals were observed. Lambs borne by ewes grazed on *C. nlemfuensis* developed goitrous and skeletal abnormalities and the rate of birth was reduced; this was ascribed to the effect of high contents of HCN, 125 p.p.m. and up to 149 p.p.m. in grass heavily fertilized with nitrogen. *Cynodon aethiopicus*, also heavily fertilized and irrigated, contained 247 p.p.m. of HCN and produced still stronger detrimental effect on lambing and lamb abnormalities (Rodel, 1972). Oestrogenic activity in *C. nlemfuensis* (referred to as *C. dactylon*) was observed in Malawi for a relatively short period at the end of the dry season; the content of stilboestrol reached 4.5 μg equivalents per 100 g of freshly cut grass and this might have been connected with a decline of heat in heifers (Millar, 1967).

Reproduction. Clayton (*Flora of East Tropical Africa*, 1974) states that *C. aethiopicus* 'is fairly well isolated genetically, crossing reluctantly with *C. dactylon* and *C. nlemfuensis*. There are two ploidy levels ($2n = 18$ and 36), which cannot be distinguished morphologically'. Observations in Kenya have shown that at least some forms of *C. aethiopicus* are probably apomictic; they formed abundant seed under isolation or in clonal colonies grown from a single root. On the other hand *C. nlemfuensis* appears to reproduce itself sexually and in Kenya selfing was difficult and a few selfed plants that were obtained were dwarfed and did not persist, whereas the progenies of single plants grown in a polycross arrangement were healthy and vigorous and differed in a number of characters showing their cross-pollinating habit and considerable heterogenity. Both diploid ($2n = 18$) and tetraploid ($2n = 36$) were found in *C. nlemfuensis*.

There are no bred cultivars of *C. aethiopicus* or *C. nlemfuensis* but some outstanding clones were isolated and multiplied vegetatively; these are: 'Star grass No. 2', a clone of *C. aethiopicus* grown in Rhodesia under the name of *C. plectostachyus*, 'Muguga star grass', a good type of *C. nlemfuensis* found in Kenya, and '*Cynodon* IB-8' (Ibadan 8), a clonal selection made by H. R. Chheda from *C. nlemfuensis* originating from Tanzania which in southern Nigeria proved to be much more productive than local types of *Cynodon* (Harlan, 1970).

All types of *Cynodon* can be easily propagated by stolons. Attempts made in Kenya to develop a type with good seed production had little success mainly due to poor combining ability for spikelet disease resistance. Seed of both species is in the form of spikelets with one caryopsis each and there are 2,200,000–4,000,000 such spikelets per kg.

Cynodon plectostachyus (K. Schum.) Pilger. Giant star grass; Naivasha star grass (Common at Lake Naivasha in Kenya) (Fig. 16) $2n = 18$

Stoloniferous perennial without underground rhizomes. Stolons thick, arching. Stems 30–90 cm high. Leaves 10–30 cm long and 4–7 mm wide, soft and hairy. Racemes 3–7 cm long, curling upwards at maturity, 7–20 in two to seven whorls. Spikelets 2.5–3 mm long. Glumes reduced to triangular scales up to 0.25 the length of the spikelet; the lower 0.2–0.3 mm and the upper 0.4–0.6 mm long.

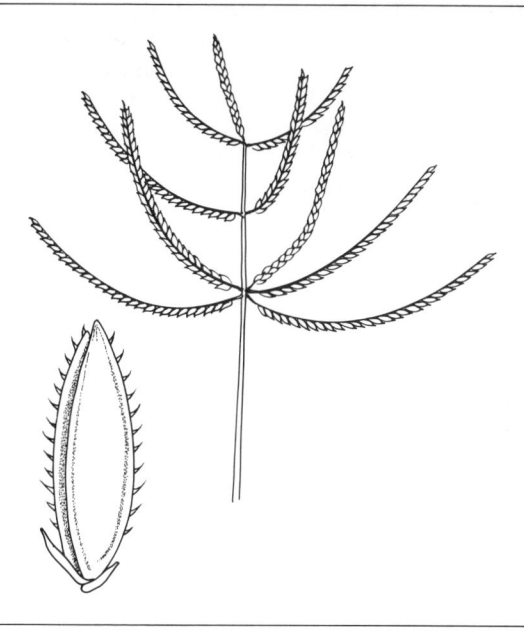

Fig. 16 *Cynodon plectostachyus.*

Occurs naturally in Uganda, Kenya, Tanzania and Ethiopia at 800–2,000 m alt, mainly in the Rift Valley, in dry areas with light-textured soil. It can easily be distinguished from the other species of *Cynodon* by the short glumes, the tendency of the racemes to curl upwards and the arching stolons.

This grass is palatable to all grazing animals but is of low competitive vigour and in mixed swards with *C. nlemfuensis* usually gives way to the latter. A good seed producer.

Information in the literature on the performance of *C. plectostachyus* in natural and sown pastures should be, in most cases, referred to *C. aethiopicus* with which it had been often confused.

Dactylis L.

Includes a few temperate Eurasian, mainly Mediterranean, species one of which is widely cultivated.

Dactylis glomerata L. Cocksfoot; Orchard grass

A temperate grass of Mediterranean origin widely grown for grazing and hay in practically all temperate countries. In the tropics and subtropics it is cultivated, often on a large scale, in Africa (Kenya, Tanzania, Rhodesia, South Africa), Australia (NSW, Victoria and other areas of southern Australia) and America (Brazil, Colombia, Venezuela, Argentina) in the areas where climatic conditions approach or are similar to temperate conditions. Seed is usually imported from UK, Denmark, Netherlands, USA, partly or mainly because the short summer photoperiods of the tropics reduce seed setting and some cultivars do not flower at all. Danish (e.g. S.37) and British (S.24) cultivars are favoured in Africa.

Dactyloctenium Willd.

Tufted or stoloniferous annuals or perennials. Inflorescence digitate of two (sometimes one) to several spikes terminating in a pointed extension of the flattened rhachis. Spikelets elliptic to ovate, broad, laterally compressed, falling off as a unit but without glumes, with a few to several florets most of them bisexual. Glumes and lemmas boat-shaped. Grain angular, variously reticulate or corrugated and enclosed, when young, in a free hyaline pericarp. The 13 species so far known occur mainly in Africa, India, Bangladesh, Burma, and on the Indian Ocean shores and islands. The most common, *D. aegyptium*, is widely distributed in the tropics and subtropics of the Old World and introduced to America where it has naturalized.

Dactyloctenium aegyptium (L.) Willd. Crowfoot grass (Fig. 17)

An annual of varying habit, with much branched stems up to 70 cm tall, or occasionally taller, sometimes erect but mostly geniculately ascending and rooting from lower nodes, or shortly stoloniferous. Leaves 3–25 cm long and 3–15 mm wide, somewhat crisp. Inflorescence digitate or subdigitate with three to nine spikes radiating from the top point of the stem, often with one or two spikes below the rest. The spikes are 1–7 cm long and have protruding bare tops. Spikelets three- to four-flowered, broadly ovate, compressed from side. Florets 2.5–4 mm long. Grain angular, transversely rugose, white or brown, about 1 mm long.

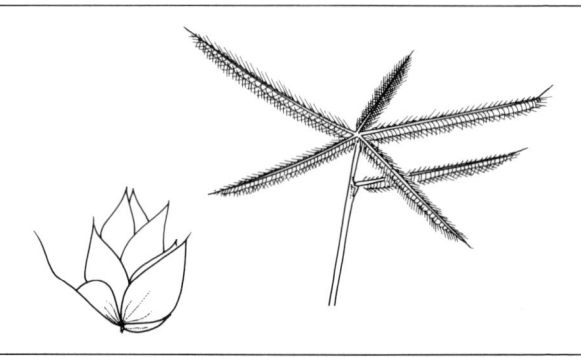

Fig. 17 *Dactyloctenium aegyptium* (from Edwards & Bogdan, 1951).

Occurs naturally in the tropics, subtropics and warm temperate regions of the Old World, from sea level to 2,000 m alt, on roadsides, in fallow and waste lands, as a common weed in cultivation and in pastures of relatively dry areas where it can tolerate a considerable degree of salinity. It has been introduced, apparently accidentally, to America, and in South American states it has spread as a weed in maize and other crops and even penetrated into semi-natural grasslands of less humid areas of north-eastern Brazil.

Dactyloctenium aegyptium is one of the best grazing annual grasses in semi-arid areas, mainly in overgrazed pastures, where it is readily eaten by stock, but it dies out with the onset of the dry season. It produces good quality herbage with CP content ranging from 6 to 18 per cent (Edwards & Bogdan, 1951; Long *et al.*, 1969) and is valued as a grazing grass in north-eastern Brazil. With the increase of grazing pressure the plants acquire more spreading habit and produce more seed, the seeds are however smaller than those formed on taller plants grown under light grazing (Sant, 1964).

Various numbers of chromosomes have been found, and $2n = 20$ and 40 represent perhaps diploid and tetraploid forms of the same plant type; other chromosome numbers found were $2n = 34$, 36 and 48.

Dactyloctenium giganteum Fisher & Schweick. is closely allied to *D. aegyptium* from which it differs taxonomically by the size and colour of anthers. It also differs by the larger size of the plant and of its parts and by the shape of the inflorescence in which the spikes are suberect and not horizontally spreading as in *D. aegyptium*. *Dactyloctenium giganteum* occurs in East and South Africa, at altitudes of 200 to 2,000 m above sea level, and in Rhodesia this giant form was used for increasing the productivity of Kalahari sandveld pastures. Seed broadcasting into a disced ground resulted in stands yielding 650 kg DM/ha. The application of fertilizer N at 67 kg/ha increased the yield to 2,570 kg/ha and of 200 kg N to 5,200 kg DM. P and K fertilizers were less effective (Rhodesia Report, 1966-67).

Dactyloctenium bogdanii S. M. Phillips

A stoloniferous perennial with upright, unbranched stems; the floral parts are similar to those of *D. aegyptium* but the spikes and spikelets can be longer and the grain is slightly smaller. Occurs in north-east Africa: Ethiopia, Sudan, Kenya and Tanzania, at 280-1,700 m alt, in dry open plains where it can form frequent spreading colonies, especially on alluvial soil and can tolerate a considerable degree of soil alkalinity. In Kenya *D. bogdanii* is considered to be one of the most drought-resistant perennial grasses (Bogdan & Pratt, 1967) and is perhaps suitable for reseeding denuded grazing land in arid areas receiving less than 600 mm annual rainfall, although its suitability for the purpose has yet to be proved (in Bogdan & Pratt this species is referred to as *Dactyloctenium* sp.). There are about 2,400,000 seeds (caryopses) per kg.

Dactyloctenium radulans Beauv. Button grass

Differs from *D. aegyptium* in having shorter spikes, up to 1 cm long, and the usually shorter herbage. It is a short-lived annual common and often numerous in dry parts of Australia, in poor pastures and on denuded land where it is well eaten by grazing animals, particularly sheep. Also a weed of dry-area cultivation (Tothill & Hacker, 1973).

Dichanthium Willemet

Tufted or spreading perennials. Panicle usually subdigitate with a short primary axis. Racemes on short, fine pedicels. Spikelets dorsally compressed, in pairs, one sessile the other on a fine pedicel, all similar in shape but the sessile spikelets are bisexual and awned, and the pedicelled spikelets are male or sterile and have no awn. One or two pairs of spikelets at the bottom of the raceme are male or sterile. The 15 species of this genus are distributed in the Old World tropics and subtropics, mainly in moderately moist areas and they are mostly good grazing grasses. This genus is closely allied to *Bothriochloa* and *Capillipedium*.

Dichanthium annulatum (Forsk.) Stapf (*Andropogon annulatus* Forsk.) Marvel grass; Sheda (León & Sgaravatti); Apang; Karad (India); Hindi grass (Philippines)

Tufted perennial with the stems up to 1 m high, erect or geniculately ascending to form broad, somewhat spreading tufts. Leaf sheaths cylindrical; ligule large, membraneous, leaf blades up to 30 cm long and 3–4 mm wide. Inflorescence subdigitate with three to nine pale green or purplish racemes arising from a fine, glabrous common rhachis up to 2 cm long. Racemes 4–6 cm long, densely beset with spikelets arranged in two rows and slightly overlapping each other. Peduncles of the racemes filiform, glabrous, up to 6 mm long. Spikelets 3–4 mm long. Lower glume not winged. Sessile spikelet bisexual fertile (except in one to two lowermost pairs), with a fine awn 15–20 mm long. Grain compressed, oblong, 2 mm long.

Occurs throughout Africa and eastward to China, Malaysian region, Australia and Fiji and is very common in India, Pakistan, Burma and Thailand. In India it occurs in a great variety of forms and Chakravarty (1971) writes that the basal crowns can spread and reach, in spaced plants, over 2 m in diameter; this may, however, actually refer to *D. caricosum*.

Dichanthium annulatum grows in moderately dry to moist areas with an annual rainfall ranging from 300 to 1,500 mm and on a variety of soils; it tolerates soil salinity but not acidity. In India and Burma *D. annulatum* is well liked by the grazing animals and is regarded as one of the best pasture grasses although the content of CP in the herbage is usually low to medium. Sen & Ray (1964) report it to be 3.1–4.1 per cent, Chakravarty (1971) 3.9–7.0 per cent and Dougall and Bogdan (1960), in

Kenya, 8.5 per cent. At CP content of 4.6 per cent its digestibility was 28 per cent and the content of DCP 1.3 per cent (from Butterworth, 1967). In the same sample CF content was 39 per cent, NFE 46 per cent and their digestibility 60 and 42 per cent, respectively. The content of P is given for India as 0.10–0.11 per cent and for Kenya 0.15 per cent, but Chakravarty also reports it to be 0.40–0.50 per cent. *Dichanthium annulatum* is cultivated in India, mainly on an experimental scale, and is established either from seed or from tuft splits, in rows 50–80 cm apart. Two to three cuts per year can be obtained in semi-arid areas, about eight cuts in moist areas and ten cuts under irrigation. Herbage yields range from 2 to 20 t/ha (30 per cent moisture content) depending on the moisture regime of the soil. In Cuba, however, where *D. annulatum* is considered to be indigenous (Pérez Infante, 1970), it yielded in a semi-arid area 17 t DM/ha, 46 per cent of the yield being produced during the dry season under sprinkler irrigation. Under these conditions the production was spread relatively evenly throughout the year and the grass is recommended for young stock and breeding cows. Good responses to fertilizer N were observed in India and Cuba. Drilling annual legumes (*Vigna radiata*, *V. aconitifolia*, *Cyamopsis tetragonoloba* or *Atylosia scarabaeoides*) between the rows of *D. annulatum* increased total yields of herbage over those of grass alone. In semi-arid areas sown *D. annulatum* pastures can support five sheep/ha compared with two sheep in natural grassland dominated by the same species (Chakravarty, 1971).

In India, a superior cultivar, **Marvel** 8, was selected which yielded 25 t green fodder/ha under irrigation and 12 t under dry-land conditions outyielding local unimproved types which produced 10 and 6 t/ha, respectively (Dabadghao & Shankarnarayan, 1973).

Dichanthium annulatum is essentially a pseudogamous apomict. Tetraploids with chromosome number $2n = 40$ prevail in the main area of its distribution but diploids have also been found in India. In South East Africa the plants are hexaploids. Diploid races of India are genetically isolated but tetraploids cross relatively easily and crosses with *D. aristatum* and *D. caricosum* have been obtained (De Wet & Harlan, 1970a). Chakravarty reports a low seed productivity – about 15 kg seed/ha.

Dichanthium aristatum (Poir.) C. E. Hubbard (*D. nodosum* Willem., *Andropogon aristatus* Poir.). Angleton grass; Alabang

Racemes two to six, with hairy peduncles; spikelets 3.5–5 mm long, much overlapping, otherwise as *D. annulatum* but larger and more vigorous. S. C. Pandeya regards this species to be merely an ecological form of *D. caricosum* (Bor, 1960).

This densely tufted perennial occurs in South East Africa, southern India and the Indian Ocean Islands, and is said to have been introduced to Polynesia and West Indies (Hubbard & Vaughan, 1940) and also to the Philippines and South America; it is reputed to be an excellent grazing grass (Bor, 1960). In the Philippines, where it became common and is used for pasture improvement (Farinas, 1970) it is described as being a short, creeping grass with the stolons reaching up to 3 m in length and may actually be *D. caricosum*. In Colombia *D. aristatum* is grown from seed, seed yields reaching 330 kg/ha; best germination was achieved after the seed has been stored for about 7 months (Alarcón et al., 1969). *Dichanthium aristatum* is a tetraploid ($2n = 40$) which can be crossed with *D. annulatum*, but there are also diploid and hexaploid forms.

Dichanthium caricosum (L.) A. Camus (*Andropogon caricosus* L.). Nadi blue grass; Angleton grass

Annual or creeping perennial; leaf sheaths compressed; ligule short, ciliate; racemes solitary, or occasionally two or three; spikelets much overlapping; keel of the lower glume winged; otherwise as *D. annulatum*.

Distributed naturally in India, Burma and the Malaysian region, and introduced to the Philippines (see under *D. annulatum*), Polynesia, West Indies and South America. This creeping grass is palatable and well liked by cattle. In Cuba, *D. caricosum* yielded 15 t DM/ha, 40 per cent of the yield being produced in the dry season under sprinkler irrigation (Pérez Infante, 1970) and it was also tried with some success in Fiji. Butterworth (1967) quotes in his review 6.1–7.0 per cent CP, 37–40 per cent CF and 40–43 per cent NFE, their digestibility being 43–59 per cent, 56–78 per cent and 60–67 per cent, respectively. Dabadghao & Shankarnarayan (1973) state that J. G. Oke, who studied *D. cariscosum* in India, distinguished three forms: spreading, bushy and semi-erect, and selected two promising types: **Marvel** 93 and **Marvel** 40; the latter has an erect habit and gave over 11 t fresh herbage/ha, whereas Marvel 93 is a spreading plant; its yields were slightly lower, some 10 t fresh material/ha. However, *D. caricosum* is supposed to be of a creeping habit, and the two selected cultivars may perhaps belong to more than one species. *Dichanthium caricosum* is a tetraploid with $2n = 40$ chromosomes or hexaploid ($2n = 60$) although $2n = 50$ has also been reported (from Bor, 1960) and crosses with *D. annulatum* have been obtained.

Dichanthium fecundum S. T. Blake

Tufted perennial up to over 1 m high. Stems slender, usually branched. Leaves up to 5 mm wide. Inflorescence consists of one, or less commonly two or three pedunculate racemes, 4–6 cm long. Sessile spikelet 4–5 mm long, with an awn 1.5–2.8 cm long (awnless in the lowermost pair of spikelets). Pedicelled spikelet male or bisexual, 5–6 mm long, awnless. Grain 2 mm long.

Differs from *D. annulatum* in having only one to three racemes and in the pedicelled spikelet exceeding in size the sessile spikelet. Closely resembles *D. caricosum*. An Australian species common in sub-humid tropical grasslands where it can be a dominant or a co-dominant grass often associated with *Eulalia fulva*. Produces good grazing.

Dichanthium sericeum (R.Br.) A. Camus. Queensland blue grass

$$2n = 20$$

Tufted perennial 30–80 cm high; inflorescence of two to four racemes which are 4–7 cm long. Differs from *D. annulatum* in having the racemes beset with spikelets up to the very base whereas in *D. annulatum* and *D. caricosum* the raceme bases are free from spikelets.

An important grass in arid areas of Australia, mainly in its northeastern parts. On black heavy soil it can be locally dominant. The grass is well grazed and is perhaps one of the most palatable local Australian grasses which also produces a fair bulk of herbage (Tothill & Hacker, 1973).

Digitaria Haller

Annual or perennial grasses. Spikelets variously arranged but mostly in loose or dense and then spike-like racemes which are gathered in panicles or are occasionally solitary. The spikelets are usually dorsally compressed and so orientated that *the upper glume faces the raceme rhachis*. Lower glume is a minute scale or absent; upper glume from one third to the full length of the spikelet. Florets two, the lower reduced to an empty lemma, the upper bisexual; its *lemma is folded over the palea and covers it completely from both sides*. Both lemma and palea papery or membranaceous and not hard or crustaceous.

A large genus with over 300 species, mainly tropical and subtropical but also of warm temperate areas. Section *Erianthae* is almost exclusively of African origin, and although not very large is the most important as it includes the widely cultivated Pangola grass (*D. decumbens*) and a group of closely allied species, often known as woolly finger grasses, which are of importance in natural African grasslands and also of a considerable potential value for introduction into cultivation. The more valuable species of this group are *D. eriantha, D. pentzii, D. milanjiana, D. macroblephara, D. nodosa, D. setivalva, D. smutsii* and *D. valida*. They have clearly digitate or subdigitate panicles and are tufted or stoloniferous perennials (Fig. 18); at least in some of them stoloniferous and tufted forms are found in the same species. These species grow in relatively dry parts of South and tropical Africa with an annual rainfall ranging from 500 to 1,000 mm and with well-pronounced one or two dry seasons. Each species varies; there are numerous transition forms and plant naming is not easy partly because the taxonomy of the group is confused and requires a revision. Some of these species (apart from cultivated *D. decumbens*) have been tried in

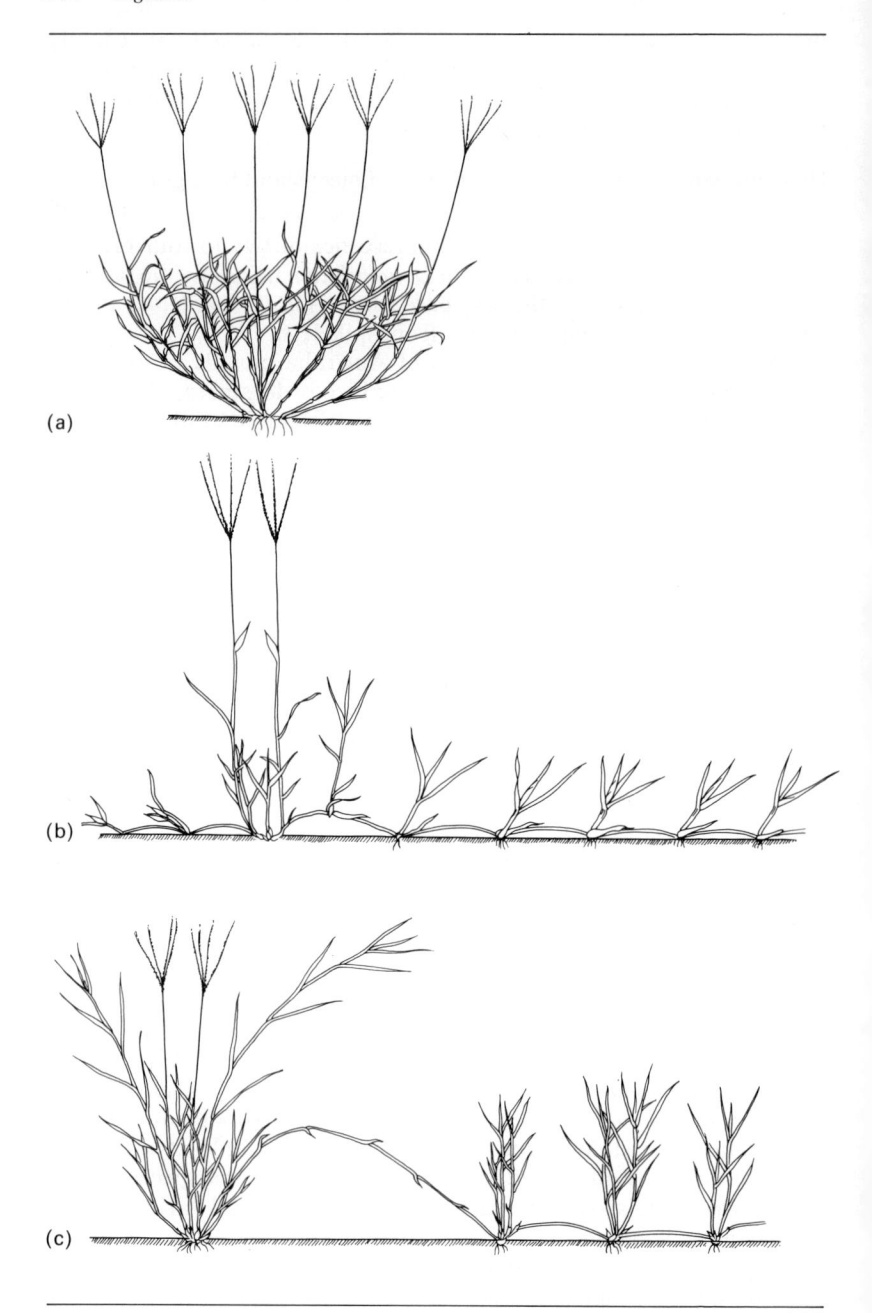

Fig. 18 Habits of perennial species of *Digitaria*: (a) *Digitaria nodosa*, (b) *D. milanjiana*, (c) *D. macroblephara* (from Edwards & Bogdan, 1951).

cultivation in Africa, but mainly in the Caribbean area of America. The data concerning their performance should however be accepted with caution as the names attached to species and varieties under trial may not necessarily be correct. Some varieties of this group which are under trial have not been botanically named and are grown under local names as e.g. **Slender-stem digitgrass**, which has been approved for farm-scale cultivation in the USA, **Pretoria** digitaria in South Africa or **Umfolozi** and **Tsotsoronga** in Zaire.

Three growth forms in plants of the *Erianthae* section have been distinguished in Kenya (Edwards & Bogdan, 1951): (a) tufted type without stolons typical of *D. nodosa*; (b) stoloniferous form with long horizontal runners forming even stands typified by *D. milanjiana*; (c) a type developing stolons which grow first vertically and resemble young fertile shoots but have more numerous nodes and grow taller; having reached a certain height they bend down, root when they reach the ground and then grow horizontally as usual stolons. This last kind of stolon behaviour can be regarded as an adaptation reducing the competition between the main tuft and its secondary tufts developed from stolons; similar stolons have been observed in *Sporobolus helvolus*.

There is a number of annual species such as *D. sanguinalis* (L) Scop., *D. adscendens* (H.B.K.) Henr., *D. longiflora* (Retz.) Pers., *D. ternatea* (A.Rich.) Stapf and many others, mostly arable weeds. They are eaten by stock grazed on stubble or at roadsides and can be nutritious but are only of a temporary nature and not productive enough to be of any use in cultivation, either as grazing or as fodder grasses.

Digitaria californica (Benth.) Henrard (*Panicum californicum* Benth., *Trichachne californica* (Benth.) Chase) $2n = 54$

Tufted perennial. Stems 20–100 cm high. Leaves from almost glabrous to pubescent, 2–15 cm long and 2–5 mm wide. Panicle 5–20 cm long, with four to eight distant erect or suberect racemes 2–8 cm long. Spikelets 3–4 mm long, woolly and with long, silvery white or purplish marginal hairs which are longer than the spikelet. Lower glume a minute triangular scale.

Distributed from southern USA to Argentina, in arid and semi-arid grasslands where it is common and sometimes abundant. Well liked by stock before flowering but becomes less palatable at later stages of growth. It forms medium-size tufts with the leaves high on the stem, often reaching the panicles.

Digitaria decumbens Stent. Pangola grass; Pongola grass (Fig. 19)

Perennial forming tufts from long creeping stolons rooting from nodes. Stems up to 120 cm high but usually much shorter, often branched. Leaves numerous, glabrous, linear-lanceolate to linear, 10–25 cm long and 2–7 mm wide. Inflorescence a terminal digitate panicle of 5–10 spikes (racemes), usually arranged in one whorl but the more vigorous plants may have two to four spikes forming a second whorl above or below the main one. Spikes up to 13 cm long, densely beset with paired spikelets, one sessile the other on a short pedicel. Spikelets 2.7–3.0 mm long, with two florets. Lower glume a minute scale, upper glume three quarters the length of the spikelet, silky hairy with minute adpreesed hairs.

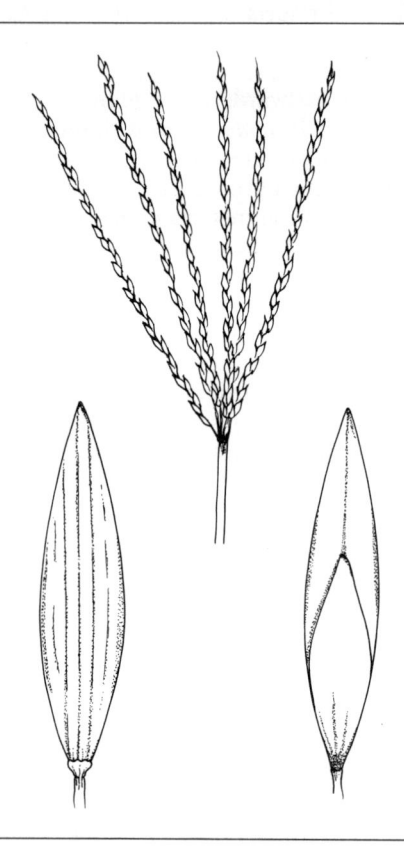

Fig. 19 *Digitaria decumbens.*

The botanical name *D. decumbens* is now traditionally attached to Pangola grass but is not necessarily correct and Chippindall (1955) considers that Pangola grass is more likely to be one of the numerous varieties of *D. pentzii*, a stoloniferous grass common in South Africa. The section *Erianthae*, to which *D. pentzii* (and *D. decumbens*) belongs, is taxonomically difficult mainly because of the numerous transitional forms between the species of the section which are mostly apomictic and often produce little or no seed. Moreover herbarium specimens on which the taxonomy is normally based, often lack basal plant parts important for determining plant identity. It is more or less generally considered that for the time being the cultivated Pangola grass should remain known as *D. decumbens* pending a thorough revision of the *Erianthae* section, a revision which should include a study of living plants.

The name Pangola grass has been derived from the Pongola River in the Piet Retief district of eastern Transvaal in South Africa or in adjacent Zululand districts, and some authors suggested the name Pongola grass, possibly with a view of avoiding confusion with any other forms of *Digitaria* which may come from the Pangola River area of western Transvaal. It has also been suggested (Oakes, 1969a) that the name Pangola grass should be applied only to those clones which, under the USA introduction number P.1.111 109–110, were originally brought to USA in 1935 and which are now widely grown in a number of countries, but not to any later African introductions of the same or closely allied species.

Introduction. *Digitaria decumbens* was first grown near Pretoria in South Africa and then introduced to USA in 1935; it reached Florida in 1937 where it was thoroughly tried, found to be an outstanding pasture grass and released to farmers in 1944. In the late 1940s and early 1950s, *D. decumbens* was introduced into the West Indies, Central America and the northern parts of South America and later to some countries outside America: Australia, West and East Africa, the Philippines, Hawaii, India, Pakistan, Malaysia and some others. In nearly all areas of introduction Pangola grass has shown good or outstanding results and is now cultivated on a farm, often large scale, in the majority of subtropical, tropical and even temperate warm countries.

Pangola grass is valued for the ease of establishment, the ability to grow under a variety of climatic and soil conditions, tolerance to almost any kind of management, good herbage production, high nutritive value and palatability. In spite of its vigour and aggressiveness Pangola grass can be relatively easily eradicated, is suitable for arable rotations and is e.g. grown in rotation with sugar cane in some Caribbean islands. In the last decade a stunting virus disease was found spreading over large areas and threatening the use of Pangola grass in cultivation. Substitutes for Pangola grass resistant to stunting virus are being sought and *Cynodon nlemfuensis*, *Brachiaria decumbens* and some other grass species have begun to replace Pangola grass; resistant types of *Digitaria* allied to Pangola grass are also under trial.

Environment. High air temperatures stimulate the growth and development of Pangola grass and Schroder (1971) found the maximum shoot growth at 43°C, a temperature higher than that required for the best growth of the other grass species he tested. The roots grew best at soil temperatures of 27°/30°C but very little growth was observed below

16°C and above 41°C. Low night temperatures produce an adverse effect on grass growth and herbage yields; this effect is now usually ascribed to the accumulation of starch grains in the leaf chloroplasts in cool or cold nights. The adverse effect of starch accumulation can be explained by the blocking of the system transmitting carbohydrates which for its normal functioning requires the reduction of starch to water-soluble sugars. Chatterton et al. (1972) have shown, however, that starch accumulation occurs mainly in non-tillering plants, whereas in the leaves of actively tillering plants starch is removed even during cool nights with the temperature around 6°/10°C and the flow of the products of photosynthesis is not blocked. In spite of high temperature requirements Pangola grass can grow well in the subtropics, often even better than in the tropics, but at high altitudes, above 1,200–1,500 m, its productivity drops, possibly because of cool nights. Pangola grass originates from a country with an annual rainfall ranging from 500 to 900 mm and with a well pronounced dry season and can withstand droughts. It grows better under more humid conditions and is normally cultivated in the areas where the annual rainfall reaches or exceeds 1,000 mm. Pangola grass can grow on various types of soil, from acid with 4.5 pH to alkaline with pH values exceeding 8.0 and can tolerate slight waterlogging although drainage is required for the sites where waterlogging is more severe. Clay soils are less suitable than loams. Pangola grass is tolerant to the presence of aluminium in the soil or to small or moderate amounts of sodium or sodium chloride.

Establishment. No viable seed is practically formed and Pangola grass is established only vegetatively. The simplest and most used way to establish the grass is to cut the herbage when it is stemmy, spread it on the prepared seedbed and disc in; 0.5 to 2 t fresh material are required for planting a hectare. To economize planting material the stems can be chopped to pieces with about three nodes each. If the ground is too wet for tractor discing, cattle trampling can press the stems into the soil but this method is said to result in an uneven establishment (Nestel & Creek, 1962). More precise planting can result in quicker and more uniform establishment; the stems or pieces of stolons are then planted by hand, in furrows or holes made by hoes and various spacings have been suggested from 25 × 25 cm to 1 × 1m or 1 × 1.2 m; spacings 50 × 50 cm or 100 × 50 cm may be adequate; closer spacing, resulting in more rapid establishment, can be used if labour is not a problem. Planting by stems which had been kept watered, protected from drying and direct sun, and begun rooting can be done with advantage; this method can be used, however, for relatively small areas or for experimental plots. When planting with particular care the young fresh tops of shoots should be cut off and the stems stuck into the soil up to about 5–10 cm deep although about half the length of the cutting should be above the soil level. An application of selective herbicides on weed-infested soil is often recommended. Successful attempts have been made in the Virgin Islands (Oakes, 1968b) to sod-plant Pangola grass into an established sward of hurricane grass (*Bothriochloa pertusa*), a pasture weed in the Islands, or on to subsoiled or ploughed swards; other grasses tried for the same purpose failed to establish or were later replaced by the same *B. pertusa*, whereas Pangola grass persisted.

Management, fertilizing. One of the advantages of Pangola grass is its satisfactory growth and performance under almost any density of grazing. Sward management trials have been mainly confined to the height and frequency of grazing usually studied by clipping technique. As expected, herbage yields increased with the increase of intervals between cuttings and in a trial by Pérez Infante (1970) in Cuba the yields were under 5 t DM/ha in the dry season under sprinkler irrigation when the herbage was cut at 36-day intervals and about 6 t when cut at 60-day intervals; in the wet season yield increases were more substantial and the grass yielded 10 and 14 t DM/ha, respectively. In unfertilized grass the effect of the length of intervals between the cuts was negligible but it was well expressed in grass stands given fertilizer N. It seems that the increased intervals between grazings or cuts increased herbage yields only or mainly under the conditions favourable for plant growth. Yields of consumable herbage may, however, decrease with the increased intervals between grazings as it was shown by Vicente-Chandler *et al.* (1972) in Puerto Rico. In their trials, the amounts of DM consumed by grazing cattle decreased from 14.1 t/ha when the animals were grazed every 14 days to 13.0 t under grazing with 21-day intervals and to 11.1 t at 28-day intervals. In the same trials, grazing to about 15 cm from the ground level resulted in larger amounts of consumed herbage than low grazing to 5 cm. These results are at variance with the earlier data by Caro-Costas & Vicente Chandler (1961) who obtained 32.8 t DM/ha/year under low clipping of up to 7 cm and 22.3 t under clipping to 15–25 cm from the ground level. Pangola grass forms a dense growth and can be to a certain extent sodbound or rootbound and occasional cultivation, discing or rotovation can improve the growth and herbage yields. Sprinkler irrigation is sometimes applied in the dry season and in Brazil (Ladeira *et al.*, 1966) irrigation during the 5 months of the dry winter season increased the yields of fresh herbage in this cool season from 3.5 t to 5.3 t/ha unfertilized, and from 4.7 to 7.1 t when given moderate amounts of NPK fertilizer. Cool season yields of Pangola grass were in this trial considerably lower than those of *Panicum maximum* or *Melinis minutiflora* and this was attributed to its high sensitivity to low temperature.

Fertilizer N produces a considerable effect on the yields and CP content of Pangola grass herbage. The rates of application varied widely in various experiments, from under 100 to 1,800 kg N/ha given mainly in the form of ammonium sulphate; other forms of N usually had similar or slightly smaller effects. Although in trials in Cuba (Crespo 1972a,b), urea applied at 60 kg N/ha after each cut as a foliar spray increased DM yields from 5.7 to 17.2 t DM/ha, soil (broadcast) application increased the yields to 19.2 t. Under soil application the invasion of other species increased by 35 per cent, whereas no increase of pasture weeds was observed under foliar application and a combination of foliar and soil applications has been recommended. Even in the dry season considerable increases of DM yields following a foliar application of urea was observed and 60 kg N/ha resulted in 606 kg DM/ha, whereas unfertilized grass yielded only a fraction of this amount. Applied N increases the content of N or CP in the herbage and in Cuba (Crespo, 1972b), fertilizer N increased N content in dry herbage from 0.97 per cent to 2.14 per cent, which corresponds to 6.1 and 13.4 per cent CP, respectively. Less spectacular increases were however observed in the majority of trials and e.g. in a trial by Oakes (1969a) the application of 670 kg N/ha increased CP content of the herbage from 8.02 to only 9.20 per cent. Fertilizer N decreases Pangola grass tolerance to low temperatures and especially to frosts and it is not recommended to apply N when a cool or cold season approaches. Yield responses to fertilizer N are usually linear up to about 300–350 kg N/ha, but decrease at heavier applications and, economic considerations permitting, 300 kg is a rate which can be recommended in most cases. Generous application of N can reduce the content of dry matter in fresh herbage and this may possibly result in decreases of grass intake by the grazing animals. Split application after each grazing or cutting gives normally better results than single application. The percentage of utilized or recovered N ranges widely from 20 to 70 per cent depending on the locality, the application of other fertilizers but mainly on the rates of applied N, the highest per cent of recovery of N being at low or moderate rates when the recovery can reach 60–70 per cent and it decreases at higher rates of application, although Oakes (1969a) reported only 27–29 per cent recovery at 170–340 kg N/ha.

Phosphorus as fertilizer is usually less effective than nitrogen although Chesney (1969a) observed, in small plots, considerable responses to P and obtained 11 t DM/ha after an application of 33 kg P_2O_5/ha, compared with 7.6 t for the grass receiving no P. In most cases, however, P applied alone gave little or no increase of herbage yields but it can be highly effective if applied together with high rates of N and a considerable interaction between the two nutrients has been observed. Moreover, on soil deficient in P the grazing cows sometimes show symptoms of P deficiency which can be remedied by fertilizer P.

Potassium fertilizers are not often used but in some experiments responses to K were obtained, especially in conjunction with high rates of fertilizer N and an interaction between the two nutrients was noted. K increases to some extent Pangola grass tolerance to low temperatures. A portion of K fertilizer can be replaced by sodium, to which Pangola grass responds positively, and as most of the other tropical grasses do not respond to Na it has been suggested that Na can be used to stimulate selectively the growth of Pangola grass in a mixture with other grass species (from Nestel & Creek, 1962). Chesney (1969a) obtained yield increases from 7.1 to 12.5 t DM/ha when 66 kg K_2O were applied but Mg produced no effect. Pangola grass is reputed to be indifferent to liming but in Hawaii liming increased liveweight gains of cattle grazed on grass fertilized with N.

Chesney (1969b) obtained considerable yield increases by the application of a mixture of micronutrients containing B, Cu, Fe, Mn, Mo and Zn. Of these nutrients Cu is perhaps the most important because Pangola is sensitive to Cu deficiency and responds well to copper oxide or copper sulphate; the application of P can increase Cu content in the plant. Sulphur can also improve Pangola grass yields and especially Pangola/clover mixtures and is usually applied in the form of gypsum (from Nestel & Creek, 1962).

Association with legumes. Pangola is an aggressive grass and can suppress the companion legume in grass/legume mixtures. This aggressiveness can be increased by the application of N and pure stands of N-fertilized Pangola grass often give higher yields of herbage than its mixtures with legumes although CP yields and its content in the herbage are usually higher in the mixtures. The application of nitrogen, especially at high rates, decreases or even eliminates the legumes from the mixtures and Whitney (1970) reports from Hawaii a decrease of the percentage of *Desmodium intortum* grown in mixture with Pangola grass from 50 per cent to below 10 per cent when 410 kg N/ha were applied and to below 1 per cent when the intervals between the cuts were reduced from 10 to 5 weeks. Several herbage legumes have nevertheless been grown with good success in mixtures with Pangola grass, provided that little or no nitrogenous fertilizer is applied. Young & Chippendale (1970) in Queensland, Australia, grew Pangola grass with *Trifolium repens* and *Lotononis bainesii*, low-growing perennial creeping legumes, but it was not possible to maintain them in the sward for the 5 years of trial; however, in another trial *L. bainesii* grew reasonably well up to the sixth year (Bryan, 1968). A *Centrosema pubescens* mixture was grown with good success in Brazil (Aronovich *et al.*, 1970) and gave 410 kg/ha of liveweight increase of calves grazed on the mixture, whereas pure stands of Pangola grass gave 349 kg liveweight increase/ha. An application of 100 kg N to pure Pangola grass resulted, however, in a still higher liveweight increase of 531 kg. In Florida a Pangola/*C. pubescens* mixture yielded 9.5 t DM/ha/year compared with 5.2 t obtained from Pangola grass alone, and the presence of the legume increased CP content in the herbage from 6.0 to 9.6 per cent and trebled the yields of CP (Kretschmer, 1970); *Macroptilium atropurpureum* (Siratro) has also shown good performance in Pangola grass mixtures, yielding 11.6 t

DM/ha of herbage containing 9.6 per cent and 1,130 kg CP/ha; *Desmodium intortum* was still another outstanding legume and its mixture with Pangola grass yielded 13.7 t DM in a trial in which unfertilized Pangola grass grown alone yielded 3.3 t. In Hawaii (Whitney & Green, 1969) a Pangola grass/*D. intortum* mixture did well and gave 10.8–12.0 t DM/ha compared with 7.5 t obtained from a Pangola/*D. canum* mixture and with 3.8 t from the grass alone; yields of the two mixtures were equivalent to those of Pangola grass fertilized with 450–550 kg and 240 kg, N/ha, respectively. Other species of *Desmodium* which performed satisfactory in Pangola grass mixtures were *D. sandwicense*, *D. heterocarpon* and *D. uncinatum*. In Brazil, a mixture with *Glycine wightii* increased liveweight of steers by 208 kg/ha/year, whereas pure unfertilized Pangola grass gave 139 kg liveweight increase and Pangola fertilized with 100 kg N/ha gave 239 kg; liveweight gains per steer per day were 316, 231 and 294 g, respectively (Buller *et al.*, 1970). In the same trial a Pangola/*Stylosanthes guianensis* mixture gave still better results than the *G. wightii* mixture but *S. guianensis* did not persist more than 2 years, possibly because of the year-round close grazing. In other trials *S. guianensis* and *S. humils* have also shown good performance in Pangola grass mixtures. Other legumes recorded to do well in association with Pangola grass are *Pueraria phaseoloides*, *Calopogonium mucunoides*, *Teramnus labialis*, *Alysicarpus vaginalis* and *Macroptilium lathyroides*, and in cooler situation or seasons also *Lotus uliginosus*, and *Vicia dasycarpa*. It should be noted however that in Pangola grass mixtures most of the legumes under trial did not last more than 2 or 3 years.

Pests and diseases. Yellow sugar-cane aphid, *Sipha flava*, seems to be the most important insect pest in the Caribbean area and has been reported from Florida, Trinidad, Puerto Rico, Jamaica and Cuba. This insect, which causes grass stunting, is not specific to *C. decumbens* but occurs on other grasses; in cases of heavy infestation parathion at 0.4 kg/ha can control the insect. Occasional severe damage is also caused by army worms of the genera *Laphigma*, *Spodoptera* and *Mocis*; these insects are again non-specific and damage other grasses, especially low growing or creeping and spraying with DDT at about 1 kg/ha is recommended in case of severe attacks (Nestel & Creek, 1962). Of other insect pests damaging Pangola grass *Taxoptera graminis*, *Blissus leucopterus*, a coccid and *Antonina graminis* have been mentioned by Nestel & Creek and severe attacks by *A. graminis* badly damaging Pangola grass pastures were reported from Paraguay (Giménez Ferrer, 1970); the insects attack the basal parts of plants causing necrotic lesions; correct management with rest and the use of fertilizers reduces the damage. Pangola grass is susceptible to some root nematodes. In Florida, the roots are attacked by *Belonolaimus longicaudatus* and in Brazil by *Platylenchus brachyurus*.

The most damaging and destructive is stunting virus. The virus was first reported from Surinam by Dirven and Van Hoof (1960). The first symptom of the disease is slow recovery after cutting and the affected plants do not develop runners. The plants become dwarfed and have small or minute leaves crowded at stem bases with much shortened internodes. The virus is transmitted by *Sogata fructifera*, an aphid; it spreads to adjacent plants and can affect large areas greatly reducing herbage yields and eventually destroying the plants. The stunt virus was later found in Guyana, Jamaica, Malaya, Borneo, Taiwan and Fiji although it is not quite certain yet whether the virus in these countries is identical or only similar to that found in Surinam, although its effect is similar. Although the virus spreads rather slowly, it presents a real danger for Pangola grass cultivation and in some countries Pangola grass is being replaced by resistant grass species such as *Cynodon nlemfuensis* or *Brachiaria decumbens* in Jamaica. Other grass species can, however, also be susceptible to the same or a similar virus and Revilla (1966) found in Peru that *B. mutica* and *Paspalam conjugatum* can be affected but to a lesser degree than Pangola. Varieties of *D. decumbens* resistant to stunt virus were found and a resistant variety grown in Taiwan is under trial in several countries, but even in this type the resistance is not complete and the infected plants develop somewhat deformed flowering parts; moreover this variety may not necessarily belong to typical Pangola grass but can be a variety of *D. pentzii*. Schank et al., (1972) in the USA have shown that susceptible types of *D. decumbens* and of its hybrids can perhaps be recognized by the presence of very thick walls of the bundle-sheath cells of the leaf not found in the virus-resistant hybrids.

Herbage yields. Nestel & Creek (1962) give average yields of herbage as ranging from 11 to 22 t DM/ha (10,000–20,000 lb/ac) for moderately to well fertilized Pangola grass and later trials support this view. In Hawaii (Whitney & Green, 1969) 18 t DM/ha was obtained from grass given 640 kg N/ha; in French Caribbean islands Pangola yields varied from 5 to over 20 t and in Sarawak (Ng, 1972) 6.0 t DM/ha were obtained without fertilizers and this increased to a maximum of 17 t at 896 kg N/ha, and CP yields ranged from 298 to 1,731 kg/ha. There is other evidence of similarly high yields. Exceptionally high yields were reported from Puerto Rico where Sotomajor-Rios et al., (1971) obtained up to 36 t DM/ha/year in six cuts, and from Australia where 28.0 t DM were reported by Grof & Harding (1970) for Queensland and 28.4 t by Hendy (1972) for northern Australia. Much lower yields were also reported, e.g. in Colombia, where Durango & Padilla (1972) obtained only 3.28 t DM/ha, or in the Caribbean area (Richards, 1970), where Pangola grass yielded 8.26 t DM when fertilized with 364 kg N/ha and 8.51 t when given 728 kg N. Yields of unfertilized grass usually range from 3 to 7 t DM/ha. It should be remembered that the indicated yields have been obtained from experimental plots and may be higher than those obtained under farm conditions. Moreover yields in the areas affected by stunting virus have been drastically reduced.

Conservation. Pangola is essentially a grazing grass but herbage surplus at the time of peak growth is sometimes cut and made to hay or silage. Satisfactory hay can be prepared but drying and fermenting is not easy because of the seasonal high air humidity; nevertheless haymaking is common in Florida and good hay has been prepared. Ensiling is also difficult and seldom successful mainly because pH values of the ensiled material are high and although reduced from 6.2 to 5.6 by adding zinc bacitracin they remained still too high. Consequently, poor results have often been obtained from feeding Pangola grass silage to cattle (Catchpoole & Henzell, 1971) although remarks can be found in the literature about good silage prepared in Florida.

Chemical composition, digestibility. CP content in the DM of Pangola grass ranges within wide limits, from 3 to 24 per cent, although the 24 per cent reported from Surinam by Appelman & Dirven (1962) is an exception and CP content seldom exceeds 14 per cent even for grass cut at short intervals and well supplied with fertilizer N. CP of unfertilized grass usually ranges from 4 to 8 per cent and of grass well supplied with fertilizer N from 8 to 13 per cent, and is normally lower than CP content in a number of cultivated tropical grasses as has been shown by Gomide *et al.* (1969a) in Brazil or by Bryan (1968) in Australia. Fertilizer N increases the content of CP and Rivera-Brenes *et al.* (1961) gave its content as 6.93, 8.02, 11.09 and 12.06 per cent in grass given, respectively, 224, 448, 896 and 1,344 kg N/ha; similar or even more marked increases of CP contents with the increased rates of fertilizer N have been observed in numerous other trials. Increasing the intervals between grazing or cutting decreases CP content as e.g. in a trial by Appelman & Dirden (1962) in which CP content increased from 4.8 per cent in grass cut every 8 weeks to 11.2 per cent when cut every 3 weeks. It has also been observed that first or early cuts in a season contain more CP than subsequent, and especially last, cuts. CF content is not particularly high and usually ranges from 28 to 35 per cent, increasing with longer intervals between cuts which results in more mature grass and low leaf: stem ratios. The contents of EE range from 1 to 3 per cent and of NFE from 41 to 63 per cent (Butterworth, 1967). The contents of Ca and P vary and are usually sufficient for cattle requirements although the content of P can sometimes be too low in grass given no fertilizer P and grown on soil deficient in phosphorus. In trials by Gomide *et al.* (1966b) it ranged from 0.11 to 0.16 per cent depending on plant age and the frequency of cutting, but the content of Ca was high and ranged from 0.50 to 0.76 per cent; the same authors also recorded a high content of Mn, 248 p.p.m., and moderate contents of Zn and Cu.

Total DM digestibility is reasonably high and varies from 47 to 67 per cent and the content of TDN accordingly from 44 to 66 per cent. CP digestibility varies very considerably, from 7 to 66 per cent largely, but not always, depending on the content of CP and N balance can occasionally be negative. Digestibility of CF ranges from 53 to 73 per cent, of NFE from 41 to 71 per cent and of EE from 2 to 50 per cent (Butterworth, 1967). It should be noted that in digestibility trials quoted by Butterworth hay or dried grass were much more mature and less digestible than the grass usually obtained in field trials. In general Pangola grass, given moderate rates of N or NPK, has a nutritive value sufficiently high for cattle and sheep requirements but unfertilized grass, especially when grazed at long intervals, is usually deficient in protein although its energy content would be adequate. The content of vitamin A is apparently adequate and Kirk *et al.* (1972) report from Florida that cows grazed for 9 years entirely on Pangola grass have shown no symptoms of vitamin A deficiency.

Animal production. Animal production from Pangola grass is usually high. Evans (1969) in southern Queensland, Australia, obtained 1,275 kg liveweight increase/ha from cattle-grazed Pangola grass fertilized with 450 kg N/ha/year and Caro-Costas *et al.* (1972) in Puerto Rico obtained 1,060 kg/ha. Such high yields are however rare although Plucknett (1970) reported from Hawaii liveweight gains of 700–850 kg/ha from Pangola grass given N and limed. Beef yields reported from Mexico by Garza *et al.* (1970) – 175 kg/ha from unfertilized Pangola grass and 344 kg from grass receiving 100 kg N/ha – are also on the high side as liveweight gains would be about twice as high. More usual liveweight gains range normally from 100 to 400 kg/ha and daily gains of steers from 0.2 to 0.8 kg but can occasionally be higher. Milk production is usually high and Blydenstein *et al.* (1969) in Costa Rica obtained 6,014 kg milk/ha/year and an average milk production of 6–8 kg/animal/day from cows grazed at 2.57 head/ha on a Pangola grass pasture fertilized with NPK. The animals were, however, given concentrates and it was estimated that from the total amount of nitrogen, 233 kg/ha/year, consumed by the cows 44 kg were obtained from the supplement concentrates. Carrying capacity of Pangola grass can be very high and Caro-Costas *et al.* (1972) in Puerto Rico report 6.5 steers/ha on a heavily fertilized pasture but it is usually lower, two to four cows or steers per ha or sometimes lower. Pangola grass is well liked by all grazing animals and the intake of grass in terms of kg DM per 100 kg bodyweight is satisfactory or good but it varies at various seasons and is relatively low during cool or dry seasons and during the mass flowering when numerous stems reduce the palatability of the grass. Blydenstein *et al.* (1969) observed fluctuations from 1.8 to 3.6 kg/100 kg bodyweight.

Reproduction and cytology. It has already been mentioned that

Pangola grass produces no, or only occasional, seeds although flowering heads are well-formed and can be numerous. It seems that only one viable seed was recorded until 1956 and Sheth *et al.* (1956) state that Pangola grass is highly male sterile and also highly female sterile. The only seed obtained was grown into an adult plant with 27 $2n$ chromosomes, several of which were univalent, and this plant can be regarded either as a triploid or as an apomictic plant with an irregular number of chromosomes; the normal number of chromosomes in Pangola is $2n = 30$. Although some authors expressed an opinion that irradiation or strong chemical reagents, e.g. colchicine, can result in more regular chromosome behaviour and in the formation of viable seed, it seems that no such plants have so far been produced.

Digitaria didactyla Willd. Blue couch $\qquad 2n = 18, 36$

Stoloniferous perennial forming low dense mats 2–15 cm high; flowering stems 10–40 cm in height. Racemes two to three, occasionally four, terminal, slender, 3–6 cm long. Spikelets 2.5 mm long. Upper glume one half to two thirds the length of the spikelet. Originates from Mascarene Islands in the Indian Ocean and introduced throughout the tropics and subtropics as a lawn grass. Henrard (1950) states, however, that southern Asia, Australia and New Zealand can be regarded as another area of its natural distribution. The grass is used to a limited extent for sheep grazing. R. J. Jones *et al.* (1969) report that in Queensland *D. didactyla* can be a weed in grass plots, replacing weak grasses in the year of establishment but giving reasonably high yields of its own herbage; the authors remark that this grass is, agronomically, unjustifiably neglected, whereas it could be a useful grazing species.

Digitaria eriantha Steud. Woolly finger grass

Tufted perennial sometimes sending out short runners. Stems up to 80 cm high. Leaves 8–20 cm long and up to 4 mm wide. Racemes three to nine, dense, 6–15 cm long. Spikelets 3–3.5 mm long; lower glume up to 0.5 mm long, upper glume about two thirds the length of the spikelet. Marginal hairs of spikelets spreading. Widely distributed in South Africa extending to East Africa and India.

A good grazing grass productive in cultivation. In a small-plot trial by Stomayor-Ríos *et al.* (1971) in Puerto Rico it outyielded, when heavily fertilized, all other species of *Digitaria* under trial, including Pangola grass, and produced, in six cuts, over 45 t DM/ha/year and 2,740 kg CP. Seed formation is however poor and planting by splits is hardly a practical proposition for a non-stoloniferous grass. Diploids and tetraploids with respective $2n$ chromosome numbers 18 and 36, and also plants with the irregular chromosome number of $2n = 40$ have been found.

Digitaria gazensis Rendle $\quad 2n = 18$

Tufted perennial 50–120 cm high with the spikelets on 10–18 cm long racemes gathered in panicles. Spikelets 2 mm long. Occurs in East and South tropical Africa. In Kenya the herbage analysed at the beginning of flowering contained over 13 per cent CP. There are a number of other tufted perennial species rather similar in habit to *D. gazensis* of limited local importance for grazing.

Digitaria macroblephara (Hack.) Stapf. Woolly finger grass

Stoloniferous perennial. Stems up to 80 cm high with hairy nodes. Stolons often grow first vertically and then bend down and function as normal stolons. Leaves 3–15 cm long and 2–5 mm wide. Racemes 3–11, 2–20 cm long, not dense. Spikelets 2.5–3.5 mm long; lower glume a minute scale, upper glume 2–2.5 mm long. Marginal hairs of spikelet long, spreading. Occurs in tropical East and North-East Africa in arid and semi-arid areas, mainly on sandy loams and does not withstand waterlogging. It is one of the most drought-tolerant species of *Digitaria*; in Kenya it was observed to remain green almost throughout the dry season in areas with an annual rainfall of 500–700 mm and two dry seasons a year. A good grazing grass; analysed in Kenya at the flowering stage, it contained 12.3 per cent CP, 32.2 per cent CF, 42.5 per cent NFE, 0.44 per cent Ca and 0.18 per cent P (Dougall & Bogdan, 1960).

Digitaria milanjiana (Rendle) Stapf. Woolly finger grass

Stoloniferous perennial with flowering stems 50–120 cm high. Leaves up to 30 cm long and 4–12 mm wide. Racemes 4–15, 8–20 cm long, dense. Spikelets 2.5–3 mm long. Lower glume a narrow-triangular scale, upper glume one third to two thirds the length of the spikelet. Marginal hairs of the spikelet relatively short and not spreading. Henrard (1950) states that *D. milanjiana* differs from *D. pentzii* mainly in being stoloniferous and by the leaves not crowded at the base. A variable species distributed in tropical, mainly eastern Africa from Rhodesia to Kenya and also in Angola, and introduced to some other countries including Puerto Rico, Ghana and Fiji.

Digitaria milanjiana grows in the areas of moderate rainfall, mostly on sandy loams and does not tolerate waterlogging. In Rhodesia it has been tried with some success in pure stands and in mixtures with *Trifolium repens* and *Lotononis bainesii* (Clatworthy, 1970). In Puerto Rico it yielded in six cuts over 34 t DM/ha and 2,860 kg CP in small, heavily fertilized plots (Sotomayor-Ríos *et al.*, 1971), the yields being slightly lower than those of the best yielding types of *Digitaria* under trial. In a trial in Fiji (Roberts, 1970) it produced 38 t fresh grass/ha and was the best yielding species. Seed production is uncertain but when seed is produced it requires over 5 months of storage before germination can begin. The removal of floral scales improves germination and dehulled seeds can germinate soon after harvest, the germination reaching its maximum after about 5 months of storage. Gibberellin can stimulate germination of dehulled but not of intact seeds (Baskin *et al.*, 1969). *Digitaria milanjiana* is a diploid with 18 $2n$ chromosomes.

Digitaria nodosa Parl. Woolly finger grass

Tufted perennial 20–80 cm high. Leaves 5–30 cm long and 2–5 mm wide. Racemes 4–12, 5–15 cm long. Spikelets 2.5–3 mm long; lower glume a triangular scale, upper glume from three-quarters the length of the spikelet to almost its full length. Marginal hairs on spikelet not spreading sideways. Distributed in Kenya, Sudan, Eritrea in Ethiopia, Somalia, tropical Arabia and further north-east to Punjab (India) and Afghanistan. An excellent grazing grass which often occurs on black waterlogged clays.

Digitaria pentzii Stent. Woolly finger grass

Tufted or stoloniferous perennial. Stems up to 120 cm high. Leaves up to 30 cm long and 6 mm wide. Racemes 3–14, up to 18 cm long. Spikelets 2.5–3.5 mm long. Lower glume a minute scale, upper glume about three quarters the length of the spikelet. Marginal hairs erect, not spreading. Var. *stolonifera* (Stapf.) Henr. has long creeping stolons and *D. decumbens* is believed by some botanists to be a form of this variety. *Digitaria pentzii* is a polymorphic species and apart from *D. decumbens* it is also interconnected with *D. valida* which can be regarded as a robust type of *D. pentzii*. Polymorphism of *D. pentzii* can be partly explained by the varying ploidy which ranges from diploids to hexaploids with $2n$ chromosome numbers forming a seris of 18, 27, 36, 45 and 54; irregular chromosome numbers were also observed. *Digitaria pentzii* is widely spread in South Africa and extends to adjacent tropical countries.

Digitaria pentzii has been introduced to Puerto Rico where in small, well-fertilized plots it produced 27–31 t DM/ha and 1,690–1,970 kg CP, but these yields were much lower than those obtained from other species and forms of *Digitaria* under trial (Sotomayor-Ríos *et al.*, 1971). In Brazil, cv. A-24 yielded in 13 months 89 t green fodder/ha, unfertilized, and 109 t when given moderate rates of NPK, or over 25 and over 30 t DM, respectively (Zúñiga *et al.*, 1967). In the Cameroun *D. pentzii* yields were lower and 19–22 t fresh herbage or 5.7–6.4 t hay/ha were harvested in South Cameroun and 16.2 t fresh and 4.2 t hay/ha in the drier North Cameroun (Borget, 1968). High yields were also reported from Samoa. Other countries to which *D. pentzii* was introduced and where it was tried on a small scale include USA, Trinidad, Venezuela, Fiji, Philippines and Taiwan and in the last named country A-24 was found to be resistant to the stunt virus destroying *B. decumbens* in Brazil; A-24 suffered however from heavy attacks of Rhodes-grass scale, *Antonina graminis* (Buller *et al.*, 1970). *Digitaria pentzii* is a good grazing grass comparable with Pangola grass but less palatable during the growing season although it is well grazed by the animals towards the end of the season and during South African winters (Whyte *et al.*, 1959). Analyses in South Africa, Tanzania and Trinidad showed a good nutritive value of herbage in which CP content ranged from 6.7 to 14.5 per cent and of DCP from 3.7 to 11.7 per cent. CF content varied from 28.4 to 37.3 per cent, NFE from 40.3 to 48.4 per cent and EE from 1.9 to 3.0 per cent. CP digestibility was high and ranged from 53 to 82 per cent, depending on its content in the herbage, CF digestibility from 59 to 82 per cent, NFE from 54 to 81 per cent and EE from 37 to 56 per cent. The content of TDN varied from 55 to 73 per cent (Butterworth, 1967). Seed production is doubtful but when formed some 5 months of storage are required before germination could begin. However, seed with removed floral scales can germinate soon after harvest and good germination has been obtained after 5 months of storage; germination of dehulled, but not intact seed was stimulated by gibberellin (Baskin *et al.*, 1969).

Digitaria scalarum (Schweinf.) Chiov. African couch grass; Thangari (East Africa) $2n = 36$

Rhizomatous perennial with fine stems medium in height or short. Spikelets in small panicles, plump, not compressed. Common in tropical Africa, mainly in its eastern part, as an arable weed. It is also common and often numerous in highland grasslands of East Africa where it is usually well grazed. An analysis of herbage which showed 14.7 per cent CP and only 29.0 per cent CF indicates a high nutritive value.

Digitaria setivalva Stent. Woolly finger grass $2n = 18$

Resembles *D. macroblephara* and may be a geographical form of the latter. Occurs in East and South tropical Africa. Introduced into Puerto Rico where in small, well-fertilized plots its yields were unfavourably compared with those of other forms of *Digitaria* and ranged from 17 to 19 t DM/ha (Sotonmayor-Ríos *et al.*, 1971). *Digitaria setivalva* was also tried in Guyana where on sandy loam of low fertility it responded well to applied N which increased annual yields of DM from 2.5 to 8.7 t/ha; applied P raised DM yields from 4.5 to 8.7 t and K from 5.8 to 7.4 t, whereas there was little or no response to Mg and Ca. The recovery of applied N, P and K was at a low and variable level. N content in the herbage ranged from 0.56 to 1.53 per cent depending on the time of cutting and the application of nitrogenous fertilizer corresponding to CP contents of 3.50 to 9.56 per cent. Fertilizer N slightly increased the content of nitrogen in the herbage, but the application of P, K or Mg usually decreased it (Chesney, 1972).

Digitaria smutsii Stent $2n = 18, 36$

Large, robust perennial forming broad tufts. Stems stout, 1.5 m high. Leaves up to 50 cm long and 8–12 mm wide. Racemes 4–10, up to 13 cm long, often almost horizontally spreading. Spikelets 3.5 mm long; lower glume very small, upper 2 mm long. Marginal hairs of spikelets short and not spreading sideways. Occurs in South Africa, mainly in Transvaal. Introduced to a few countries including Puerto Rico, Ghana and Kenya but without much success. In Puerto Rico, in small, heavily fertilized plots, DM yields ranged from 27 to 33 t DM/ha but the grass was outyielded by several smaller types of *Digitaria* (Sotomayor-Ríos *et al.*, 1971).

Digitaria swazilandensis Stent. Swazigrass $2n = 18$

Stoloniferous perennial similar in habit to *D. didactyla* from which it differs by slightly taller herbage, slightly smaller spikelets and shorter upper glume. Distributed in South Africa, Swaziland and adjacent territories. It forms dense low swards and is used for sheep grazing in Swaziland and in South Africa where it was also grown in mixtures with *Trifolium repens* in which the grass did well (Whyte *et al.*, 1959). The herbage analysed by R. C. Elliott in Rhodesia contained 13.2 per cent CP, 7.9 per cent DCP and 58.9 per cent TDN (from Butterworth, 1967).

Digitaria valida Stent. Giant pangola grass; Pangola gigante

$2n = 18, 27, 36, 30$

Tufted or stoloniferous large perennial with the stems reaching up to 130 cm in height. Resembles *D. eriantha* but is larger and has longer leaves and larger spikelets; some botanists regard *D. valida* as a robust form of *D. pentzii*.

A South African species introduced to the Caribbean area of America and also to Ghana. In Puerto Rico, in small, heavily fertilized plots, it was outyielded by nearly all other forms of *Digitaria*; nevertheless it yielded about 19 t DM/ha and 1,390 kg CP (Sotomayor-Ríos *et al.*, 1971). Lower yields, 25.6–42.7 t fresh material or 9.2–12.4 t hay/ha, were obtained in South Cameroun and 5.4 t hay/ha, in the drier North Cameroun. In earlier trials in Puerto Rico, DCP contents were determined to be 2.8 and 2.4 per cent in herbage cut at the 50th and 80th days after planting, respectively, and TDN contents 59.2 and 72.9 per cent (Arroyo & Brenes, 1961). Viable seed is produced and the progenies vary in habit, and tufted, readily flowering plants, and stoloniferous forms developing few if any flowering heads were observed which shows that tufted and stoloniferous forms can belong to the same species (Virkki & Purcell, 1967); this may also be true in regard of some other species of the section *Erianthae*.

Diheteropogon Stapf

A small African genus with four species, closely allied to *Andropogon* from which it differs by the large pedicelled spikelet which is larger than the sessile spikelet, long pungent callus and hairy awns (Clayton, 1966). It has also at least one homogamous pair of spikelets at the base of the peduncled raceme. Racemes are paired at the top of stem or its branches and this separates it from *Heteropogon* in which the racemes are single.

Diheteropogon amplectens (Nees) W. D. Clayton (*Andropogon amplectens* Nees). $2n = 40$

Tufted perennial 60–150 cm high. Leaves 2–20 mm wide. Inflorescence usually simpler than of *D. hagerupii* but other characters resemble those of the latter species, except that the awn is 3–5 cm long. Two varieties are distinguished:
1. Var. *amplectens* with the leaves 2–5 mm wide, rounded at the base.
2. Var. *catangensis* (Chiov.) W. D. Clayton with the leaves 5–20 mm wide, broadly cordate at the base.

Var. *amplectens* occurs in East Africa (Kenya, Tanzania, Rhodesia) and in South Africa and var. *catangensis* west of this area – in Senegal, Zaire, Tanzania. Both varieties grow on poor sandy soil or on rocky slopes and can be numerous or even subdominant in natural grasslands; both are of medium grazing value.

Diheteropogon grandiflorus (Hack.) Stapf (*Andropogon grandiflorus* Hack.)

A tufted perennial. Differs from the two other species of *Diheteropogon* by the presence of three to nine homogamous pairs of spikelets at the raceme bottom, by larger spikelets, the pedicelled spikelet being 16–25 mm long, and by longer awns which are 9–15 cm long. Leaves and stems vary considerably and some types are good grazing grasses. Occurs mostly on poor sandy soil, often in temporary wet situations, in central parts of Africa: Nigeria, Zaire, Tanzania, Malawi, Zambia and Angola. Common in grassland and can be abundant.

Diheteropogon hagerupii Hitchc.

Annual with the stems up to 1.5 m high. Leaves broad, cordate at the base. Inflorescence spathate, the rays terminating in pairs of racemes which are 4–7 cm long; one of the racemes sessile, the other on a short peduncle; the peduncled raceme has one pair of homogamous, male spikelets. Other pairs of spikelets are heterogamous: the sessile spikelet bisexual, fertile, 9 mm long, with a twisted and geniculate hairy awn up to 8 cm long. Pedicelled spikelet awnless, larger than the sessile spikelet.

Occurs naturally in West Africa from Mauritania in the north to Nigeria in the south, in semi-arid areas, on poor sandy or gravelly soils, Very common in the Sahelo–Sudan ecological region with an annual rainfall of 400–500 mm, and is one of the important grasses of the region, a grass which fully develops only relatively late in the season (Valenza, 1970). Late burning encourages the spread of this grass. It often forms associations with *Andropogon pseudapricus* and *Zornia glochidiata*. A grass of medium grazing value. Bartha (1970) states however that *D. hagerupii* is well liked by animals. The plants he analysed contained 1.6 per cent EE, 46.6 per cent NFE, 38.2 per cent CF, 0.57 per cent Ca, and only 6.9 per cent CP and 0.11 per cent P.

Echinochloa Beauv.

Spikelets oblong to ovate or elliptic, in groups of two or three on short thick pedicels. Upper glume often attenuated into a short awn. Florets two, the lower male or empty, its lemma often terminates in an awn. Upper floret bisexual, fertile, flat or concave on one side and convex on the other, with the lemma and palea crustaceous at maturity. Inflorescence a panicle, usually dense and compact. Annuals or perennials of wet situations or arable weeds. Perennial species are mostly useful fodder grasses in seasonal swamps of Africa or America.

Echinochloa colona (L.) Link. Jungle rice; Shama millet; Jharua grass
$2n = 54, 72$

Annual up to 90 cm tall. Ligule absent. Leaves up to 25 cm long. Panicle narrow, erect, up to 15 cm long. Spikelets 2.5–3 mm long, awnless.

Occurs throughout the world tropics, up to 1,500 m alt, in swampy situations, at river banks and as a weed of arable land. It is well eaten by cattle and is reported to increase milk output of buffaloes (N. C. Das Gupta, from Whyte, 1964). Sen & Mabey (1966) analysed 4-, 8- and 12-week-old herbage and found that its quality was high at the three stages although CP content decreased from 13.8 per cent in the 4-week growth to 10.3 per cent in the 12-week growth; CF content was low, 22.6 and 25.6 per cent, respectively, and the contents of P and Ca were adequate for cattle requirements; Vonesch & Riverós (1967–8) obtained comparable results in regard of CP and CF contents.

Other well-known annual weeds which are grazed by cattle or fed as cut grass are *E. crus-pavonis* in the tropics and *E. crus-galli*, which is distributed, mainly outside the tropics, in the northern hemisphere.

Echinochloa frumentacea (Roxb.) Link. Japanese barnyard millet; Sanwa millet $2n = 36, 54, 56$

A robust annual with branched stems up to 1 m high. Leaves up to 40 cm long and 25 mm wide. Panicle with up to 15 branches. Spikelets awnless. Cultivated as a millet in India, South East Asia, China and Japan but is not of any great economic importance. *Echinochloa frumentacea* has been noted for its fast growth and quick ripening. The straw is often fed to cattle and other farm animals. In Australia (Wheeler & Hedges, 1971) grazing *E. frumentacea* herbage resulted in liveweight gains (in wethers) of 242 kg/ha. There is a report by Healey (1969) that grazing young herbage before it reached a height of 60 cm caused photosensitivity disorders in sheep, young lambs being more susceptible than adult animals; the effect was of the primary nature and caused no lasting harmful effect.

Echinochloa 129

Echinochloa haploclada (Stapf) Stapf $2n = 36, 72$

Tufted perennial. Stems thin, up to 2.5 m high but usually much shorter. Ligule usually absent. Leaves 15–30 cm long. Panicle 10–25 cm long. Spikelets 2–3 mm long, awnless or with minute awns.

Occurs in tropical East, North East and South East Africa, from sea level to about 1,800 m alt and it forms extensive colonies in seasonal swamps or waterlogged ground usually on black clays, or in other wet habitats. It is well grazed when young.

A dry-land type found on Mt Marsabit in northern Kenya was cultivated with some success in grazed plots in Kenya. *Echinochloa haploclada* is a good seed producer with 7–30 per cent spikelets containing caryopses, and can be easily established from seed. Field tests (Bogdan, 1963a) showed the plants to be cross-fertilized and attempts were made to apply mass selection. In grazed plots the herbage contained 10–12 per cent CP, 0.37 per cent Ca and 0.21 per cent P. *Echinlochloa haploclada* is normally a tetrapoid with $2n = 72$ but diploids, with $2n = 36$, have also been reported.

Echinochloa polystachya (H.B.K.) Hitche. Pasto alemán

Robust perennial. Stems 1–2.5 m high, thick in the lower part. Ligule a rim of stiff hairs. Leaves 20–60 cm long and 10–25 mm wide. Spikelets 5–7 mm long, lanceolate. Upper glume aristulate, with an awn 5–7 mm long; the awn of the lower-floret lemma 7–17 mm long. Fertile floret 5–6 mm long. The two better known forms, var. *polystachya* and var. *spectabilis* (Nees) Mart. Crov. differ only in small details.

Echinochloa polystachya occurs in tropical and subtropical America, from the southern United States to Buenos Aires in Argentina. It forms extensive colonies in seasonal swamps and on less wet ground, and is particularly common in South American countries. This grass is much used for grazing or soilage and for haymaking, and is of importance for the animal industry of South America, especially of Brazil. *Echinochloa polystachya* has been introduced into cultivation and grown with good success in South America, Mexico and some West Indies islands. In trials in Venezuela (Novoa & Rodríguez-Carrasquel, 1972) DM yields of irrigated *E. polystachya* were considerably lower than those of *Brachiaria mutica*, some 8–12 t/ha when fertilized with PK and given 100 kg N/ha, but doubling the rates of N increased the yields to 20–30 t in five or seven cuts. In Surinam, plant communities in rice fallows with the dominance of *E. polystachya* yielded 29 t green fodder/ha (Dirven *et al.*, 1960). CP content in trials by Dirven (1962), carried out in Surinam, was very high in the leaves, 23.5 per cent, and high in the stems, 16.8 per cent, and its digestibility reached 80 and 86 per cent, respectively. CF content was 30.4 per cent in the leaves and 33.2 per cent in the stems. Much lower contents of CP have however been reported by Combellas & Gonzáles (1973) from Venezuela; in their trials CP content of 10.3 per cent was determined in herbage cut 41 days after the previous cut and it was reduced to 8.2 per cent in herbage harvested 21 days later. Digestibility of DM ranged from 60 to 63 per cent and of CP from 72 to 73 per cent, and the animals consumed 2.4–2.7 kg of grass/100 kg bodyweight. Yields of beef obtained from steers grazed at two head/ha during the grazing season from July to January were reported from Mexico (Teunissen *et al.*, 1966) to be 245 kg/ha and higher gains of 280 kg/ha were recorded earlier by Arroyo & Teunissen (1964).

Echinochloa pyramidalis (Lam.) Hitchc. & Chase. Antelope grass

Tall reedlike perennial, often rhizomatous. Stems up to 4 m high, thick. Ligule a rim of hairs. Leaves 30–60 cm long and 6–25 mm wide. Panicle erect or slightly nodding, 15–30 cm long. Spikelets 3–4 mm long, awnless.

Distributed throughout tropical Africa, in seasonal swamps where it forms extensive colonies, and at stream banks. It can however withstand long periods of drought and in Kenya it was grown on dry land with satisfactory results. In its natural habitats *E. pyramidalis* is usually burnt towards the end of the dry season and fresh growth, which can contain at this stage up to 15 per cent CP (Edwards & Bogdan, 1951), is then grazed. Young stems, although thick, are spongy and grazed together with the leaf. At later stages of growth, even at the early flowering stage, it becomes less palatable and its CP content was determined to be 7.03 per cent (Dougall & Bogdan, 1965); the content of other nutrients were: EE 1.14 per cent, CF 30.77 per cent, NFE 51.93 per cent, Ca 0.34 per cent and P 0.13 per cent.

Echinochloa pyramidalis is cultivated in tropical South Africa and in Rhodesia and is usually established from tuft splits or pieces of rhizome because of the low percentage of spikelets containing caryopses.

The plants are usually tetraploids with $2n = 72$ but diploids with $2n = 36$ and triploids ($2n = 54$) have also been found. *Echinochloa pyramidalis* is variable and types differing in habit and herbage quality have been observed.

Echinochloa stagnina (Retz) Beauv. Bourgou.

Perennial with stems up to 2.5 m high arising from long rhizomes or creeping and rooting stem-bases often floating in water. Ligule a rim of hairs. Leaves 15–50 cm long and 5–20 mm wide. Panicle usually narrow and nodding, 4–10 cm long. Spikelets 4–6 mm long with an awn 4–24 mm long. Fertile floret 3–5 mm long.

Occurs throughout tropical Africa and also in India, Sri Lanka, other parts of Asia and the Philippines at river banks and lake shores, often in water where it is up to 2 m deep. It can be very productive and its biomass can reach 17 t DM/ha or 130 t green; aerial stems and leaves can produce up to 3.5 t DM/ha (Toutain, 1973). The herbage is highly palatable at nearly all stages of growth and is reputed to be of high nutritive value. It is utilized when the water recedes and the animals can reach the plants; it is also used for making hay. Butterworth (1967) quoting the data obtained in the Philippines gives the contents of CP as 6.7 per cent, CF 33.7 per cent and NFE 45.6 per cent, their digestibility in trials with sheep being 52.3, 62.5 and 51.8 per cent, respectively. Plants of different ploidy, with chromosome numbers $2n = 36$, 54, 108 and 126 have been found.

Eleusine Gaertn.

Annuals or densely tufted perennials with flattened stems. Inflorescence digitate, subdigitate or sometimes an elongated panicle with dense racemes. Spikelets with several florets, ovate or oblong, laterally compressed, disarticulating between the florets at maturity. Grain is enclosed in a free hyaline pericarp. The nine species so far known are distributed mainly in eastern Africa, but there are also indigenous species in southern Asia and in Australia, and *E. indica* has become a weed of warm, mainly tropical countries throughout the world. Mostly tough grasses, useless or of low quality for grazing but some species produce satisfactory grazing.

Eleusine compressa (Forsk.) Asch. & Schweinf. (*E. flagellifera* Nees, *Panicum compressum* Forsk.) $2n = 45$

Stoloniferous perennial 30–70 cm high. Leaves up to 10 cm long and 3 mm wide, flat but later convolute; basal leaf-sheaths short, velutine. Inflorescence digitate and consists of three to five spikes which are 2–3 cm long. Spikelets 5 mm long, three to eight flowered. Occurs in dry grasslands of Pakistan and northern India, partly in the tropics but mainly in subtropical areas where it is common and can be a subdominant or locally dominant grass. It is one of the most drought-resistant perennial grasses of India which occurs under an annual rainfall of 125–250 mm, and is palatable to stock (Rao, 1970).

Eleusine coracana (L.) Gaertn. Finger millet; African millet; Koracan; Ragi (India); Wimbi (Swahili); Bulo (Uganda); Telebun (Sudan)
$2n = 36$

Differs from *E. indica* in having wider spikes which curl inward at maturity and larger and almost orbicular grain which is non-shattering, and, in some varieties, not entirely covered with floral scales. Developed probably as an autotetraploid ($2n = 36$) from *E. indica* ssp. *indica* or, more likely, from ssp. *africana* without chromosome doubling. There is also an opinion that it could have arisen as a hybrid between ssp. *africana* and an unknown species (from Purseglove, 1972). Cultivated as a cereal in Africa and India. In India the straw is often used as a low-quality fodder which contains 3–4 per cent of CP, the digestibility of which has been determined to be as low as 6 per cent (from Narayanan & Dabadghao, 1972) but Mahudeswaran (1973) reported CP content from

7.5 to 10.0 per cent. The contents of CF and NFE are given by Narayanan & Dabadghao as 36 and 51 per cent, respectively, their digestibility being 69 and 58 per cent. P, Ca and K contents were determined to be 0.16, 1.11 and 1.50 per cent, respectively, i.e. adequate for animal needs. It is said that ensiling can improve the feeding value of straw. Straw yields from 1.12 to 2.24 t/ha for dry-land crops and up to 8.96 t for irrigated crops have been reported. In the northern, subtropical India (Punjab), *E. coracana* is grown for green fodder or hay and then the herbage is reputed to be of high nutritive value and can yield 15 t green fodder/ha in three cuts (Narayanan & Dabadghao, 1972).

Eleusine indica (L.) Gaertn. Goosegrass; Fowl-foot grass; Rapoka grass (Rhodesia); Gondirimi (Mali).

Annual 15–50 cm high. Stems slender, glabrous and smooth, but tough. Leaves linear, usually folded, 5–30 cm long and 3.5–5 mm wide, glabrous, smooth, tough. Inflorescence digitate or subdigitate with up to 12 slender spikes 5–12 cm long and 3–5 mm wide. Spikelets 3–4 mm long, elliptic, three to nine flowered; florets 2.5–4 mm long. Grain 1.0–1.3 mm long. Spikelet easily shatter at maturity. There are two main forms: ssp. *indica*, as described, and ssp. *africana* (Kennedy-O'Byrne) Phillips which is more robust, taller, and larger in all parts. Florets 3.5–5.0 mm long and grain 1.2–1.6 mm long. It is a tetraploid with the chromosome number $2n = 36$, whereas ssp. *indica* is a diploid ($2n = 18$).

Ssp. *indica* occurs naturally throughout the tropics, subtropics and warm areas of the Old World as a weed of arable land, on roadsides, wasteland and sometimes in overgrazed pastures. It has spread to tropical and subtropical America and became a common weed in cultivation and in some South American countries penetrated into semi-natural grasslands. Ssp. *africana* is distributed in Africa, mainly in its eastern part, from Ethiopia to Cape province in South Africa; in the tropics it occurs at higher altitudes than ssp. *indica*.

Eleusine indica is reported to be palatable to cattle in Fiji, Hawaii, South America and in other areas, and Derbal *et al.* (1959) state that in Mali *E. indica* gives 30 t fresh herbage/ha, is very palatable and can be grown for hay. However, the leaves and stems are tough and the actual palatability of *E. indica* and its value as fodder or grazing require confirmation. Analysed in West Africa (Sen & Mabey, 1966), *E. indica* contained reasonable but not very large amounts of CP; its content decreased from 12.8 per cent in 4-week-old herbage to 8.2–8.4 per cent in 8- to 16-week-old growth and to 6.0 per cent when the herbage was 16 weeks old. The content of CF ranged from 25.1 to 30.2 per cent and of P from 0.06 to 0.15 per cent, mostly inadequate for animal requirements. Ssp. *indica* is a troublesome weed of arable land in many countries and so is ssp. *africana*, especially in newly sown grasses as it has been observed in Kenya. Incidentally, both subspecies, especially ssp. *africana*, are the plants which, in contrast with the majority of other annual grasses, are difficult to pull out from soil if the plants are not too old.

Eleusine multiflora A. Rich.

Annual with ascending stems up to 45 cm high. Leaves 5–25 cm long and 3–6 mm wide, flat. Panicle of two to eight spikes which are up to 3 cm long and distant on the panicle axis, except some upper spikes which can be crowded on top of the stem. Spikelets 7–11 mm long, broad, with 5–15 florets which are 3.5–5 mm long. Grain laterally compressed, 1–1.2 mm long.

Occurs in Ethiopia and in tropical East Africa, from 1,500 to 2,700 m alt, where it is common on disturbed places and as a weed of cultivation. It is particularly characteristic of cattle-herding places where, on the well-manured ground, *E. multiflora* is one of the first plants which colonize such places. Very palatable to cattle.

Eleusine tristachya (Lam.) Lam.

Tufted perennial 10–45 cm high. Leaves linear, 6–25 cm long and 1–2 mm wide. Panicle digitate with one to three racemes which are 1–4 cm long. Spikelets 5–9 mm long and 4–6 mm wide, with 6–13 florets.

Distributed in South America, mainly in southern Brazil, Uruguay and Argentine and naturalized in some other countries. Often numerous in grassland and is valued as fodder (Burkart, 1969).

Enneapogon Beauv.

About 30 tufted perennial or annuals of dry open habitats, mainly in Australian and African tropics.

Enneapogon polyphyllus (Domin) Burbidge. Leafy nineawn (Lazarides, 1970); Whitetop (Siebert *et al.*, 1968)

Annual or short-lived perennial forming dense, somewhat viscous tufts 20–40 cm high. Stems numerous, slender, terminating in dense panicles 5–9 cm long. Spikelets 7–10 mm long with about three florets of which only the lowermost is fertile; lemma terminates in nine slender awns.

A species widely spread in the drier parts of Australia, except in the south, and particularly abundant in central Australia where it often dominates over extensive areas and thrives under a wide range of soil conditions. It responds rapidly even to moderate falls of rain and can produce fresh growth in 4–5 days after the rain has fallen. It often covers the ground in *Spinifex* stands when the latter are burnt. *Enneapogon polyphyllus* is a valuable pasture grass palatable at all stages of growth, and can even be used for fattening stock, but it lasts only for short periods and disappears after the rains have stopped (Lazarides, 1970). Crude protein content in green herbages was determined as 10.3 per cent (Siebert *et al.*, 1968) but reduced to 7.5 per cent at maturity and to 5.0 per cent in completely dry herbage; CF content is low and ranges from 23 to 29 per cent.

Enteropogon Nees

Some six tufted perennial species of the Old World tropics.

Enteropogon macrostachyus (A. Rich.) Benth.
Tufted perennial up to 1 m high with narrow flat leaves. The slender stems terminate in a single, rarely paired, one-sided dense spike 8–20 cm long. Spikelets three-flowered, but with only the lowermost floret (or occasionally the second one also) being fertile. Fertile floret 7–10 mm long and terminates in an awn 10–18 mm long.

Occurs in semi-arid areas of tropical Africa; in bush, and on forest margins and clearings, often on rocky ground. It has been tried with moderate to good success for reseeding denuded pastoral land in Kenya (Bogdan & Pratt, 1967), under an annual rainfall of 550–800 mm. In further trials (Fredenslung & Cassady, 1969) it contributed 33–73 per cent to a sward of mixed sown local grasses. *Enteropogon macrostachyus* yielded 880 kg air-dry matter/ha when sown in intact bush, and 2,650 kg in plots where the bush was slashed. Seed is large, 1 kg containing about 175,000 spikelets with one caryopsis each, and can be easily harvested by hand or by machinery if the grass is grown on flat land.

Other species promising for reseeding denuded land are *E. rupestris* (J. A. Schmidt) A. Chev. (*E. somalensis* Chiov.), which occurs in dry areas of northern Kenya and *E. sechellensis* (Baker) Th. Dur. & Schinz., a grass of coastal areas of East and South East Africa, and also of Seychelles and Madagascar.

Entolasia Stapf

A monotypic genus with one species.

Entolasia imbricata Stapf. Bungoma grass (Bungoma is a town in western Kenya from which *E. imbricata* was first brought to trial). Tufted perennial 40–80 cm high, with rather soft erect or ascending stems. Leaves linear, up to 20 cm long and 6–8 mm wide, somewhat fleshy and succulent. Panicle 10–15 cm long with erect branches densely beset with spikelets which are 5 mm long, broadly lanceolate in shape and dorsally compressed. The lower floret is reduced to a lemma, the upper floret bisexual and fertile.

Entolasia imbricata occurs occasionally in eastern tropical Africa, at about 1,500 m alt, in swamps with poor soil deficient in N. When grown in grass nursery at Kitale, Kenya, this grass showed outstanding palatability and high CP content which ranged from 15 to 30 per cent (Bogdan, 1963). Isolated plants formed no seed which was well produced in mixed populations and the progeny varied in habit, leafiness, vigour and some other features; samples originating from four sites were mixed to form cv. Nzoia. *Entolasia imbricata* flowers in the afternoon and seed is normally well formed. The weight of 1,000 seeds is about 3 g which corresponds to about 330,000 seeds/kg. The optimum emergence was observed from a depth of 2–3 cm and only 9 per cent emerged from a depth of 7–8 cm (Mwakha, 1971). In small-scale trials herbage yields per plant were progressively decreased during the growing season when the plants were cut at 2-week intervals, but increased when cut at 4–8-week intervals. At Kitale *E. imbricata* was one of a few grass species undamaged by the army worm (*Laphygma exempta*); the caterpillars tried to eat the leaves but soon abandoned them. Basic information from Mwakha (1970).

Eragrostis Wolf

A large genus with more than 300 species, mostly subtropical or tropical. Spikelets laterally compressed, awnless, with several (3 to over 60) florets. Caryopsis small, globose to elliptic, easily detached from the floral scales. An important character for recognizing the numerous species is the pattern of breaking the spikelets at maturity: breaking can begin from the base or the top of the spikelet; rhachilla can break or remain intact when the florets break off; the palea can remain on the rhachilla or fall off together with the rest of the floret.

Annual or tufted perennials, rarely rhizomatous perennials. Annuals are common in arid and semi-arid areas where they can be numerous and provide good grazing in wetter years. A number of annuals are also common, but not very serious weeds of arable land.

Eragrostis caespitosa Chiov. Cushion lovegrass
Tufted perennial often forming low loose cushions. Stems 15–50 cm high, slender, wiry. There is no basal leaves but the stems are densely beset with short leaves 3–12 cm long. Panicle 3–13 cm long, fairly dense. Spikelets 2.5–5.5 mm long, with 4–15 florets, breaking up from the apex; rhachilla fragile. Florets 1.3–1.7 mm long with ciliate palea. Caryopsis very small dark reddish-brown.

Distributed in semi-arid areas of East Africa including the Somali Republic, from sea level to 1,600 m alt, in grasslands where it can be a co-dominant grass or form large colonies on poor sandy soil, often in slightly overgrazed pastures. *Eragrostis caespitosa* is well grazed by stock and is a valuable grass on poor soil where other grasses are sparse. Analysed at the flowering stage, the herbage contained 8.7 per cent CP, 37.5 per cent CF, 1.8 per cent EE and 47.2 per cent NFE (Dougall & Bogdan, 1958). The contents of Ca (0.49 per cent) and P (0.17 per cent) were adequate for stock requirements. The seed, in the form of naked caryopses, is extremely small – there are up to 40 million seeds per kg.

Eragrostis cilianensis (All.) Mosher (*E. major* Host). Grey lovegrass
$$2n = 20 \text{ or } 40$$

Annual. Stems erect or ascending, 10–90 cm high. Leaves flat, up to 15 cm long and 3–8 mm wide, usually with a row of warty glands on the margins. Panicle ovate, 4–30 cm long, moderately dense and with stiff branches; branchlets and pedicels usually with scattered glands. Spikelets with 5–60 florets, ovate to oblong, 3–20 mm long and 2–4 mm wide, greenish-grey in colour, breaking up from the base. Florets 2–3 mm long, often glandular. Caryopsis sublobose, reddish brown, 0.5 mm long.

Occurs in all tropical, subtropical and warm temperate areas of the Old World and accidentally introduced to some areas of the American continents were it has naturalized. It can grow under various annual rainfall and in Africa is often a dominant or co-dominant grass in the areas with about 250 mm of annual rainfall where perennial grasses cannot survive. *Eragrostis cilianensis* is also a pioneer grass which colonizes denuded land in the higher rainfall areas and is a common weed of arable land predominantly in semi-arid areas. Plant size varies considerably: under poor rainfall the plants of 10 cm in height flower and produce seed, whereas in moist years they grow to a height of 50–90 cm and produce a considerable bulk of greyish-green herbage. The herbage is rather stemmy but the stems are not hard and the whole plant is readily eaten by stock. Plants collected on rich alluvial soil in dry parts of Kenya contained 15.3 per cent CP, 2.3 per cent EE, 29.0 per cent CF and 42.7 per cent NFE (Dougall & Bogdan, 1965). Ca content was high (0.70 per cent) and the content of P (0.20 per cent) satisfactory. Seed is abundantly produced and in Kenya *E. cilianensis* is recommended for reseeding denuded pastoral land in the arid, annual-grass zone (Bogdan & Pratt, 1967). Seed is in the form of naked caryopses and 1 kg contains about 10 million such seeds.

Eragrostis curvula (Schrad.) Nees. Weeping lovegrass; Pasto lloron.

Densely tufted perennial, stems slender to robust, 30–120 cm high. Leaves up to 30 cm long and 1–5 mm wide, often convolute or filiform. Panicle variable; loose and spreading to dense and narrow. Spikelets linear, 4–10 mm long and 1–1.5 mm wide, greyish-green, breaking up from the base, with 4–13 florets. Rhachilla persistent. Lower glume 1–1.8 mm long, slightly shorter than the upper glume. Florets 1.8–2.2 mm long. Caryopsis ellipsoid, 0.7 mm long.

The taxonomy of *E. curvula* and its relationship with allied species has not been sufficiently sorted out and Leigh & Davidson (1968) and also Clayton (1974) treat *E. curvula* in a wide sense, as a complex of species which, apart from *E. curvula* in the narrow sense, includes *E. chloromelas* Steud. and *E. robusta* Stent and transitional forms between these three taxa. *Eragrostis curvula* in a narrow sense can pehaps be identified with the Ermelo type or cultivar which has filiform drooping leaves and forms dense and broad tufts.

Eragrostis curvula occurs naturally in South Africa, Rhodesia, Botswana and (possibly) in Mozambique, its area of distribution extending from South African subtropics deep into the tropics. In the countries where it is indigenous, *E. curvula* seldom occurs in natural grasslands in any appreciable numbers but is common in fallow land, other disturbed situations and in overgrazed pastures where the balance between the local species has been disrupted. In fallow land it appears some 2–3 years after the cessation of cultivation and can form almost pure stands or mix with *Cynodon dactylon*. As a pasture species *E. curvula* has been introduced to a number of tropical, subtropical and warm temperate countries: Australia, Kenya, Tanzania, Argentina, USA and some others, where it has naturalized to a limited extent. In South Africa and in adjacent territories, it is valued for the ease of establishment, reasonable yields of herbage, especially when fertilized, and where it is reputed to be palatable and of considerable importance for the animal industry. However, in the countries of its adoption *E. curvula* is often regarded as a second-rate species inferior to the best grazing grasses and it has not spread to any appreciable extent beyond the experimental areas. However in south-western states of the USA *E. curvula* has been well received and is used for grassing the land cleared from bush or forest and is recognized as a valuable species.

Eragrostis curvula is a drought-resistant plant which in South Africa grows mainly in the areas with an annual rainfall ranging from 500 to 750 mm. It thrives under high air temperatures ($25°/35°C$) and can withstand low night temperatures without accumulating excessive amounts of starch. It is indifferent to daylength in regard to its effect on flowering and its shade tolerance is poor. *Eragrostis curvula* grows usually on light-textured soils or sands and can withstand considerable soil salinity but not waterlogging. It was observed (Visser, 1965) that root exudates of *E. curvula* can adversely affect germination of maize and wheat seeds but stimulate it in sunflower and cowpea.

Cultivation. Establishment is effected by seed and 0.5 to 1 kg seed/ha is required for drilling in rows or 1–2 kg for broadcast sowing, although some authors suggest higher rates. Shallow drilling to a depth of 1 or maximum 2 cm is essential. In the USA it is sometimes sown with *Lespedeza sericea* but in the subtropics and tropics *E. curvula* is grown pure or in mixture with lucerne; little information is available on mixtures with other legumes. In Kenya a locally found form has been tried with some success for reseeding denuded range land in moderately dry areas with an annual rainfall of 550–850 mm. Good seedbed preparation was essential and pelleted seed gave better establishment than non-pelleted pure seed; seed dressing with aldrin (against harvester ants) can be useful (Bogdan & Pratt, 1967).

Yields and nutritive value. *Eragrostis curvula* is grown for grazing and for hay and herbage yields of unfertilized stands range mostly from 1 to 2 t DM/ha. Much higher yields were however recorded and in the Bombay area of India the grass yielded 11.8 t fresh fodder/ha in the first year of growth, 27.2 t in the second year, 22.0 t in the third year and only 5.3 t in the fourth; yields of *E. chloromelas* were much lower and ranged from 3.5 t in the fourth year to 12.6 t in the third (Whyte, 1964).

Responses to fertilizers, especially to fertilizer N, are very marked and in various trials, in which 140–450 kg N/ha were applied, yields of 6–11 t DM/ha of grazed herbage or 6–9 t of hay were obtained. In a South African trial, 160 kg of applied N/ha increased herbage yields from 2.0 t DM/ha to 10.3 t (Birch, 1967). In the same trial fertilizer N increased CP content in the herbage from 6.6 to 11.8 per cent. Crude protein content seems to be below average but young 2-week-old plants can contain up to 20 per cent CP. Vonesch & Riverós (1967–68) determined CF content as ranging from 28 to 37 per cent and the content of NFE 43–50 per cent. *Eragrostis curvula* palatability is a controversial issue. It is generally accepted in South Africa and in Rhodesia that this grass is palatable and readily grazed by all kind of stock and gives substantial liveweight gains. Rodel (1970) in a trial in Rhodesia obtained up to 550 kg/ha liveweight gains in heifers grazed on *E. curvula*; in another Rhodesian trial 320 kg/ha liveweight increases of steers were reported. On the other hand Nel et al. (1964) write that in their trials *E. curvula* pastures or hay were unsuitable for fattening sheep, and later Leigh & Davidson (1968) remarked that *E. curvula* is a grass species most unpalatable amongst grasses recommended for sown pastures. Organic matter and CP digestibilities were however found to be satisfactory. *Eragrostis curvula* tolerates close grazing probably because the growing points of its tillers are situated close to the ground level.

Seed production. Seed setting and production are good. Seed is small and 1 kg can contain 3 to 5 million seeds. No past-harvest maturation was necessary and good seed germination was obtained 1 month after harvesting in trials quoted by Whyte (1964), although in *E. chloromelas*, which is a part of *E. curvula* complex, seed germination improved after 2 years of storage. Seed of *E. chloromelas* germinated well, to 68–82 per cent, at constant temperatures of 16° to 37°C and alternating the temperature did not increase the percentage of germinated seeds (Knipe, 1967).

Reproduction. All species and varieties of the *E. curvula* complex are apomicts although sexual plants have occasionally been found (Voigt, 1971). The plants are pseudogamous, i.e. pollination is needed for the initiation of endosperm and the development of seed (Streetman, 1963). The basic chromosome number (x) is 10 and $2n = 20, 40, 50$ and 60 have been found in *E. curvula*, 40 and 60 in *E. chloromelas* and 70 and 80 in *E. robusta*. Aneuploid (irregular) chromosome number of 63 has also been found in *E. curvula*.

Agricultural grouping in the E. curvula complex. Six main agronomically different types have been recognized in the *E. curvula* complex by Leigh and Davidson (1968):
1. Robusta blue – vigorous plants with tall, thick stems and relatively broad, bluish-green leaves and large panicles; a leafy type.
2. Robusta green – stems shorter and finer than in robusta blue; leaves dark green; moderately leafy plants.
3. Robusta intermediate – intermediate between 1 and 2 types.
4. *Eragrostis curvula* of 'weeping' habit with long, arching, very narrow leaves. Leafy, with the predominance of leaves; here belongs the widely known cultivar Ermelo, seed of which is commercially produced on a large scale and which has spread to a number of countries outside the area of its natural distribution, and cultivated, mainly on an experimental scale. Leigh and Davidson consider it to be the least palatable of the six types.
5. Tall chloromelas – stems long and slender; leaves long, bluish green.
6. Short chloromelas – stems short, slender; leaves short, bluish green; the basal leaves with curved dead tips.

These six types are easier to distinguish on the second year of growth when they are fully developed, than in the first year.

Apart from the well-known cv. **Ermelo**, references to cvs. **American Leafy**, **Kromkraal** and **Witbank** can be found in the literature and for the areas outside the tropics cv. **Morpa** and **Catalina** have been selected. Cv. Morpa, developed in Oklahoma, USA, is more palatable than the common type and gives higher liveweight gains of cattle but is somewhat less winterhardy. Cv. Catalina, developed in Arizona, USA, is valued for the high drought resistance of seedlings and for its general ability to withstand droughts, and is recommended for oversowing denuded pastures.

Eragrostis eriopoda Benth.

Tufted perennial 30–40 cm high with wiry, many-noded stems covered at the bottom with woolly scales. Leaves convolute, short and narrow, suberect or spreading. Panicle 12–18 cm long, erect, with occasional short horizontal branches, but mostly with almost sessile spikelets, single or grouped on the main axis. Spikelets linear to lanceolate with up to 24, but usually fewer florets. Glumes subequal, ovate. Florets 2–2.5 mm long with the lemmas falling off with the grain, whereas the paleas remain on the spikelet axis for a longer time. Grain ovoid, under 1 mm long.

A common and often dominant grass on sandy soils of Mulga land in south-western Queensland and also a characteristic species of an extensive arid area around Alice Springs and of other parts of Northern Territory of Australia. It forms associations with *Aristida browniana* and provides some grazing in areas of 150–350 mm of annual rainfall. The grass is fibrous and of low palatability but is grazed when more palatable grasses are scarce.

Eragrostis lehmanniana Nees $2n = 40, 60$

Tufted perennial 30–60 cm high. Stems usually geniculately spreading at the base, occasionally erect. Leaves narrow linear. Panicle lax, open, 6–20 cm long, with the lowest branches single or in pairs. Spikelets spreading, 4–8 mm long and 1–1.5 mm wide, linear, with 4–13 florets which are greyish-green to dark olive-grey in colour. Lemmas 1.5 mm long. Grain about 0.6 mm long.

Occurs naturally in tropical and subtropical South Africa and tried in cultivation, mostly in the subtropics or warm temperate areas.

Eragrostis lugens Nees. Pasto mosquito (Argentina)

$2n = 40, 80$ and about 108

Tufted erect perennial 15–70 cm high with fine stems and narrow leaves 1–3 mm wide. Panicle large loose. Spikelets 2–5 mm long, olive-grey in colour, with two to nine florets. Glumes ovate, unequal; florets 1–2 mm long.

Widely spread in tropical and subtropical America from southern USA, through Mexico to Brazil, Uruguay and Argentina. Occurs often in degraded pastures replacing more palatable species. Burkart (1969) states that *E. lugens* is not very productive and that its nutritive value is not high but indicates that because of the wide spread it has a certain importance for grazing. Earlier authors had considered this grass to be of a higher value and an analysis has shown CP content of 8.82 per cent (Pio Corrêa, 1926) and later (Vonesch & Riverós, 1967/68) determined the content of CP as 16.8 per cent, EE 1.7 per cent, CF 30.0 per cent and NFE 50.7 per cent. *Eragrostic polysticha* Nees is a more robust plant rather similar to *E. lugens*; occurs in the same areas except that it is absent in the USA and Mexico.

Eragostis superba Peyr. (*E. superba* var. *contracta* Peter). Masai lovegrass (Masai is the name of an African tribe in Kenya and Tanzania). $2n = 20$ or 40

Tufted perennial 20–120 cm high with flat leaves up to 10 mm wide. Panicle 10–30 cm long, sparsely branched. Spikelets flatly compressed from the sides, 6–20 mm long and 3–10 mm wide, with 6–30 florets which remain attached to the spikelet at maturity, the spikelet falling off as an unbroken unit. Caryopsis elliptic, 1–1.5 mm long.

Occurs mainly in East Africa, from Sudan to South Africa, under an annual raifall of 500–900 mm, in grassland and bush, and is often abundant on sandy soil but does not grow under waterlogged conditions. Mucilagenous layers on root surface to which sand grains adhere and form a protective cover were observed in grass grown on sandy soil. The root system is relatively shallow (Taerum, 1970). In Kenya, *E. superba* has been used for reseeding denuded pastoral land (Bogdan & Pratt, 1967; Pratt, 1963, 1964; Pratt & Knight, 1964; Jordan, 1957) and found to be one of the best grasses for the purpose. Introduced to Senegal and tried in small plots, it gave satisfactory production, was leafy and competed well with weeds; yields of 3,310 kg DM/ha/year without weeding and 3,939 kg when weeded three times were obtained. On sandy soil it produced 3,946 kg DM/ha/year and 5,494 kg on alluvial soil. In Senegal the persistence was poor and *E. superba* yielded 4,793 kg in the first year of production and only 405 kg in the second year (Nourrissat, 1966). In Kenya however it was more persistent and lasted more than 2 years when heavily grazed and to over 4 years in another trial with lighter grazing (Pratt, 1963). *Eragrostis superba* is readily grazed by cattle and other stock and at the early flowering stage it contained about 12 per cent CP and 30–35 per cent CF, and Nourrissat reports just under 4 per cent DCP in the herbage. It was tentatively grown in semi-arid areas of Kenya as a ley grass with a moderate success.

Seed is abundantly produced and can easily be collected by stripping the ripe panicles. Each spikelet has numerous florets but contains only one to three well-developed caryopses, and there is no point in breaking the spikelets when sowing. There are about 100,000 spikelets per kg. In laboratory trials seed germinated better under light than in darkness, and at 30°C than at 20°/25°C (Al-Ani & Ouda, 1969). *Eragrostis superba* is essentially a cross-pollinating plant, and 674 seeds per panicle were found in freely pollinating plants, compared with 14 seeds in isolated plants (Streetman, 1963); single plant progenies revealed a considerable variability in habit, leafiness and the resistance to rust.

Eragrostis tef (Zucc.) Trotter (*E. abyssinica* (Jacq.) Link; *Poa tef* Zucc.). Teff; T'ef $2n = 40$

Annual 20–90 cm high with fine stems. Leaves up to 30 cm long and 1–4 mm wide. Panicle 10–40 cm long, loose or contracted, with the lowermost branches arranged in a whorl. Spikelets 5–9 mm long and 1.5–2 mm wide, with 4–12 florets remaining intact at maturity. Glumes slightly unequal. Florets 2–3 mm long. Grain 1–1.5 mm long.

Known only in cultivation or as an escapee from cultivation. *Eragrostis tef* is widely grown in Ethiopia as a cereal, mainly in two forms: with red and white grain (Purseglove, 1972). There are, however, several cultivars which differ in the shape of the panicle, its colour which can be straw-coloured to purple, grain colour and other characteristics.

Eragrostis tef has been introduced to other tropical and subtropical, mostly African, countries, as a hay crop which has an advantage of rapid growth and can be harvested in 9–12 weeks after sowing. It was also tried with some success as a nurse crop for perennial pasture or fodder grasses. Whyte et al., (1959) state that E. tef is widely grown for hay in the northern states of the Union of South Africa and Butterworth (1967), quoting from South African sources, gives CP content of hay as ranging from 8.2 to 11.4 per cent, DCP of 4.3 to 6.4 per cent, CF content 30 to 35 per cent and DM digestibility of about 60–65 per cent. When cultivated experimentally in Kenya it produced about 2 t DM/ha in 9 weeks after sowing and the herbage contained 15.5 per cent CP. In Ethiopia E. tef is sown at a rate of 10–12 kg seed/ha although Whyte et al. (1959) quote sowing rates of 15–20 kg/ha. Seeds germinate and seedlings emerge fast, the plants grow rapidly and E. tef can be grown in dry areas because of its short lifespan which allows the plants to escape droughts. Eragrostis tef is said to be a self-pollinating species. Seed is well produced and there are 2,500–3,000 seeds/g or 2.5–3 million seeds/kg.

Eriochloa H.B.K.

A small genus of Paniceae with up to 20 species distributed mainly in the tropics and subtropics. The plants resemble some species of *Brachiaria* but usually have less dense racemes and can be best recognized by the presence of a small, bead-like swelling at the base of the spikelet.

Eriochloa meyeriana (Nees) Pilger (*Panicum meyerianum* Nees)

$2n = 36$

An intermediate form between *Eriochloa* and *Panicum*. The racemes are less dense than those of *Eriochloa* and the lower glume is better developed and is in the form of a truncate scale about one quarter the length of the spikelet; the bead-like swelling below the spikelet is present. A strong perennial which can produce long creeping stems when grown on bare ground. A good grazing grass common on swampy ground, river banks and other wet situations in semi-arid areas of tropical Africa.

Eriochloa punctata (L.) Desv. ex Hamilt. (*E. polystachya* H.B.K. – a synonym under which this species is better known to pastoralists). Carib grass; Janeiro; Angolinha; Malohilla (León & Sgaravatti) $2n = 36$

Annual or short-lived perennial. Stems erect or geniculately ascending, up to 150 cm high, but usually shorter, with four to six nodes. Leaves glabrous, 10–30 cm long and 4–12 mm wide. Panicle erect, about 15 cm long, with a few erect or suberect racemes. Spikelets about 3 mm long, broadly lanceolate. Lower glume clasps firmly the spikelet base which is in the form of a bead-like swelling and hardly extends beyond the swelling. Upper glume acute, as long as the spikelet. Lower floret reduced to an empty lemma resembling the upper glume. Upper floret white, shiny, 2 mm long.

Occurs in tropical and subtropical America and West Indies and there is a local South American form – *Eriochloa montevidensis* Griseb. (*E. punctata* var. *montevidensis* (Griseb.) Osten). Often found in wet situations – on river banks, lake shores, seasonally flooded ground – where it can form fairly large colonies. The soft herbage is well grazed by cattle and has medium to high nutritive value. In Surinam, CP content was determined (in *E. polystachya*) by Dirven (1962) to be 14.4 per cent in the leaves and 7.1 per cent in the stems or 10.3 per cent in the whole herbage, and its digestibility was higher in the stems than in the leaves, 61 and 53 per cent, respectively. The leaves contained 27.5 per cent CF and the stems 37.1 per cent, the total content in the whole herbage being 32.9 per cent.

Eriochloa procera (Retz.) C. E. Hubbard is a closely allied Old World type, morphologically similar to *E. punctata* and with the same number of chromosomes ($2n = 36$) but is essentially a perennial grass. It occurs in India and other parts of South and South East Asia and in Africa, in habitats similar to those of *E. punctata*.

Eriochloa nubica (Steud.) Hack. & Stapf ex Thell is an annual species allied to *E. procera* but usually smaller; it grows in comparable situations and in the same areas as *E. procera*. Its chemical composition was found to be close to that determined by Dirven for *E. polystachya* and the contents of Ca and P, 0.51 and 0.30 per cent, respectively, were more than adequate for cattle requirements. Sen & Mabey (1966) report however low contents of P in plants analysed in West Africa; they also give 13.2 per cent CP content for 4-week-old herbage which decreased to 4.2 per cent in 12-week-old herbage and CF contents were 21.2 and 31.3 per cent, respectively.

Euchlaena Schrad.
Two species of Mexican origin. Related to maize.

Euchlaena mexicana Schrad. Teosinte; Makchari (India) (Fig. 20)
Robust annual resembling maize. Stems thick, many-noded, branched at the base, 1.5 to 4, or occasionally 5 m high. Leaves about 70–90 cm long and up to 8 cm wide. Flowers unisexual: male (staminated) flowers in terminal panicles (tassels), female (pistillate) in ears in leaf axils, the tassels are similar to those of maize, with numerous unbranched or sparsely branched racemes beset with numerous paired spikelets, one sessile, the other on a short pedicel, both with two staminate florets. Ears of pistillate flowers are covered with leaf-like spathes or husks, two to a few ears are enclosed in the same sheath. Rhachis of the ear flat, obliquely articulated into joints. Each joint has a spikelet on one of the flat sides; in adjacent joints the spikelets are situated on opposite sides of the rhachis. The spikelets are sunk into joints and covered with the lower glume. The stigmas (silks) are long and protrude from husks as in maize. At maturity the joints disarticulate easily together with the spikelets

which contain a grain or are sometimes empty. Empty spikelets are creamy-white whereas those containing a grain are brown.

Outwardly *E. mexicana* resembles maize but differs by abundant tillering which results in broad tufts and by an ability to recover and produce new growth after cutting.

Fig. 20   *Euchlaena mexicana.*

Occurs naturally in Mexico and Central America and is apparently the nearest to maize of all living wild plants; there is also an opinion that *E. mexicana* could have originated from natural crosses between the primitive forms of maize and *Tripsacum dactyloides* (Relwani, 1968a). *Euchlaena mexicana* grows in humid tropics under an annual rainfall over 1,000 mm and requires fertile soil for vigorous development. It can withstand temporary flooding and excessive soil moisture content but does not grow under long-term waterlogging.

Introduction.   *Euchlaena mexicana* is grown, to a limited extent, as a fodder crop in Mexico and in Central American countries. It has also been introduced to the USA where it is occasionally cultivated in the south, and has spread, as a cultivated plant, to the Caribbean islands and to South America, reaching Argentina; in these countries it did not however acquire any considerable importance for animal husbandry. It has been tried in some African countries but the greatest success with *E. mexicana* has been obtained in India. It was first introduced to Madras in 1881 (Relwani, 1968a); at present it is grown mainly in the northern subtropical India although it is also of some importance in the Indian south and west. *Euchlaena mexicana* is also grown in Pakistan. The grass is valued for its productivity, the ease of establishment and for its tolerance of flooding and excessive soil moisture.

Establishment. Good seedbed should be prepared by ploughing and harrowing and/or discing. In the drier areas *E. mexicana* is sown early in the rainy season and in almost any month in the areas with more abundant and evenly spread rains. In India it is sown between early March and the end of July. In the subtropics with relatively cold winters, good field emergence of seedlings can be expected only when soil temperature reaches 15°C. Row sowing is much preferred to broadcasting and the distance between the rows is usually 50–60 cm, or wider on poor soil. Seed rates vary from 35–40 kg/ha on good fertile soil to 15–20 kg on poor soil (Relwani, 1968a). However, in western India much lower seed rates are normally used. Establishment by transplanting 1-month-old seedlings is also possible and it may be useful under the shortage of irrigated land which may still be under another crop at the time of sowing of *E. mexicana*. A heavy basic application of f.y.m. is often given, sometimes in combination with moderate rates of mineral N fertilizer. Applications of P and K depends on the fertilizers given to preceding crops. During the first 2 months the plants grow slowly and only later do they begin growing fast and tiller, the tillers, usually 8–10 and up to 20 or more per plant, forming broad tufts. Weeding at the early stages of growth is necessary, especially if the weeds are annuals with the early and rapid growth or rhizomatous perennials. Weeding can be reduced by the application of herbicides and Panday *et al.* (1969) in India obtained good results with the application of 1.5 kg simazine/ha which, combined with hand weeding, increased the yields of herbage from 5.78 t DM/ha for untreated plots to 14.01 t.

Management. *Euchlaena mexicana* is mainly used for feeding freshly cut grass to cattle, but hay or silage are sometimes prepared. One to four cuts are taken depending on the time of sowing, rains, irrigation and soil fertility. Irrigation is needed if the soil becomes dry and the crop is irrigated at 15–20 day intervals (Relwani, 1968a). The plants are usually harvested when they reach some 2 m in height and are cut at a height of 25–30 cm from the ground to allow further tillering from the lower parts of the tuft. In Japan, the first cut is made at 10 cm from the ground level and the next cut at 25 cm (Relwani, 1968a). In general, cutting at a greater height usually results in higher yields of herbage. It is recommended to apply some 75–100 kg/ha of ammonium sulphate or equivalent amounts of other nitrogenous fertilizers after each cut.

Association with legumes. Annual legumes such as soyabean, *Vigna radiata*, *V. aconitifolia*, *Mucuna* (*Stizolobium*) sp. (Relwani, 1968a) and also *Vigna unguiculata* can improve the nutritive value of herbage. In northern India *Trifolium alexandrinum*, *Medicago sativa* or *Melilotus parviflora* can be undersown to *E. mexicana* with advantage in September if the grass is to be cut in November.

Herbage yields. Average yields of green fodder as estimated by Relwani (1968a) for India approach 30 t/ha and are higher than those of maize or sorghum; yields up to 70 t/ha have been obtained on fertile or well-manured soil, whereas on poor soil they can be about 15 t. Figures given by Narayanan & Dabadghao (1972) are within these limits and range from 17.8 t green fodder/ha to 44.8 t for irrigated land and from 16.8 to 22.4 t for dry land. Yields of some 50 t fresh fodder have also been reported from Louisiana, USA.

Chemical composition, nutritive value. Crude protein content of *E. mexicana* herbage, as quoted from various sources by Relwani (1968a), ranges from 4.47 per cent to 11.99 per cent, EE from 1.20 to 2.34 per cent, CF 19.57 to 32.20 per cent, NFE 51.33 to 55.32 per cent, Ca 0.65 to 0.91 per cent and P 0.16 to 0.28 per cent, which in most cases are sufficient to satisfy cattle requirements. Apparent digestibility for cattle was determined to be 63 per cent for CP, 46 per cent for EE, 62 per cent for CF and 51 per cent for NFE. The forage is usually coarser than that of maize or sorghum but is well eaten by cattle. Whyte (1964) reports, however, that at Poona, in India, *E. mexicana* herbage was, as fodder, inferior to that of maize and affected adversely milk yields.

Reproduction, hybridization. *Euchlaena mexicana* flowers later than maize and requires 90–100 days from seedling emergence to flowering. It is a cross-pollinated plant and, as in maize, the tassels flower earlier than the silks, at a time when the latter are not yet ready to accept pollen, the difference in time being greater than in maize. *Euchlaena mexicana* can be crossed with maize which has the same number of chromosomes, $2n = 20$. Chaudhuri & Prasad (1969), in India, found that hybrids can be more easily obtained by pollinating maize silks with *E. mexicana* pollen than by reciprocal pollination. Six hybrids obtained were intermediate between the parent plants in a number of characters although plant height and sugar content were higher in the hybrids than in the parental plants. Some hybrids were backcrossed to maize, the backcrosses being the most promising for the development of cultivars suitable for fodder production. In western Pakistan (Ahmad *et al.*, 1968) maize × *E.*

mexicana hybrids combined the prolific tillering of *E. mexicana* with the rapid early growth of maize and yielded 42.07 t green fodder/ha compared with 35.64 t for *E. mexicana* and 27.03 t for maize.

Seed production. For seed production *E. mexicana* can be grown in wider rows than for herbage, with the seed rates reduced to 20–25 kg/ha on fertile soil, or even to 5–7.5 kg (Relwani, 1968a) when seeds are placed 40 × 40 cm. In India seed usually ripens in October–November and seed yields normally range from 0.7 to 1.3 t/ha. Seed, as harvested, can contain a considerable proportion of creamy-white seeds which are empty or have unripe caryopses; these are lighter than the brown, fully developed seeds and can be separated by seed immersion into a 2 per cent solution of common salt.

Eustachys Desv.

Leaf-sheaths strongly keeled, flagellate. Panicle digitate, of two to many one-sided racemes densely packed with spikelets. Spikelets golden-brown or dark brown, with two florets, the lower bisexual and the upper empty; the lower floret with a very short awn or with no awn. Ten to twelve species of the tropics and subtropics, mainly in America.

Eustachys distichophylla (Lag.) Nees (*Chloris distichophylla* Lag.). Capim cebola, Cocorobo (Brazil); Paraguillas, Pasto bora (Argentina).

$2n = 20$ or 40

Tufted perennial 70–140 cm high. Leaves folded or flat, abruptly narrowed and pointed at the apex, 4–30 cm long and 10–15 mm wide. Racemes 10–30, 6–17 cm long. Lower floret 2.5–3 mm long with numerous long spreading hairs and no awn or with a short point. Second floret 1.5–2 mm long. Caryopsis 1.2 mm long.

Occurs naturally in Brazil, Uruguay and Argentina. Pio Corrêa (1926) writes that this grass is of considerable value as forage and deserves introduction into large-scale cultivation. It is nutritious and being a local Brazilian grass is well adapted to the environment and to poor soils exhausted by previous crops. He reports that before flowering *E. distichophylla* contained 11.0 per cent CP, 3.9 per cent EE, 41.2 per cent NFE and 32.6 per cent CF. Another analysis of the herbage cut at the flowering stage gave comparable contents of nutrients and 7.4 per cent of DCP was determined in hay. Analysed in Uruguay, it gave lower content of CP – 7.0 per cent. It was suspected that the herbage may contain HCN but no cattle poisoning was reported from Brazil.

Eustachys paspaloides (Vahl) Lanza & Mattei $\qquad 2n = 36$
Perennial forming spreading clumps and colonies by means of slowly spreading stolons. Stems 20–80 cm high. Leaves 2–18 cm long and 2–5

mm wide, abruptly acute. Racemes mosly 4–10. Lower floret 1.5–2.5 mm long, with or without a mucro; second floret 1.2–1.5 mm long.

Occurs naturally in Arabia and in tropical and subtropical Africa from Zaire and Ethiopia in the north and throughout the rest of Africa to the south, in semi-arid areas, on various soils including seasonally waterlogged situations and rocky ground. It forms dense leafy herbage 15–30 cm high and provides a useful grazing. Analysed at the early flowering stage, the herbage contained 9.7 per cent CP, 2.6 per cent EE, 33.8 per cent CF and 55.0 per cent NFE. Ca and P contents were adequate, 0.49 and 0.24 per cent, respectively (Dougall & Bogdan, 1958).

Eustachys petraea (Sw.) Desv. (*Chloris petraea* Sw.) $2n = 40$

Perennial with slender stems 50–100 cm high, more or less decumbent and rooting from lower nodes or producing stolons. Leaves 3–8 mm wide. Racemes four to six, mostly 4–10 cm long. Spikelets 2–2.5 mm long; lemma of the lower (main) floret shortly ciliate on the nerves. Rather similar to *E. distichophylla* but differs in being smaller and having narrower leaves, fewer and shorter racemes and smaller spikelets.

Widely spread in tropical and warm America from Coastal Plains, Texas and Florida in the USA, to West Indies and to South America, including Argentina. Grows mostly on sandy soil. A useful, well-grazed grass.

Burkart (1969) has not included this species in his Flora of northern Argentina (Entre Rios province) and probably regards it as belonging to *Chloris capensis* (Houtt.) Thell, or perhaps as still another species.

Festuca L.

About 80 species of temperate and subtropical regions. Two–three species are widely cultivated in temperate areas and *F. arundinacea* also in the subtropics and at high elevations in the tropics.

Festuca arundinacea Schreb. Tall fescue.

This temperate grass is grown on a farm scale at altitudes over 2,000 m in the tropics of South America (Bolivia, Venezuela) and Africa (Kenya, Tanzania) and at lower altitudes in irrigated winter pastures of South Africa and Rhodesia. In a few other tropical countries *F. arundinacea* is grown mainly on an experimental scale. Kentucky 31 and Alta, both of USA breeding, are the most popular cultivars. Seed is imported from Europe or from the USA and in Kenya over 12 t seed were imported during 5 years, from 1966 to 1971. *Festuca pratensis* Huds. is little grown in the tropics.

Hemarthria R.Br.

About ten species of tropical Africa and Asia which grow mainly in wet habitats.

Hemarthria altissima (Poir.) Stapf & Hubbard (*H. compressa* (L.f.) R.Br.; *Rottboellia altissima* Poir., *R. compressa* L.f. var. *fasciculata* Hack.). Limpograss, Capim gamalote (Brazil); Pasto Clavel, Gramilla canita (Argentina); Baksha, Panisharu (India) $2n = 20, 18, 36, 40$

Perennial with short rhizomes and long spreading, decumbent branched stem bases rooting at the nodes; the upper parts of stems ascending to erect or suberect position reaching up to 150 cm in height, but usually to 30–80 cm. Leaves up to 20 cm long and 6 mm wide. Inflorescence consists of several spikes (racemes) appearing singly or in groups of two to four from axils of the upper leaves. Spikes almost cylindrical, 6–10 cm long, with thick joints. Spikelets in pairs: the sessile spikelet bisexual, 5–6 mm long, with a coriaceous lower glume; lemma of the fertile spikelet smaller than the glume. Pedicelled spikelet male, much shorter than the sessile one.

Distributed in tropical, subtropical and warm temperate areas of the world, including North and South America where it might have been introduced from the Old World and naturalized. Occurs in wet situations such as river banks, seasonally flooded river valleys, seasonal swamps, etc.

Hemarthria altissima is a leafy grass which provides grazing and fodder much valued in South America, Australia, India, Rhodesia, South Africa and perhaps in some other countries. It is sometimes grown from cuttings on seasonally wet soil and about 2,000 ha of limpograss were under production in Florida in 1971–2 (Schank *et al.*, 1973). This grass requires at least temporary wet ground but can also withstand seasonal droughts. It is well grazed by cattle, but Pio Corrêa (1926) writes that because of its prostrate habit the animals cannot utilize fully this high-quality grass. The value of *H. altissima* for the autumn or early dry-season grazing has been recognized because grazing during that season is often scarce and dry land grasses are then of low palatability and nutritive value. In Argentina Vonesh & Riverós (1967/68) determined the content of CP at the flowering stage as only 7.1 per cent, EE 2.3 per cent, CF 27.8 per cent and the content of NFE was high, 56.7 per cent.

Diploid and tetraploid forms have been found and genetical variation in herbage digestibility was observed: DM digestibility *in vitro* of diploids and hybrids was 60–63 per cent in 5-week-old growth and around 55 per cent in mature plants, but in tetraploids it was 68 and 66 per cent, respectively, showing the possibility of herbage quality improvement. In a bunch type from Argentina, possibly another species of *Hemarthria*, DM digestibility was much lower: 53 and 38 per cent, respectively (Schank *et al.*, 1973).

Heteropogon Pers.

Some ten to twelve perennial or annual tropical species.

Heteropogon contortus (L.) Beauv. ex Roem. & Schult. Spear grass; Black speargrass; Tangle grass (Fig. 21)

Tufted glabrous perennial with slender, often branched stems 40–100 cm high. Leaf sheaths compressed. Leaf blades 7–20 cm long and 3–7 mm wide. Inflorescence a single raceme, or a scanty spathate panicle with slender rays 4–15 cm long, narrow spatheoles 7–10 cm long and slender penduncles supporting dense cylindrical racemes 3–8 cm long. The lower part of the raceme bears 3–10 pairs of male spikelets which are flat

Fig. 21 *Heteropogon contortus.*

and green; the upper part bears up to 12 pairs of heterogamous spikelets, one of the pair being flat and green, the other having a cylindrical, dark fertile floret provided with a stout, kneed and spirally twisted awn 5–10 cm long. The awns of the whole raceme, and also often of adjacent racemes or even plants usually twist together.

Widely spread throughout the tropics and subtropics of the whole world, and also in the Mediterranean and some other temperate areas; it is common in India and West Africa and particularly in Australia where it covers some 30 million ha and is of considerable economic importance. It often dominates the herbaceous cover in woodlands and bush and less often in open grassland. *Heteropogon contortus* thrives under light grazing, whereas heavy grazing can suppress or eliminate it; there is however conflicting evidence about the effect of heavy grazing. *Heteropogon contortus* is highly resistant to grass fires. Due to the twisting movement of the seed caused by the movment of the awn in response to changes in air humidity, the seed is buried in the soil, thus contributing to the fire tolerance of the grass; moreover, high soil temperature stimulates germination. Fire can also accelerate the development of reproductive tillers (Lazarides *et al.*, 1965).

Heteropogon contortus is very polymorphic, and in Australia numerous introductions were classified into geographical types also differing in morphology (Tothill, 1967), and further classification within the types has been attempted. It is maintained, however, that formal intraspecific taxonomy is premature. Plants of a few ploidy levels with $2n = 20$, 40 60 and 80, and also with $2n = 44$ or 50 were found. The chromosome numbers are little connected with the types defined by Tothill, but a trend was observed that the further from India the larger the number of chromosomes is likely to be found. Reproduction is apomictic. In Australia the populations grown north of 20°S. flower late in the season, when the daylength is under 12 h, but no such trend was observed south of 20°.

Heteropogon contortus is seldom cultivated, apparently because of the difficulties with seed production; moreover there are other grass species more valuable and suitable for cultivation. Nevertheless it is sometimes cultivated experimentally, and in India under an annual rainfall of 1,150 mm sown *H. contortus* yielded 15.6 t fresh herbage/ha, and in mixture with *Stylosanthes humilis* 19.5 t; mixtures with *S. guianensis* and *Centrosema pubescens* yielded less than *H. contortus* alone. In another year *H. contortus* alone yielded 342 kg CP/ha, its mixture with *Stylosanthes guianensis* yielded 435 kg and with *S. humilis* 535 kg (Chatterjee *et al.*, 1969). Analysed in West Africa (Sen & Mabey, 1966) *H. contortus* contained 9.4 per cent CP in 4-week-old herbage, the content of CP decreasing to 6–7 per cent in 8- to 16-week-old herbage and to 2.5 per cent 9 months after the last cut. CF content ranged from 31.2 to 33.7 per cent and the contents of Ca and P were low, the content of the latter ranging from 0.13 per cent in young herbage to 0.05 per cent in old plants. A much lower CP content is however quoted by Butterworth (1967) from Indian data: 3.5 per cent, and that of DCP only 0.8 per cent. CF content is given as 40.1 per cent, its digestibility 63.4 per cent and DM digestibility in trials with cattle was about 50 per cent.

Heteropogon contortus is normally grazed in natural stands, and in Australia and to a lesser extent in other countries, is also oversown to natural grasslands with *S. humilis*, and occasionally with some other legumes. Stocking rates and animal production vary. In a trial in Queensland (Shaw & 't Mannetje, 1970) stocking rates for natural *H. contortus* pastures ranged from 1.8 to 3.6 ha per head of cattle. Liveweight gains per ha were in general low but slightly greater, 29 kg, at the higher rate of 1.8 ha per animal than at the lower rate, which amounted to 25 kg. In dry years, however, the higher rate resulted in liveweight losses. Liveweight gains per animal were greater at the lower than at the higher stocking rates, 83 and 47 kg, respectively. Grass fertilized with PK gave liveweight increases of 100 kg per animal and 62 kg per ha. Undersowing with *Stylosanthes humilis* increased the liveweight gains to 148 kg per animal and per ha when the mixture was fertilized and grazed at 1 ha per head. Under particularly favourable conditions *H. contortus/S. humilis* mixtures can be grazed at still higher stocking rates up to 0.4 ha per head. Daily gains of 27 g per head of sheep liveweight were recorded in Queensland (Playne, 1969), but normally sheep are not grazed on *H. contortus* pastures because the sharp seeds and twisted awns contaminate the wool. Sheep consume only moderate amounts of *H. contortus* herbage and digest it poorly.

Hymenachne Beauv.

Tropical grasses, mostly of wet situations; this genus contains eight species

Hymenachne amplexicaule (Rudge) Nees (*Panicum amplexicaule* Rudge); Canutillo; Carrizo chico.

Rhizomatous perennial often spreading on the ground or floating in water. Stems glabrous, up to 160 cm high erect or ascending from a prostrate or creeping base. Leaf-sheaths often spongy. Leaf blades mostly lanceolate, markedly wider in the basal than in the upper half, cordate at the base, 10–45 cm long and up to 3 cm wide. Panicle narrow, spikelike, cylindrical, 20–40 cm long; sometimes with two to a few long upright branches. Spikelets lanceolate, upright, 4–5 mm long. Lower glume shorter than the spikelet; upper glume acute or mucronate as long as the spikelet and similar to the lower lemma. Upper floret with herbaceous lemma and palea and easily detached caryopsis.

Distributed in South and Central tropical America, in swampy situation, at river banks and on land seasonally flooded with river water where it can form extensive colonies, and where it is grazed by cattle even at advanced stages of growth when flood water recedes. In Surinam, CP content was found to be high, 15.8 per cent in the whole plant, the leaves containing 22.6 per cent CP and the stems 8.9 per cent. High CP digestibility 66–80 per cent was observed and it was higher in stems than in leaves (Dirven, 1962). Much lower CP content was however observed for the grass grown in rice-field fallows where it can be abundant. Dirven *et al.* (1960) report that in rice-field fallows of Surinam plant communities dominated by *H. amplexicaule* can be very productive but the yields vary within very wide limits and range from 3.7 to 20.6 t green fodder/ha.

Hymenachne pseudo-interrupta C. Muell.

Perennial similar to *H. amplexicaule* in habit and in floral parts but the leaves are narrower, linear to linear-lanceolate, not cordate at the base. Distributed in the Indo-Malayan region in habitats similar to those of *H. amplexicaule*, often as a lower storey in reed (*Phragmites*) communities. This grass is well liked by buffaloes.

Hyparrhenia Anderss. ex Fourn. Hoodgrasses

Perennials or annuals usually forming tufts of erect, stout or slender stems and long linear leaves. Spikelets in paired articulated racemes, each pair on a common fine peduncle, which is supported by a leaf-like spatheole ('hood'), and a few to many pairs form a variously branched 'spathate' panicle, the branches being again supported by leaf-life spathes. Each raceme-joint bears two spikelets, one sessile and the other on a short stout pedicel. The spikelets have two florets each: the lower one reduced to a lemma, while the upper one is bisexual and fertile in the sessile spikelet, and male or barren in the pedicelled spikelet. The fertile floret has a kneed awn spirally twisted below the knee. There are 52 species in the genus (40 perennial and 12 annual) confined mainly to tropical Africa, but some species are also found in the Mediterranean region, India, Indo-China, Indonesia, Australia and America (Clayton, 1969). Various species of *Hyparrhenia* occur and often dominate in savanna type grasslands, and one species, *H. rufa*, has been introduced into cultivation, and is of considerable importance in Latin America and especially in Brazil. All the species are grazed to a certain extent, especially the most leafy and less robust types or at the younger stages of growth. The more common and relatively more palatable perennial species are listed below.

Hyparrhenia collina (Pilger) Stapf

Loosely tufted perennial. Stems slender, 30–130 cm high. Spatheoles 2–4 cm long. Racemes 1–2 cm long with four to seven awns per pair of racemes.

Distributed mainly in East and Central Africa from Sudan to Natal, predominantly in wet situations, on alluvial or waterlogged soil but sometimes also on dry land. When closely grazed *H. collina* forms dense, fairly even swards and is valued as a grazing grass.

Hyparrhenia cymbaria (L.) Stapf $\qquad$ $2n = 20, 30$ or 40

Perennial forming tufts from creeping rhizomes. Stems 2–4 m high. Spatheoles boat-shaped, red at the later stages of flowering, 0.8–2 cm long. Racemes 0.7–1.3 cm long with three to six awns per pair of racemes.

Occurs mainly in East Africa, from Ethiopia to Natal and reaches West Africa. Common in grasslands and in bush and tolerates shade better than other species of *Hyparrhenia*. Young stems are slender and when kept short the plants form leafy and even swards and are well grazed. An analysis of this grass by Dougall *et al.* (1964) showed 7.32 per cent CP, 30.3 per cent CF, 44.6 per cent NFE, 0.18 per cent P and 0.32 per cent Ca.

Hyparrhenia filipendula (Hochst.) Stapf $\qquad$ $2n = 40$

Tufted perennial. Stems slender to moderately robust, 60–120 cm high. Spatheoles 5 cm long. Racemes 1–1.2 cm long with two to four awns per pair of racemes.

Occurs throughout tropical Africa but mainly in its eastern part as a common and sometimes dominant species in savanna grasslands. Well grazed by cattle and other stock when not too mature and Dougall *et al.* (1964) determined 9.17 per cent CP, 30.0 per cent CF, 46.0 per cent NFE, 3.16 per cent EE and the contents of P and Ca as 0.18 and 0.30 per cent, respectively. The nutritive value can be however lower and the contents of CP and P of 6 and 0.10 per cent, respectively, were also recorded. A certain amount of oestrogenically active substances (up to 9 μg stilboestrol equivalent/100 g fresh grass) were found in Rhodesia towards the end of the dry season (Millar, 1967).

Hyparrhenia hirta (L.) Stapf

Tufted perennial. Stems slender, 30–120 cm, but mostly up to 60 cm high. Spatheoles 3–8 cm long. Racemes 3–4 cm long with 8–16 awns per pair of racemes.

Occurs in East and South Africa and in the Mediterranean region and is the only species of *Hyparrhenia* reaching that far north. Common and can be numerous in relatively dry grasslands. It was tried under cultivation in Australia but without much success. $2n$ chromosome number can be 30, 40 or 60 but irregular numbers of 44 and 45 were also encountered.

Hyparrhenia anamesa W. D. Clayton, which is distributed mainly in East Africa, is a type somewhat intermediate between *H. hirta* and *H. filipendula*.

Hyparrhenia nyassae (Rendle) Stapf

Loosely tufted perennial. Stems 50–150 cm high. Spatheoles linear, 3–6 cm long. Racemes brown or rufous, 2–3 cm long with 8–13 awns per pair.

This species is closely allied to *H. rufa* from which it can be distinguished by the presence of dense white hairs on the lower leaf sheaths which are hairless in *H. rufa*. There are diploid and tetraploid forms with $2n = 20$ and 40, respectively. Occurs mainly in East, South-East and West Africa and has also been recorded for Vietnam and Thailand. A grass of medium grazing value.

Hyparrhenia papillipes (Hochst. ex A. Rich.) Anderss. ex Stapf (*H. lintonii* Stapf).

Perennial 20–100 cm high, usually with numerous slender stems often branched in the lower parts. Basal leaves often absent and those on well-developed stems are 3–15 cm long and 2–4 mm wide. Panicles with a few (2–10) pairs of racemes. Spatheoles 4–7 cm long. Racemes 2–4 cm long, with 9–19 awns per pair.

Occurs on high ground in Ethiopia, Kenya, Uganda and Tanzania, on rocky slopes, where it often forms loose cushions, and also on waterlogged black clays. It is well grazed when young but not at later stages of growth. Analysed in Kenya at the early flowering stage it contined 11.4 per cent CP, 31.8 per cent CF, 43.2 per cent NFE, high contents of EE, 5.0 per cent, and of P, 0.26 per cent, were recorded (Dougall & Bogdan, 1958). In another analysis more mature plants contained however only 4.7 per cent CP and 0.09 per cent P (Dougall *et al.*, 1964).

Hyparrhenia pilgerana C. E. Hubbard

Loosely tufted perennial. Stems slender, 30–100 cm high, erect or rambling. Spatheoles 2–3 cm long, reddish brown. Racemes 1–1.5 cm long, with six to seven awns per pair of racemes.

Occurs mainly in East Africa, from Ethiopia to Malawi, in seasonal swamps, where it often forms dense growth and provides reasonable grazing after the old growth is burnt or otherwise removed. In Kenya CP content in the herbage was determined as 6.7 per cent, CF 33.7 per cent and NFE 40.8 per cent; P content was low, 0.12 per cent, and that of Ca adequate, 0.50 per cent (Dougall *et al.*, 1964).

Hyparrhenia rufa (Nees) Stapf (*Trachypogon rufus* Nees). Jaragua (Yaragua, Faragua) which can be interpreted as 'master of the field' (Parsons, 1972); Yayale; Veyale (Mali, Derbal *et al.*, 1959) (Fig. 22).

Perennial or sometimes annual with slender to robust stems 0.3–2.5 m high. Leaves 30–60 cm long and 2–8 mm wide. The spathate panicle is of various size, with the spatheoles 3–5 cm long, usually shorter than the raceme-pair peduncles. Racemes fulvous or rufous, 2–2.5 cm long, with 9–14 awns per pair of racemes. Spikelets 3–5 mm long.

Var. *rufa* is common throughout tropical Africa, including Madagascar, and in America.

Var. *siamensis* W. D. Clayton, which differs in having six to seven awns per pair of racemes and in some other minor characters, occurs naturally in Burma, Thailand and Laos.

Hyparrhenia rufa is widely cultivated in South and Central America, and Parsons (1972) considers it to be an exotic, accidentally and repeatedly introduced to America by slave ships, and then naturalized; some botanists are, however, of the opinion that eastern tropical

Fig. 22 *Hyparrhenia rufa.*

America is a part of the area of natural distribution of *H. rufa*. At present *H. rufa* is grown in Mexico, Cuba, Nicaragua, Puerto Rico, Costa Rica, Brazil, Venezuela and Colombia; it is often naturalized and constitutes a considerable part of herbage in local grasslands. It is also cultivated on an experimental scale in Uganda. *Hyparrhenia rufa* is particularly important in Brazil and Nicaragua, and it is estimated that about two-thirds of the steers in central Brazil are fattened on *H. rufa* pastures, and that in the milk-producing areas of central–southern Brazil *H. rufa* is grown on 21–73 per cent of the farms, depending on the district. In Africa *H. rufa* occurs at altitudes up to about 1,800 m, and in Brazil and other American countries it is grown in low and relatively dry areas below 2,000 m. It can withstand burning but seeds lying on the ground cannot survive fires. In America *H. rufa* is usually grown in pure stands, but in Uganda it is experimentally cultivated in mixtures with *Stylosanthes guianensis* or *Centrosema pubescens*.

Establishment. In farm practice *H. rufa* is usually established from seed locally collected by hand, and sown uncleaned and unprocessed as a mixture of chaff with some seed, in which viable seeds constitute about 5 per cent. The chaffy seed is spread over a roughly prepared seedbed, or on burnt grass or bush without any soil preparation, and when sown in this way stands of *H. rufa* take some 2 years to establish (Crowder *et al.*, 1970). If clean seed is sown into a well-prepared seedbed in 25–40 cm rows and fertilized, the grass can be ready for grazing in about 5 months' time.

Management, fertilizing, herbage productivity. *Hyparrhenia rufa* is sometimes cut for hay, but is usually grazed. To maintain leafy swards it should be grazed frequently or continuously, although grazing below 30 cm is usually not recommended. In Uganda a grazing rotation under which a *H. rufa/Stylosanthes guianensis* mixture was grazed by steers for 7 days and rested for 14 days gave satisfactory results, similar to those obtained in a rotation with twice as long periods of grazing and rest. Average yields of herbage range from 3 t DM/ha/year un-irrigated to 13 t under sprinkler irrigation, but Crowder *et al.* (1970) report an unusually high yield of 45 t/ha. In Costa Rica herbage yields were increased by the application of nitrogenous fertilizers, and in Uganda (Wendt, 1970) good responses were obtained to 34–67 kg P/ha, DM yields being increased from 24 to 33–35 t/ha (two-year totals); responses were also obtained from applied S but not from K.

Chemical composition, nutritive value. The contents of nutrients in the herbage vary within wide limits, and CP content usually ranges from 3 to 15 per cent, but Daubenmire (1972) reports the highest content of CP in natural stands of *H. rufa* in Costa Rica as 7.4 per cent which, in the dry season, can fall to 1.4 per cent, far below the subsistence level. CF contents from 28 to 38 per cent have been reported and P content is often on the low side, 0.06–0.16 per cent. Nitrogenous fertilizers increase CP content in the herbage but their effects can be of only temporary nature and in trials by Tergas et al. (1971) in Costa Rica, 150 kg N/ha increased CP content from 5 to 9 per cent 48 days after the application, but this returned to the initial level in 84 days after the application. Digestibility of DM, as quoted from various sources in Butterworth's review (1967) ranged from 43 to 67 per cent – depending mainly on the stage of growth – of CP from 16 to 60 per cent, CF from 47 to 70 per cent and NFE from 50 to 71 per cent. Experimental ensilage in Nigeria showed slow fermentation of *H. rufa* silage and silages prepared in Brazil and Nigeria had very low DCP content, from negative values to 2.1 per cent.

Animal production. *Hyparrhenia rufa* is used mainly for beef production and 1 ha of pasture is usually considered adequate for one steer. In Colombia steers grazed at this stocking rate gained 0.37 kg liveweight per head per day but only 0.28 kg at 2 steers/ha (Crowder et al., 1970). In Mexico 2-year average yields of beef amounted during the wet season to 147 kg/ha for steers grazed on unfertilized pasture and 205 kg when 100 kg N/ha were applied; these gains were lower than those obtained from *Digitaria decumbens* (Garza et al., 1970), and in an earlier trial in the same country liveweight gains from *H. rufa* reached 190 kg/ha (Arroyo & Teunissen, 1964). Mixtures with legumes give higher liveweight gains and in Uganda steers grazed on a *H. rufa/Stylosanthes guianensis* mixture at 0.2 ha/head gained 0.25 kg per animal per day or 422 kg/ha/year, and 0.29 kg and 165 kg, respectively, when grazed at 0.6 ha/head; grazing at 0.2 ha/head was however too heavy, and in 3 years' time the paddocks were invaded by grass weeds (Stobbs, 1969a). In another trial the same mixture grazed rotationally yielded 710–780 kg liveweight per hectare in 420 days. In Uganda liveweight gains were increased by the application of fertilizer P to a *H. rufa/Centrosema pubescens* mixture.

Variability. *Hyrraparrenia rufa* varies considerably in vigour, leafiness and other important characteristics, and there is a scope for selection of superior types. The chromosome number was determined mainly as $2n = 36$, but there were also chromosome counts giving $2n = 20$, 30 or 40, which indicates deeper differences in plant karyotypes than the usual ploidy differences.

Seed production. Seed is usually collected by hand from stands used for fodder or grazing, and the resulting chaffy mixture is sown. When *H. rufa* is grown mainly for seed the best distance between the rows can be 50 or 60 cm. The period between flowering and seed ripening is some 35 days, and in a trial in Colombia (Alarcón et al., 1969), 277 kg seed/ha were harvested. The highest germination was obtained after 100–129 days of storage, but Daubenmire (1972) reports no post-harvest dormancy, and the retention of seed viability for at least 14 months. Seed is best stored at a temperature of 10°C and under low relative humidity. (Botanical information mainly from Clayton, 1969.)

Hyperthelia W. D. Clayton

This genus which resembles *Hyparrhenia* includes a few tropical African species.

Hyperthelia dissoluta (Steud.) W. D. Clayton (*Hyparrhenia dissoluta* (Steud.) Hutch.; *Hyparrhenia ruprechtii* Fourn.) $2n = 40$

The small genus *Hyperthelia* with its four species is closely allied to *Hyparrhenia* from which it differs mainly by a groove on the back of the lower glume. *Hyperthelia dissoluta* is better known to pastoralists as *Hyparrhenia dissoluta* or *Hyparrhenia ruprechtii*.

Tufted perennial 1–3 m high. Stems moderately stout. Leaves up to over 30 cm long and 4 mm wide, with a ligule up to 25 mm long. Panicle spathate, narrow, not large; spathes narrow, resembling the leaves but shorter. Spatheoles narrow, 5–8 cm long. Racemes up to 3 cm long, with an appendage at the base and with two awns per pair of racemes. Sessile (fertile) spikelet 4–7 mm long, pedicelled spikelet 12 mm long. Awns 5–9 cm long.

Common throughout tropical Africa, in open grasslands mostly resulting from periodical fires, and in bush or woodland, preferably on light loams or sandy soils and avoids waterlogged black clays. Well grazed when young but may soon become stemmy although some forms of *H. dissoluta* are leafy and one of them analysed in Kenya contained at the early flowering stage 12.9 per cent CP, 3.0 per cent EE, 33.7 per cent CF, 41.6 per cent NFE, 0.25 per cent Ca and 0.19 per cent P (Dougall & Bogdan, 1960).

Ischaemum L.

Annuals or perennials. Spikelets in dense spikelike racemes single, paired or subdigitate on top of the stems. Racemes with several to many joints easily disarticulated at maturity. Joints and pedicels stout. Sessile and pedicelled spikelets equal or subequal except that the sessile spikelet is bisexual and the pedicellate spikelet bisexual or male. Each spikelet with a fine, geniculate and spirally twisted awn. Glumes flat or slightly convex or concave. Fifty tropical or subtropical species mainly in South and South-East Asia and a few species in America and Africa. Some species, especially *I. timorense*, are well grazed by cattle, horses and sheep.

Ischaemum goebelii Hack. (*I. aristatum* Hack. ssp. *imberbe* Hack. var. *imbricatum* Hack.; *I. imbricatum* (Hack.) Stapf) $2n = 24$

Tufted perennial up to 100 cm high, with single or paired racemes.

Lower glume of sessile spikelet with faint knobs. Awn 1–1.5 cm long. Common in grasslands in relatively humid areas of Assam, Burma, Sri Lanka, India, Malaya and South-East Asia in general. A robust leafy grass well grazed by stock and considered to be of value in pastures. In a trial in Sarawak (Ng, 1972) *I. goebelii* (under the name *I. aristatum*) only relatively slightly responded to low rates of fertilizer but considerable responses to high rates were noted. It yielded 3.1 t DM/ha, unfertilized, 5.7 t when given 122 kg N/ha, 9.3 t at 448 kg N and 14.0 t at 896 kg N/ha. Herbage cut every 6 weeks contained 7.25 per cent CP when the grass was given no fertilizer N, 9.62 per cent at 112 kg N/ha and 15.75 per cent at 896 kg N; however CP content of herbage cut every 14 weeks ranged only from 4.87 to 6.69 per cent even under high rates of N. CP yields ranged from 192 to 940 kg/ha, depending mainly on the fertilizer rates.

Ischaemum indicum (Houtt.) Merrill (*I. aristatum* auct. var. *indicum*)
$2n = 36$ or 72

Annual with lanceolate or linear leaves. Racemes paired or subdigitate. Spikelets about 4 mm long; glumes glabrous or hairy. Awn 1–1.5 cm long. Occurs in South and South-East Asia, on disturbed land and as an arable weed. It is well grazed by cattle and Butterworth (1967) quotes an analysis of herbage showing 9.2 per cent of CP, 30.7 per cent CF, 1.8 per cent EE and 48.8 per cent NFE. In trials with sheep the digestibilities of these nutrients were 55, 71, 51 and 62 per cent, respectively.

Ischaemum pilosum (Klein ex Willd.) Wight (*Andropogon pilosus* Klein ex Willd.)

Rhizomatous perennial up to 120 cm high. Rhizomes densely covered with white scales (leaf sheaths). Leaves hard, bluish green. Racemes paired or subdigitate. Glumes hairy. Awns under 7 mm long. Occurs in India on black, heavy, sometimes waterlogged soil, in natural grasslands. Grazed when young. *Ischaemum afrum* (J. F. Gmel.) Dandy (*I. brachyatherum* (Hochst.) Hack.), a tropical African species, is similar in appearance and ecology but seems to have much longer awns. In Sudan (Whyte *et al.*, 1959) 2-week-old herbage of the latter species contained 8.7 per cent CP which was reduced to 4.0 per cent when cut 9 weeks later and mature plants contained 2.8 per cent CP. CF content was high even in young herbages and ranged from 42 to 44 per cent during the plant growth.

Ischaemum rugosum Salisb. Bher (India) $2n = 18, 20$

Annual with lanceolate or linear leaves. Spikelets under 5 mm long with markedly transversely rugose lower glume; awn over 1.5 cm long. A variable species which occurs in dry and in wet habitats, often in arable land as a weed and in natural grasslands where it is well grazed by stock. Sexual reproduction was reported by Brown & Emery (1958).

Ischaemum timorense Kunth. Loekoentoegras (Surinam, Dutch)
$2n = 20, 36$
A slender creeping perennial. Stems 20–45 cm high with silky nodes. Leaves narrow-lanceolate 2–10 cm long and 3–6 mm wide. Inflorescence consists of two racemes about 7 cm long. Spikelets 2.5 mm long. Awn slender.
Distributed naturally from India to Malaya and in Polynesia. Common in grasslands and on roadsides in wet areas. A useful grazing grass which seeds well and can be established from seed, but is not particularly vigorous or productive. Introduced to Surinam where a 5 m^2 plot yielded 15.3 kg green herbage which corresponds to about 30 t/ha. When cut every 3 weeks it contained 10.7 per cent CP which gradually decreased with the decrease in cutting frequency to 5.3 per cent in grass cut every 8 weeks; the digestibility of CP was low and ranged from 32.7 to 36.7 per cent (Appelman & Dirven, 1962). Leaves and stems analyzed separately gave 15.3 and 6.2 per cent CP, respectively and CP digestibility ranged from 41 to 47 per cent. CF content was 30 per cent in the leaves and 35 per cent in the stems. Young, 3-week-old herbage well fertilized with NPK had high CP content, 20.2 per cent which was reduced to 16.0 per cent in 6-week-old herbage (Dirven, 1971). *Ischaemum timorense* was also tried with some success in small observational plots in Kenya and in Trinidad where it was established by surface spreading of grass cut at the hay stage and disced in.

Iseilema Anderss.

Contains some 20 species of tropical Australia and southern and southeastern Asia. Resembles *Themeda* but the plants are usually smaller and the stems finer.

Iseilema laxum Hack. Musal (India) $2n = 24, 28, 36$
Tufted perennial with the stems 20–60 cm high, erect or ascending, often from semi-creeping many-noded bases. Leaves 8–15 cm long and 1–3 mm wide. Panicle 15–30 cm long, with the branches fascicled from leaf axils. Ultimate racemes arise from the axils of boat-shaped spatheoles which are 8–10 mm long. The racemes are on very short peduncles and are scarcely exerted from the spatheoles. Four lowermost empty spikelets form an involucre resembling that of *Themeda triandra* but the spikelets are provided with pedicels. Above the involucre the raceme has one joint with two pedicelled male spikelets and one sessile bisexual fertile spikelet; some racemes have two joints and two fertile spikelets. The spikelets are 5 mm long and fertile spikelet has an awn 1.3 cm long.
Distributed in southern India, Burma and Sri Lanka, often on 'black cottom soil' of seasonally waterlogged flats and depressions and is in general a common species in natural grasslands of India including those of the *Sehima/Dichanthium* zone and occurs under an annual rainfall of

500 to over 1,000 mm. *Iseilema laxum* is considered to be one of the most palatable natural grasses of India well eaten by cattle but less so by sheep. The herbage is grazed or cut for haymaking or occasionally ensiled. The highest yields of herbage have been obtained by cutting the grass at 30- to 60-day intervals. Hay produced from herbage cut at the preflowering stage contained 6.5, 38 and 43 per cent of CP, CF and NFE, respectively, and CP and CF digestibilities were determined to be 53 and 73 per cent (Chandra, 1964).

Iseilema prostratum (L.) Anderss., a species allied to *I. laxum*, is also a leafy perennial much similar to *I. laxum* but slightly larger. It also differs in having longer peduncles so that the racemes are clearly exserted from the spatheoles and in having smaller spikelets which are 3–3.5 mm long. It is a species with a wider distribution than *I. laxum* and occurs in most of India, Sri Lanka and Burma, and some, if not most, information attached to *I. laxum* can perhaps be referred to *I. prostratum*; this is however difficult to establish from the published data.

Ixophorus Schlechtd.

A monotipic genus closely allied to *Setaria* but differing in having the palea of the lower floret much wider than the rest of the spikelet.

Ixophorus unisetus (Presl) Schlechtd. (*Urochloa uniseta* Presl.). Hierba blanca Honduras; Pasto Honduras; Pasto Hatico $2n = 34$ Perennial forming large tufts. Stems 0.75–2 m high. Leaves of varying length and 0.5–3 cm wide. Panicle mostly 15–20 cm long, with 5–25 moderately dense, distant racemes. Spikelets lanceolate, 4–5 mm long, each supported by a single bristle which is usually longer than the spikelet and remains on the raceme after the spikelet had fallen off. Lower glume short, less than one quarter the length of the spikelet; upper glume three-quarters the length of the spikelet. Lower, sterile floret longer than the rest of the spikelet. During the late flowering and ripening stages the palea of the lower floret develops wings and becomes much broader than the rest of the spikelet. Fertile floret small.

Occurs naturally in Mexico, Central America and in the northern parts of South America, in humid areas and on moist fertile soil. This grass is palatable to stock but cannot withstand grazing and is usually cut for feeding green. It is utilized in natural stands or planted by seed in rows 75–100 cm apart. It can also be established from tuft splits or stem cuttings (Whyte *et al.*, 1959). In Central America a mixture of *I. unisteus* with *Pueraria phaseoloides* was grown with good success and lasted for several years. However, in Brazil it did not perform well and, when cultivated, yielded in the wet season about 20 t fresh herbage, unfertilized, and 40 t/ha when fertilized with moderate amounts of NPK + B + Zn, and was one of the poorest yielding grasses in the trial.

In the dry and cool season the yield was about 2 t green herbage/ha when given no fertilizers; fertilizing alone produced no or little yield increase, but fertilizers combined with irrigation increased the yield to some 7 t/ha (Pereira *et al.*, 1966).

Lasiurus Boiss.

Contains two or three species which occur in tropical and subtropical Africa, India and Pakistan.

Lasiurus hirsutus (Forsk.) Boiss. (*Elyonurus hirsutus* Forsk.). Sewan.
This includes *L. sindicus* Henrard separated from *L. hirsutus* by the presence of dense hairs on the stem, just below the inflorescence and on parts of internodes, the character which was thought to be confined to the eastern part of the area of *L. hirsutus* distribution. Hairy stems are however occasionally found in other areas of *L. hirsutus* distribution and it is hardly necessary to give *L. sindicus* a species status. Nearly all information on the pastoral value of *L. sindicus*, found mainly in the Indian literature, should be referred to *L. hirsutus*. $2n = 56$

Strong perennial up to 1 m high. Stems much branched and often subwoody at the base, glabrous to hairy. Leaves up to 30 cm long and up to 6 mm wide, flat or convolute. Stems and their branches, if any, terminate in a single, densely villose articulate raceme 10–15 cm long. Joints 4–8 mm long and bear awnless spikelets, usually in threes of which two are sessile and one pedicelled. Sessile spikelet 6–9 mm long, with two florets of which the upper is fertile; lower glume is often elongated and terminates in a two-toothed appendix. Pedicelled spikelets male or barren.

Occurs naturally in dry areas of Mali, Niger, North Africa, Egypt, Somalia, Ethiopia, Iraq, South Pakistan (Sind) and north-west India, partly in the tropics but chiefly in subtropical areas.

Lasiurus hirsutus is one of the most important grazing grasses of north-west India, at 25–27°N. and under an annual rainfall below 250 mm. It grows best on alluvial soil or on brown sands. In north-west India *L. hirsutus* is often used for the establishment of permanent pastures; the land is ploughed and 3–6 kg of unhusked seed/ha are drilled in rows some 75 cm apart, up to a depth of 1–1.5 cm. However, in trials by Chakravarty & Verma (1972) 30 cm between the rows resulted in better plant tillering than wider spacing. In the same trial hand weeding increased DM yields from 2.23 t/ha to 5.00 t in the second year of growth and from 1.87 to 3.86 t in the third year; it also increased the number of tillers/plant. Average DM yields of sown *L. hirsutus* are around 1.5 t DM/ha, but Verma & Chakravarty (1969) reported 2.7–10.5 t fresh forage/ha and Whyte (1964) 4.9 to 12.0 t, depending on the frequency of cutting. Drilling annual legumes (*Vigna radiata*, *V. aconitifolia*, *Cyamopsis tetragonoloba*) every year between the rows of *L. hirsutus* increased the yields of herbage by 20–30 per cent (Daulay *et al.*, 1968). Under natural conditions CP content of *L. hirsutus* usually ranges from 5.9 to 6.7 per cent but when cultivated and fertilized with NPK CP content can reach 13 per cent and even up to 15 per cent; CF content varies from 24 to 38 per cent, Ca content is usually high, 0.76–1.11 per cent and P content ranges from 0.15 to 0.44 per cent (Gupta & Saxena, 1970). These authors also estimate that three hectares of *L. hirsutus* pasture are sufficient to support an adult animal (cattle) throughout the year, and a 3.6 kg liveweight increase in 14 days was recorded when heifers were grazed at 2–4 ha per animal. Continuous grazing is preferred to deferred grazing.

Leersia Sw.

Contains twenty mostly tropical species of wet situations.

Leersia hexandra Sw. (*L. abyssinica* A. Rich.; *L. capensis* C. Mueller). Arroz bravo; Barit (León & Sgaravatti) $2n = 48$

Perennial 30–100 cm high, with numerous fine stems arising from long and much branched rhizomes. Leaves 10–20 cm long and 4–8 mm wide, retrorsely and sharply scabrid on the midrib of the lower surface. Panicle small, 5–15 cm long. Spikelets 3.5–5 mm long and 1.2–1.4 mm wide, laterally much compressed, with one floret. Glumes in the form of minute scales or absent. Lemma pectinate-ciliate on the keel.

Leersia hexandra occurs throughout the world tropics in shallow water of flood plains and swamps or at river banks, and forms extensive colonies and carpets of herbage. In spite of scabrid leaves it is well grazed by cattle especially when kept low. Dirven *et al.* (1960) also report it to be one of the main grasses in fallows after the cultivation of irrigated rice in Surinam where plant communities dominated by *L. hexandra* yield 7–14 t green fodder/ha. In Surinam, and perhaps in other countries too, it can be an important weed of rice and is grazed after the rice has been harvested. Butterworth (1967) quotes CP contents of 5.8–6.3 per cent for India, 7.3 per cent for the Philippines and 10.1 per cent for Tanzania, and the contents of DCP 2.3–2.4, 4.9 and 7.0 per cent, respectively. TDN content varied from 46.7 to 61.3 per cent. Higher contents of CP, 12.0–12.2 per cent were determined in Argentina by Vonesch & Riverós (1967–8); the contents of CF were 27.7–28.3 per cent and of NFE 43.6–44.2 per cent.

Leersia denudata Launert, which occurs in East, Central and South Africa and forms extensive colonies at 1,500–2,300 m alt, is a non-rhozomatous perennial rather similar in appearance to *L. hexandra*. The leaves are not sharply scabrid and the plant is more readily grazed than *L. hexandra*, from which it can also be distinguished by the ciliate and not pectinate keel of the lemma (Clayton, 1970).

Leptochloa Beauv.

Contains some 25–30 annual or perennial species which occurs throughout the tropics and in warm temperate regions.

Leptochloa obtusiflora Hochst. $2n = 20$

Tufted perennial. Stems 40–150 cm high, erect or geniculately ascending. Leaves 10–35 cm long and 2–15 mm wide, flat. Inflorescence consists of 5–20 distant, slender, suberect, spreading or pendent spikes. Spikelets 4–7 mm long with 5–15 florets easily disarticulating at maturity, Glumes subequal; florets 1.7–2 mm long, with blunt, often two-lobed lemma. Grain oblong, 1–1.2 mm long, usually remaining enclosed between the lemma and palea at maturity.

Occurs in the northern half of tropical Africa, from Tanzania and Angola to Sudan and Ethiopia, and again in Arabia and India, from sea level to 1,800 m alt, in bushland, woodland and in grassland, in semi-arid or moderately humid areas. *Leptochloa obtusiflora* is a variable grass and leafy forms are well grazed. A leafy form analysed at the early flowering stage had a very high content of CP – 18.4 per cent, 1.7 per cent EE, 28.0 per cent CF and 42.4 per cent NFE, and the contents of P and Ca, 0.22 and 0.62 per cent, respectively, were adequate (Dougall & Bogdan, 1960). Seed setting is outstanding and 1 kg of seed can be collected from about 100 plants. Seed can be easily harvested by hand-stripping the panicles, but cutting with sickles or harvesting can result in seed losses. The seeds are in the form of florets with one caryopsis each and there are 3–3.5 million of such seeds per kg (Bogdan & Pratt, 1967). In Kenya *L. obtusiflora* is regarded as potentially suitable for reseeding denuded pastoral land in not very dry areas, but no experimental evidence is available.

Leptothrium Kunth

Two species: one in tropical Africa and south-western Asia, the other on the Caribbean islands.

Leptothrium senegalense (Kunth) W. D. Clayton (*Latipes senegalensis* Kunth)

Short-lived perennial forming small tufts. Stems 10–60 cm high, erect or ascending. Leaves 2–10 cm long and 1–3 mm wide. Inflorescence a raceme 2–17 cm long with a few to many pairs of spikelets (some spikelets can be single) on flat common peduncles. The first spikelet has an upper glume 2.5–8 mm long, mostly longer than the rest of the spikelet, flat, lanceolate and usually recurved, often forming a hook. Upper glume is 3–3.5 mm long and encloses the rest of the spikelet which consists of a single floret. The glumes and the flat peduncle are pectinate-spinulose on the margins and the upper glume is also tuberculate. The lower glume of the second floret is erect and does not exceed the upper glume in length.

Distributed in tropical Africa, mainly in its northern half, from Senegal to Somalia and further east to Pakistan. Occurs in arid and semi-arid bushland from sea level to about 1,500 m alt. The grass is stemmy and rather harsh but is surprisingly well eaten by stock at almost any time of the year. The bulk of herbage is never great even when the plants are numerous. *Leptothrium senegalense* is a good seeder and was successfully used for reseeding denuded pastoral land in central Kenya and tried further north, also in Kenya, and although better grasses are usually used for reseeding it is a useful species for the purpose when only a minimum seedbed preparation can be provided. Another valuable feature is a relatively good palatability in the dry season although outwardly the grass looks as though it would not be accepted by the animals. A Kenya analysis of herbage harvested at early flowering showed a satisfactory content of CP – 9.3 per cent, high content of CF – 38.9 per cent, and the content of NFE was 43.0 per cent. Ca and P contents were moderate, 0.41 and 0.15 per cent, respectively (Dougall & Bogdan, 1965). Seed is in the form of a pair of spikelets attached to the supporting flat peduncle and if each spikelet contains a caryopsis then there are about 350,000 such 'seeds' per kg (Bogdan & Pratt, 1967).

Lolium L.

About 12 species of temperate Europe and Asia. Two or three species are widely cultivated as pasture grasses.

Lolium perenne L. Perennial ryegrass

Tufted perennial widely cultivated in Europe, eastern and south-eastern states of the USA and in most temperate countries of all continents and in both northern and southern hemispheres. It is also grown in the subtropical areas of Australia, South America and South Africa; in the tropics perennial ryegrass is grown to a limited extent at high altitudes, the elevation depending mainly on the latitude; in the equatorial zone, e.g. in Colombia or Kenya, ryegrass is grown above 2,400–2,500 m alt, but the altitudinal level of its cultivation lowers as the latitude increases. More vigorous and upright cultivars, e.g. S24, are preferred at high altitudes to the pasture types mainly because of the low temperatures in the growing seasons which result in a slow growth. Perennial ryegrass is often grown in mixtures with white clover (*Trifolium repens*).

Lolium multiflorum Lam. or Italian ryegrass resembles perennial ryegrass except that the florets, awnless in perennial ryegrass, are provided with awns and the plants are less persistent, dying out after 2–3 years of productive growth. Italian ryegrass withstands drier and hotter conditions better than *L. perenne* and in the tropics it can be grown at somewhat lower altitudes than the latter.

Melinis Beauv.

About 15–20 species of tropical and South Africa.

Melinis minutiflora Beauv. Molasses grass; Herbe du Brézil; Wynne grass (Jamaica); Gordura; Melado; Calinguero. In Colombia, Venezuela and Puerto Rico the name Yaragua, usually associated with *Hyparrhenia rufa*, is also used as a result of an original error (Parsons, 1972) (Fig. 23)

A viscous, usually tufted perennial. Stems up to 1.5 m in height, many-noded, erect or geniculately ascending, often much branched. Leaves linear to linear-lanceolate or lanceolate, 5–20 cm long and 4–13 mm wide, hairy, the hairs are glandular and exude a sticky, sweet-smelling substance. As growth progresses, the basal leaves die out and adult plants have no basal leaves but only those on stems and branches at some distance from ground level. Panicle linear to ovate, 10–30 cm long, dense and contracted but opening during flowering, pale green or usually purple. Spikelets on fine pedicels, narrow, oblong 1.8–2.4 mm long, glabrous or hairy, strongly groved and nerved. Lower glume a minute scale up to 0.3 mm long; upper glume as long as the spikelet, two-lobed. Lower floret empty, with the lemma as long as the upper glume or slightly shorter, two-lobed, with a fine awn 6–15 mm long, or occasionally almost awnless. Upper floret bisexual, fertile, 1.5–1.8 mm long. Caryopsis narrow oblong, 1.5 mm long.

Melinis minutiflora is indigenous in nearly all parts of tropical Africa at elevations of 800–2,500 m and is also recorded from Madagascar and Comoro Islands; it is widely spread as a naturalized grass in South America. Occurs naturally in bush, at forest edges and in open grassland, often on steep, rocky slopes.

Environment. *Melinus minutiflora* requires a cool and moderately moist climate with an annual rainfall over 900 mm although it can remain green and productive well into the dry season. It grows well on acid or poor leached soils where it can compete with other grasses but does not tolerate heavy clays or waterlogging. Parsons (1972) indicates that in America *M. minutiflora* thrives 'on thin soils especially in the more temperate "coffee climates" '.

Introduction. *Melinus minutiflora* is cultivated, mainly on an experimental scale, in some countries of its origin and on a farm scale in northwestern Kenya. It has been tried with varying success in a number of other tropical countries, but only in tropical South America and some West Indian islands has it been cultivated for a considerable period of time on a large scale producing a great impact on the animal industry, both as a cultivated and a naturalized grass.

Melinus minutiflora was apparently accidentally introduced into Brazil early in the nineteenth century and it was from there that it was

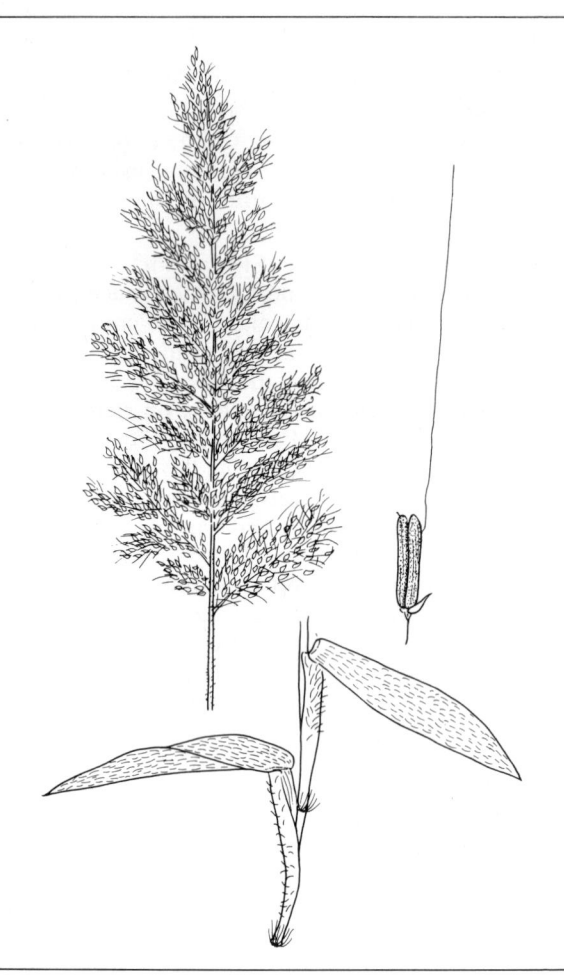

Fig. 23 *Melinis minutiflora.*

first described botanically (Parsons, 1972). It soon reached Venezuela and later, in the twentieth century, spread to Peru, Colombia, Cuba, Puerto Rico and some other parts of the West Indies, Central America and Mexico. In South America, *M. minutiflora* has found particularly suitable conditions for growth, being perfectly naturalized and now covering large areas, especially in the cooler regions. It has been introduced into cultivation and is at present grown on a large scale. In Brazil and in other countries of tropical South America and in some West Indian islands it now plays an important role in the beef and milk industry. Parsons (1972) writes that 'Molasses grass is the basis of much of the modern beef cattle industry of Minas Gerais and Goias' of Brazil.

Milk production of the south central, milk-supplying area of Brazil also largely depends on molasses grass. It is valued as an easily established, palatable and productive grass of satisfactory nutritive value and also for soil conservation on steep slopes with poor soil in sufficiently wet areas.

Establishment. Seeds of *M. minutiflora* are small and a reasonably well prepared seedbed is required; however, in Brazil, it was successfully established when sown on ash of burnt forest or bush without any soil tillage. Seed is usually broadcast but sometimes drilled and no cover is required after sowing though rolling is useful. Although very shallow or surface sowing is recommended, in Kenya, on good soil, the seedlings emerged best from the soil depth of 2–2.5 cm (Bogdan, 1964a). The seedlings can be easily recognized by the ovate-orbicular first leaf lying flat on the ground. Sowing rates vary and Whyte *et al.* (1959) cite 2 to 4.5 kg/ha for Queensland and 20–25 kg/ha for Brazil, the sowing rate variation apparently depending on seed quality. If seed quality is known then 0.5–1 kg or 3 to 6 million of 100 per cent pure germinating seeds (PGS) per ha or 300 seeds per m^2 can be adequate; the actual amount of seed sown would depend on seed purity. It was found in Australia (Prodonoff, 1967) that presowing chilling of seed for 5 days at 8°C increases the germination by 4–21 per cent. P, and in some areas K, are the fertilizers usually applied at the establishment, but N is unnecessary at this early stage. The seedlings emerge on the 6th–9th day after sowing, and although they are weak at the start, the establishment does not normally present any difficulties. In Kenya *M. minutiflora* is usually established under maize, the seed being broadcast after the last weeding, when maize plants reach a height of 60–80 cm; no detrimental effect of *M. minutiflora* on maize was observed. If the young plants are overgrown with weeds, an early single cut, or grazing if the weeds are palatable, help the establishment. Establishment from splits is almost impossible and even under ideal conditions hardly more than 10 per cent of plants would survive.

Management, fertilizing. *Melinus minutiflora* is used for grazing and only seldom for haymaking but the height and frequency of grazing has been studied using mainly the cutting technique. In the majority of trials, cutting at about 10 cm from ground level resulted in higher herbage yields than closer cutting and frequent close cutting damaged the plants. Cutting at 12 cm encouraged the lateral spread of plants, produced a good ground cover and delayed flowering (Paula *et al.*, 1967). It is recommended to graze *M. minutiflora* rotationally, at 40- to 60-day intervals, repeating the grazing when the herbage reaches a height of 35–45 cm (Whyte *et al.*, 1959). Haymaking is practised to some extent in South America but it is believed in Kenya that cutting *M. minutiflora* for hay at the flowering stage can considerably damage the plants. *Melinus minutiflora* is not resistant to burning and can be destroyed by grass fires. Its responses to fertilizer N varies: N can double the herbage yields in some areas and habitats and produce no response in others. Good responses to moderate rates of P, and sometimes to K, were reported, and considerable yield increases from the application of NPK have been reported from Costa Rica and elsewhere. Andrew & Robyns (1971) found that 0.18 per cent of P in the herbage was critical for the plant and P content below this level indicates that P fertilizer is needed. In a trial in Brazil, sprinkler irrigation during the five months of dry season increased green herbage yields in this period from 5.7 to 7.2 t/ha, unfertilized, and from 7.4 to 11.0 t when given moderate amounts of NPK (Ladeira *et al.*, 1966).

Association with legumes. *Melinus minutiflora* is a grass of considerable competitive vigour and only a few legumes can withstand its pressure in mixtures. The best results have so far been obtained from mixtures with stylo (*Stylosanthes guianensis*), mainly in Africa, and with tropical kudzu (*Pueraria phaseoloides*). In Kenya (Strange, 1961) an admixture of *Trifolium semipilosum* increased the yields of CP but not of DM and a *Desmodium uncinatum* mixture did well in small plots. In Australia, a *Stylosanthes humilis* mixture was tried with some success.

Diseases. In north-west Kenya, the main *M. minutiflora*-growing area, a 'small-leaf' or stunting virus disease appeared in 1955–60 and greatly contributed to the decrease in *M. minutiflora* popularity. The affected plants produce minute leaves and panicles and the disease can spread rapidly. The insect vector has not been discovered so far, although some attempts in this direction were made. The cultivated commercial type is particularly susceptible to the virus.

Herbage yields. Crowder *et al.* (1970) estimate that in Colombia DM yields reached 6–8 t/ha per year and can be doubled by the application of 150 kg N/ha. Yields obtained in other countries (Fiji, Puerto Rico, Costa Rica, Kenya, Brazil) range mostly from 3 to 8 t DM/ha. Unusually high yields of fresh herbage, 44.3 t/ha for unfertilized stands, and 78.3 t for those receiving NPK, were reported by P. J. Mata from Costa Rica.

Chemical composition, palatability, nutritive value. CP content in the *M. minutiflora* herbage usually ranges from 6 to 10 per cent but it can be as low as 4 per cent and as high as 14–18 per cent; the last figures are for young, grazing-stage herbage of the year of establishment. Gomide *et al.* (1969a) determined CP content every 4 weeks; it decreased from 17.2 per cent in the 4-week-old herbage to 11.3 per cent in the 8-week-old herbage and then gradually to 6.2 per cent in the herbage uncut for 32 weeks, and they consider 9.4 per cent CP as an average for the species. CF content is on the high side and usually ranges from 30 to 40 per cent. It is interesting to note that the content of lignin in *M. minutiflora* plants does not increase from the flowering stage onwards (Lewin & Melotti, 1965–66). High content of EE has often been observed, which is not uncommon for aromatic grasses. In spite of relatively low CP content and a high percentage of CF, molasses grass is readily grazed by cattle if the animals are given sufficient time to get used to the sweet smell of the herbage. Good palatability has been reported from various countries and *M. minutiflora* is highly valued by graziers, especially in Latin America. Gomide *et al.* (1969b) have also shown the decrease of P content in the herbage from 0.30 per cent in a 4-week-old growth to 0.12 per cent 32 weeks later, whereas the content of Ca fluctuated from 0.38 to 0.46 per cent without any clear relationship with the age of herbage. According to the same authors, the content of Mn varied from 102 to 137 p.p.m. and of Zn from 27 to 47 p.p.m., their contents in a young growth being only slightly higher than in older plants, but the content of Cu was markedly higher in the early than late cuts and ranged from 185 to 981 p.p.m.

Herbage digestibility is usually on the low side and Melotti (1969) reports from Brazil a DM digestibility of 54 per cent and Butterworth (1967) in his review quotes 42 to 51 per cent. Crude protein digestibility in Melotti's work is given as 38 per cent and it ranged from 18 to 46 per cent according to Butterworth's data, who gives DCP contents from 0.1 to 2.8 per cent; it should be taken into account, however, that in most of the trials he quotes the herbage examined had passed the prime grazing stage. Butterworth also quotes the digestibility of CF as ranging from 44 to 66 per cent and of NFE from 42 to 62 per cent. The digestible energy of 2.465 Kcal/g was determined by Melotti.

Animal production. In spite of the usually moderate yields of herbage, stocking rates reported in the literature are relatively high and mostly range from 0.5 to 2 ha of *M. minutiflora* pasture per head of cattle and liveweight gains of 0.4–0.5 kg per day per animal.

Flowering, reproduction. The narrow and contracted panicles open during the flowering period and close again at ripening. Flowering (in Kenya) occurs 166–202 days after sowing, depending on the variety. In field trials (Bogdan, 1960a), practically no cross-pollination was observed and only 0.1 per cent of plants failed to breed true to type; it was presumed that the species can well be an apomict. Its chromosome number is $2n = 36$.

Variability, cultivars. The types cultivated in various countries (USA, Uruguay, Malawi, Kenya) were practically uniform and identical when compared in Kenya and only the Nigerian cultivated type differed from the others. The wild East African types varied, however, quite considerably in habit, vigour, plant height, hairiness, degree of viscousness, presence or absence of awns and their length, time of flowering and seed setting, and combinations of valuable features were found in some introductions. The types cultivated or naturalized in Brazil are not completely uniform. No breeding work has been undertaken so far and the commercial type which is widely grown and which is known in Kenya as **Kitale commercial** was probably picked up in south-eastern Africa as a good, well-performing variety, and may well be similar to the **Roxo** (red) type grown in Brazil. Other recognized Brazilian types are **Cabelo de negro** (negro's hair) which is smaller and more resistant to grazing than the Roxo, the **Francano**, similar to Roxo but more vigorous, and the **Branco** (white), which has light-green leaves and in which the quality of herbage is not as high as in Roxo (Whyte et al., 1959). In Kenya, cv. **Chania**, originating from the Chania River in the Thika area, and cv. **Mbooni hill**, also from Kenya, were found to be worth further trials. Chania molasses grass has a more prostrate habit than the commercial type and the Mbooni hill variety is a creeping stoloniferous type which forms a continuous sward; seed yields of the two new varieties are however lower than those of the Kitale commercial cultivar. The two introductions showed a considerable, although not complete, resistance to the small-leaf virus to which the commercial type is highly susceptible.

Seed production. According to Whyte *et al.* (1959) seed yields can reach 280 kg/ha; this is however seldom achieved and in Kenya, in small plots, seed yields of 165–250 kg/ha were recorded for commercial types of different origin and from 12 to 108 kg for wild varieties (Bogdan, 1960b). Mechanical harvesting is not easy, mainly because of the large bulk of viscous herbage, and seed is usually harvested by hand. Seed is in the form of spikelets containing one caryopsis each and there are about 6 million of such 'seeds' per kg. Commercial types of *M. minutiflora* have long awns which make the seed fluffy; hammer milling breaks off the awns and the processed seed flows easily. In sealed containers seed can retain its viability for a long time but when stored in cloth bags it ceases to germinate on the fifth year of storage (Da Rocha *et al.*, 1965).

Panicum L.

Perennials or annuals of various habit. Panicles loose or contracted, branched. Spikelets on fine to stout pedicels at the ends of ultimate branches, awnless, slightly dorsally compressed, flatter on the side of the lower glume and convex on the other side. Lower glume small, upper glume of the size and shape of the spikelet. Florets two, the lower male or sterile, with the lemma similar to upper glume, the upper bisexual, its lemma and palea hard, crustaceous, tightly enclosing the grain, the margins of lemma overlapping the palea. A very large genus with up to

500 species which occur in warm countries, mainly in the tropics, throughout the world. This genus includes a number of useful grazing species but only a few are of major importance for the animal industry. *Panicum maximum* is one of the most important grazing and fodder grasses, followed by *P. coloratum*. Genus *Panicum* was more widely applied in the past but its limits were narrowed when species formerly included in *Panicum* were transferred to the newly created genera of *Paspalum*, *Brachiaria* and a number of others.

Panicum antidotale Retz. Blue panic; Giant panic; Barwari; Ghamur; Gift (India)

Vigorous tufted perennial. Stems thin, wiry, bulbous at the base, 1–2 m high or occasionally taller. Leaves glabrous, bluish-green, 15–60 cm long and 4–12 mm wide. Panicle up to 30 cm long, loose, with numerous, fine, ascending branches. Spikelets 2.5–3 mm long, on fine pedicels. Lower glume half the length of the spikelet; upper floret 2.0–2.5 mm long with shine crustaceous lemma and palea which are beige in colour with darker spots. Grain ovate, 1 mm long.

Indigenous in subtropical northern India, Pakistan, Afghanistan and Iran, and also in southern tropical India. Narayanan & Dabadghao (1972) state that it can grow under an annual rainfall of 130 mm although Relwani & Bagga (1969) gave a higher rainfall, 500–700 m, as suitable for the species. The drought resistance of *P. antidotale* can perhaps be attributed to a deep penetration and vigorous development of roots. *Panicum antidotale* grows on flat land but also in hilly areas, on a variety of soils but avoids waterlogged ground. Anderson (1970) observed, in Queensland, 100 per cent survival after a 5-day flooding but only 35 per cent of plants survived a 10-day and none a 20-day flooding, a relatively poor performance compared with other grasses he tested. *Panicum antidotale* is also a frost-tolerant species.

Panicum antidotale was first introduced into cultivation in the USA, where it was grown in the south-western states of Arizona, Texas and California early in this century, and in 1930 was established on an appreciable scale in Queensland for grazing and seed production (Barnard, 1969). It was later 're-introduced' to India where local types were then also collected and grown in experimental plots, the best of them on a farm scale. Introduced to other tropical and subtropical countries where it has so far been grown mainly in experimental plots.

Cultivation. *Panicum antidotale* can be established from tuft divisions and by raising seedlings in nurseries; this method requires, however, a considerable labour force and the grass is more commonly grown from seed. Relwani & Bagga (1969) consider cross discing to be an adequate seedbed preparation but more thorough soil tillage is also recommended. Seed is sown at 3 to 7 kg/ha, preferably in rows spaced 30–60 cm apart, to a depth of about 1 cm. Seedlings are weak and the establishment is slower than in the majority of drought-tolerant grasses and weeding is essential. Chakravarty & Verma (1972) obtained fourfold higher herbage yields in the second year of growth when the sown grass was weeded once a year compared with unweeded plots; two weedings a year resulted in relatively small further yield increases. Manuring with 12–15 t f.y.m./ha 2–3 weeks before sowing plus some 100 kg superphosphate, or more if the soil is particularly deficient in P, and 100 kg ammonium sulphate after each cut is recommended. The grass responds well to fertilizers in years with good rainfall or under irrigation. The application of 2–2.5 t lime is recommended for acid soils (Relwani & Bagga, 1969). The fertilizers mentioned have been applied in experiments but rarely in farm practice. Irrigation greatly increases the yields but is usually reserved for more valuable crops. The first cut can be made 50–60 days after sowing and cutting at 20–25 cm from the ground level is recommended.

Herbage yields and quality. Herbage yields range from 10 to 50 t fresh fodder/ha, and Whyte (1964) quotes yields of 18.50 t, 36.54 t, 17.46 t and 4.74 t green fodder/ha for the first and three subsequent years of growth, respectively, obtained in three to four cuts a year in the Bombay area of India; these yields were lower than those of the best producing grasses in the trial. Under sewage irrigation the yields can exceed 150 t green fodder/ha.

Chemical analyses show a satisfactory nutritive value of the herbage: the content of CP ranges from 7 to 13 per cent, depending on the stages of growth and fertilizers, CF content from 34 to 40 per cent, NFE from 30 to 43 per cent, EE from 1.2 to 2.4 per cent, P_2O_5 from 0.20 to 0.30 per cent and CaO from 0.4 to 0.5 per cent (Relwani & Bagga, 1969); Vonesch & Riverós (1967–8) give however lower content of CF and higher content of NFE for plants grown in Argentina. Relwani & Bagga also state that DM intake by sheep reaches 3 kg per 100 kg liveweight and that DM contains 58 per cent TDN. Although the grass is reputed to be well eaten by stock, its palatability was estimated to be only 11 to 31 per cent of that of *Dichanthium annulatum*, one of the most palatable Indian grasses (Dabadghao & Marwaha, 1962), and other sources give only slightly better palatability estimates. The grass is said to acquire a bitter taste at the late flowering stage and to contain appreciable amounts, up to 4–5 per cent, of oxalates which may result in kidney disorders as stated by Relwani & Bagga (1969). The same authors report that leguminous plants (*Cyamopsis tetragonoloba, Vigna unguiculata, V. radiata, Stylosanthes guianensis, Clitoria ternatea, Centrosema pubescens* and some others) grown with *P. antidotale* can increase the nutritive value and palatability of the mixtures compared with grass alone.

According to cytological evidence (Brown & Emery, 1958), the reproduction is by sexual seed and the plants are diploids with the chromosome number $2n = 18$ (Burton, 1942) but tetraploids with

$2n = 36$ have also been found. In Arizona, USA, superior types of *P. antidotale* were developed and cv. A-130 was released in 1950; cv. T-15327 was later selected based on A-130 (Barnard, 1969). Considerable work has also been done in India where, amongst other good types, cv. 297 was developed at Jodhpur.

Seed production. Seed is well formed, and although flowering takes a long time to complete and seed sheds easily, yields of 100–180 kg/ha from dry-land crops and up to 370 kg from irrigated stands have been obtained (Relwani & Bagga, 1969). Post-harvest maturation up to about a year is required for optimum germination after which 50–70 per cent germination can be obtained and maintained for as long as 5 or even 6 years. Seed can germinate at temperatures of 20° to 35°C, better at 30°C than at 20°C, and although at optimum temperatures the germination is about equal in darkness and under light, at marginal temperatures of about 20°C darkness can reduce germination by half (Mukherjee & Chatterji, 1970).

Panicum bulbosum H.B.K. Bulb panicum $2n = 36, 54$ or 72

Essentially as *P. maximum* from which it differs mainly by thickened lower stem internodes forming cormlike swellings 1 or occasionally up to 2 cm in diameter. According to Hitchcock (1950), it occurs in Mexico and in adjacent parts of the USA in moist places of canyons and valleys. In pastoral and agricultural literature it is often referred to as *P. maximum* var. *gonylodes* (Jacq.) Doell or as cv. gonylodes of Guinea grass. The performance and the value of this form is close to that of other types of *P. maximum*.

Panicum coloratum L. Coloured Guinea grass; Kleingrass; Makarikari grass.

Tufted perennial, erect, ascending or spreading by long creeping stems. Stems 40–150 cm high. Leaves 5–40 cm long and 4–14 mm wide, flat, often fleshy, from green to bluish in colour. Panicle loose, erect or nodding; the lowermost branches single or paired although in some vigorous types they tend to form a whorl. Spikelets oblong, acute, 2–3 mm long, glabrous and green, bluish or tinged with purple. Glumes unequal, the lower one a quarter to one third the length of the spikelet, the upper as long as the spikelet. Lower floret male, upper floret bisexual, fertile, its lemma and palea smooth and glossy, from pale yellow to brown in colour, depending on the variety and the stage of maturity. Grain whitish, broadly ovate, with a large embryo and a dark circular scar on the opposite side.

According to Hutchison & Bashaw (1964), *P. coloratum* is a cross-pollinating, sexual species possibly with some degree of apomictic reproduction. In Kenya the apparent uniformity of cv. Solai swards established from seed suggest, however, apomixis, rather than sexual reproduction, at least in this variety. Apart from 2*n* chromosome numbers 18, 36 and 54, which indicate normal diploids, tetraploids and hexaploids, irregular numbers of chromosomes, 2*n* 32, 44 and 56, have been found, and occasional univalent chromosomes were present in tetraploids (from Bolkhovskikh *et al.*, 1969) which suggest irregularities in sexual reproduction. No field or green-house trials on reproductive habits seem to have been undertaken so far and breeding behaviour of *P. coloratum* requires further investigation which may reveal apomixis at least in some forms of the species.

Panicum coloratum is a variable species some forms of which may deserve a species status. The main form, var. *coloratum*, occurs in tropical Africa, mainly in its eastern parts. The tufted plants are erect and seldom exceed 120 cm in height. Leaves glabrous, but those of seedlings and young plants are often hairy. Occurs at the altitudes between 500 and 2,000 m, mainly in seasonally waterlogged flats and depressions but avoids particularly heavy clays waterlogged for long periods of time. Annual rainfall under which this grass occurs ranges from 600 to 1,200 mm and the grass is suitable for cultivation within these rainfall limits. The Kabulabula type originating from Botswana is perhaps only a more vigorous form of var. *coloratum*; it is tall and, perhaps because of its vigour, tends to form a whorl of two to three lowermost branches of the panicle. This form or cultivar is sometimes referred to as *Panicum kabulabula* but no such name seems to be recognized in the taxonomic literature.

Var. *makarikariense* Goossens (*Panicum makarikariense* (Goossens) Van Rensb.), also known as Makarikari grass, is a distinct type forming broad suberect tufts or spreading by creeping stems; it has bluish, glaucous, usually fleshy leaves with strong white midribs. Occurs in the Makarikari swampy area of Botswana in tropical South Africa and in adjacent territories, on heavy clays of seasonally waterlogged or flooded plains, can withstand low-rainfall conditions and thrives under an annual rainfall from 400 mm, but mostly from 600 to 900 mm. It also differs from the main form in other ecological and agricultural characteristics, and is better known to the pastoralists than the main variety. Cultivation and agricultural value of the two types are considered separately.

Var. **coloratum**. A typical form of this variety, cv. Solai, is grown mainly in Kenya where it was selected from a local wild type. A giant cv. Kabulabula ('*Panicum kabulabula*'), which can provisionally be included in var. *coloratum*, but can also belong to var. *makarikariense*, originates from Botswana, South Africa. It is better known and more widely spread in experimental cultivation than cv. Solai; the plants are tall, upright and vigorous and give high yields of herbage especially in the more moist areas (Bryant, 1967a, b). Both cultivars, and especially cv. Solai, are good seeders and can be relatively easily established from seed. Various sowing rates have been suggested and in Kenya one or, preferably, 2 kg of pure germinating seed (PGS)/ha is recommended, the actual rate depending on the percentage of PGS which in cv. Solai is usually high. In early trials with cv. Solai, low germination of untreated seed was reported but in the more recent years all Kenya commercial seed has been in the form of ripe fertile florets, with the soft floral scales (the glumes and the lower floret) removed, and no scarification or other additional seed treatments seem to be necessary for seed stored for 6 months. Shallow sowing is essential on soils with poor physical properties and easily packing, in which only a small proportion of seed germinated from a depth below 1 cm, whereas in soils with good structure sowing to a depth of 2–3 cm results in good seedling emergence.

Herbage yields and quality. Herbage yields vary and depend on the cultivar or variety soil fertility and the height and frequency of cutting or grazing. In a clipping trial in Kenya at 2-week intervals between cuttings the yields increased sharply from 2.53 t DM/ha to 6.15 t when the level of cutting increased from 5 to 15 cm above the ground level and to 9.81 t when cut at 25 cm. When cut at 8-week intervals the yields ranged from 18 to 19 t DM/ha and little depended on the height of cutting, and intermediate reaction to the cutting height was observed at 4-week intervals between the cuts (Kenya Report, 1970). In the same trial herbage yields of *P. coloratum* cv. Solai were of the same order as those of *Setaria anceps*, but much lower yields were also reported; cv. Kabulabula being a more vigorous plant can produce higher yields. Very high yields of herbage were obtained in Rhodesia (Rodel & Boultwood, 1971) from cv. Bushman Mine, which however, may perhaps belong not to the main variety but to var. *makarikariense* It yielded 17.89 t DM/ha/year in a 4-year trial, was the highest yielding out of 30 grasses in the trial and showed very good drought resistance: the grass recovered to almost 100 per cent in the year that followed a severe drought. Two other varieties of *P. coloratum* grown in the same trial yielded considerably less – 7–10 t DM/ha. In another trial in Rhodesia (Mills *et al.*, 1973) with a few grasses, cv. Bushman Mine almost doubled the yields in the first year when it received 675 kg N/ha compared with plots fertilized with 225 kg N but the high rates of N produced little after effect. Irrigation increased the yields and especially the ability of grass to utilize high amounts of fertilizer N.

In cv. Solai, CP content in plants cut at 8-week intervals was 13.8 per cent and it was even higher in herbage cut every 2 weeks, but much lower CP contents were also reported. In a trial in Rhodesia already mentioned, the average content of CP in Bushman Mine herbage was 13.0 per cent. The content of P in cv. Solai was 0.18 per cent, an amount sufficient for animal requirements. Cv. Solai is not an aggressive grass and forms well-balanced and persistent mixtures with pasture legumes; *Desmodium uncinatum* and *Glycine wightii* did particularly well in Kenya and the percentage of the two legumes in the mixtures ranged from 30 to 60 per cent and those of *D. sandwicense* and *Trifolium repens* from 10 to 50 per cent. The grass varies in palatability and, depending on the variety and the stage of growth, is satisfactorily to well eaten by livestock. However, in trials in Rhodesia (Rodel, 1972) ewes grazed on *P. coloratum* produced two goitrous and one deformed lambs out of eight born and this is ascribed to the effect of HCN content in the herbage, although it was not particularly high -51 p.p.m.

Seed production is normally good in comparison with the majority of other tropical grasses. In trials by Boonman (1971b) in Kenya, cv. Solai produced numerous flowering heads, over $500/m^2$, and yielded 30–78 kg PGS/ha/harvest, the average being 52 kg, or twice this amount per year. The best results were obtained when seed was harvested 7 weeks after the initial panicle emergence, or when about 50 per cent of spikelets have shed. Hearn & Holt (1969) found that the percentage of seed set in the *P. coloratum* type grown in Texas, USA, is an unheritable feature and so is the number of fertile florets/panicle but seed retention in ripening panicles was not inheritable. Seed, as harvested, is in the form of spikelets; the soft glumes are easily removed during seed processing in seed cleaning installations and seed as sold is in the form of smooth and glossy fertile florets. The weight of 1,000 spikelets, each containing a caryopsis, is about 1.2 g (800,000–850,000 seeds/kg) and of processed seeds, i.e. fertile florets, 1.0–1.1 g (900,000–1,000,000 seeds/kg). Seed of var. *coloratum* require post-harvest maturation; the length of the maturation period seems to be of the order of a few months.

Var. **makarikariense** or **makarikari grass** (Fig. 24) has been known for a considerable time as a good pasture grass with outstanding potentials for cultivation. Originating in South Africa, it was introduced to several tropical and subtropical countries, including Australia, where it was first brought in 1936 or perhaps in 1920 (Bryant, 1967a) and where it has been extensively studied; numerous trials were also carried out in Rhodesia and South Africa. This variety is even more suited for waterlogged black clays than var. *coloratum* and can withstand prolonged flooding; Anderson (1970) observed 100 per cent survival of young plants after 20 days of flooding, whereas the other grass species under trial either completely died out or survived up to 55 per cent at most. Var. *makarikariense* is also highly resistant to frost; R. M. Jones (1969) observed 7–85 per cent survival after a winter with the lowest temperature of $-10°C$, the survival depending on plant size and the cultivar. It can, therefore, be grown in those subtropical areas where winter frosts are common (Lloyd & Scateni, 1968).

Cultivation. The establishment is less easy than of var. *coloratum*: the seedlings are small and weak and grow slowly, and relatively poor stands are formed in the first year of growth when the grass can be easily

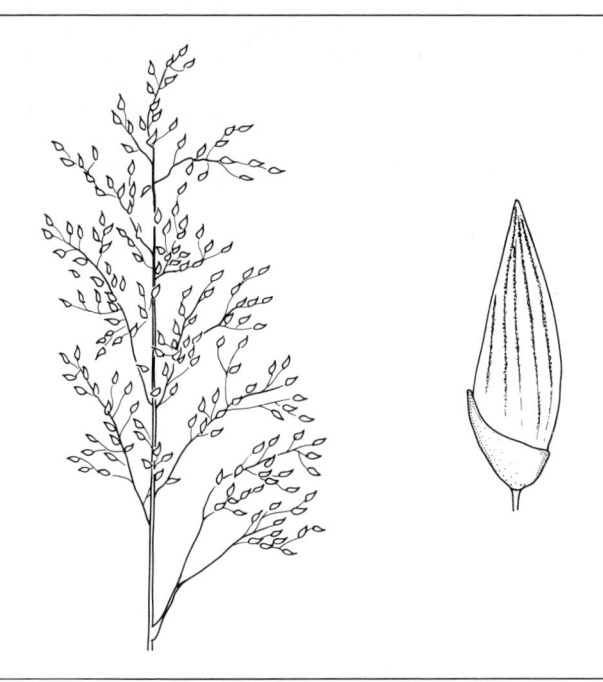

Fig. 24 *Panicum coloratum* var. *makarikariense*.

invaded by weeds; the application of herbicides to weedy stands have been recommended but some of the herbicides (e.g. Tordon or 2,4-D amine) may produce an adverse effect on makarikari seedlings. Seed production is not so satisfactory as in var. *coloratum* and seed is often in short supply and expensive. Sowing rates of 5–7 kg clean seed/ha have been suggested for close swards and 1–2 kg for sowing in 60 or 90 cm rows but Lloyd & Scateni (1968) recommend lower rates, 1–2 kg seed of 50 per cent germination, or even smaller amounts when drilled in rows, especially if mixed with gypsum. Seed should be sown into dry soil, to a depth of about 2 cm, a depth at which seed can contact firm soil. On light soil rolling, broadcasting and chain-harrowing is recommended. Tillering in young plants of var. *makarikariense* begins slowly but lasts for a considerable period of time, longer than in var. *coloratum*, and tillers may develop even during the flowering period. Of the grass/legume mixtures a good success has been obtained, mainly in Australia, from mixtures with lucerne and with *Macroptilium atropurpureum*.

Herbage yields and quality. Yields of herbage can be considerable and in Australia up to 23 t DM/ha/year have been obtained from some selected lines but they are usually of the order of about 12 t DM/ha as it was e.g. observed for cv. Bambatsi. Responses to fertilizers vary and linear yield responses to up to 450 kg applied N/ha were observed in some cultivars but not in the others. Herbage quality is usually high, CP content averaging 10.3–10.7 per cent (Bryant, 1967) and varying from 4.7 to 18.1 per cent. Butterworth (1967), quoting S. J. Myburgh of South Africa, gives CP content from 7.1 and 7.6 per cent and CF content of 34.1–35.2 per cent in samples used for digestibility trials with sheep. In these trials CP digestibility ranged from 45.1 to 50.6 per cent, DCP content 3.2–3.9 per cent and the content of TDN 56.2–56.3 per cent. However, in a herbage sample from Australia, CP and DCP contents were very low, 3.9 and 0.8 per cent, respectively. The herbage is very palatable and is readily eaten by both cattle and sheep.

Seed production. This is relatively poor in var. *makarikariense* although yields of 45 kg clean seed/ha have been obtained from cv. Bambatsi and well-fertilized stands yielded up to 145 kg seed/ha. Seed shattering, which reduces the yields, is considerable, especially in the more moist areas. When harvesting, Lloyd & Scateni (1968) recommend cutting flowering heads when one quarter to one third of seed has shed, and to spread the cut panicle on the floor in a 15–20 cm thick layer for 14–17 days 'to sweat' so that immature seed can ripen and then clean the seed. For obtaining good germination seed should be stored for up to 3 years but scarification, i.e. loosening or removal of hard floral scales, encourages germination in seeds stored for a relatively short period of time, and Bryant (1967a) reports that scarification of seeds stored for 12 months increased the germination from 20 to 75 per cent. Lloyd & Scateni (1968) remark, however, that untreated seeds can germinate much better in soil than under laboratory conditions.

Cultivars. Plants of the makarikari variety can differ in habit and other morphological and agronomic features and two main types have already been recognized at the early stages of introduction into cultivation: Zhilo, which has a spreading habit and Bambatsi which is erect or suberect.
1. Cv. **Zhilo**, known under this name in South Africa, has spreading and creeping stems terminating in panicles. When grown widely spaced single plants can cover an area of two metres in diameter.
2. Cv. **Bambatsi**, first grown in Rhodesia, has an erect or suberect habit with the stems reaching up to 150–180 cm in height. It is less adapted to waterlogging but is a relatively good seeder. In Australia swards sown from seed usually contain about 5 per cent of spreading or creeping plants of the Zhilo type.

Cv. **Pollock** has been developed in Australia from selected plants of the Zhilo type and has essentially similar growth characteristics. It is the best performer under severe waterlogging but a relatively poor seeder (Barnard, 1972).

Cv. **Burnett** contains plants of both types – upright and spreading in about equal porportions. This cultivar may be a mixture of the bambatsi and pollock plants although a possibility of hybridization with subsequent segregation into the two types cannot be excluded.

In Rhodesia, local cv. **Bushman mine** has been grown with an outstanding success. This cultivar may however belong to var. *coloratum*.

Panicum elephantipes Nees. Gamalote; Canutillo (Argentina); Paja de agua (Venezuela) $2n = 30$

Robust glabrous perennial floating in water with the help of slightly spongy lower parts of stems, which can be up to 2 cm in diameter, and spongy lower leaf-sheaths. Stems 1–1.5 m long, ascending, with numerous roots arising from the lower nodes. Leaves up to 35–45 cm long and 1.2 cm wide. Panicle loose, large, 30–50 cm long. Spikelets oblong-lanceolate, 4.5–5 cm long and 0.7–1.2 mm wide. Lower glume very short, 1 mm long; upper glume as long as the spikelet. Lower floret sterile. Fertile spikelet shiny, 3.5–4 mm long. Grain 2.3–2.5 mm long.

Widely spread in tropical and subtropical America from the Caribbean area in the north to La Plata in Argentina, mostly in the lower reaches of large rivers, sometimes very numerous and forming extensive 'floating meadows'. An excellent fodder (Burkart, 1969). In Venezuela, *P. elephantipes* is common in seasonally flooded areas where flood water comes from rivers; in such areas the grass floats only during the floods (Ramia, 1967).

Panicum maximum Jacq. Guinea grass; Capim coloniao, Sempreverde (Brazil); Hierba de India (Venezuela); Privilegio; Zacaton (Mexico); Fataque (Mauritius); Herbe de Guinée; Pasto Guinea. (Fig. 25)

Tufted perennial 0.5 to 4.5 m high. Stems mostly erect but can be ascending, glabrous or hairy, stout to slender, with 3–15 nodes. Leaves linear to linear-lanceolate, 15–100 cm long and up to 35 mm wide. Panicles loose, 15–60 cm long and up to 25 cm wide, much branched, the lowermost branches in a **distinct whorl**. Spikelets 3–4 mm long, green or purplish, glabrous or sometimes hairy. The lower glume one quarter to one third the length of the spikelet, broad, clasping the spikelet base; upper glume as long as the spikelet. Lower floret male or empty, depending on the variety; upper floret up to 3 mm long and has finely transversely wrinkled lemma and palea. The elliptic grain is about 2 mm long.

The species is native of tropical Africa extending to the subtropics of South Africa. It occurs at forest edges, in bush, as a pioneer grass covering the land cleared from forest and in grassland with scattered trees where it tends to grow in light shade, under trees. From sea level to 1,800 m alt, rarely slightly higher.

Panicum maximum is a very variable species; innumerable distinct types occur naturally in Africa, especially in East Africa, and about a

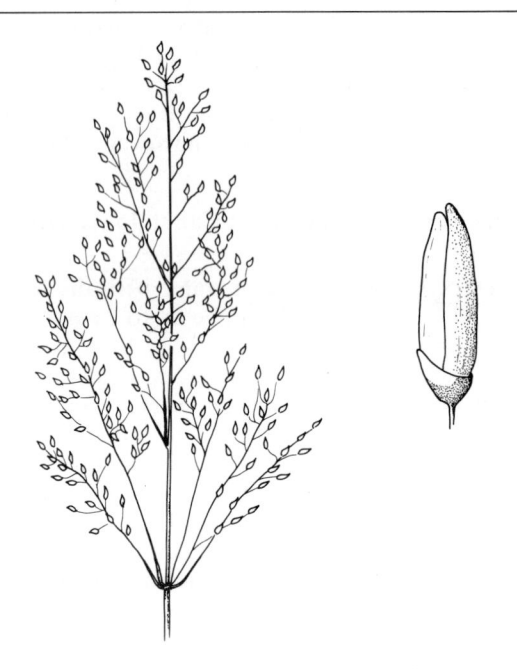

Fig. 25 *Panicum maximum.*

dozen botanical varieties have been named. The great variability observed in this species requires a thorough intraspecific revision which should include a study of living material and the use of botanical varietal names in pastoral work seems to be premature. Nevertheless, two distinct, untypical forms should be mentioned: var. *trichoglume* is widely grown under the common names slender guinea, green panic or Petri and although the name var. *trichoglume* is taxonomically invalid it is perhaps advisable to retain it for the time being; the name var. *pubiglume* may not necessarily be applied to green panic only as there are other distinctly different cultivars or wild plants with hairy spikelets. Another easily recognizable type found in Kenya and known as cv. Embu differs from all other varieties of *P. maximum* in having long creeping stem bases, many-noded stems and almost lanceolate leaves although its floral parts are typical for the species. This type may be eventually transferred to an allied species.

Environment. *Panicum maximum* grows in warm, frost-free tropical climate and R. M. Jones (1969) observed that nil to 6 per cent of plants survived a winter with the minimum temperature of $-10°C$. It tolerates shade and can be grown in tree plantations, and is reported to grow well under mango trees in India (Narayanan & Dabadghao, 1972). Some varieties can grow in semi-arid tropics under an annual rainfall of 650–800 mm but most types perform better in more humid areas with over 1,000 mm of rain. *Panicum maximum* occupies mainly well-drained, lightly-textured soils – preferably sandy loams or loams – and does not tolerate heavy clays or prolonged waterlogging or flooding; short-term flooding is tolerated only if the top parts of grass are exposed to the air. In trials in Australia (Anderson, 1970) green panic (var. *trichoglume*) survived 5–10 day flooding to 90–100 per cent but none survived 20 days of flooding. Cutting the grass before flooding reduced the survival.

Introduction. *Panicum maximum* is cultivated to a moderate extent in Africa and has been introduced to other tropical and subtropical countries; India, Sri Lanka, Australia, Malaya, New Guinea, Sarawak, the Philippines, Hawaii and some other areas where it is grown on farms but mostly on an experimental scale. In India, however, where it was first introduced in 1793, it became one of the important fodder grasses. The main areas of *P. maximum* cultivation are, however, South and Central America, the West Indies and, to a lesser extent, the south-east of the USA. The grass was introduced to America a long time ago, probably in the eighteenth century and certainly early in the nineteenth century (Parsons, 1972), at first accidentally, in slave ships, from West Africa. It soon spread throughout the Caribbean Islands to the northern parts of South America, Central America and the USA and, independently, from south-east coastal Brazil inland. In America this grass was soon recognized as an excellent fodder and pasture species; it is now grown on a large scale and has also spread to natural and semi-natural vegetation where it has perfectly naturalized and now constitutes a considerable portion of certain types of natural grasslands, especially those with trees or bush, and often dominates the ground where natural woodland had been cleared. As a cultivated grass *P. maximum* is much valued for its high productivity, palatable herbage and good persistence. Extensive experimental work has been done, especially in Puerto Rico, where high herbage and animal production from *P. maximum* has been achieved.

Establishment. *Panicum maximum* can be established either by seed or vegetatively, by tuft splits. Vegetative propagation can be applied in practice only to the large types, single plants of which occupy considerable space. The splits can be spaced about 2 m by 0.5–1 m and 5,000–10,000 splits are required to plant one hectare. When planted in wet weather or on irrigated land, the splits survive reasonably well and the grass can be first cut or grazed 3–4 months after planting. Seed sowing can also give good results but seed production is difficult and seed is in short supply; nevertheless, small types, such as slender guinea, cv. Sabi and cv. Makueni, are better seeders than the majority of large types and are actually established from seed.

The seed bed is prepared in the usual way although spreading uncleaned seed on the soil surface of arable land has been reported from South America as a frequent practice. In Hawaii (Motooka et al., 1967), sowing from air (in mixture with other grasses) at 2.4 kg/ha into burnt bushland without soil tillage gave satisfactory germination and good early growth. Seed is usually sown at a rate of 4 to 10 kg, depending on seed quality which is often low although in the earlier years sowing rates up to 110 kg/ha were sometimes recommended (Motta, 1953). Seed with 100 per cent purity and germination can be sown at 1–2 kg/ha. Germination improves with storage and fresh seed should not be sown. Seed sowing to a depth of 1–2 cm can be recommended; seeds of *P. maximum* can germinate at a greater depth than those of other tropical grasses with small seeds and in trials in which the soil was periodically dried and moistened to a depth of 2–3 cm (Bogdan, 1964a) the best emergence was observed from 2.5–3 cm. Mulching with straw which retains moisture near the ground surface can improve the establishment. Germination and emergence are slow and uneven and the seedlings need some protection from weeds. It is recommended in Rhodesia to let weed seedlings emerge, to destroy them and then plant *P. maximum* seed. Herbicides can also be used and Bailey (1967) obtained good results with a presowing application of a series of herbicides including different forms of 2,4-D. On weedy land, an early cut, or grazing if the weeds are palatable, can be recommended. Of the fertilizers, P is usually applied at sowing, the rates depending on soil fertility, and N can be applied later.

Management, fertilizing. Panicum maximum herbage is grazed or cut for soilage or hay but ensiling is seldom practised. The utilization depends on the type of agriculture of the area and in India the grass is almost invariably cut and fed green. Cutting or grazing to 15–20 cm from the ground, or even higher, has usually no advantage over low cutting to about 5 cm but Whyte et al. (1959) recommend cutting at 15–20 cm. The general trend is that with frequent cutting or grazing the increase in cutting level increases the yields, whereas the reverse was observed for infrequent cutting. Rotational grazing with 3- to 9-week intervals between the grazings during the growing season often increases the herbage and animal yields compared with continuous grazing and is usually recommended, but there are also reports that continuous grazing can yield as much, if not more, grass as rotational grazing; it can, however, have a detrimental effect on grass which becomes evident only after a few years of utilization. Picard & Fillonneau (1972) noticed that cutting in a period between an advanced stage of stem elongation and the beginning of flowering results in poor meristematic activity immediately after cutting which reduces the regrowth, and suggested that cutting at this particular period should be avoided.

Great responses to N fertilizer expressed in DM and CP yield increases and also in the increases of animal production were obtained in a number of countries, the responses being usually the highest at moderate rates of N, 100–250 kg/ha, decreasing gradually at further application to a maximum of about 600 kg N/ha. An increase of 38 kg DM production per kg of applied N was obtained in Queensland by Grof & Harding (1970) at 140 kg N/ha, but reduced to 27.6 kg at 280 kg N/ha and to 18.4 kg at 420 kg/ha. *Panicum maximum* usually responds to fertilizers, and especially to fertilizer N, better than a number of other tropical grasses, and in Uganda (Stobbs, 1969d) the application of N increased its proportion in a sown mixture with *Hyparrhenia rufa* to 25–30 per cent compared with an almost complete dominance of *H. rufa* in paddocks receiving no N. Increases in *P. maximum* proportion in natural grasslands under the effect of N were also observed. N is often applied at 50 kg/ha after each cut or grazing or at 100 kg after every second cut but single application can occasionally be as effective as split application. Responses to P depend to a great extent on the content of available P in the soil and very considerable increases in DM yields, sometimes even greater than from fertilizer N, were obtained on P-deficient, e.g. granitic, soils and in a trial on relatively poor soils in Uganda the application of 34 and 68 kg P/ha increased the total 3-year DM yields from 26.6 t to 33.5 and 35.7 t/ha, respectively (Wendt, 1970). In the same trial S applied at 22 kg/ha increased DM yields from 28.3 t to 33.4 t/ha, and responses to S were also observed in Australia and Kenya on soils deficient in this nutrient. Responses to K were even more erratic than to P although in some trials considerable gains in DM yields were obtained on soils well supplied with other mineral nutrients. The effect of minor nutrients was little studied but responses to Mo are known. In India f.y.m. is often used and a basic application of 6 to 25 t/ha followed by further dressings of 6 to 10 t is recommended and can be combined with the application of mineral fertilizer N (Narayanan & Dabadghao, 1972).

Supplementary sprinkler irrigation during the dry season is sometimes applied and in Brazil (Ladeira *et al.*, 1966) it increased herbage yields during the five dry and cool months from 6.7 to 8.0 t/ha, unfertilized, and from 9.1 to 10.9 t when fertilized with moderate amounts of NPK. In irrigated trials the effect of irrigation itself is however seldom reported. Sewage water irrigation is practised in India and high yields up to over 100 t fresh material/ha were obtained.

Association with legumes. Mixtures with *Centrosema pubescens* and with *Stylosanthes guianensis* were successfully established and grown in Australia and elsewhere. Other legumes grown with *P. maximum* with varying, but usually reasonable or good, success are *Glycine wightii*, *Macroptilium atropurpureum*, *Pueraria phaseoloides*, *Desmodium uncinatum* and *D. intortum*, and occasionally also *Stizolobium deeringianum*, *Calopogonium orthocarpum*, *Indigofera spicata* and *Leucaena leucocephala*. In trials in Queensland (Grof & Harding, 1970)

Stylosanthes guianensis formed the most balanced mixture compared with the other four leguminous species under trial, the legume constituting 47 per cent of the herbage, the total yield of the mixture being 5.77 t DM/ha and 66 kg of N/ha to which the legume contributed 46 kg and the grass 20 kg. Mixtures with *Macroptilium atropurpureum* and *Centrosema pubescens* yielded 4.43 and 3.98 t DM/ha, respectively, in which the legumes constituted 25 and 43 per cent. In Kenya (Kenya Report, 1970) mixtures with *Desmodium uncinatum* and *D. intortum* yielded, in 1969–70, 7.6 t and 9.1 t DM/ha, respectively, compared with 4.1 t yielded by *P. maximum* grown alone. There were other trials in which an admixture of legumes increased DM yields up to 100 per cent, but usually the increases are much smaller, sometimes negligible, and in some rare cases the yields can be even lower than those of pure grass. Crude protein yields increase almost invariably by the presence of legumes in the sward; the increases can be as high as two- to four-fold. Animal production from *P. maximum*/legume mixtures is as a rule much higher than from the grass alone.

Diseases. *Panicum maximum* is relatively free from leaf diseases but cv. Gamalote is almost invariably affected by leafspot caused, at the later stages of growth, by *Cercospora fusimaculosus*. The spikelets are often subjected to smut or *Fusarium* diseases and some varieties are much more susceptible to spikelet diseases than the others. A bunt disease, apparently similar to that affecting *Setaria anceps* and probably caused by the same pathogene (*Tilletia echoinosperma*), has been observed in certain areas of Kenya.

Herbage yields. Numerous and controversial data on *P. maximum* productivity can be found in the literature. Crowder *et al.* (1970) state that in Colombia well fertilized and irrigated *P. maximum* can produce 40–50 t DM/ha and in India a yield of 226 t fresh herbage/ha/year in 12 cuts was recorded for the sewage-irrigated grass (Narayanan & Dabadghao, 1972). In Puerto Rico 46.72 t DM/ha were recorded for a crop given about 900 kg N/ha and in other trials over 35 t/ha (Little *et al.*, 1959; Vicente-Chandler *et al.*, 1959). Fairly high yields were also obtained in Thailand, 20 t DM/ha/year in the first 2 years of growth when the grass was well fertilized with NPK and irrigated during the dry season (Holm, 1972). Lower yields were obtained by Borget (1966) in French Guiana, 14.4 t DM/ha, but more realistic yields range mostly between 4 and 12 t DM/ha or between 15 and 50 t fresh herbage and can sometimes be still lower. The yields depend on the cultivar, soil fertility, the fertilizers applied, the rainfall and the management.

Chemical composition, nutritive value. *Panicum maximum* is reputed to be very palatable to all kinds of stock, at least at reasonably early stages of growth, a few weeks after the last cut or grazing. At the later stages of growth the leaves remain highly palatable, whereas the thick stems of robust varieties are left uneaten, but the herbage of finer varieties, containing at the more mature stages numerous thin stems, is eaten less willingly. According to a number of analyses quoted by Butterworth (1967), for the grass cut at different stages of growth, CP content ranges from 4 to 14 per cent, but higher contents of CP were also recorded. In Venezuela var. *trichoglume* contained over 20 per cent CP 28 days after the last cut which was reduced to 8.8 per cent in 56-day-old herbage (Combellas & Gonzáles, 1973). Similarly in Brazil (Gomide *et al.*, 1969a) 4-week-old grass contained 22.6 per cent CP and 12.4 and 8.5 per cent CP 12 and 32 weeks after the last cut, respectively. High contents of CP were also reported from Puerto Rico, Kenya and occasionally also from other areas, for the young growth heavily fertilized with N. The effect of fertilizers is however not always consistent and although considerable increases of CP content from applied N have been obtained, Grof & Harding (1970) report from Australia that in their trials the content of CP in *P. maximum* herbage remained almost unchanged when N rates were increased from 140 to 420 kg/ha. The content of CF ranges from 28 to 36 per cent and depends mainly on the frequency of cutting, i.e. the age of herbage. P content usually exceeds 0.15 per cent and is normally adequate for cattle requirements although there are reports from Ghana on 0.11–0.15 per cent P in relatively young herbage (Sen & Mabey, 1966) and similar data were obtained by Gomide *et al.* (1969a). The content of NFE usually varies from 40 to 50 per cent and seems to be slightly lower than in a number of other grasses, and EE content varies from 0.6 to 2.8 per cent. A relatively high content of Na was also recorded. Data obtained by several authors and quoted by Butterworth (1967) give DM digestibility as varying mostly from 40 to 62 per cent and the content of TDN from 38 to 61 per cent, accordingly, although in one case very old herbage had only 31 per cent TDN. Crude protein digestibility ranges from 15 to 73 per cent depending on plant age and on CP content, and the content of DCP from 0.6 per cent in herbage containing about 4 per cent CP to 71 per cent when the content of CP was over 15 per cent. According to Butterworth CF digestibility ranged from 40 to 70 per cent, NFE from 26 to 67 per cent and these figures show a tremendous variation of nutritive value of *P. maximum* herbage, the variation depending mainly on the stage of growth at which the herbage was harvested.

Animal production. *Panicum maximum* can be utilized intensively and high stocking rates of 2 steers/ha, and sometimes even higher, were reported from Puerto Rico, Venezuela and Brazil. However, in Australia and in some other areas, stocking rates in the less humid tropics and subtropics are usually of the order of 2–4 ha per head of cattle. Grass/legume mixtures allow higher stocking rates than pure grass and in Hawaii six cows were supported on one hectare of a *P.*

maximum/*Leucaena leucocephala* mixture (Plucknett, 1970). Liveweight gains are usually of the order of 200–400 kg/ha, depending on herbage yields and on the content of DCP, and in Mexico the liveweight gains of Zebu steers were reported to be 190 kg/ha during the growing season lasting from July to January. However, much higher gains were sometimes obtained as e.g. 722 kg/ha in humid tropics of Queensland (Grof & Harding, 1970). Liveweight gains per animal per day also vary and gains of 0.5 kg in the dry season and 0.8 kg in the wet season are considered normal gains for Columbia (Crowder *et al.*, 1970). Milk production was reported to vary from 1,000 to 8,000 kg/ha/year and 7,790 kg, or 15 kg per day per cow, were obtained from a well-fertilized pasture in Puerto Rico in which *P. maximum* contained 18.2 per cent CP (Caro Costas & Vicente-Chandler, 1969). In Hawaii, cows grazing on a *P. maximum*/*Leucaena leucocephala* mixture gave 9,770 kg milk/ha or 12.5 kg per cow per day, and in Queensland, Australia, cows grazed on a mixture of cv. Petri (var. *trichoglume*), receiving no fertilizers, produced 3.1 t milk per cow in one lactation period (Tucker *et al.*, 1972).

Flowering, reproduction. Javier (1970) observed in the Philippines that 25 per cent of spikelets flowered 6 days after the panicle emergence and that 50 per cent of spikelets shed 15 days later. Alarcón *et al.* (1969), in Colombia, found, however, that the spikelets mature in about 32 days after anthesis. In the Philippines the flowering occurred early in the evening, with its peak at 6 p.m. to 10 p.m., depending on the variety, and lasted for some 80 minutes. *Panicum maximum* is an aposporous and pseudogamous facultative apomict with about 2–3 per cent of sexual reproduction which can be effected by cross- or self-pollination and this rate is normally maintained in the progenies of sexual plants (Combes & Pernès, 1970). Javier (1970) accepts the predominately apomictic reproduction of *P. maximum* but states that sexuality, according to cytology, should range from 22 to 53 per cent; Combes & Pernès note, however, that the 2–3 per cent of sexual reproduction actually observed is very far from that estimated by the frequency of reduced embryo sacs. Under pseudogamie, pollination is necessary for the formation of the endosperm which is needed for the normal growth of the embryo. Apomixis and pseudogamie in *P. maximum* were discovered by Warmke (1954), who observed that the spikelets in which the stigmas were cut off produced no seed. If, however, the stigmas were cut off 3 hours after the anthesis or later, seed was formed normally which indicates that pollen tubes reach the ovule in less than 3 hours. The majority of plants and populations are tetraploids, normally, allotetraploids, with $2n = 32$ although hexaploids ($2n = 48$) occur fairly frequently and they can produce more numerous plants, about 14 per cent, by sexual reproduction than the tetraploids (Combes & Pernès). Other ploidy levels were also reported: triploids ($2n = 24$), pentaploids ($2n = 40$), octoploids ($2n = 64$), nonaploids ($2n = 72$) and also plants with irregular chromosome numbers ($2n = 31, 36, 37, 38$). There are also diploid forms ($2n = 16$) which differ from the higher ploidy types in being entirely sexual and Combes & Pernès found a concentration of purely sexual diploid populations at Korogwe, in the Tanga area of Tanzania, and another less clearly defined diploid centre in an area near Dar es Salaam. These sexually reproducing diploids serve as a centre of variability. However, the sexually obtained progenies of tetraploids and hexaploids also contribute to a considerable extent to the variability of the species and various forms that arise are then fixed by apomixis.

Seed production and germination. Seed is in the form of a spikelet with a single caryopsis and varies in size depending on the variety. In large-spikelet varieties, 1,000 such spikelets weigh 1.40 g (700,000 per kg) and 1,000 scarified seeds, i.e. those with removed soft glumes, weigh 0.85 g (1,200,000 per kg). Unprocessed seeds of small-spikelet cv. Sabi weigh 0.75 g per 1,000 or 1,350,000 per kg (Bogdan, 1966a). Whyte *et al.* (1959) give higher numbers of seeds per kg, 1,760,000–3,100,000, possibly because large numbers of empty spikelets could also be present.

Flowering in *P. maximum* lasts for a considerable period of time (except perhaps when it begins only at the end of the season), and this affects seed harvesting and yields. Seed damage by birds and poor seed formation are the two other important factors adversely affecting seed production. In a number of varieties a great majority of spikelets can be empty but in some types the percentage of spikelets containing caryopses can be reasonably high as e.g. in cv. Makueni of Kenya origin or in cv. Sabi from Rhodesia, which are relatively good seed producers. Little information is available about seed yields and old Kenya reports about Slender guinea producing some 300 lb seed/ac (about 335 kg/ha) refer perhaps to seed of poor quality in which the percentage of spikelets containing caryopses may be as low as 1 to 5 per cent. For cv. Makueni, Boonman & Van Wijk (1973) give realistic 25 kg of pure germinating seed/ha. Javier (1970) in the Philippines reports 48–51 kg/ha of seed with 5–7 per cent of good full seeds for three large cultivars and 99–156 kg for two selections from cv. Sabi. Harvesting is done by hand, stripping the panicles of large varieties or cutting the herbage near the ground in smaller varieties, stooking it, and thrashing later by beating with sticks. To prevent excessive seed shedding and bird damage, the panicles can be tied together before harvesting (Fig. 26) or the stooks are piled together and covered with another, smaller stook placed upside-down. Seed requires post-harvest maturation which, according to various authors, may take 6 to 18 months. Seed longevity depends on storage conditions, and in sealed containers and at a temperature of 10°C germinability can last for a long time but is lost much earlier when stored in bags at higher or alternating temperatures. In Rhodesia (Smith, 1970), freshly harvested seed germinated to 5 per cent, the germination increased to 24 per cent by storage for about a year and to 40 per cent when seed was treated

Fig. 26 *Panicum maximum*. Tied up plant tops prior to ripening to prevent bird damage and seed shedding.

for 10 min. by concentrated sulphuric acid. The soft glumes of the spikelet may inhibit the germination but hard glumes (lemma and palea of fertile floret) should not be removed, and in Smith's trials the naked caryopses did not germinate. Germination can also be increased by soaking in KNO_2 solutions or by seed leaching in running water. Alternate wetting and drying at the early stages of germination can increase seed germinability; and so can also alternating temperatures, e.g. from 10 to 40°C, during the germination.

Variability and cultivars. *Panicum maximum* is a very variable species, especially in East Africa, and numerous natural types exist, some of which have been described as botanical varieties. The plants differ in habit, height, stem thickness, the degree of branching, etc, and two main groups differing in their agronomical characteristics can be distinguished:

1. Large or medium types suitable for both soilage and grazing; they can be economically established from tuft splits at 5,000–10,000 splits per ha.
2. Small, low-growing types suitable mainly for grazing, which should be grown closer and can be established on a farm scale only or mainly from seed; some of them are, fortunately, reasonably good seeders.

The cultivars that are grown in various countries represent apomictic or vegetative clones with little or no variability within the cultivar. When the cultivars are transferred to different environments the vegetative clones can, within certain narrow limits, be more variable than the progenies of apomictic seeds (Pernès et al., 1970). To the large (1) or intermediate types belong most of the known cultivars: **Coloniao (Colonial)** grown in Brazil and introduced to other countries, **Boringuen** and **Broadleaf** (both grown in Puerto Rico), **Guinea** (Venezuela), **Hamil** which outyielded a number of other cultivars in trials by Grof & Harding (1970) in Australia, **Gatton** (Australia), **Semper verde** (Brazil) the name also used in Brazil to denote *P. maximum* as a species, **Sigor** and **Nchisi** (Kenya), **King ranch** (USA) and a number of other, less-known cultivars mentioned by Motta (1953) and various authors. Of the small (2) type cv. **Green panic** or **Slender guinea** (var. *trichoglume*), named cv. **Petrie** in Australia, is widely distributed in cultivation, mainly in the Old World; it forms numerous fine stems with the leaves situated higher on the stem than in most varieties. It produces large amounts of seed but the percentage of seeds containing caryopses is usually small. Cv. **Sabi** is of Rhodesian origin; it has bluish leaves and grows to a height of about 1.5 m; it is of medium drought resistance and a good seeder. Cv. **Makueni**, which grows up to 1 m in height, originates from dry south-eastern parts of Kenya, is drought resistant and its seed production is reasonably good. Cv. **Embu** (Kenya), as has already been mentioned, is a distinct creeping type which forms continuous swards. It is not a particularly good producer and in Australia it yielded much less than the ordinary types in the wet and warm season when the main bulk of herbage is produced, but gave 5.14 t DM/ha during the dry and cool season when other cultivars yielded 1.89 to 3.46 t (Grof & Harding, 1970).

Panicum repens L. Torpedo grass (USA)

Perennial up to 1 m high with long stout rhizomes which spread rapidly and can reach up to 7 m in length. Stems erect or ascending, manynoded. Leaves firm and rigid, 7–15 cm long and 3–6 mm wide. Panicle contracted or loose, 7–20 cm long, with the lower branch distant from the rest. Spikelets ovate-oblong, acute, glabrous, pale, 2–3 mm long. Lower glume broad, clasping the base of the spikelet. Lower floret male. Upper floret bisexual, white, smooth and glossy, 2 mm long.

Widely distributed in the tropics and subtropics throughout the world, on flat, wet or swampy ground, in coastal marshes and on sandy alluvial soil, often as a pioneer grass rapidly spreading by means of rhizomes. It can withstand flooding but is also drought-resistant to a certain extent as its rhizomes can survive long spells of drought. It is reported from India (Narayanan & Dabadghao, 1972) to be palatable to stock, to form good pastures under irrigation and to tolerate close grazing, and Rhind (1945) considers it to be a good fodder. In some other countries the opinion is less favourable; the grass is reported to be not too palatable because of its tough leaves and Stewart and Stewart (1970) report from East Africa that *P. repens* is mostly ignored by wild ungulates. CP content of the herbage was determined to be 5.8 per cent, CF 27.4 per cent, NFE 47.0 per cent, Ca 0.19 per cent and P 0.14 per cent (Sen & Ray, 1964), indicating a low nutritive value. Whyte *et al.* (1959) quote CP content of 7.48 per cent in 4-week-old herbage which gradually decreased to 3.31 per cent in mature plants and CF content varied from 36.6 to 42.7 per cent. A higher CP content, 14.5 per cent average for the first year of growth, indicating a high nutritive value at this stage, was obtained in Rhodesia from an outstanding cultivar **Victoria Falls**. When established vegetatively, from rhizomes, it yielded 6.2 t DM/ha in the first year of growth, 20.9 t in the second year, but the yields declined to 8.4 t in the third year and to 6.1 t in the fourth. This cultivar was susceptible to droughts and after a severe drought it recovered to only 13 per cent (Rodel & Boultwood, 1971). Because of its rhizomatous habit *P. repens* can be a serious weed of arable land, difficult to eradicate and is unsuitable for arable rotations.

Seed formation is poor and seed sheds easily. Chromosome numbers $2n = 36$, 40, 45 and 54 and these irregular numbers may suggest apomictic reproduction.

Panicum trichocladum K. Schum. Donkey grass; Ikoka (Tanzania)

$2n = 32$

Trailing perennial. Stems slender, many-noded and branched, ascending from the prostrate, often creeping bases rooting from nodes. Leaves almost horizontally spreading, narrow-lanceolate, rounded at the base, 4–15 cm long and 6–15 mm wide. Panicle loose, 5–15 cm long and 5–8 cm wide; stems under the panicle and ultimate panicle branches with numerous long spreading hairs. Spikelets oblong, 2.5–3 mm long; lower glume under one quarter the length of the spikelet, upper glume as long as the spikelet. Lower floret male, upper floret bisexual, fertile, with glossy lemma and palea.

Occurs in East, South and Central Africa in the areas with an annual rainfall exceeding 850–900 mm, in light forests, forest edges and stream banks. Palatable to cattle; in some areas of Kenya it is considered particularly suitable for calves and is grown to a limited extent by small farmers. The leaves, as picked up by the animals, can contain 17.5 per cent of CP and 0.29 per cent P in the wet season when it is nutritious enough to secure the growth of young cattle; in the dry season, with the increase of plant age, its nutritive value rapidly decreases and the leaves contain 8.5 per cent CP and 0.10 per cent P, the contents inadequate for the growth of young animals (Van Voorthuisen, 1971). When cultivated, *P. trichocladum* sometimes suffers from the attacks of a small-leaf virus.

There is a large number of other rambling species of *Panicum* confined mainly to light forests, forest edges and swamps. *Panicum hochstetteri* Steud. and *P. transvenulosum* Stapf are examples of such species in East Africa; they are however of less importance for grazing than *P. trichocladum*.

Panicum turgidum Forsk. Morkuba (Mauritania); Thomam (Arabia); Tumam (Egypt, Sudan); Du-ghasi (Somalia). $2n = 18$
Perennial up to 100–150 cm high with glabrous, erect, ascending or prostrate, variously branched stout stems. Leaves rigid, smooth, glaucous, up to 15–20 cm long and 7 mm wide but often much reduced in size. Panicle of different size and shape, compact to loose and can reach up to 20 cm in length, but mostly much smaller. Spikelets on relatively thick branches and pedicels, glabrous, gaping, 4 mm long. Glumes subequal, often as long as the spikelet. Lower floret male; upper floret bisexual 3 mm long; grain 2 mm long.

This often suffruticous grass occurs in northern tropical and subtropical Africa from Mauritania in the west to Somalia in the east and also in tropical and subtropical Saudi Arabia, and southern subtropical parts of Iran and Pakistan (Williams & Farias, 1972). *Panicum turgidum* grows under about 250 mm of annual rainfall and penetrates deep into Sahara and Arabian deserts but avoids particularly dry areas with an annual rainfall below 200–250 mm. It can survive severe drought, mostly in a dormant state and recover even when little rain has fallen. Seedlings can only seldom be found and it was suggested that they are far less drought resistant than the adult plants. When green the grass is well eaten by all stock, possibly because of the lack of softer grasses and is of importance in North African and south-west Asian deserts and semideserts; in a dry, dormant state it is however browsed only by camels and donkeys.

Paspalidium Stapf

Tufted or stoloniferous perennials. Panicle long, narrow, with short dense racemes appressed to panicle axis, at least in their lower parts. Racemes with the spikelets turned to the raceme axis by the upper glume and the convex side of fertile floret. Spikelets with two florets, the lower male and the upper bisexual, awnless. There are about 20 species distributed mainly in warm areas of the Old World. Good grazing grasses.

Paspalidium desertorum (A. Rich.) Stapf (*Panicum desertorum* A. Rich.)
Stoloniferous perennial forming small tufts of erect or ascending shoots mostly from rooting nodes of long stolons. Differs from *P. geminatum* in having firm, not spongy stems, narrower leaves and longer spikelets which are 2.5–3 mm long and often have purple tips and also by its ecology.

Occurs in north-eastern tropical and subtropical Africa including northern Kenya, Sudan, Ethiopia and Somalia and also in Arabia and India in semi-arid or arid areas, mostly on alluvial soil or on loams but avoids heavy clays or sandy soils. It is one of the perennial grasses which penetrate into the annual grass zone where only a few perennials can survive long dry seasons. It is a good grazing grass and an analysis of herbage grown in nursery plots in Kenya and harvested early in the season had a high content of CP, 13.8 per cent, and a moderate content of CF, 29.4 per cent. The content of NFE was 42.6 per cent and high contents of Ca (0.97 per cent) and a satisfactory content of P (0.18 per cent) were also recorded (Dougall & Bogdan, 1965). A good seeder, *P. desertorum* is considered in Kenya to be a promising grass for reseeding denuded pastoral land, receiving 400–600 mm of annual rainfall. There are 1.2 million seeds (spikelets containing one ripe caryopsis each) per kg (Bogdan & Pratt, 1967).

Paspalidium geminatum (Forsk.) Stapf (*Panicum geminatum* Forsk.)

$2n = 18$

Stoloniferous or rhizomatous perennial often floating in water. Stems firm or more usually spongy, erect or ascending from a long, rooting base, five- to many-noded. Lower leaf-sheaths of floating or submerged shoots are spongy. Leaves 10–25 cm long and 3–8 mm wide, flat or convolute towards the apex. Panicle slender, narrow, 10–30 cm long with short suberect or erect dense racemes which are 2–5 cm long. Spikelets 2–2.5 mm long, ovate, glabrous and pale. Lower glume truncate, 0.5 mm long, clasping the spikelet base. Upper glume almost as long as the spikelet. Lower floret male, as long as the spikelet and so is the upper, fertile spikelet; its lemma and palea crustaceous, finely transversely wrinkled.

Occurs in the tropics and subtropics of Africa and Asia and also in warm temperate areas of Africa, Asia and North America, in moist situations, mostly alongside river banks, often in shallow water and in seasonally flooded grasslands. A good grazing grass well liked by stock.

Paspalidium paludivagum (Hitchc. & Chase) Parodi (*Panicum paludivagum* Hitchc. & Chase) $2n = 36$

Similar to *P. geminatum* but differs in slightly shorter racemes, longer and more narrow spikelets, which are 2.5–3 mm long and have purplish tips, and by shorter glumes, the upper glume being about two thirds the length of the spikelet.

Occurs in southern parts of USA, Mexico, Central America and South America where it reaches Buenos Aires in the south. Forms colonies on river and stream banks and provides good natural fodder; tends to become a weed in rice fields (Burkart, 1969). In Argentina, Vonesch & Riverós (1967–8) determined the contents of CP in plants analysed at the vegetative to flowering stages to range from 10.6 to 21.3 per cent, EE 1.0–2.1 per cent, CF 21.3–32.1 per cent and NFE 36.9–51.4 per cent, showing good nutritive value of relatively young plants.

Paspalum L.

Predominantly perennials, tufted, stoloniferous or rhizomatous. Panicle of two to over 30, mostly dense, spreading or suberect racemes. Racemes with fine, narrow or flattened rhachis and the spikelets are turned to the rhachis by their convex side. Spikelets flat on one side and convex on the other, lanceolate to orbicular, with two florets: the lower male or sterile, the upper bisexual, awnless and with hard lemma and palea firmly clasping the grain. Lower glume small or, in most species, absent. A large genus of up to 250 species distributed in the tropics, subtropics and, to a lesser extent, in warm temperate areas throughout the world but mainly in America. Most of the species are good grazing grasses, the better known species are *P. dilatatum* and *P. notatum* introduced to a number of countries and often grown on a farm scale.

Paspalum almun A. Chase $2n = 24$

Tufted perennial 10–60 cm high with very short rhizomes. Stems fine, erect or geniculate and rooting at the base. Leaves flat, 4–20 cm long and 3–6 mm wide. Racemes terminal, paired but occasionally up to seven; 2–9 cm long. Spikelets 2.4–3.5 mm long and 1.3–2 mm wide; fertile floret 2.2–2.8 mm long and 1.2–1.5 mm wide.

Distributed from Texas, USA, in the north to northern Argentina in the south, in savannas with fertile soil. An excellent grazing grass for cattle.

Paspalum commersonii Lam. (*P. scrobiculatum* L. var. *commersonii* Stapf) Scrobic paspalum (Australia). Closely related to *P. scrobiculatum* L., an annual grown in India as a millet. $2n = 20$ or 40

Tufted, usually short-lived perennial. Stems soft and succulent, up to 100 cm, but mostly 40–60 cm high. Leaves glabrous or hairy, up to 30 cm long and 12 mm wide. Panicle consists of one to four racemes 4–9 cm long. Rhachis flat, about 2.5 mm wide. Spikelets 2 mm long, almost orbicular. Seeds (ripe fertile florets) light-brown, about 660,000 per kg. Indigenous throughout tropical Africa.

Paspalum commersonii has been tried in cultivation in Kenya, Rhodesia and Australia. In Kenya and Australia herbage yields were lower than those of the majority of other grasses. A grass of low competitive vigour easily overgrown by weeds. It requires fertile soil and is highly palatable and digestible, probably because of its soft stems, and retains these qualities at maturity better than other grasses, but its CP content is relatively low and in a trial by Bryan (1968) it ranged from 5.2 to 9.7 per cent (N × 6.25); Sen & Mabey (1966) gave however a higher content of CP: 12.8 per cent in 4-week-old herbage which declined later to 7–8 per cent and remained at this level for a considerable period of time; P content was very low. 0.06–0.09 per cent in both young and more advanced herbage. In a trial in subtropical Queensland (Bryan, 1968) a

P. commersonii/Desmodium intortum/Trifolium repens mixture yielded 1.08–2.20 t DM/ha; *P. commersonii* strongly dominated in the first year but was gradually reduced to 66 per cent of the sward and gave some room to *D. intortum* but not to *T. repens*. On the fifth year the mixture was overgrown by weeds. *Paspalum commersonii* is sexually reproduced, and is a predominantly cross-fertilized species with a high degree of fertility. In Australia cv. **Paltridge** was registered in 1966 (Barnard, 1972).

Paspalum conjugatum Swartz. Sour grass; Sour paspalum; Bitter grass; Grama de antena; Pasto amargo, Capim amaroso; and there are numerous other common names listed by Beetle (1974) $2n = 40$

Stoloniferous perennial with the stems 20–60 cm high. Leaves linear-lanceolate, 4–20 cm long and 5–10 mm wide. Spikes two, or occasionally three, widely diverging, often almost horizontal, very narrow and 5–12 cm long. Spikelets flat, ovate to almost orbicular, 1.4–1.8 mm long. Lower glume absent. Upper glume and the lower lemma similar, both with scattered long hairs on margins. Caryopsis about 1 mm long.

This description refers mainly to var. *conjugatum*; var. *parviflorum* Doell has smaller spikelets and glabrous leaves and var. *pubescens* Doell is a coarser and larger plant with longer, pubescent leaves, longer spikes and larger spikelets (from Beetle, 1974).

Originates from American tropics where it is common, mainly in eastern regions, but has now spread to the tropics and subtropics throughout the world. *Paspalum conjugatum* grows under humid climates, mostly on acid wet soil and tolerates shade. It can invade old sown pastures planted with low-growing grasses and is also common in fallows, representing, together with some other stoloniferous species, the second stage of succession. *Paspalum conjugatum* is usually of low productivity although Dirven (1971) reports a yield of 27 kg from a 5 m^2 plot which corresponds to 54 t/ha. Tried in Sarawak (Ng, 1972), it showed almost a linear response to applied N yielding 2.9 t DM/ha/year when given no N, 5.7 t at 112 kg N/ha and 14.0 t at 900 kg N with intermediate yields at intermediate rates of N. Crude protein content of herbage cut every 6 weeks ranged from 6.0 to 9.2 per cent, and was 16.1 per cent at 900 kg N/ha. Herbage cut every 14 weeks contained 5.3 to 6.2 per cent CP and CP content little depended on the rates of applied N. Yields of CP ranged from 174 to 1,541 kg/ha. Dirven (1962) determined the content of CP to be 12.2 per cent in the whole herbage (15.6 per cent in the leaves), but the digestibility of CP was low – 41 per cent. *Paspalum conjugatum* is a grass of low palatability and is grazed only when young and less willingly than other species of *Paspalum* or *Axonopus compressus* with which it often associates, and is hardly eaten at all at later stages of growth when it develops numerous wiry stems and Rhind (1945) reports that in Burma cattle eat the grass but horses dislike it. Mowing or burning are therefore required to maintain the grass in a palatable condition. There is an old report that eating large amounts of *P. conjugatum* herbage disturbs the digestion of cattle (Beetle, 1974). Whyte *et al.* (1959) also mention that 'seeds tend to stick in throats of livestock and choke the animals'.

Paspalum dilatatum Poir. (*P. pauciciliatum* (Parodi) Hester (*pro parte*). Dallis grass; Pasto Dalis; Pasto miel (Fig. 27).

A leafy, somewhat coarse and strong perennial forming broad spreading tufts. Stems 50–150 cm tall. Leaves glabrous, except a few long hairs at the base, up to 45 cm long and 3–13 mm wide. Ligule membranaceous, 2–3 mm long. Panicle with 3–11 spike-like distant racemes 3–10 cm long. Rhachis of the racemes flat, narrow, up to 2 mm wide, densely beset with paired spikelets one of which is sessile, the other on a short pedicel. Spikelets 2.5–4 mm long, broadly ovate, with only one (upper) glume which is of the shape of the spikelet but sharply pointed and with a fringe of long hairs on margins. Florets two: the lower reduced to an empty lemma similar to the glume, the upper bisexual, suborbicular, dorsally flattened, with almost white, shiny lemma and palea.

Fig. 27 *Paspalum dilatatum.*

Rosengurrt et al. (1970) distinguish two subspecies: ssp. *dilatatum*, the main and widely spread type with purple anthers, and ssp. *flavescens* Roseng. with yellow anthers and slightly larger spikelets than in the main form, and much more restricted in its distribution. They treat *P. pauciciliatum* (Parodi) Hester as a distinct species although some other authors consider it as only a variety of *P. dilatatum*. It is possible that *P. pauciciliatum* is the 'prostrate' type recognized by pastoral agrostologists (Hart & Burton, 1966; Barnard, 1969) and this view is supported by Fernandes et al. (1968). Pending a revision of the genus, it is perhaps advisable to treat *P. pauciciliatum* as a part of *P. dilatatum* in a broad sense. Morphologically *P. pauciciliatum* is rather similar to *P. dilatatum* s.str. except that it has slightly smaller spikelets and its lemma and palea have three veins compared with nine veins in typical *P. dilatatum*.

Occurs naturally in south-east Brazil, northern Argentina, Uruguay and adjacent, mostly subtropical, but also tropical, parts of the neighbouring states, mainly in the areas with an annual rainfall over 900 mm. *Paspalum dilatatum* has been introduced to a number of other countries, mainly subtropical and tropical, but also with warm temperate climate where it is grown mainly in the coastal areas. Introduced to Australia in 1870 (Barnard, 1969), it has become a grass of considerable importance in southern Australia and in some tropical parts of Queensland, especially in the coastal belt, where it is now one of the main cultivated perennial grasses. In the USA, where it was introduced in 1875 (Narayanan & Dabadghao, 1972), *P. dilatatum* is of considerable importance in south-eastern states. It has also reached India, some African countries, Madagascar, the Philippines, Hawaii and Fiji where it is grown predominantly on an experimental scale; it acquired some importance in the northern island of New Zealand and in Japan and is also under trial in South Korea and Jordan. *Paspalum dilatatum* is valued mainly for its vigour, persistence, the ability to withstand high grazing pressure, high herbage yields and a relatively high CP content.

Environment. Similarly with other tropical grasses with the C_4 pathway of photosynthesis, high light intensity, up to 6,400 lux, benefits the net photosynthetic rate in *P. dilatatum* (Cooper & Tainton, 1968), the rate of tillering and leaf appearance. Daylength produces some effect and a 14 h day seems to be the most suitable for the species. High air temperatures have a beneficial effect on photosynthesis and tillering. *Paspalum dilatatum* can withstand frosts and after a winter with the lowest temperature reaching $-10°C$, a 100 per cent survival was observed in Australia by R. M. Jones (1969), and Hacker et al. (1974) also found it to be the most frost-resistant grass of the species they tested. *Paspalum dilatatum* grows best under an annual rainfall of over 1,000 mm; it shows, however, some drought resistance and can withstand, often in a dormant state, relatively long dry periods. It requires fertile soil with high moisture content and yields well on loams, alluvial soil and moist heavy soil. Light-textured soils are, in general, less suitable although the 'prostrate' variety has shown good performance on light soils in the extreme south-east of the USA (Hart & Burton, 1966).

Establishment. The early growth is slow and a well-prepared seedbed, with little or no weed infestation, is required. *Paspalum dilatatum* is often established mixed with rapidly growing grasses, such as *Chloris gayana* or *Melinis minutiflora* which can compete with weeds but are less persistent, whereas *P. dilatatum* can take over from *C. gayana* and even eliminate it in the later years. Sowing rates of 9 to 14 kg seed/ha for the establishment with legumes and 4 to 7 kg when sown with other grasses are usually recommended (Whyte *et al.*, 1959). Sowing is done at the beginning of rains or in spring in the subtropics, but in the south-east of the USA autumn sowing is advised (Wheeler, 1950). Seeds eaten by cattle and found in dung germinate to 20–30 per cent (Yamada *et al.*, 1972) and can contribute to the spread of *P. dilatatum* outside the established swards.

Management, fertilizing. *Paspalum dilatatum* is a vigorous and deep-rooted grass which tolerates high stocking rates and frequent defoliation; e.g., in Queensland, stocking of 0.4 ha/head of cattle increased the percentage of *P. dilatatum* in a mixed sward, whereas 0.8 ha/head decreased it (Bryan, 1970). The grass may become dormant during southern Australian winters but Wheeler (1950) states that it remains green throughout the winter in the south-eastern USA and is a good winter-grazed grass. Fresh growth formed after cutting or previous grazing is readily eaten by cattle but when the herbage exceeds the height of some 30 cm the grass becomes much less palatable; grazing at 4- to 6-week intervals is therefore advised but more frequent grazing usually results in considerable decreases of herbage yields and animal production. The grass can be grazed low without any marked detrimental effect on the sward but grazing to higher levels may result in higher yields. In South Africa, cutting at 2.5 cm from the ground in the first year and at 10 cm in the second resulted in high yields in both years (R. I. Jones, 1967). *Paspalum dilatatum* responds well to fertilizer N. In Queensland, the application of 225 kg N/ha increased the yields of *P. dilatatum* herbage in a grass/legume mixture from 816 to 5,107 kg/ha, increased its proportion in the herbage and also the rate of its growth from 8 to 56 kg/ha/day in the middle of the growing season (Cassidy, 1971). In South Africa, three-fold yield increases of pure *P. dilatatum* were obtained under the application of 434 kg N/ha, ammonium nitrate being more effective than the other nitrogenous fertilizers used in the trial (Haylett, 1970). *Paspalum dilatatum* is however not so demanding for N as *Pennisetum clandestinum*; in a mixed pasture in Queensland, *P. dilatatum* dominated over *Pennisetum clandestinum* when given 168–224 kg N/ha but the reverse was observed under the application of over 224 kg N. P and K fertilizers usually produce little effect on pure *P. dilatatum* but increase the yields of legume mixtures. Responses to micronutrients vary and Mo, Co and B, and sometimes also Zn, can stimulate the growth of *P. dilatatum*. Responses to spraying with gibberellic acid (GA), expressed in yield increases of over 200 per cent in glasshouse trials (Lester & Carter, 1970), were obtained only under relatively low temperatures, whereas under warmer conditions the effect of GA was negligible.

Association with legumes. Paspalum dilatatum is often grown in association with other grasses and, in the main area of its cultivation, with legumes. *Trifolium repens*, either Ladino or an ordinary small type of New Zealand or Louisiana breed, is most commonly used. The clover can be sown together with *P. dilatatum* or oversown later into an established sward and in Queensland oversowing with 5.5 kg/ha of white clover seed resulted in 22 per cent of clover in the sward 14 months after oversowing and in 40 per cent when fertilized with P, Cu and Mo (Bryan, 1967). It should be noted that *P. dilatatum* roots contain substances which can have a detrimental effect on clover seed germination. Other perennial legumes tried were *Desmodium intortum* in Swaziland, and *Centrosema pubescens* and *Macroptilium atropurpureum* in Australia. An unfertilized *D. intortum* mixture yielded 3,860 kg DM/ha and a mixture fertilized with 25 kg P/ha 4,720 kg (I'Ons, 1969). An annual legume, *Macroptilium lathyroides*, has also been grown in mixture with *P. dilatatum*, and established swards of *P. dilatatum* were oversown with *Vigna unguiculata* and *Lablab purpureus* with a reasonable success.

Herbage yields. Roberts (1970) reports from Fiji that in 1950–2 trials yields of fresh herbage ranged from 39 to 65 t/ha. In trials in Rhodesia (Mills *et al.*, 1973) DM yields ranged from 1.7 to 26.0 t/ha, depending on plant age, the rates of fertilizer N and irrigation. Grass fertilized with 225 kg N/ha yielded 6.1 t DM/ha in the first year of growth but the yields decreased to 5.5 t in the second year and to 1.7 t in the third year. Increasing the rate of fertilizer N to 675 kg/ha resulted in respective yields of 14.9, 10.4 and 4.6 t and irrigation plus the high rate of N gave 26.0 t DM/ha. In general, yields of pure *P. dilatatum* given moderate amounts of N can be expected to range from 3 to 10 t DM/ha. Yields of some mixtures with legumes have already been quoted.

Ensiling. Grass surplus is sometimes cut to maintain a certain height of grass sward and its palatability, and attempts have been made to use the surplus for making silage. The ensiled material contains insufficient amounts of sugars for the production of good silage which can be partly remedied by adding molasses, the best results being obtained from adding some 40 kg molasses per ton of green herbage. DM losses during the ensiling are low, averaging about 9 per cent, and pH of the silage is about 4.8, somewhat high for good silage. DM digestibility in experimentally made silages only slightly decreased compared with fresh grass. Silage fed during the dry season reduced the losses of weight in cattle but did not produce any liveweight gains (from Catchpoole & Henzell, 1971).

Chemical composition, nutritive value. Narayanan & Dabadghao, (1972) gave CP content of 6.7 per cent for the grass at the flowering stage which also contained 40.9 per cent CF and the contents of Ca and P were 0.27 and 0.14 per cent, respectively. However, *P. dilatatum* analysed in Argentina (Vanesch & Riverós, 1967–8) at various stages of growth showed a better nutritive value: 13.4–18.5 per cent CP, 1.3–2.4 per cent EE, 24.4–34.8 per cent CF and 40.1–48.6 per cent NFE. Graham (1964), working with sheep, reports high digestibility for *P. dilatatum/Trifolium repens* mixture containing 23 per cent CP and 22 per cent CF: 75 per cent for organic matter and 82 per cent for CP; digestible energy reached 73 per cent of gross energy and metabolizable energy 82–84 per cent of the digestible energy; the digestibility of hay was only slightly lower. Grass palatability, although satisfactory or good for regenerating herbage, decreased rapidly at later stages of growth.

Animal production. *Paspalum dilatatum* can withstand heavy grazing and high stocking rates are usually practised. Animal productivity can be high and Bryan (1968) reports annual liveweight gains of 312 kg/ha for cattle grazed at 0.6 ha/head. In Rhodesia the grass has been reported to produce a harmful effect on sheep: two lambs with neonatal goitre and two with skeletal deformities out of nine born to ewes grazed on *P. dilatatum* receiving high rates of fertilizer N were reported (Rodel, 1972). These abnormalities were attributed to the effect of cyanogenetic glucosides in the grazed grass although the content of HCN in the herbage was relatively low, 42 p.p.m. Grazing *P. dilatatum* infected with ergot can also result in cattle poisoning, and Rhind (1945) writes that excess of intake may cause diarrhoea.

Reproduction, variability. *Paspalum dilatatum* flowers freely and Knight (1955) reports from Mississippi, USA, that the most abundant formation of panicles takes place at a day length of 14 h; at 8 to 12 h the flowering and seeding are erratic and at a day length of 16 h the plants produce smaller numbers of panicles than at 14 h. He also observed that low night temperatures and cool days inhibit the flowering. Reproduction is predominantly apomictic, the apomixis being of aposporic and pseudogamous type. There is, however, a yellow-anthered wild erect type which reproduces sexually; it has 40 2n chromosomes and has been described as ssp. *flavescens* Roseng. The yellow-anthered sexual type was crossed with the common apomictic type, having 50 2n chromosomes. The univalent chromosomes were lost during meiosis in F_1 and

F_2 and further generations were sexual. In the types selected for seed production from the progenies of the crosses seed setting was increased twice from F_2 to F_4 (Bennett et al. 1969). Two types of the purple-anthered plants have been distinguished in cultivated *P. dilatatum*: 'prostrate' with $2n = 40$ and 'common' with $2n = 50$ although plants with irregular numbers of chromosomes ranging from 50 to 63 were also found. Of the prostrate, '*P. pauciciliatum*', type two varieties – B230 and B430 – have been selected in Louisiana, USA, and they are admittedly better seed producers than the common type. Prostrate and common types were also compared in Georgia, USA (Hart & Burton, 1966), and the prostrate varieties gave higher yields of herbage, especially at high rates of applied N, and were recommended for light-textured, poorly drained soils of the extreme south of the USA.

Seed production. Seed of *P. dilatatum* is abuntantly formed but easily shed before or during the harvest. Moreover the shed seed contains much higher proportion of good, full seeds than those persistent on the racemes as it was shown by Knight (1955). In Knight's trials shattered seeds of plants grown under a 14 h photoperiod contained 47–56 per cent seeds with caryopses and non-shattered sees only 1.3–1.5 per cent. Seeds harvested 14, 21 and 28 days after peak flowering were dried at temperatures ranging from 38° to 60°C and the temperature of 60°C produced no adverse effect on germination. Early harvesting with subsequent drying at 60°C produced the best results, although in seed harvested 14 days after flowering and containing a large proportion of unripe caryopses the percentage of germination was lower than in those harvested later (Bennett & Marchbanks, 1969). Germination of good seed is of the order of 75–85 per cent (Wheeler, 1950). According to Wheeler, normal yields of seed are of the order of 110–220 kg/ha but can occasionally reach 750 kg and in Louisiana, USA, seed yields ranged from 250 to 500 kg/ha. Seed fed to cattle can germinate in dung to 20–30 per cent. Seed is in the form of spikelet containing one caryopsis each, and, according to various sources, there are 480,000–650,000 such seeds per kg. *Paspalum dilatatum* seed are very sensitive to CH_3Br (methyl bromide) and the slightest fumigation with this chemical can be damaging (Johnston & Miller, 1964).

Paspalum distichum L. Gramilla blanca; Pata de gallina; Gramilla; Pasto dulce. $2n = 40$ or 60 and also 48

Rhizomatous and stoloniferous perennial. Stems 20–30 cm high. Leaves linear, 2–12 cm long and 2.5–6 mm wide, mostly glabrous. Racemes (spikes) two, or occasionally three, on stem top. Spikelets ovate-lanceolate, 2.6–3.2 mm long. Lower glume often present and then can be up to three-quarters the length of the spikelet. Upper glume of the size and shape of the spikelet. Lemma of the lower floret similar to the upper glume. Fertile floret 2.5–2.7 mm long and 1.2 mm wide.

Occurs naturally over a wide area in America, from the southern USA in the north to central Chile and northern Argentina and Uruguay in the south. It is an adventitious naturalized plant in southern Europe, mainly in the Mediterranean area, and also in Asia where it is common in Burma (Rhind, 1945) and particularly in Japan; in those areas it often becomes a serious weed of fertile lowland soil.

Paspalum distichum is well adapted to various soils but prefers wet, fertile soils rich in humus and can withstand waterlogging and flooding. Its rhizomes can spread at different layers of soil, and when reaching soil surface they can grow as stolons, i.e. above-ground creepers. It can withstand a considerable degree of grazing and trampling and on hard soil can form a low dense carpet 5–10 cm high (Burkart, 1969). *Paspalum distichum* is particularly well adapted to the conditions of shallow river banks and small river islands. Herbage yields are not high but the plant is highly palatable to grazing animals. According to Vonesch & Riverós (1967/8) it contained at the preflowering vegetative stage 24.3 per cent CP and only 21.7 per cent CF.

Paspalum fasciculatum Willd. ex Flügge. Gamalote; Sorgo amargo; Venezuela grass (Puerto Rico)

Robust perennial 1–2 m high forming broad tufts and clumps sometimes up to 4 m in diameter and also spreading by stolons. Stems erect or ascending, branched, compressed in the basal parts and up to 10 mm wide. Leaves 20–40 cm long and 15–20 mm wide. Panicle with 12–15 racemes; racemes 6–14 cm long, the lower longer than the upper ones thus forming obconical, V-shaped panicles. Spikelets 4–4.5 mm long and 1.5–1.8 mm wide, with long hairs on margins. The lower glume absent, the upper of the size and shape of the spikelet and similar to the lemma of the lower floret. Upper, fertile floret slightly shorter than the lower floret.

Occurs in a wide area extending from Mexico in the north to subtropical Argentina in the south, on river banks, small riverine islands and in seasonally flooded areas. In Venezuela it often dominates or even forms pure stands in savannas seasonally flooded up to 20 cm of water from the ground level where it is grazed mainly in the dry season and, covering large areas, is of considerable importance for the cattle industry (Ramia, 1967). In South and Central America *P. fasciculatum* is used as a natural fodder or grazing grass and has also been introduced for trials into other countries, e.g. to Tanzania. Butterworth (1967) quotes in his review CP contents ranging from 6.4 to 8.3 per cent and DCP from 3.0 to 4.1 per cent; the content of TDN varies from 44.5 to 55.3 per cent of NFE 51.2–48.8 per cent and CF from 28 to 30 per cent.

Paspalum guenoarum Arachevaleta. Pasto rojas; Wintergreen paspalum. $2n = 40$

Robust perennial up to 1–2 m high, with or without rhizomes. Stems with three to four nodes. Leaves glabrous, 6–18 mm wide. Panicle of 2–13 distant racemes which are 7–17 cm long. Spikelets obovate, 3–3.5 mm long. Rather similar to *P. plicatulum* but more robust in all parts. Burkart (1969) distinguishes two varieties:

1. Var. *guenoarum*, provided with short rhizomes; racemes 4–13, spikelets almost glabrous.
2. Var. *rojasii* (Hack.) Parodi (*P. rojasii* Hack.) without rhizomes; racemes two to four, spikelets hairy; a less robust plant than var. *guenoarum*.

Both varieties occur in southern Brazil, southern Paraguay, Uruguay and northern Argentina, under subtropical or tropical humid climates, on fertile soil. In Argentina and Paraguay it is used for green fodder, hay and sometimes for making silage and has been successfully cultivated. Seed is well produced and the sowing rates used are 12–15 kg/ha when broadcast or 10–12 kg for sowing in rows, usually 20–40 cm apart (Whyte *et al.*, 1959), but Barnes in Rhodesia recommends 6–7 kg seed/ha. In a comparison of five pasture grass species in Rhodesia (Barnes, 1968), *P. guenoarum* provided the best grazing and the highest liveweight gains in steers, 370 kg/ha in 222 grazing days, but in a more recent trial it showed poor performance and yielded only 5.7 t DM/ha/year, compared with 15–17 t obtained from the best grasses in the trial (Rodel & Boultwood, 1971). According to Barnes the grass is leafy, with only a small proportion of stem and the herbage remains green and grazable in winter. Newly harvested seed has to be stored for at least 6 months before sowing. Rodel & Boultwood (1971) give an average content of CP in the herbage as 11.5 per cent, which is lower than the CP content in all other grasses they tried and Vonesch & Riverós (1967–8) determined the content of CP as ranging from 8.8 to 12.1 per cent.

Paspalum maritimum Trin.

Perennial creeping by rhizomes covered with densely pubescent scales. Stems up to 100 cm high. Leaves glabrous except scattered long hairs at the base. Racemes four to seven, suberect, 4–8 cm long. Spikelets obovate-elliptic, convex on one side and flat on the other, about 2 mm long, loosely hairy with the hairs arising from dark bases. Fertile floret pale, smooth and shiny.

Distributed naturally in British and Dutch Guiana, north-eastern Brazil and Cuba. An important grazing grass in the forest zone, in open glades and in grasslands developed after forest clearing.

Paspalum nicorae Parodi (*P. plicatulum* Mishx. var. *arenarium* Arachevaleta). Brunswick grass (USA) $2n = 40$

Perennial 20–70 cm high forming tufts and also spreading by rhizomes which are 5–25 cm long. Stems fine. Leaves 4–5 mm wide, often folded or convolute. Panicles of two to five distant racemes which are 2–7 cm long. Spikelets 2.7–3.3 mm long and 1.6–1.8 mm wide, lanceolate or obovate, without long hairs on margins. Fertile floret 2.5–3 mm long and 1.4–1.7 mm wide. Caryopsis dark, 1.8 mm long and 1.4 mm wide.

Occurs naturally in Argentina and Uruguay. Introduced to south-eastern USA where it showed good performance in fodder grass trials. In small-plot trials at Athens, south-eastern USA, unfertilized *P. nicorae* gave 4.24 t DM/ha when cut low at intervals from 1 to 6 weeks, and cutting frequency little affected the yields but fertilizer N increased them to 7.36 t at 112 kg N/ha, 9.24 t at 224 kg N and to 10.572 at 336 kg N. At the high rates of fertilizer N the effect of cutting frequency was evident and the plants cut every week produced at 224 kg N 7.73 t DM/ha and those cut every 6 weeks 10.77 t with intermediate yields for intermediate cutting frequencies; at 336 kg N the corresponding yields were 9.29 and 12.63 t/ha. The weight of roots and rhizomes were little affected by fertilizer N or by cutting frequency (Beaty *et al.*, 1970).

Paspalum nicorae has been reported to be a pseudogamous apomict but in trials by Burson & Bennett (1970) 18–48 per cent spikelets developed caryopses under open pollination and 2–26 per cent in isolation.

Paspalum notatum Flügge (*P. uruguayense* Arachevaleta). Bahia grass; Grama dulce; Forquinha; Gengibrillo (León & Sgaravatti, 1971); Pasto horqueta (Fig. 28) $2n = 20, 30$ or 40

Creeping perennial with stolons and rhizomes; the stolons are firmly pressed to the ground, have numerous very short internodes and root freely from nodes which also develop leaves and shoots. Fertile stems two to four noded, 15–70, occasionally to 100 cm high. Leaves linear, mostly 5–20, sometimes to 50 cm long, gradually tapering to a fine point, flat, convolute or folded, 2–10 mm wide. Racemes (spikes) terminal, green, mostly 5–10 cm long, paired or in three to five, densely beset with spikelets on one side. Spikelets glabrous, elliptic-obovate, 2.5–4 mm long and 1.7–2.2 mm wide. Glume one, of the size and shape of the spikelet. Florets two; the lower reduced to an empty lemma similar to the glume. Upper floret slightly smaller than the spikelet; caryopsis 1.8 mm long and 1.2 mm wide.

A polymorphic species of which three main varieties are usually distinguished:

1. Var. *notatum* with short racemes and small spikelets 2.5–3 mm long. Leaves short, narrow. A tetraploid apomict with the chromosomes number $2n = 40$. Distributed in the northern part of the species area, mainly in southern USA.

2. Var. *latiflorum* Doell. Vigorous plants 30–100 cm high; leaves broad, often hairy; racemes two to five, 5–16 cm long. Spikelets 3.3–4 mm long and 2.3 mm wide; fertile floret 2.7–3.2 mm long and 1.7–1.8 mm

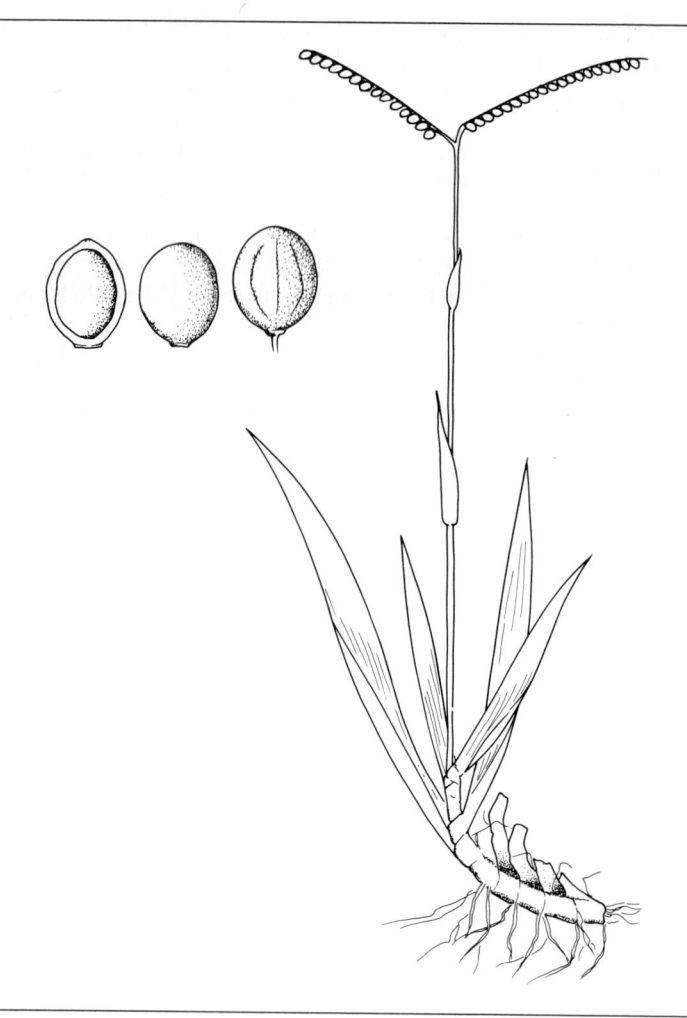

Fig. 28 *Paspalum notatum.*

wide. Tetraploid with $2n = 40$; most types are apomicts. Occurs in tropical and subtropical South America, Mexico and West Indies, mostly in open grassland. Provides good palatable fodder.

3. Var. *saurae* Parodi. An endemic of northern Argentina where it can be locally abundant. Burkart (1969) considers it to be a distinct species, *P. saurae* (Parodi) Parodi. Parodi, who described it first as a variety, raised this taxon later to the species rank. From the pastoral point of view it is perhaps wiser not to separate the two species but to include *P. saurae* in *P. notatum* as a variety, mainly because in the

agricultural literature varietal names are not always given and some information ascribed to *P. notatum* may actually refer to *P. saurae*. Var. *saurae* has long narrow leaves and small spikelets, 2.8–3.2 mm long and 1.8–2 mm wide, slightly larger than those of var. *notatum*. A sexual diploid ($2n = 20$) which can intercross freely and produce morphologically and biologically variable progeny; the well known cv. Pensacola has been developed from such progenies. The material from which cv. Pensacola was bred had been found near Pensacola, a town on the coast of the Gulf of Mexico, where it is believed to be accidentally introduced from Argentina by ships importing Argentinian cattle (Burton, 1967).

Paspalum notatum occurs naturally in a wide area extending from the southern USA and Mexico to Buenos Aires in Argentina and to the West Indies. It has been introduced to other tropical, subtropical and warm temperate areas, in the first instance to USA where it is now widely cultivated in the southern states, especially in Florida, and where extensive research with this grass has been done. It is also grown in southern Japan where its physiology, has been studied. *Paspalum notatum* is grown to a lesser extent in some countries of its origin: Argentina, Brazil, Bolivia and on some West Indian islands and has been introduced to India, Australia, some African countries, Taiwan, Hawaii and New Zealand.

Paspalum notatum forms a dense cover and is valued for its productivity, relative ease of establishment and persistence. Some authors consider, however, that *P. notatum* is more suitable for beef than for milk production. If grown in arable rotation, it is not easily eradicated after the end of the grass break and the crop that follows can be poor. *Paspalum notatum* is grown mainly for grazing, to a lesser extent for hay and also for soil conservation, and is particularly suitable for the protection of sloping ground against erosion.

Environment. Similarly with other tropical grasses, *P. notatum* can develop a considerable photosynthetic activity under high-intensity light, but can also tolerate a high degree of shading, and when used for lawns can grow under trees better than a number of other tropical grass species. Photoperiodically it is a long-day plant (Knight & Bennett, 1953). Temperature requirements for normal development are high although the maximum tillering was observed at 20–25°C (Kawamura & Yamasaki, 1972) and *P. notatum* can withstand cold weather better than *P. dilatatum*. *Paspalum notatum* tolerates some degree of frost, and in Queensland, Australia, survived to 90 per cent a winter with the lowest temperature of about $-10°C$ (R. M. Jones, 1969). Night temperatures below 13°C can inhibit flowering. It has often been mentioned in the literature that sandy or light-textured soils are the best for this species, but it can also grow well on wet clay soils where it develops better than

Cynodon dactylon or some other species of *Paspalum*. *Paspalum notatum* is tolerant to flooding and continuous flooding for 36 days produced no harmful effect on the grass as reported by Schroder (1966) who, in the same trials in Florida, also showed that this grass, can tolerate up to 4,500 p.p.m. of NaCl in irrigation water; higher salinity, 9,000–27,000 p.p.m. NaCl greatly reduced photosynthesis and transpiration but had little effect on respiration. *Paspalum notatum* occurs naturally in the tropics and subtropics with a moderate or high annual rainfall but it can also tolerate droughts, apparently because of its long roots which can penetrate deep into the soil.

Establishment. Establishment is effected by seed but *P. notatum* can also be planted by pieces of rhizomes; the rhizomes take well but spread slowly and should be planted densely, some 15–25 cm apart. For seed sowing a well-prepared seedbed is needed and seed is sown to a depth of 1–2 cm and at a rate of 10–20 kg/ha. Germination is slow, mainly because the hard scales, lemma and palea of the fertile floret firmly clasping the caryopsis, do not allow rapid penetration of water and seed treatment with 60 per cent sulphuric acid for 23 min followed by submerging in water for 15 min has been recommended by Gamboa & Guerrero (1969); this treatment resulted in the separation of seed scales and 60 per cent of germination. Early development of young plants is slow.

Fertilizing, Management. As shown mainly with cv. Pensacola, *P. notatum* normally responds well to fertilizer N but the maximum rates of applied N beyond which little or no response is observed are usually lower than for *Cynodon dactylon* and some other tropical and subtropical grass species and are usually around 100–200 kg N/ha, although responses to up to 600 and even 900 kg Na/ha were sometimes observed (see p. 210). On the other hand there was no response to fertilizer N in a trial in Uganda (Harker, 1962). In Alabama, USA, four forms of applied N were equally effective and resulted in herbage yield increases of 48 kg DM/kg applied N (Scarbrook, 1970). The application of N can also increase the digestibility of DM, CP and CF. Yield increases have also been observed from the application of fertilizer P and this nutrient has also promoted tillering. Potassium applied together with N and P can increase DM yields and herbage digestibility but it can also decrease the content of CP in the herbage. In Uganda, in a trial in which N and P fertilizers produced no effect, gypsum increased herbage yields, the increase being attributed to the effect of sulphur (Harker, 1962).

The main bulk of *P. notatum* herbage is concentrated at a relatively low level above the soil surface and Beaty *et al.* (1968b), working with cv. Pensacola, determined the average height of the sward as being 14–16 cm. When the grass was cut at a height of 5 cm, about 60 per cent of the herbage was recovered but only 22–44 per cent was recovered when cut at a height of 6.25 cm (2.4 in). Grass fertilized with 336 kg N/ha had the average height of 19–29 cm and a larger proportion of herbage was recovered when it was cut at both levels. However, to achieve still fuller utilization low cutting or grazing are needed. It is interesting to note that carbohydrates of *P. notatum* stubble can contribute more to the formation of new leaves after cutting or grazing than the carbohydrates of roots and low cutting can be harmful to plant recovery after grazing or cutting (Ehara *et al.*, 1966). It is usually recommended to graze or cut the grass at 6-week intervals. In trials with cv. Pensacola in USA, the grass cut every week yielded 2,260 kg DM/ha and 3,880 kg when cut every 6 weeks (Beaty *et al.*, 1963).

Irrigation during dry periods increases herbage yields, and maintaining soil moisture content at a high level, about 85 per cent of field capacity, gives the highest yields, whereas in some other tropical grasses the highest yields are usually obtained at a lower level of soil moisture content.

Association with legumes. Paspalum *notatum* is an aggressive grass and when fully established leaves little room for the development of legumes; mixtures with leguminous plants had, therefore, only a moderate success and fertilizer N can be more suitable and even profitable than N fixed by the legumes, especially as the demand for N is normally not high.

Trifolium repens and *Lotononis bainesii*, both low creeping perennials, increased the yields of total herbage of mixtures and the contents of CP compared with *P. notatum* grown alone. In Japan and Australia it has been grown with some success in mixtures with lucerne and with siratro (*Macroptilium atropurpureum*); in the USA, the latter mixture yielded up to 8 t DM/ha (Kretschmer, 1972). In USA a mixture with annual *Arachis monticola* yielded 730 kg of herbage, the yields decreasing in the following 2 years to about 200 kg/ha; the same legume did however better in a mixture with *Cynodon dactylon*, yielding 1,850 kg/ha in the first year of growth (Beaty *et al.*, 1968a). Creeping perennial *A. glabrata* gave satisfactory results in trials in Florida, USA (Prine, 1964). In those areas of the USA where *P. notatum* is dormant or near dormant for 6 months or even for a longer period, during the autumn and winter, a hybrid vetch from a cross between *Vicia sativa* and *V. cordata*, bred for the purpose, was grown together with *P. notatum* and yielded well in cool months when *P. notatum* produced little or no growth (Donnelly & Hoveland, 1966).

Association with bacteria and fungi. In 1960s a close association of *P. notatum* with nitrogen-fixing soil bacteria of the genus *Azotobacter* was discovered and the species associated with *P. notatum* was described by J. Döbereiner (1966) and named *A. paspali*. *Azotobacter paspali* survives and fixes atmospheric nitrogen only when it grows in close vicinity to *P.*

notatum, practically on its roots and rhizomes. *Azotobacter paspali* can exist, though much less frequently, on roots of a few other species of *Paspalum*: *P. dilatatum*, *P. plicatulum*, *P. vaginatum*. *Azotobacter paspali* associates only or predominantly with the tetraploid forms of *P. notatum*, and in trials in Brazil (Döbereiner, 1970) *A. paspali* was found on 98 per cent plants belonging to a local Batatai form of var. *latiflorum* and only on 3 per cent of plants of the narrow-leaved diploid Pensacola type originating from Argentina. Considerable fixation of atmospheric nitrogen observed in soil under *P. notatum*, was ascribed to the close relationship between the grass and the bacteria (Kass *et al.*, 1971) and the roots of plants associated with *A. paspali* and their rhizomes contained more N than those of the plants grown in the absence of *A. paspali*; in the above-ground plant parts the content of N was not however increased. In early field trials N fixation was estimated to be up to 10 kg N/ha and never more than 20 kg. However, Döbereiner *et al.* (1972), who studied the activity of nitrogenase in the rhizosphere of *P. notatum*, maintain that N fixation by the *P. notatum/A. paspali* system is more effective and that the amounts of fixed N can reach as much as 90 kg/N/ha.

A symbiosis of *P. notatum* cv. Batatai with an endogenous mycorrhizal fungus of the genus *Endogene* resulted in more vigorous growth of grass seedlings some 4 weeks after the inoculation with the fungus than that of non-inoculated seedlings (Mosse, 1972). In pot trials the growth improvement of inoculated seedlings was clearly evident and comparable with that produced by the application of fertilizer phosphorus.

Pests and diseases. A sting nematode, *Belonolaimus longicaudatus*, can attack the roots of *P. notatum* and various cultivars or types differ in their resistance to this nematode; e.g. cv. Paraguay was more resistant than cv. Pensacola, which was, however, not susceptible at soil temperatures above 39–40°C (Boyd & Perry, 1970). Ergot is not uncommon on the spikelets of *P. notatum*.

Herbage yields. Herbage yields of *P. notatum* are usually not particularly high although 17 t DM/ha were obtained in Australia (Shaw *et al.*, 1965) from heavily fertilized and irrigated grass. In Cuba (Pérez Infante, 1970) yields of 14 t DM/ha were recorded from stands sprinkler irrigated in the dry season; the dry season production contributed however only 19 per cent of the total yield, whereas other grasses in the trial produced in the dry season 33–46 per cent of their total yields. Particularly high yields of herbage were obtained in trials in Rhodesia (Mills *et al.*, 1973), where 900 kg applied N/ha nearly doubled the yields obtained under the application of 225 kg N/ha and resulted in a yield of 20.7 t DM/ha in the first year of utilization, 20.4 t in the second year and 15.7 t in the third year. Under this fertilizer rate plus irrigation yields of DM reached 37 t/ha and were higher than those of three other grasses under trial. On the low side, yields of 4–5 t DM/ha were obtained in India and in Uganda which were considered as maximum yields of *P. notatum* in these countries. The yields depend mainly on the rates of fertilizers, especially of N, as it has been shown in the Rhodesian trial, and also in Georgia, USA, where herbage yields were increased from 1,400 kg DM/ha to 5,670 kg when 220 kg N/ha were applied (Beaty *et al.*, 1963). In general, 3 to 8 t DM/ha can be expected from moderately to well fertilized grass grown under suitable climatic conditions.

Chemical composition, nutritive value. CP content in *P. notatum* herbage is usually not high although Rodel & Boultwood (1971) give the average content of CP as being 14.0 per cent and Montgomery *et al.* (1972) also reported from Louisiana, USA, 14 per cent in young herbage cut in May; CP content decreased to 8–10 per cent in July–August but increased again in autumn to 10–14 per cent depending on the frequency of cutting. McCormick *et al.* (1967) gave lower contents for *P. notatum* hay, 6.7–8 per cent, and Butterworth (1967) quotes comparable figures obtained by R. Milford in Queensland, Australia, which range from 4.0 to 9.0 per cent CP depending on the stage of growth and the cultivar. In trials with wethers, Milford determined the digestibility of CP as ranging from 15 to 49 per cent in samples containing 5.8 per cent or CP or over and DCP content ranged accordingly from 0.9 to 4.4 per cent. A negative value of CP digestibility was however determined in the herbage containing 4.0 per cent CP. Total DM digestibility as determined by Milford ranged from 40 to 53 per cent although other authors give higher figures reaching 65–75 per cent. In a trial in Japan (Ehara & Tanaka, 1961), CP content was higher in plants subjected to lower (15°C) than to higher (25°C) temperature and the reverse was observed for the content of lignin. The CF content is usual for a tropical grass, about 30 per cent in the younger herbage and 35 per cent in more mature plants. High content of P, up to 0.35 per cent, was reported. The content of HCN, 28 p.p.m., was the lowest for the five grass species examined by Rodel (1972), in which the content of HCN ranged from 28 to 247 p.p.m.; Rodel ascribes the absence of neonatal goitre and skeletal deformities in lambs born to ewes grazed on *P. notatum*, but evident when ewes were grazed some other grasses, to the low content of HCN. *Paspalum notatum* is palatable to all kind of stock but frosting decreases its palatability.

Animal production. Animal production is reasonable and can often reach the level of production observed for the grasses yielding more herbage or containing more CP, or, occasionally exceeds this level. In trials in Rhodesia (Clatworthy, 1968), liveweight gains of steers grazed at 4.6 head/ha on *P. notatum* fertilized with NPK and limed, were 560 kg/ha in one year and 320 kg in another year; the gains were slightly lower than those from *P. plicatulum*. In Florida, USA, liveweight gains were 670 kg/ha/year but were again lower than from the three other grasses in the trial (Haines *et al.*, 1961). High liveweight gains were also obtained in Georgia, USA (McCormick *et al.*, 1967), and they ranged from 0.3 to 0.5 kg/calf/day and were higher than the gains obtained from *Cynodon dactylon* pastures, where the herbage contained more CP than that of *P. notatum*. Keeping the animals housed on wet mornings, when the herbage contained high percentage of water, can help to increase the DM uptake by the animals as it was shown by Dirven & Ehrencron (1963b) in Surinam.

Seed production. Seed yields are relatively low and can be decreased still more by ergot infestation. Seed is in the form of the fertile floret, the caryopsis being enclosed between hard floral scales; there are some 350,000 such seeds per kg.

Cultivars. Several cultivars of *P. notatum* are known. Some of them have been developed by combined intercrossing and selection and this refers to the sexual var. *saurae*. Cultivated types of other varities, mainly of var. *latiflorum*, are apomictic.

Of var. *saurae*, **Pensacola**, selected by G. W. Burton, is the best known cultivar widely grown in the USA and introduced elsewhere. It has long narrow leaves, spreads more rapidly than most other types, is reasonably tolerant to frost (Wheeler, 1950) and susceptible to nematode infestation. Pensacola provides good grazing for a long period with the peak of growth and production late in the season: late summer and early autumn. Seed is commercially available.

Tifhi-1 is a hybrid cultivar of the same variety *saurae*. Herbage production can be similar to that of cv. *Pensacola* but it provides more grazing early in the season, in spring and early summer than later in the season, a reverse performance compared with Pensacola.

The better known cultivars of other varieties are:
Common, a cultivated type of an early introduction.
Argentina.
Batatai, a local Brazilian type in general use.
Paraguay, a selection from material introduced to USA from Paraguay. Leaves relatively narrow; good seed production and good resistance to nematodes.
Wilmington is perhaps the most frost-resistant cultivar which can be grown in more northern states of the USA than other cultivars.
Wallace and **Tamba** are quoted by Wheeler (1950) but seldom mentioned in more recent publications.
Andre da Rocha and **Capivari** are Brazilian types regarded by E. R. Prates as ecotypes. In a trial in Brazil their yields were similar to those of cv. *Pensacola* but the contents of CP in the herbage was lower.

Paspalum orbiculare Forst. $2n = 40$ or 54

Tufted and possibly also creeping perennial 30–60 cm high with long leaves. Panicle usually with three to five racemes. Spikelets pointed, broadly ovate, about 2 mm long. Occurs in grasslands of India, South-East Asia, New Guinea, Australia, Polynesia and also in Africa to where it was possibly introduced. Jacques-Félix (1968) reports that *P. orbiculare*, together with other sward-forming grass species, can, under an increased density of grazing, replace tufted grasses in the Cameroons.

Paspalum plicatulum Michx. Plicatulum (Australia)

Tufted perennial up to 1.2 m high, Leaves up to 90 cm, but mostly about 40 cm long, slightly hairy. Panicle consists of 5–15 racemes, 2–6 cm long; rhachis narrow. Spikelet ovate-elliptic, 3 mm long and up to 2 mm wide. Seeds (ripe fertile florets) dark brown and shiny.

Occurs naturally in South and Central America and was introduced to Australia in 1932, firstly from Guatemala and later from other sources. It has also been tried with encouraging results in Kenya and the Philippines. In Australia *P. plicatulum* is a summer-grown grass grown in south-eastern and eastern Queensland, in areas with an annual rainfall over 750 mm. It is highly adaptable and can grow on relatively poor soil, and tolerate waterlogging and short-term floods but is frost-susceptible, although some frost resistance was observed in types introduced from relatively high altitudes of America (Hacker *et al.*, 1974). A nitrogen-fixing bacterium, *Azotobacter paspali*, normally associated with *P. notatum*, has been found in the rhizosphere of a small number of *P. plicatulum* plants (Döbereiner, 1970). In Australia *P. plicatulum* was successfully tried in mixtures with *Macroptilium atropurpureum*, *Desmodium uncinatum* and *Stylosanthes guianensis* but *Trifolium repens* and *Lotonis bainsii* showed poor performance in mixtures with *P. plicatulum*. Cattle grazed on *P. plicatulum*/legume mixtures at 1 beast to 0.4–0.6 ha gained up to 220 kg liveweight/ha (Barnard, 1972); however, in trials by Bryan (1968) in Australia *P. plicatulum*/legume mixtures were less productive, in terms of liveweight increases/ha, than the three other grasses grown with the same legumes, and the content of N in *P. plicatulum* ranged from 0.73 to 1.56 per cent, which corresponds to 4.6–10.5 per cent CP (N × 6.25). Experimental ensiling of *P. dilatatum* herbage was not successful: fermentation was slow, pH of the silage remained high and the concentration of lactic acid was below 0.5 per cent (Catchpoole & Henzell, 1971). In plants grown in boxes near Brisbane, the panicles emerged 142 days after sowing and seed was ready for harvesting 21 days later (Chadhokar & Humphreys, 1970). Seed production of *P. plicatulum* is normally good and there are 780,000–950,000 seeds per kg. Fertilizer N increases seed yields as it was observed in field trials in Australia when N in the form of urea was applied at 50 to 400 kg/ha. The increases ranged from 5.6 kg crude seed/kg applied N at 50 kg N/ha to 1.2 kg at 400 kg N/ha. Fertilizer N improved seed viability but the crop given 200–400 kg N/ha tended to lodge in wet years and can adversely affect water regime of the plant in dry years (Chadhokar & Humphreys, 1973). Two cultivars have been selected in Australia: **Rodd's Bay** (Rodd's Bay is on the eastern coast of Queensland at about 24°S.) and **Hartley**. They belong respectively to var. *plicatulum* and var. *glabrum* Arech. Cv. Rodd's Bay differs from cv. Hartley in having narrower leaves, somewhat larger seeds, earlier flowering and better seed production.

Paspalum plicatulum is an apomictic tetraploid with $2n = 40$, but diploids ($2n = 20$) have also been recorded as well as plants with the chromosome number $2n = 30$.

Paspalum pulchellum Kunth

Tufted perennial. Stems simple, erect, up to 80 cm high. Upper leaf sheaths often without blades. Leaves flat or convolute, up to 25 cm long and 2–4 mm wide, covered with a few to numerous stiff hairs. Racemes two to three, approximate, 2–8 cm long. Spikelets solitary, elliptic, glabrous, 2 mm long. Sterile lemma usually dark purple. Fertile floret pale, smooth and glossy.

Occurs in the West Indies and also in Brazil and neighbouring territories in seasonally waterlogged savannas where it can be numerous and of importance for grazing. It often dominates in the grass cover in savannas of Trinidad and Surinam. It recovers rapidly after grass fires and burning greatly encourages its flowering (Richardson, 1963).

Paspalum urvillei Steud. (*P. vaseyanum* Scribn.). Vasey grass

$$2n = 40 \text{ or } 60$$

Perennial forming large tufts mostly 1–2 m and occasionally 2.5 m high. Leaves up to 50 cm long and 5–15 mm wide, pilose near the base but otherwise glabrous. Panicle 10–40 cm long, narrow-pyramidal, with 12–25 suberect racemes which are 7–14 cm long. Spikelets 2.2–2.7 mm long, obovate, pointed, fringed with long white hairs; the glume is adpressedly silky. This species is allied to *P. dilatatum* and belongs to the same group *Dilatata*.

Occurs naturally in southern Brazil, Uruguay and Argentina where it is used for grazing or hay. Introduced accidentally to the USA where it naturalized in the southern states under subtropical and warm temperate conditions, mostly in the Coastal Plain Region. In the USA it was used mainly for making hay possibly because heavy continuous grazing can destroy this grass. Similarly with *P. dilatatum*, *P. urvillei* can grow on wet soil but also withstands droughts. Seed setting is good but flowering usually continues over a long period of time and this affects seed yields. Wheeler (1950) recommends cutting the first crop for hay and reserve the second cut for seed. There is just under 1 million seeds (spikelets) per kg (440,000 per lb) and Wheeler suggests sowing rates of

10–18 kg seed/ha. Good results with *P. urvillei* were obtained in Rhodesia, mainly on vleis, i.e. seasonally wet lowland.

This grass has also been introduced to other warm countries – South Africa, Portugal, Japan. In studies in Japan, *P. urvillei* grew best under temperatures around 25°C. Under excessive rates of fertilizer N herbage yields declined but increases in yields with the increased frequency of cutting were observed (Ehara & Tanaka, 1972). A chemical analysis has shown a good quality of herbage which contained 11.7 per cent CP, 27.8 per cent CF and 47 per cent NFE (Abusso, 1970).

Paspalum vaginatum Sw. Salt water couch; Sea-shore paspalum

$2n = 20$

Perennial with long creeping stolons and rhizomes producing erect or suberect many-noded stems up to 60 cm tall. Leaves 3–15 cm long and 2–8 mm wide. Inflorescence of two (rarely three) spike-like racemes beset with paired spikelets. Spikelets oblong, acute, 3–4.5 mm long.

Occurs naturally in tropical Africa, possibly also in South America in saline or alkaline marshes at sea shores and at higher ground up to 800 m alt. It forms dense, often pure stands and can be grazed if the animals have access to it. *Paspalum vaginatum* has been introduced to Australia where it is popular, mainly as a green cover for saline seepages and seed has been produced. A few clones have been selected.

This species is closely allied to *P. paspaloides* (Michx.) Scribn., formerly known as *P. distichum*, which usually grows on fresh-water swampy ground.

Paspalum virgatum L. $2n = 40$ or 80

Perennial, forming dense clumps 1–2 m high. Leaves flat, 10–25 mm wide. Panicle 15–25 cm long, slightly nodding, with 10–16 racemes, ascending or drooping, 5–15 cm long. Spikelets obovate, 2.2–2.5 mm long, brownish, hairy on margins towards the top end. Distributed in a wide area from southern Texas in the USA through Central America to South America and the West Indies. Introduced to other tropical and subtropical areas in which it is mostly considered to be a satisfactory grass suitable for soilage (Madagascar) or grazing (Congo People's Republic). However, in countries where better grasses are grown (Puerto Rico) *P. virgatum* is undesirable in pastures. Nitrogen-fixing *Azotobacter paspali*, normally associated with *P. notatum*, was also found in the rhizosphere of some plants of *P. virgatum* (Döbereiner, 1970).

Paspalum wettsteinii Hack. Broadleaf paspalum.

Semi-prostrate tufted and stoloniferous perennial forming spreading clumps up to 1 m in diameter. Stems up to 90 cm high, with two to five nodes. Leaves lanceolate-linear, rounded at the base, up to 40 cm long and 30 mm wide, with wavy margins, sparsely hairy or glabrous. Panicle with 4–10 distant racemes which are 3–10 cm long. Spikelets ovate, blunt, 2.2–2.5 mm long, strongly convex on one side, brownish-green. Glume one (upper) of the size and shape of the spikelet; lower, sterile, and upper, fertile florets about as long as the spikelet.

Occurs naturally in southern Brazil, Paraguay and northern Argentina, and is grown as a pasture grass in Brazil (Barnard, 1969). Introduced to Australia where it has been cultivated predominantly on an experimental scale for some years mainly in south-eastern Queensland and north-eastern New South Wales, and seems to be more suitable for warm temperate and subtropical conditions than for true tropics. *Paspalum wettsteinii* proved to be very palatable and grazed selectively when grown together with other grasses of good quality. It also associated well with legumes such as *Glycine wightii*, *Desmodium intortum*, *D. uncinatum*, *Macroptilium atropurpureum* (Siratro) and some other species, and produced good yields of herbage. Cultivar **Warral** has been registered in New South Wales – the seed is well produced and the grass is believed to be an apomict.

Pennisetum L. Rich.

Annuals or perennials, tufted or creeping, often with branched stems. Panicles dense, spikelike, sometimes reduced to a few spikelets. Spikelets solitary or in groups of two to seven, surrounded by a few to numerous bristles which can be plumose, and one bristle is usually longer than the rest. Spikelets lanceolate to ovate, lower glume shorter than the spikelet or absent. Florets two: the lower male or sterile, the upper bisexual, or can be male in the outer spikelets of a group. Lemma of the fertile spikelet coriaceous, usually smooth and shiny at maturity. A large genus of some 120–130 species which occur in the tropics and subtropics, and to a lesser extent in warm temperate areas, throughout the world. A number of species are excellent fodder and grazing grasses, such as elephant grass (*P. purpureum*), pearl millet (*P. americanum*) and their hybrids, Kikuyu grass (*P. clandestinum*) and several others. On the other hand there are fibrous, hard, hardly palatable species (e.g. *P. schimperi*) widely spread but undesirable in pastures.

Pennisetum americanum (L.) K. Schum. (*P. typhoides* (Burm. f.) Stapf & Hubbard; *P. typhoideum* Rich.; *P. glaucum* (L.) R. Br.; *P. spicatum* (L.) Koern.). Pearl millet; Bulrush millet; Bajra (India), Mwele (East Africa). Kala-sat (Burma) (Fig. 29)

Tufted or sometimes single-stemmed erect annual 1–3 m, occasionally to 4 m, in height. Stems solid (not hollow), glabrous except below the spike and at the nodes which are hairy. Leaves 20–100 cm long and 5–50 mm wide. Inflorescence a dense spike (spike-like panicle) 10–50 cm long and 0.5–4 cm in diameter, cylindrical or tapering to the apex or to both ends. The hairy axis of the spike beset with numerous peduncles 2–25

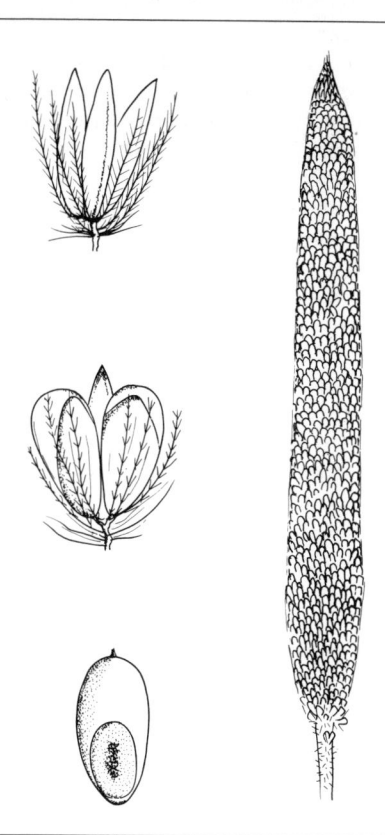

Fig. 29 *Pennisetum americanum.*

mm long, each bearing 25–90 bristles, some of them plumose, and single, or grouped in two to five, spikelets. Spikelets 4 mm long, with the lower floret male. In grain cultivars the caryopses are large, about 4 mm long and of different colour (grey, pale-yellow, white or slightly bluish in different types or cultivars, protruding from the floral scales or, mainly in wild or naturalized types, they can be much smaller and are then enclosed between lemma and palea.

Being a cross-pollinated plant, *P. americanum* is very variable and some taxonomists regard certain types of it as distinct species, whereas others, including Bor (1960), bulk the minor species together into a single species and I am inclined to agree with this latter treatment clearly and convincingly outlined by Purseglove (1972) in his brief review of *P. americanum* taxonomy and origin. As a cultivated cereal, *P. americanum* has most probably developed from wild plants and was apparently cultivated first in West Africa in its northern, Sahel zone bordering the Sahara. It penetrated to East Africa and the Sudan and then to India perhaps some 2,000 years ago where it has been cultivated on a large scale for several centuries. It reached Europe in the sixteenth century and penetrated to Spain and North Africa where it is grown for grain. In the last century it reached the USA where it has been grown for fodder and, also as a fodder plant, to Australia where it has found particularly suitable conditions of growth in the Northern Territory and met with an outstanding success. In the USA its importance as a fodder grass was for some time overshadowed by the spread of fodder sorghum, Sudan grass and their hybrids, but the interest for this grass has been recently revived with the development of new and productive forage cultivars, mainly in India, USA and Australia. Pearl millet is valued as a forage grass for the ability to grow under low rainfall using short wet seasons, rapid growth, good quality of fodder, easy seed production and the ability to regenerate and to produce new growth after grazing or cutting.

Environment. Pennisetum americanum is a facultative short-day species; under a 12-hour photoperiod the temperature and the genotype do not affect flowering and the plants flower early, whereas under longer photoperiods both genotype and temperature can affect flowering and in some but not all genotypes the flowering is delayed under 14- and 16-hour photoperiods. Photoperiods can also affect herbage production and in three out of five genotypes tested by Begg & Burton (1971), DM yields were doubled, or more than doubled, under 14- and 16-hour photoperiods compared with the 12-hour photoperiod, and the number of tillers and of leaves on the main stem increased, showing that some genotypes and cultivars can produce more herbage under subtropical or sometimes warm temperate conditions than in the tropics. However, Australian cultivars Katherine and Ingrid yielded more DM in the tropical Northern Territory of Australia with shorter photoperiods than in subtropical New South Wales. High temperatures are required for rapid growth especially at later stages of plant development. *Pennisetum americanum* tolerates low-rainfall conditions and in the Sahel zone of West Africa early maturing cultivars can be grown for grain under an annual rainfall of 250–300 mm, whereas in the same area grain sorghum requires at least 350 mm of rain. Tolerance to low annual rainfall depends mainly on the fast growth and early maturation so that the plants escape the drought and complete their development before the rainy season ends. *Pennisetum americanum* is, however, not drought resistant when it actively grows and if a drought occurs during the growing season the plants would not normally survive it in a dormant state as the sorghums can do. Vigorous and late maturing cultivars can only grow under higher rainfall but a humid tropical climate is unsuitable and although pearl millet yielded in New Guinea 5.9 t DM/ha (Hill, 1969) the yields were much lower than those obtained from other fodder grasses.

Pearl millet can grow on a variety of soils, except those waterlogged or seasonally flooded during the plant growth, and favours light loams and sandy soils in preference to heavy soils on which it develops more shallow roots than on light soils. The plants are relatively tolerant of low soil fertility and can produce at least some crop on soils on which some other cereals would fail.

Establishment. Establishment does not present any particular difficulties but a dry spell after the first rains can kill the seedlings in which case reseeding is necessary. No particular care for seedbed preparation is needed. Various sowing methods have been used by small farmers in Africa and India for the establishment of pearl millet as a cereal and it is sometimes planted together with sorghum. Seed rates of 5–9 kg/ha are usually adequate but for growing pearl millet as a fodder crop the rates should be higher and Metcalfe (1973) gives sowing rates as 22–34 kg/ha. Pearl millet grown for fodder is drilled in rows and satisfactory results were obtained from the rows 25–30 cm apart. Mulching with straw can be useful, and in trials in India it reduced water requirements of pearl millet by 25–30 per cent and increased fresh fodder yields from 49.4 to 53.9 t/ha (Pal & Pandey, 1969). Pearl millet responds well to nitrogenous fertilizers which are normally used in moderate amounts; fertilizer P increases the effect of the larger rates of N but is seldom effective when applied alone. Young plants grow fast and weeding is seldom necessary. Pearl millet is sometimes irrigated and the plants withstand small amounts of salt in the irrigating water.

Pearl millet can be grazed or cut for hay, silage or soilage and cutting or grazing can begin as early as 4 or 6 weeks from sowing. The plants regenerate well after cutting or grazing and up to four or sometimes even six cuts can be taken, but a single cut per season just before or during the flowering or at the milk-ripe stage of maturity is a common practice in some areas.

Herbage yields. Herbage yields vary very widely from below three to over 20 t DM/ha, depending on the climate, soil fertilizers and cultivar, and yields of 7–10 tons/ha can be accepted as average for experimental fields or well managed farms. In the Katherine area of the Northern Territory of Australia particularly high yields were occasionally recorded, 20–21 t DM/ha, and in Rhodesia even up to 25 t or up to 50 t fresh material. Yields of 100 kg nitrogen/ha, and occasionally to over 125 kg were reported which corresponds to 600–900 kg CP/ha.

Cut at the early flowering stage and analysed in India (Goswami, *et al.*, 1970) the herbage of pearl millet contained 6.8–12.8 per cent CP, 0.9–1.8 per cent EE, 29–34 per cent CF and 41–52 per cent NFE; the same plants showed high contents of Ca, 0.29–0.69 per cent and still higher contents of P, 0.47–0.84 per cent. In Mexico, however, low contents of CP, around 2–3 per cent, were reported. At those stages of growth, when the grass is normally cut for forage, herbage digestibility can be high, and vary from 63 to 82 per cent for DM, 62–80 per cent for CP, 60–75 per cent for CF and 69–80 per cent for NFE as it was shown in trials with cattle and sheep (from Butterworth, 1967). The straw, often used as roughage in India and Africa where pearl millet is grown as a cereal, is of poor quality and although CP content can reach 4 or even 5 per cent its digestibility is low and the content of DCP can be below 1 per cent. As in other fodder and pasture grasses, the content of CP decreases with the increase of intervals between grazing or cuttings, whereas total yields of fresh herbage and of DM increases.

Goswami *et al.* (1970) found 1.19–2.16 per cent oxalic acid in the DM but no harmful effect of pearl millet herbage on the animals has so far been reported.

Animal production. Animal production seems to be little covered in the literature, although it has been reported from Florida (Dunavin, 1970) that cattle grazed on the early sown cv. Gahi-1 gained 3.7 kg/ha/day, exceeding the gains obtained from grazing Sudan grass hybrids; when grazed the late-sown crop of the same cultivar the liveweight gains were 2.6 kg/ha/day, similar to those obtained from Sudan grass.

Diseases and pests. Of the major diseases mentioned by Purseglove (1972), the 'green ear', a downy mildew caused by *Sclerospora graminicola*, which damaged not only the ears but the whole plant, is possibly the only serious fungus disease affecting forage yields; it is common in India and in Africa and breeding resistant cultivars would perhaps control it. Other diseases: rust (*Puccinia penniseti*) and leaf spot caused by *Curvularia penniseti*, *Helminthosporium turcicum* and *Pyricularia grisea* are less important. The diseases of reproductive plant parts are smut caused by *Tolyposporium pennicillaria*, honeydew caused by *Sphacelia sorghi* and ergot caused by *Claviceps microcephala* can damage seed crops. Of the pests, caterpillars and grasshoppers can be damaging to herbage but the major pests are granivorous birds, chiefly *Quelea quelea*, which can inflict considerable losses of seed in Africa and India. The use of 'bearded' cultivars, in which numerous long and sharp scales protruding from between the grain can reduce the losses or even prevent the damage.

Flowering, reproduction, improvement. Ten to twelve weeks after seedling emergence are required before the first spikes appear. The plants are markedly protogynious and the styles appear first, 2–3 days after the spike has emerged, starting from the top of the spike and spreading down until the zone of the emergence of styles reaches the spike bottom in about 24 hours. About 2 days later, when the stigmas begin to dry up, the anthers of bisexual florets appear in the same order and about 2 days later the anthers of the male florets emerge and repeat the pattern (Krishnaswamy, 1962). This sequence of flowering practically excludes cross-pollination within the same spike but some cross-pollination may occur between different spikes of the same plant if they flower at slightly different times. Nevertheless *P. americanum* is essentially a cross-pollinated plant, the xenogamy resulting in a considerable variability of all plant characteristics and new combinations of characters may appear in each generation. There are many local types or cultivars in Africa and India, some of them improved by natural selection or by local farmers who usually select the best spikes for seed for the next year's planting. A number of cultivars grown for grain have also been improved by plant breeders in India and to a lesser extent in some African countries, such as Rhodesia. Crosses between the plants originating from distant localities may show a considerable hybrid vigour in the first generation (F_1) and the technique of growing hybrid seed, similar to that of maize, has been developed in India when male sterile plants were found or developed and selfed lines possessing desirable genes were selected and used for the production of hybrid seed. As in maize each generation should be grown from hybrid seed resulting from crossing selfed lines.

Pennisetum americanum is a diploid with the chromosome number of $2n = 14$, although occasional abnormal chromosome numbers, $2n = 15$ to 17, have also been found.

For fodder production grain cultivars can be used but better results are obtained from specially bred forage cultivars produced in the USA, India, Australia and also in Rhodesia. Cultivars developed for forage should have numerous tillers, be vigorous and leafy and have an accentuated ability of recovery and the production of fresh growth after cutting or grazing. Not all these essential characteristics are always combined in one and the same cultivar; when vigour and leafiness are not incorporated in the same cultivar, the leafy one can perhaps be preferred. In Georgia, USA, a tall and a 'dwarf' near-isogenic cultivars, reaching 213 and 107 cm in height, respectively, and producing in the year of trial 7.2 and 5.6 t DM/ha, were compared. When cut 74 days after sowing, the dwarf cultivar was more leafy and contained more CP and less CF than the tall and more vigorous cultivar, and cattle fed on dry herbage of the dwarf cultivar gained about 50 per cent more weight than those fed with the tall cultivar (Johnson *et al.*, 1968). *Pennisetum americanum* has been crossed with *P. orientale* (Willd.) L.

Cultivars. The better known fodder cultivars are those developed in the USA, mainly at Tifton in Georgia, and in India and Australia. A USA hybrid, cv. **Gahi-1**, has shown good performance in Florida outyielding Sudan grass hybrids. **Tift 23 B** and **Tift 23 DB**, the two near-isogenic cultivars mentioned above, gave a good performance in Georgia, especially the Tift 23 DB which has herbage of outstanding quality.

In India, cv. **BH-4**, a tall and vigorous type, is suitable for the more wet areas where it produces large bulks of herbage.

In Australia (Barnard, 1972) Cv. **Katherine pearl**, selected from material received from Ghana, showed outstanding performance in the Northern Territory where it averaged almost 13 t DM/ha during 11 years of trials, the maximum yield being about 22 t/ha. High CP content in young plants and 8 per cent at maturity were recorded. Cv. **Ingrid pearl** has been selected at Katherine, Northern Territory, from material originating from Senegal. It is a very tall and vigorous type, the plants reaching 3.8 m in height and is slightly earlier than cv. Katherine pearl; in variety trials it yielded slightly less DM but more CP than Katherine pearl; the maximum yields of DM reached 20 t/ha.

Cv. **Tamworth** is a hybrid cultivar developed from Gahi-1. In New South Wales it gave an average yield of about 9.7 t DM/ha, outyielding the Katherine and Ingrid cultivars when they were grown for comparison in New South Wales.

In Rhodesia cv. **Tjolotjo bearded** gave, in 1968–9 trials, DM yields of about 25 t/ha approaching the highest yields of sorghum hybrids.

Seed production. Being a cereal, pearl millet produces seed easily and even the cultivars bred for forage production retain or only slightly reduce their seed yielding capacity. Their seed yields can be expected to be anything between 500 and 2,000 kg/ha. Seed ripens about 40 days after pollination; it varies in size, the average number of seeds per kg being, according to Metcalfe (1973), about 200,000. Seed of most pearl millets has a period of dormancy lasting several weeks after harvesting; the dormancy extends to over 12 months if seed is kept in sealed containers at temperature of $0°/6°C$ (Burton, 1969).

Pennisetum clandestinum Hochst. ex Chiov. Kikuyu (Kikuyo) grass; this widely used common name is derived from the Kikuyu people of Kenya. (Fig. 30) $2n = 36$

Creeping perennial with strong, thick stolons and rhizomes which have numerous short internodes; both rhizomes and stolons root freely from nodes. Sterile upright shoots have relatively short stems and long leaves up to 20 cm in length; fertile shoots which terminate in flowering heads, do not exceed the sterile shoots in height (as they do in most grasses) but are usually hidden beneath the sterile shoots at the bottom of the sward. The grass can usually produce fertile shoots and flower when it is closely grazed or mown. The only visible evidence of flowering in short herbage is the abundance of excerted anthers borne on upright filaments up to 5 cm long which emerge at night, and in the morning form a bluish-white tinge over the sward, not unlike ground frost. The spikelets are hidden in the leaf sheaths of the fertile shoots, except for the upper parts which protrude from the sheaths. The 10–20 mm long spikelets are supported by a few to 15 slender bristles which are shorter than the spikelet. The spikelet has two florets similar in appearance, of which the lower is empty and the upper contains a pistil with a long feathery stigma and

Fig. 30 *Pennisetum clandestinum.*

three long stamens. In some types of Kikuyu grass the stamens are absent or rudimental. The grain (caryopsis) is about 2 mm long and dark brown when ripe.

Pennisetum clandestinum is relatively uniform in appearance, although Edward (1937) distinguished three ecotypes in Kenya: 'Kabete', 'Molo' and 'Rongai', which differ in the thickness of stolons, the colour and width of leaves and in the presence or absence of stamens.

Kikuyu grass is a native of highlands and mountains of tropical East Africa: Ethiopia, Eritrea, Kenya, Uganda, Rwanda, Zaire and Tanzania, where it occurs at altitudes ranging from 1,500 to almost 3,000 m, at forest edges, on land recently cleared from forest, on roadsides and as a weed in arable land.

Environment. Kikuyu grass requires fertile soil and is tolerant to high soil acidity. Fresh land from under cut or burnt forest is quickly colonized by Kikuyu grass, immediately or after a short period of tall-weed domination. In Kenya, and apparently also in other countries, when the soil loses its fertility, Kikuyu grass is gradually replaced by a coarse-grass phase in the plant succession, mainly with the dominance of *Pennisetum schimperi*. In Kenya, natural stands of Kikuyu grass are often associated with Kenya white clover (*Trifolium semipilosum*). Kikuyu grass does not grow well under waterlogging but thrives at lake shores as it does at Lake Naivasha in Kenya.

Water requirements of Kikuyu grass can be assessed from the average rainfall for the areas of its natural distribution, which ranges from just under 1,000 mm p.a. to 1,600 mm. Evapotranspiration under a Kikuyu grass sward at Muguga, an area in Kenya with marginally low rainfall, ranged from 3.8 to 5.1 mm per day (Glover & Forsgate, 1964, from Mears, 1970); here the grass was able to extract moisture from the soil down to a depth of 120 cm, although root penetration can be even deeper. Kikuyu grass can extract water from the soil until its content is close to the wilting point.

Poor performance of Kikuyu grass at low altitudes in the tropics indicates that it is not well adapted to high temperatures, although in some types of Kikuyu grass decreases in air temperature from 27° to 10°C arrested the excertion of the stamens. Average minimum and maximum temperatures in the areas of Kikuyu grass natural distribution range from 2° to 8°C and from 16° to 20°C, respectively. Only occasional night frosts occur in these areas, but in the subtropics, where frost are more frequent though still light, the herbage can be damaged by frost; the plants are, however, not necessarily killed. In south-eastern Queensland, at 27°C, a winter with lowest temperature reaching −9°C did not kill the plants which had a survival rate of 100 per cent (R. M. Jones, 1969).

Introduction, geography of cultivation. The valuable qualities of Kikuyu grass, its persistence under grazing, the considerable bulk and good quality of herbage have been recognized for a considerable time. The grass was tried under cultivation in Kenya and introduced to the highlands of other tropical countries and to the subtropics. First introductions of Kikuyu grass were reported for South Africa in 1910 and for Australia in 1919 (from Mears, 1970). At present Kikuyu grass is cultivated experimentally or on a farm scale in many African countries, Australia, southern USA, Central and South America, southern Asia and on a number of warm islands. In most of these countries Kikuyu grass is still under experimental cultivation but in others, such as Hawaii, Sri Lanka, Australia and a few others it has become one of the most important pasture grasses, mainly at relatively high altitudes.

Establishment. No commercial seed is usually available and Kikuyu grass is normally established from pieces of stolons, the size of which depends on the material available. Sprigs containing two to three nodes are reported to establish successfully and to form swards, but usually the sprigs are larger and cut to contain more internodes. When labour is not a problem, planting is done by hand, in furrows or in holes; otherwise the sprigs can be scattered from a vehicle and disced or harrowed afterwards. The machines used in the USA for planting Bermuda grass can also be used. Planting by about 45 × 45 cm is often recommended, but closer spacing, although requiring more labour and material, would result in quicker establishment. Provided some rain falls after planting, Kikuyu grass takes well and establishes itself easily.

Kikuyu grass seeds are normally eaten by the grazing animals together with the herbage, and new plants can establish from seeds which pass undamaged through the digestive tract and germinate in dung. The position of seed close to the ground where it is bound to be eaten with the herbage is perhaps an adaptive feature facilitating the spread of the grass.

Kikuyu grass persists as long as soil fertility is maintained and is suitable for permanent pastures; it is not advisable to cultivate it as a grass break in arable rotations where it becomes a troublesome weed, difficult and costly to eradicate. Mechanical eradication by repeated ploughing during the dry season was reasonably successful but costly, and so was eradication with dalapon and other herbicides on trials in Kenya, New Zealand, USA and India.

Association with legumes. Kikuyu is a strong and aggressive grass which on fertile soil, in climatically suitable areas and under close grazing, does not normally allow any other plants into the established sward. Under severe grazing or repeated cutting the penetration of clovers is more likely, and in Kenya highlands wild *Trifolium semipilosum*, often found in Kikuyu grass swards, can make up to 15 per cent of the low grazed herbage, but disappears when the grass is allowed to grow taller. In Australia, *Glycine wightii* was grown with some success in mixtures with Kikuyu grass, but its reaction to cutting was opposite to that of clover: it was more abundant in taller swards cut every 8 or 12 weeks than in shorter swards cut every 4 weeks, and the same trend was observed in *Desmodium intortum*. In Hawaii, the application of 410 kg N/ha/year to a Kikuyu grass/*D. intortum* sward reduced the content of the legume from about 50 per cent in an unfertilized sward to some 10 per cent when cut every 10 weeks, and to less than 1 per cent when cut every 5 weeks; the legumes were apparently unable to withstand frequent cuts which suited Kikuyu grass. Ladino white clover was also reported to be successfully grown with Kikuyu grass in Hawaii, and *Desmodium uncinatum* and *Vicia sativa* in Australia. In general, however, Kikuyu grass/legume cultivation on a commercial scale is perhaps a matter for the future.

In Australia, Kikuyu grass is reported to associate with sown

Paspalum dilatatum and *Axonopus compressus*; increases in soil fertility following the application of N increased the competitive vigour of Kikuyu grass and suppressed the two other grasses.

Management, fertilizing. A plant of fertile soil, Kikuyu grass responds well to nitrogenous fertilizers by increasing herbage yields and protein content of the herbage. In southern Queensland, where Kikuyu is the only widely used grass adequately responding to fertilizer N during both summer and winter grazing, ammonium sulphate, ammonium nitrate and urea, given in equal amounts of N, produced equal linear increases in Kikuyu herbage yields up to about 100 kg N/ha, but at much higher rates, ammonium nitrate gave higher yields than urea. In various fertilizer trials, Kikuyu grass responded to N by increasing herbage DM yields by 12–27 kg/ha for each kg of applied N. The effects of P fertilizers have received little attention, although responses to triple superphosphate were reported from Kenya (Morrison, 1966) and to other forms of P from Hawaii, Zaire, Colombia and Australia. Fertilizer P was particularly useful when applied together with N and a considerable interaction between the two main nutrients was observed. In general, responses to P and K applied singly were inconsistent, and they apparently depended on the soil rather than the grass. Potassium deficiency seldom occurs, but when it does it is manifested in leaf-tip drying. In Australia Kikuyu grass was found to be very sensitive to deficiencies in S, Mg, Cu and Mn and much less so to those of B, Mo, Ca and Zn (Cassidy, 1972). A preliminary trial has shown a positive response of Kikuyu grass to spraying with gibberellin (Lester & Carter, 1970).

Kikuyu grass is unproductive on dry land and especially in dry, hot areas, but high yields can be obtained in such areas under irrigation, as it was shown in California, and there are also reports of good results from Rhodesia and Sudan.

Sod-bound and degraded Kikuyu grass pastures can be restored by heavy discing or ploughing in wet weather (Whyte *et al.*, 1959), especially when oversowing with legumes is anticipated. Under unsuitable conditions, in a dry year or dry season, renovation of old Kikuyu grass can reduce herbage yields in the year of renovation as was the case in Morrison's trial in Kenya, although herbage yields in his trial exceeded those of intact pasture in the following year (Morrison, 1966).

Pests and diseases. The only major disease of Kikuyu grass, known as 'Kikuyu Yellow' and reported from New South Wales, Australia, is manifested in spreading patches of yellow chlorotic leaves. The pathogen is apparently a soil fungus, but no definite details about it seem to be known (from Mears, 1970). In the temperate zone of the USA, Kikuyu grass grown in pots and cages was attacked by *Prosapia distanti*; 10–40 parasites per cage killed the grass (Fagan & Vargas, 1971).

Chemical composition and digestibility. Kikuyu grass is generally rich in protein, the content of which seldom falls below 12 per cent of the DM, and reaches up to 23–25 per cent in young herbage although under unfavourable conditions and in old herbage CP content can be as low as 7 or 5 per cent. The content of CP normally increases with the increase in soil fertility resulting from N and P application. In the Australian subtropics unfertilized herbage contained 11.7 to 16.9 per cent CP when given N fertilizer. There are, however, reports from Hawaii (Whyte *et al.*, 1959; Tamimi *et al.*, 1968) that CP content decreased under high rates of applied N. Similarly to other grasses, CP content decreases with the age of herbage and in Brazil (Gomide *et al.*, 1969a) the grass analysed 4 weeks after the last cut contained 21.6 per cent CP, whereas 12- and 32-week-old herbage contained 15.4 and 13.1 per cent CP, respectively, these contents being higher than those of five other grasses under trial. The content of CF in Kikuyu grass is moderate, usually about 30 per cent, as the leaves are not unduly fibrous and this low content, similarly with a high content of CP, is maintained in Kikuyu grass for a longer period than in most other tropical grasses, possibly because the herbage consists mainly of leaf with little stem, unless the cattle are forced to graze too low and eat fibrous stems below the flowering shoots.

The contents of mineral nutrients in Kikuyu grass herbage vary. The content of P can be reasonably high, 0.20 per cent or even up to 0.40 per cent but can also be as low as 0.11 per cent; the content of Ca is usually on the low side.

The digestibility of CP depends to a considerable degree on its content in the herbage and there are reports from Australia (Jeffery, 1971) that CP digestibility in Kikuyu herbage increased from 58.5 per cent, unfertilized, to 65.8 per cent when fertilizer N was applied. In India (Katiyar & Ranjhan, 1969) the average CP digestibility is reported to be as high as 81.3 per cent. Digestibility of CF, NFE and ether extract (fats) were in India 69.2, 72.5 and 67.1 per cent, respectively, and Butterworth (1967) quotes 52.5, 58.2 and 57.3 per cent for the grass 20–30 cm high, consumed by sheep.

Conservation. Kikuyu grass is almost invariably used for grazing, and attempts to prepare silage from the surplus crop resulted in considerable losses of DM, although the silage was well liked and eaten by cattle. Ensiling of the herbage reduced its digestibility to 46 per cent compared with 64 per cent for fresh herbage (Catchpoole & Henzell, 1971).

Productivity. Yields of Kikuyu grass herbage vary considerably and can be low on poor soil without fertilizers or under an unsuitable climate. Under favourable conditions yields can reach 15 t DM/ha. In Queensland, maximum average yields ranged from 10 to 12 t DM/ha. In Hawaii, Kikuyu grass mixed with undersown ladino and crimson clovers yielded 17 t DM/ha and 1,350 kg CP when fertilized with high rates of N and P and moderate amounts of K

Animal production. In Hawaii, where Kikuyu is one of the most important grasses, beef gains of 182 kg/ha/year from unfertilized grass and 437 kg from grass given NPK were reported; 102 and 257 kg/ha, respectively, were obtained in another trial. Kikuyu grass is used more often for grazing milch cows than for beef production, and in Hawaii fertilized Kikuyu grass/white clover pasture, stocked at 0.84 cow/ha, gave 2,900–3,403 kg milk/cow/year or 2,400–2,700 kg milk/ha. R. L. Colman and J. M. Holder (from Mears, 1970) obtained 99–118 kg butterfat/cow/lactation for stocking rates of 3.29–1.64 cow/ha, which corresponds to 327 and 183 kg butterfat/ha, respectively. At 4.94 cows/ha total butterfat production was increased to 447 kg/ha in the next lactation, but decreased to 261 kg/ha in the subsequent lactations. This shows that high stocking rates and intensive grass utilization are necessary for obtaining full advantages from Kikuyu grass pastures; at low stocking rates Kikuyu grass has no advantage over other grasses (Mears, 1970) but too high stocking rates can weaken it. Ungrazed Kikuyu grass which is allowed to grow freely can be invaded by pasture weeds such as e.g. *Digitaria scalarum* in Kenya (Edwards, 1940). To obtain high yields of beef or milk in the areas with well expressed dry seasons, supplement concentrates should be given to the animals during the periods of drought.

When rapid growth follows a period of drought, or when cattle have previously been on a poor diet, grazing lush Kikuyu grass can cause disorders in the animals expressed in abdominal swellings, incoordination of leg movement and other symptoms, which, in acute cases, can result in death. These disorders have so far been recorded from subtropical areas of Australia and New Zealand.

Flowering, cytology, breeding. As a rule, Kikuyu grass flowers at the bottom of the sward when the herbage is closely mown or grazed, although occasional flowering can take place on taller shoots. Fertile shoots, crowded on much branched short stems can be very numerous and they form the main bulk of the short sward. The flowers are protogenous, the stigmas emerging a day or so earlier than the stamens. The stamens do not however develop in male sterile types, one of which was found by Edwards and named Rongai ecotype. The mode of reproduction in Kikuyu grass has not been sufficiently investigated but 'it is presumed that cross pollination does occur but the possibility of apomictic reproduction cannot be entirely discounted' (Wilson, 1970), and Narayan (1955) suggests that the formation of an aposporic embryo in the ovule observed in Kikuyu grass can indicate apomixis. Apart from the three ecotypes distinguished by Edwards, a certain variability has been observed in plants grown from seed or from cuttings taken from different localities, and this variability can serve as a basis for selecting superior types or for breeding work. In Australian subtropics cv. **Whittet** and **Breakwell** have been developed. Breakwell differs from the former cultivar by a denser growth, and although it is less productive than Whittet it is more resistant to invasions by weeds or by other grasses; some 15 per cent of Breakwell plants are male sterile. Breeding for frost resistance with a view to extending Kikuyu grass cultivation still further from the tropics is also receiving attention. Wilson (1970) also reports that strain P713, an introduction to Australia from Kenya, is agronomically superior to the locally selected type; the latter gives however slightly higher seed yields.

Seed production. Flowering shoots of Kikuyu grass can be very numerous and seed is freely formed. Seed is however difficult to harvest and clean, and is not in commercial supply; farm-scale establishment is normally effected vegetatively, by sprigs. Were seed available, propagation by seed would be easier and cheaper as seed germinates freely, to 75–90 per cent (Wilson, 1970), and relatively large seeds, about 400,000 per kg, produce strong seedlings. According to Andrew & Jayawardana (1971) the seeds require a few months of post-harvest maturation. Harvesting can be done by hand and in Kenya small quantities of seed, up to 1 kg, were obtained by collecting the tops of fertile shoots, placing them in a strong bag and beating with sticks. Wilson (1970) in Australia developed a mechanical method of harvesting: the herbage is cut to a 64 mm level and removed; the resulting short sward is rotovated to a 13 mm level, the cut heads being collected in the attached seed catcher. The heads are then dried and hammer-milled at 1,255 r.p.m. Up to 482 kg seed/ha were harvested in this way. This method seems to be suitable for harvesting relatively small areas, to obtain seed for experimental work, for the establishment of nurseries to bulk vegetative planting material, or for breeding programmes.

Pennisetum pedicellatum Trin. Deenanath grass; Deenabandhu grass (India); Annual kyasuwa grass (Nigeria)

Annual 40–150 cm high with much branched stems which have up to 10 nodes. Leaves glabrous, flat, up to 40 cm long and 4–16 mm wide. Inflorescence a moderately dense spike 5–15 cm long and 8–16 mm wide excluding bristles, green, pale or purplish, and beset with clusters of two to five spikelets, or sometimes with solitary spikelets, surrounded by numerous fine bristles. The bristles are up to 12 mm long; one bristle is longer than the rest and 16–28 mm long. Solitary spikelets are on 1–2 mm long pedicels; of the grouped spikelets one is sessile and the others pedicelled. Spikelets 4–5 mm long, with two florets of which the lower is male and the upper bisexual, fertile, 2.5–3 mm long, smooth and shiny. This grass resembles large forms of *Cenchrus ciliaris*.

Occurs naturally in tropical and subtropical Africa (but practically absent in tropical East Africa) and in India, on disturbed land, at forest edges and in bush. It grows in relatively dry or moderately humid areas where annual rainfall ranges from 500 to about 1,000 mm but requires well-moistened soil during the period of active growth; it can grow on poor soil but gives much higher yields on fertile, well-drained loams.

Pennisetum pedicellatum has been under trial and also under farm-scale cultivation for a number of years, at first in Nigeria, then in other West African countries and eventually in India where it became a popular grass. It has also been introduced to Australia and the Philippines.

Establishment is effected by seed which is sown broadcast or in 45 cm rows (Narayanan & Dabadghao, 1972), the sowing rates being 2–2.5 kg/ha of cleaned seed or 8–10 kg uncleaned seed. The initial growth is fast and the grass can be utilized some 3.5 months after sowing. The main uses of the grass are for soilage and grazing but hay is frequently made and the grass can also be ensiled (Borget, 1968), although Miller *et al.* (1963), in trials with sheep in Nigeria, have found that digestibility of CP in silage is low, 13–35 per cent, depending on the stage of growth of the ensiled grass. Although *P. pedicellatum* is an annual it regenerates easily after cutting and can be cut twice, and sometimes three times, per season and gives a useful aftermath after it had been cut for fodder or after a crop of seed. Whyte (1964) even states that *P. pedicellatum* can last for up to 3 years. It can also regenerate from natural seeding if allowed to reach the seed-ripe stage. Early cutting gives smaller yields of herbage than later cuttings but of better quality and Prasad & Mukerji (1961) obtained the highest herbage yields when the crop was cut 110 days after seedling emergence and when it was cut 7–8 cm from the ground level; earlier and higher cuts resulted in lower yields. In India, a mixture with *Vigna radiata* (*Phaseolus mungo*) and *Melilotus alba* are recommended.

Herbage yields vary and Narayanan & Dabadghao quote yields of 33.6 and 45 t green fodder/ha. Bose (1965) gives a comparable yield of 37.8 t for grass grown on poor soil, and in a trial by Chatterjee & Singh (1967) the yields ranged from 54.5 t/ha when the grass was cut 6–8 weeks after sowing and again at the ear-emergence stage, to 78.8 t when it was first cut at ear emergence. In trials in Cameroun, Barrault (1973) obtained 4.48 t hay/ha from unfertilized grass, and 7.00 and 8.27 t/ha when 40 or 80 kg N/ha were applied, respectively. Yields also depend on rainfall and Barrault quotes hay yields of 6.4, 8.7, 10.2 and 14.2 t hay/ha for the areas with an annual rainfall of 750, 1,250, 1,750 and over 2,000 mm, respectively. In several trials *P. pedicellatum* outyielded *P. americanum*, fodder sorghum, Sudan grass and *Sorghum almum*.

The herbage is leafy and palatable to stock but CP content is usually reported to be not particularly high, ranging from 5 to 9 per cent. Barrault (1973) recorded however the content of N in dry herbage to be 3.03, 1.60, 0.92 and 0.55 per cent in grass cut after 30, 45, 60 and 80 days of growth, respectively, which corresponds to CP contents (N × 6.25) of

18.9, 10.0, 5.7 and 3.4 per cent. The content of CF is usually not high, especially in young grass which can contain 22–25 per cent CF, but can be 33 per cent in the more advanced herbage. Dry matter digestibility was reported by Johri et al. (1969) to be 56.3 per cent, CP digestibility 50 per cent, EE 63 per cent, CF 65 per cent and NFE 62 per cent. The digestibility of CP in northern Nigerian trials with cattle were 47.3 per cent and 65.3 per cent for later and earlier cuts, respectively (Miller & Rains, 1963).

Progeny tests have shown a strongly apomictic nature of *P. pedicellatum* and cyto-embryological investigations also suggested apomixis but with 1–2 per cent of sexual reproduction (Whyte, 1964); pseudogamic flowering was also evident. The chromosome numbers, according to Nath & Swaminathan (1957), are $2n = 36$ and 54 in tetraploid and hexaploid plants, respectively, and Joshi et al. (1959) determined the number of chromosome to be 48 in other plants of the same species. *Pennisetum pedicellatum* varies and certain superior types were selected, e.g. G.73 and T.15, both of Indian selection, and Barrault (1973) in the Cameroun selected over 16 valuable ecotypes. The pattern of internode growth can be different in different clones (Singh & Yadav, 1971) and types with one, two or more peaks of internode growth were distinguished; plants with two or more peaks were superior in forage production compared with those having only one peak of growth.

The length of the period between seedling emergence and earing can depend to a considerable degree on the rainfall and Barrault (1973) observed earing 103–112 days after sowing under an annual rainfall of 700 mm and 130 days under 1,200 mm. Seed production is often combined with herbage utilization: cutting for fodder when the plants are 60 days old and then allowing growth for seed resulted in high yields of herbage and seed. Whyte (1964) states however that *P. pedicellatum* can produce about 450–750 kg/ha of uncleaned seed if the first cuts are used for fodder and 1,100–1,350 kg when the grass is grown for seed only.

Pennisetum polystachion (L.) Schult. Thin napier grass (India); Kyasuwa; Mission grass (Fiji); N'golo (Mali). $2n = 54$

Tufted short-lived perennial or annual. Stems thin, usually branched, up to 150 cm in height. Leaves firm, up to 50 cm long and 3–15 mm wide. Inflorescence a terminal dense, cylindrical spike 3–25 cm long and 5–10 mm in diameter (excluding bristles), straw-coloured, orange-brown or purple. Spikelets solitary, supported by involucres of up to 30 bristles. Bristles slender, with long hairs in the lower part, up to 9 mm long; one bristle up to 15 mm long, longer than the rest. Spikelet 4–5 mm long, with two florets. Lower floret male or empty; upper floret bisexual, fertile, 2–3 mm long.

Distributed throughout tropical Africa and in India, Sri Lanka, Mascarene Islands and also recorded from Fiji and South America.

Occurs in open grassland, in thin bush, on fallow land, roadsides, etc., at low to medium altitudes of moderately dry areas. This grass is sometimes cultivated in the countries of its origin, especially in India, and has also been introduced to a few other countries including northern Australia. It naturalizes fairly easily and South American and Fiji records may perhaps be referred to naturalized plants. *Pennisetum polystachion* is valued for good seed production, the ease of establishment and high herbage yields.

Pennisetum polystachion is established from seed which is broadcast or sown in rows. Naranayan & Dabadghao (1972) also advise raising the seedlings in nurseries and transplanting them to the field, but this elaborate establishment can hardly be used on a large scale. Seed germination percentage is only six to eight after 1 year of storage and 14–22 per cent after 2 years as mentioned by Whyte (1964), who comments that the low germination is compensated by high seed yields. Seedlings are strong and grow fast and *P. polystachion* is sometimes used as an admixture to slow-establishing grasses to 'protect' them during the first year of growth.

Yields of herbage are usually high and Whyte (1964) gives Indian examples of 31.9 t fresh material/ha in the first year of growth, 53.8 t in the second, 44.6 t in the third and 14.8 t in the fourth year. Singh *et al.* (1968) report yields of 5.27 t DM/ha of unfertilized grass and 9.77 t DM and 1,056 kg CP/ha when given 80 kg N/ha.

The herbage can be used for grazing or hay and after cutting or grazing usually recovers slowly. Mixtures with legumes have been tried with varying, but mostly reasonable success, the legumes tried being *Centrosema pubescens*, *Stylosanthes guianensis*, *Atylosia scarabaeoides*, *Clitoria ternatea* and *Calopogonium mucunoides* in India and *S. humilis* in northern Australia. In India, an unfertilized *Centrosemce pubescens* mixture yielded 9.23 t DM/ha and a *Stylosanthes guianensis* mixture 6.45 t compared with 5.27 t from pure *P. polystachion*, the respective effects of the two legumes on the yields being equal to those of 74 and 32 kg/ha of applied N, and 97 and 41 kg/ha when related to CP yields (Singh *et al.*, 1968).

The nutritive value of *P. polystachion* herbage is generally considered to be only medium and Majumdar & Roy (1968) in India determined the content of CP as being 6.4 per cent for the herbage cut at 4-week intervals and 4.9 per cent at 6-week intervals; the contents of some other essential nutrients were 46.2 per cent NFE, 0.41 per cent Ca and 0.51 per cent P, and 49.7, 0.29 and 0.28 per cent respectively. Dougall and Bogdan (1965) reported an unusually high CP content, 17.6 per cent, for the herbage cut early in the season at the very beginning of flowering; this herbage also contained 27.1 per cent CF, 40.8 per cent NFE, 0.31 per cent Ca and 0.22 per cent P.

Seed production is good and a high yield of 417 kg/ha of uncleaned seed containing 16–27 per cent caryopses was obtained in India (Mishra

& Chatterjee, 1968) from stands receiving moderate amounts of N and P fertilizers. Cutting the herbage for fodder early in the season reduced seed yields to 240 kg/ha and to 128 kg when the plants were given no fertilizers.

Pennisetum purpureum Schumach. Elephant grass; Napier grass; Napier's Fodder; Herbe Eléphant; Elefante; Pasto gigante. (Fig. 31).

Robust perennial forming large broad clumps spreading by stem bases rooting from nodes or by short rhizomes. Stems erect, branching in the upper part, with up to 20-noded, 2–6 m tall and up to 3 cm in diameter in the lower part. Leaf sheaths glabrous or beset with hard stiff hairs. Leaves 30–120 cm long and 1–5 cm wide, glabrous or hairy, especially towards the base, with a prominent midrib on the lower side. Panicle terminal on the main stem and on side branches, spikelike, dense, cylindrical, 10–30 cm long and 15–30 mm wide (excluding the bristles), of varying colour (greenish, yellow, brownish, or purplish), with a densely hairy rhachis. Bristles surrounding the spikelets numerous, 10–16 mm long, one longer than the rest and 12–40 mm long. Spikelets 5–7 mm long, solitary or in clusters of two to five of which usually only one is fertile. The lower glume minute or suppressed, the upper 0.5–1 mm long. Lower floret male or empty, upper bisexual, fertile or sometimes male. Grain 2 mm long.

Maire (1952) distinguishes three subspecies:

1. Ssp. *benthamii* (Steud.) Maire & Weiller; stems very thick with very hairy nodes, ligule with elongated membraneous lower part, the joint between the leaf sheath and blade yellow. The two other subspecies have glabrous or slightly hairy nodes, the ligule is reduced to a rim of stiff hairs and the leaves have reddish joints:
2. Ssp. *purpureum*; panicle dense, straight.
3. Ssp. *flexispica* (K. Schum.) Maire & Weiller; panicle relatively loose, somewhat flexuose.

Whyte *et al.* (1959) mention var. *merkeri* Leeke but Maire regards this name as an error because *Pennisetum merkeri* Leeke is a synonym of *P. schimperi* which is an entirely different plant. There is however *P. merkeri* Trabut, a synonym of *P. purpureum*. Maire includes the small 'Merker' type into ssp. *flexispica*.

Pennisetum purpureum occurs naturally throughout tropical Africa, at river banks and, in more moist areas, on dry ground in savanna, often on fallow land, where it can form extensive colonies. Wild populations are often used for cutting and feeding fresh fodder to cattle, but the grass is commonly utilized in cultivation. *Pennisetum purpureum* has been introduced to practically all tropical countries and to subtropical areas, and is widely grown from sea level to 2,000 m alt for fodder and less often for grazing. The grass is valued for its high herbage yields, competitive vigour and persistence, palatability and good herbage quality.

234 *Pennisetum*

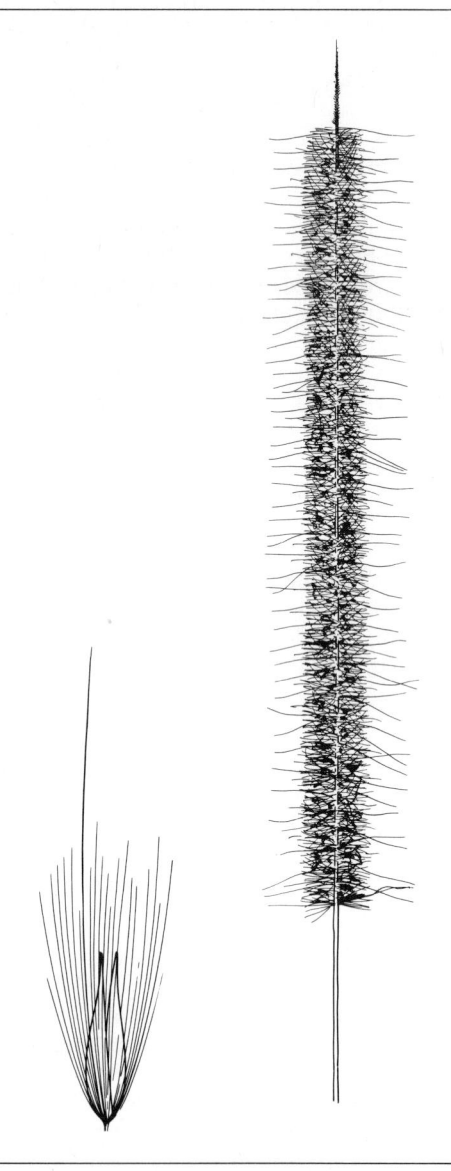

Fig. 31 *Pennisetum purpureum* (from Edwards & Bogdan, 1951).

Environment. *Pennisetum purpureum* gives higher yields under longer than shorter photoperiods (Wang, 1961). It grows best at high temperatures but can tolerate low air temperatures under which the yields are, however, reduced and the plants cease to grow at temperatures below 10°C. The herbage can be killed by light frosts but the underground parts remain alive unless the soil is frozen. Because of its tolerance to low temperatures *P. purpureum* has spread far beyond the tropics and is grown, e.g. in North Africa (Morocco) and in the warmer parts of the USA. In its wild state *P. purpureum* normally grows in the areas where the annual rainfall is over 1,000 mm but occurs also on river banks in semi-arid areas. In cultivation it can withstand considerable periods of drought although little or no growth is produced during these periods; it rapidly recovers with the onset of rains and grows fast. For high production *P. purpureum* requires fertile soil but can grow on almost any soil with reduced vigour and production. It cannot tolerate flooding or waterlogging but can grow and produce reasonably well on drained waterlogged soils on slightly elevated camber beds (Gosnell & Weiss, 1965).

Establishment. Pennisetum purpureum produces, with occasional exceptions, little or no seed, the seedlings are small and weak, and have been noted for their slow growth and the grass is normally established vegetatively, by tuft divisions or, more often, by stem cuttings. Cuttings are made from moderately mature stems and contain three nodes each. The soil is ploughed and disced or harrowed and preferably marked into lines along which the cuttings are stuck into the soil at an angle or vertically to such a depth that two nodes are in the soil and one is above the soil surface. Whenever possible the cuttings should be placed the basal end down; the basal end of the cutting can easily be recognized by the position of the bud which is situated just above (and not below) the node. The cuttings planted upside down would grow but the establishment can be delayed. The cuttings can be stored for up to 20 days, but in the subtropics with cold or cool winters the cuttings can be stored during the whole winter. When planted, the lower nodes develop roots and shoots and the upper one shoots only. Planting can also be done by placing the cuttings in furrows made by plough. The cuttings are planted in rows from 50 to 200 cm apart; the wide rows are more suitable for dry areas. The distance between the plants in rows is usually 50 to 90 cm. If available, a sugar-cane planter can be used; the machine has a tine for making a deep furrow, a hole through which the cuttings are fed into the furrow and an arrangement to cover the cuttings with soil in the rear part of the machine (Grof, 1969b). Whyte *et al.* (1959) report that in Trinidad successful establishment was achieved by ploughing cuttings into the soil. Moderate amounts of P fertilizer, some 100–200 kg superphosphate/ha, can be used at the establishment; the fertilizer is preferably placed near the rows of planted splits or cuttings. If f.y.m. is used it can be applied at ploughing. To control weeds at the first stages of establishment, inter-row cultivation and herbicides can be used; in Cuba 6 kg atrazine/ha was more effective than other herbicides under trial (Casamayor, 1970). In general, *P. purpureum* is a strong competitor and

Farinas (1970) states that in the Philippines it can suppress *Imperata cylindrica* when planted into the soil overgrown with this grass weed. Planting is done at the beginning of the rains in the areas with well-defined dry seasons and at any time if the rains continue throughout the year or the land is irrigated. The grass can be cut or grazed some 3 months after planting.

Management, fertilizing. *Pennisetum purpureum* is cut for soilage or silage, or is grazed. When grazed the plants should not be allowed to grow over 100–120 cm in height and for cutting over 200 cm. Grass productivity can differ under cutting and grazing and cutting often, but not always, results in a higher production of utilized herbage than grazing. Rotational grazing is more important for *P. purpureum* than for smaller and less clumpy grasses. The intervals between cutting or grazing usually range from 4 to 14 weeks. Yields of herbage are almost invariably higher under less frequent cutting but the content of CP in the herbage decreases with the decrease of cutting frequency. In Puerto Rico (Rivera-Brenes *et al.*, 1962) cutting the grass every 60 or 90 days gave 62 and 72 t green forage/ha, respectively, and the percentages of CP were 10.9 and 6.9. In Australia (Grof, 1969b) cutting *P. purpureum* every 4, 6 or 8 weeks resulted in DM yields of 9.16, 15.61 and 19.04 t/ha, respectively, and CP content of 11.0, 8.2 and 6.4 per cent. CP yields differed, however, only slightly and were 904, 1,155 and 1,105 kg/ha. With the decrease in the frequency of cutting the yields of TDN normally increase, the increases being somewhat slower than those of DM. The height of cutting is a controversial issue: in Puerto Rico (Caro-Costas & Vicente-Chandler, 1961) heavily fertilized grass cut at practically the ground level, 0–7 cm, gave higher yields, 31.25 t DM/ha, than when cut at 18–25 cm, which gave 26.01 t. On the other hand, in a trial in Brazil (Werner *et al.*, 1965–6), in which the grass was cut at 4-week intervals at 1–3, 30–40 and 70–80 cm above the ground level, DM yields were 4.47, 11.9 and 13.12 t/ha, and cutting at the ground level produced the poorest results. It should be noted that cutting or grazing at fixed levels is difficult to maintain because the stool increases in height, though usually slightly, from harvest to harvest. Irrigation is often applied and Pereira *et al.* (1966) report a 70 per cent increase in herbage yields obtained from sprinkler irrigation in winter months.

Pennisetum purpureum responds well to fertilizers and especially to fertilizer N especially when it is applied with basic dressings of P, K and sometimes also Mg. Grof (1969b) states that an increase of N fertilizer rate from 112 to 280 kg/N/ha doubled the herbage yields. However, even small amounts of applied N can be effective as it was shown in Uganda, where Stephens (1967) obtained an average yield increase from < 13 to 20 t DM/ha by an application of 40 kg N/ha in the first year and 80 kg yearly in subsequent years. Large yield increases were also obtained at high fertilizer rates and in Costa Rica (Guerrero *et al.*, 1970) an increase in applied N from 200 to 600 kg/ha gave an increase of herbage yield from 5,456 to 14,092 kg DM/ha. In Puerto Rico responses to N applied at 800 kg/ha and over were obtained, but such responses are not frequent. The most effective rates of N, in terms of kg DM obtained from 1 kg of applied N, have been found at much lower rates of N, and in Colombia 120 kg N/ha were about optimum (Buenaventura, 1962). Responses to N, even when applied in small dressings, usually increase with plant age. Applied N normally increases the content of CP in the herbage, often very substantially, and its percentage in DM can reach 18 per cent, and even 20 per cent, under high rates of N. P produces a less marked effect than N and increases the yields of fodder usually only in conjunction with applied N. Little information is available on the effects of K or Mg although Stephens states that K and Mg increased the yields of grass cut for soilage but had no effect on grazed grass.

Conservation. Cut herbage is usually fed fresh but silage has also been made. Although the ensiling material does not settle easily, satisfactory silages have been prepared. Losses of DM during the ensiling are usually low to moderate, about 9–12 per cent, but losses of over 20 per cent were also reported. Seven to fifteen kg of molasses per ton of grass were insufficient for lactic-acid fermentation but soaking in molasses solution resulted in a pH value just under 4, a level sufficient for correct fermentation (Catchpoole & Henzell, 1971). Low TDN content in the silage was reported but Butterworth (1967) in his review gives 55–61 per cent TDN for silages prepared in Venezuela and fed to sheep. De Lucci *et al.* (1968) report from Brazil that TDN and DM consumptions by cows fed on *P. purpureum* silage were lower than those of sorghum and maize silages. It was found in Brazil that daily intake of silage by steers was greater than of fresh herbage and daily liveweight gains were slightly higher; the silage contained less CP and digestible N than the fresh herbage and more NFE and was more suitable for steers than for lactating cows or calves.

Diseases and pests. Various fungus pathogens have been recorded on *P. purpureum*, *Helminthosporium ocillum* being perhaps the most important one and causing a serious lifespot leaf disease; some varieties, e.g. cv. Uganda Hairless, seems to be resistant to this disease. Red mite inflicted considerable damage to grazed grass grown in Kenya; the mites live on the lower side of leaves, under the protection of hairs and were not observed on varieties with glabrous leaves.

Herbage yields. On good and heavily fertilized soil, and in a warm humid climate, or under irrigation, *P. purpureum* can produce high yields of herbage. The highest yields reported in recent literature seem to be 310 t fresh material or 71.9 t dry herbage/ha from cv. Mineiro harvested in Brazil during the 13 months of trial (Zúñiga *et al.*, 1967) and

84.7 t DM obtained by Vicente-Chandler et al. (1959) in Puerto Rico; Cooper (1970) also quotes 85.1 t DM/ha obtained in El Salvador. Somewhat lower yields were obtained in Brazil by Pereira et al. (1966): in a wet summer yields of green material ranged from 170 to 220 t/ha, depending on cultivar; the highest yields were given by cv. Napper. In the dry season the yields ranged from 10 to 19 t/ha and NPK fertilizers produced no effect or slightly decreased the yields; however, when combined with sprinkler irrigation the fertilizers increased the yields to 57 t/ha in cv. Merkeron and to 30–40 t in other cultivars used in the trial. High yields, 57.7 t DM/ha/year, were also recorded in Costa Rica (Little et al., 1959). More realistic yields, up to 30 t DM/ha in the wet season and up to 6 t in the dry season, were obtained in Brazil by Carvalho et al. (1972). Yields that can be expected in farm practice may range from 2 to 10 DM/ha/year for unfertilized or slightly fertilized stands and from 6 to 30 t from grass well fertilized with N and given a basic dressing of P. Irrigation is necessary for obtaining high yields in dry areas or in dry seasons.

Association with legumes. Pennisetum purpureum is usually grown in pure stands but cultivation in mixtures with legumes on experimental and sometimes on farm scale has also been reported. Large and vigorous legumes capable of withstanding *P. purpureum* competition are normally used for the purpose. Mixtures with *Pueraria phaseoloides* had good success in Belize, formerly British Honduras, and in West Indies where *P. purpureum* is planted into established swards of the legume. *Desmodium intortum* had a reasonable success in Hawaii where it fixed about 400 kg N/ha, N fixation being only slightly reduced when it was grown with *P. purpureum* (Whitney et al., 1967). In a Kenya trial in which DM yields of *P. purpureum* were 5.74 t/ha in one year and 4.88 t in another year, its mixture with *D. intortum* yielded 8.68 t and 13.56 t, respectively, and similar results were obtained from *D. uncinatum* mixtures (Kenya Report, 1970). In another trial in Kenya (Suttie & Moore, 1966) *D. uncinatum* grown with *P. purpureum* increased the content of CP in the grass from 6.0 to 7.1 per cent, CP content of the mixture being 9.8 per cent. Mixtures with *Centrosema pubescens* have been grown in Australia and Hawaii where 11 per cent of N fixed by the legume was transferred to *P. purpureum* during the 6 months of active growth (Whitney, 1966). Other legumes tried in mixtures with *P. purpureum* were *Stizolobium deeringianum* and *Glycine wighti*.

Chemical composition, nutritive value. At the early stages of growth *P. purpureum* herbage contains a large proportion of water and only 12–18 per cent DM, less than in the majority of other tropical grasses, but the content of DM increases rapidly with plant age. CP content varies widely and depends on the frequency of cutting or grazing, soil fertility and the fertilizers applied and on the proportion of leaf in the herbage.

In trials in Venezuela by Rodríguez & Blanco (1970) the average content of CP in the herbage cut every 30, 60 and 90 days was 9.36 per cent for the leaves and 4.38 per cent for the stems, and the contents of CF 31.14 per cent and 35.62 per cent, respectively. Ware-Austin (1963) in Kenya observed seasonal variation in CP content of frequently cut herbage from 13.0 to 19.7 per cent and only in the driest month of February CP content fell below 10 per cent, and Gomide *et al.* (1969a) in Brazil gave also high CP values: 23–24, 12 and 7 per cent in 4-, 8- and 24-week-old herbage, respectively, with the average estimated content of CP about 10 per cent, comparable with CP content of five other grasses under trial except that CP content of young herbage was much higher in *P. purpureum* than in the other grass species. High CP contents, ranging from 15.6 per cent in 3-week-old herbage to 9.7 per cent in 8-week-old herbage were also observed by Appelman & Dirven (1962) in Surinam. Butterworth (1967) quotes in his review a number of analyses performed in various countries in which the contents of CP ranged from 4.4 to 20.0 per cent, of CF from 26.0 per cent in young herbage to 40.5 per cent in mature plants, NFE from 30.4 to 49.8 per cent, and of EE from 1.0 per cent to 3.6 per cent. In the Rodríguez and Blanco's trials P content was high and ranged from 0.28 to 0.39 per cent in the leaf and 0.38 to 0.52 per cent in the stem, and the content of Ca from 0.43 to 0.48 per cent and from 0.14 to 0.23 per cent, respectively. Digestibility, as quoted by Butterworth (1967) can range from 48 to 71 per cent for DM, 41 to 71 per cent for CP, 46 to 75 per cent for CF, 40 to 74 per cent for NFE and 19 to 73 per cent for EE. The content of TDN can range from 40 per cent in plants 230 cm tall to 67 per cent in young and shorter plants and those of DCP from 0.9 to 14.8 per cent. A low pH value of fresh herbage well below 6, was reported for *P. purpureum*, as well as for other allied genera and species, by Dougall and Birch (1966). Palatability of the majority of cultivated types is good and cattle eat the grass readily but at an advanced stage of growth the stems are usually left uneaten and only the leaves are stripped off and eaten. Varieties with stem nodes and leaf-sheaths covered with stiff, easily breaking hairs are avoided by the animals because eye damage can be inflicted by the hairs.

Animal production. High animal production has been obtained from cultivated *P. purpureum*. Beef cattle fed on unfertilized grass gained 328 kg/ha per year in Brazil (Lima *et al.*, 1968) and 280–323 kg in Hawaii (Plucknett, 1970) on grass fertilized with 118–200 kg N/ha; the liveweight gains were raised to 406 and 400–462 kg/ha, respectively. On heavily fertilized grass liveweight gains up to 1,200 kg/ha were obtained in Puerto Rico. In Puerto Rico, cows grazing on an old pasture given 560 kg NPK mixture/ha yielded 6,200 kg milk/ha (Caro-Costas & Vicente-Chandler, 1969). Grof (1969b) reports that liveweight gains from grazing a *P. purpureum*/*Centrosema pubescens* mixture amounted to 820 kg/ha.

Flowering, reproduction. **Pennisetum purpureum** flowers under a relatively wide range of photoperiods (Wang, 1961) although Maire (1952) states that ssp. *benthamii* never flowers in North Africa. Some varieties and cultivars develop panicles and flower more readily and earlier than the others. The flowers are strongly protogynous, the stigmas emerge 3–4 days earlier than the anthers (Grof, 1969c). Brown & Emery (1958) have found *P. purpureum* to be an apomict; their conclusion was however based on the cytological evidence only and there are indications (Hosaka & Ripperton, 1948; Grof, 1961) that at least some of the numerous varieties, and particularly diploid ones, can reproduce sexually. Grof observed that in the process of developing cv. Capricorn plants grown from seed obtained under open pollination from the Merkeron type plants varied and developed segregating populations characteristic of cross-pollinated plants and it is recommended not to propagate cv. Capricorn by seed if varietal purity has to be maintained, a measure unnecessary for the apomicts. In a later work Grof (1969c) obtained sexual progenies from a vigorous cv. Cameroon; some plants of these progenies were more vigorous than the parent plants and yielded twice more DM/plant; this performance is regarded as a manifestation of hybrid vigour which can be maintained by vegetative propagation, a usual practice in *P. purpureum* cultivation.

Seed setting in *P. purpureum* varies but is usually poor although good seed formation has been reported from the north-eastern coast of Australia. Germination percentage can reach 70 per cent in better seeding varieties and up to 15 per cent in poor seeders. No post-harvest maturation is needed and maximum germination has been observed immediately after harvesting (Grof, 1969b). After 6 months of storage germination percentage decreased from 49 to 5 per cent in better seeding varieties and from 15 to 2 per cent in poorer seeders. Seed, as harvested, is fluffy and needs processing, i.e. separation of fertile spikelets from involucres and other chaff after which the spikelets flow easily. They are small and 1 kg contains some 3 million of fertile spikelets or 'seeds'. Vegetative propagation is more reliable and quicker than the establishment from seed, and there is no demand for commercial seed.

Variability, cultivars. **Pennisetum purpureum** is a variable species and the three subspecies, recognized by Maire have already been briefly described; these subspecies are however not much known to the pastoralists. The name of var. *merkeri*, though botanically invalid, is popular and cultivars of Merker and of Merkeron groups are widely grown, and the name Merker, understood as a certain type of *P. purpureum* can better be retained.

Forms of *P. purpureum* differ in vegetative characters: the thickness of stems, the size of leaves, the hairiness of stem nodes and leaf sheaths, general vigour, the size of tufts, the number of tillers and the height of plant. Floral characters also differ and the panicles vary in size, colour and density, and in some other details of panicles and spikelets but the floral characters are rarely observed or studied because the plants are usually not allowed to grow high and to reach the flowering stage. There are also differences in chromosome numbers and diploids with $2n = 28$ and tetraploids with $2n = 56$ have been found and also a type with an irregular number of chromosomes, $2n = 27$. No information seems to exist on the linkage between the plant size and morphology and the number of chromosomes. Plants within a cultivar are usually uniform although mixtures of two or three forms also exist; the uniformity being mainly due to the fact that the cultivars are usually of clonar nature, i.e. are vegetative progenies of a single plant. The cultivars can be roughly classified into: large hairy, large glabrous or almost glabrous, and small hairy or slightly hairy and this corresponds to a certain degree to the Maire's subspecies. The well known Merker group of cultivars belong to the last named type.

Numerous cultivars exist, named or grown under numbers given by certain experimental stations. The better known cultivars are:

1. **Merker** and also rather similar **Merkeron** cultivars. This is a group which may or may not have a common origin. Selections from this type are known under numbers as e.g. Merkeron 534 (= Costa Rica 534) which is reputed to be an even better yielder than the ordinary Merkeron. Here probably also belongs a type which is known in Kenya as cv. French Cameroons (Bogdan, 1965). The Merker and Merkeron plants have numerous, relatively thin stems and narrow leaves which are glabrous or nearly glabrous; they are good yielders and are now popular in South America, West Indies and in other parts of the world, and Crowder *et al.* (1970) write that in Columbia cv. Merker, Merkeron and Costa Rica 534 produce 40–50 t fresh herbage every 35–40 days and are replacing the 'Napier' and the 'Common' types.
2. **Napier**. This name is sometimes used to denote *P. purpureum* as a species and sometimes only as a certain type of it; the characteristics of this type are however not sufficiently well defined.
3. **Capricorn**. A cultivar developed in Australia from plants of Merker type. It flowers late and is leafy and palatable (Barnard, 1972).
4. **Mineiro**. A cultivar with good records in South America because of its vigorous tillering, slow stem lignification and good production; in a variety trial in Brazil (de Carvalho *et al.*, 1972) it yielded 21.4 and 30.0 t DM/ha in two rainy seasons and 3.3 and 6.0 t DM in two dry seasons during 2 years, outyielding the other 11 cultivars under trial.
5. **Uganda hairless** or **Uganda**. This cultivar has glabrous leaves which are narrow and long; resistant to the eyespot disease caused by *Helminthosporium*.

Other cultivars worth mentioning are **Cubano**, **Domira**, **Panama**, **Ghana (Gold Coast)**, **Pungwe** and **Urukwanu**.

Pennisetum purpureum × **P. americanum** (*P. purpureum* × *P. typhoides*) Babala napier hybrid; Bana grass (S. Africa); Giant elephant grass; Pusa Giant Napier; Gajraj. $2n = 20, 21$

Pennisetum purpureum can be relatively easily crossed with *P. americanum* (= *P. typhoides* = *P. glaucum*). The first hybrid was produced in South Africa and released under the name of Babala Napier Hybrid or Bana grass. It produced very encouraging results and had a good success in the early years of testing and farm-scale cultivation, but the interest in this plant gradually faded away and in recent years little has been reported on the South African hybrid. The South African hybrid developed more tillers than *P. purpureum*, had more numerous leaves, grew faster and produced a larger bulk of fodder; the stems were, however, harder and the plants less persistent. Other *P. purpureum* × *P. americanum* hybrids were later developed in India, first at Coimbatore, in the south, and then at New Delhi, in the north, in 1961. The hybrid developed at New Delhi is well known as Pusa Giant Napier and is now grown on a farm scale. It is also grown in Pakistan and has been introduced to Thailand and to Queensland, Australia, where it is well adapted for the areas of fodder sorghum cultivation.

According to Jodhpur (1965) the leaves of Pusa Giant Napier are larger than those of elephant grass, the hairs of leaf blades and sheaths softer and less persistent and leaf edges less sharp. The stems are also less fibrous and the grass can, therefore, be cut at later stages of growth than elephant grass. The tillers of the hybrids are more numerous and grow faster.

Being a sterile hybrid (Narayanan & Dabadghao, 1972) Pusa Giant Napier is planted by splits or by stem cuttings in the same way as elephant grass and at about the same spacing. Cuttings taken from 3-month-old stems sprout better than those from older stems and they should preferably be cut from the lower two-thirds of the stem length (Khan & Syed, 1970). The hybrids yield heavily but to produce a large bulk of fodder they require considerable amounts of fertilizer N, 415–550 kg/ha according to Narayanan & Dabadghao (1972), and Chaudhry *et al.* (1969) obtained in Pakistan much higher yields from 336 kg N/ha than from the lower rates of N. Farmyard manure at 25 tons/ha before planting has also been quoted by Narayanan & Dabadghao.

The Coimbatore hybrid yielded 168 t green fodder/ha and still higher yields, 283 t/ha, were later reported for the Pusa Giant hybrid, these yields being considerably higher than the yields of *P. purpureum*. Jodhpur (1965) obtained 383 t/ha of Pusa Giant green fodder when the grass was fertilized with 112 kg N/ha, whereas under the same conditions *P. purpureum* yielded 138 t, the differences in yields being the greatest for summer cuts and the smallest for winter cuts. In Thailand (Thaiphanich, 1968, from *Herb. Abstr.*, 1972, abs. 715) the hybrid used in trials gave, however, yields similar to those of *P. purpureum*, the yields of both plants being on a low level of 34 t fresh fodder/ha. Pritchard (1971) reports that in Queensland, Australia, the hybrid yielded more fodder than fodder sorghum or *Sorghum almum*.

Herbage quality is reported to be higher than that of *P. purpureum* and it contains 25 per cent more CP and 12 per cent more sugars (Narayanan & Dabadghao, 1972). Daftardar & Zende (1968) report that CP contents ranged in their trials from 22.8 per cent in young, 14-day-old growth to 5.3 per cent in 72-day-old growth; other authors quote CP content being around 8–11 per cent. DM digestibility is reported by Pritchard (1971) to be 65.6 per cent for the leaf and 58.4 per cent for the stem plus leaf-sheaths.

Pennisetum squamulatum Fresen $2n = 54$

Tufted perennial 1–2 m tall. Leaves up to 50 cm long and 5–18 mm wide. Inflorescence a dense cylindrical terminal spike 15–25 cm long and 15 mm wide (excluding bristles). Spikelets in clusters of five to six on short common pedicels and surrounded by fine bristles which are up to 16 mm long; one bristle is longer than the rest and reaches 20 mm. Spikelets 7–8 mm long, the central spikelet in the cluster bisexual, fertile, dorsally compressed, the others male and laterally compressed.

Occurs in Ethiopia and Kenya on rocky slopes or on alluvial soil of river valleys. When tested in cultivation in Kenya (Bogdan, 1963), *P. squamulatum* was outyielded by other large grasses in the wet season, but stayed green much longer in the dry season and the leaves were readily eaten by cattle throughout the season. At the flowering stage the whole plant contained 7.0 per cent CP and a high percentage of CF but the leaves contained 15.0 per cent CP later in the wet season and 8.6 per cent in the middle of the dry season. Seed formation was poor. Introduced to India (Patil & Ghosh, 1962), *P. squamulatum* tolerated extreme temperatures, approaching freezing point in winter and 43°C in summer. Green fodder yields in India can reach over 65 tons/ha with a protein content of 10.5 per cent.

Pennisetum subangustum Stapf & Hubbard. N'golo (Mali)

An annual resembling *P. polystachion* but smaller and with smaller spikelets and shorter bristles; there are also transitional forms between the two species. The spikes are very dense and, as in *P. polystachion*, the spikelets and bristles spread horizontally at advanced stages of flowering and at ripening. The chromosome number, $2n = 54$, is the same as for *P. polystachion*, although other authors (see in Whyte, 1964) give $2n$ chromosome numbers of 24, 32 or 36.

Occurs in tropical and subtropical Africa but rare in its eastern tropical regions. In Mali (Derbal *et al.*, 1959) it produces 12 t fresh herbage/ha and is used for the improvement of natural pastures and is also grown for haymaking; its resistance to drought and to grazing pressure is poor. Boudet (1970) reports from Senegal that CP content during the rainy season can reach 15.3 per cent.

Phalaris L.

A genus of some 15 species which occur in temperate areas. *Phalaris aquatica* L. (*P. tuberosa* L.) is a tufted perennial of Mediterranean origin now widely grown in a number of countries including temperate southern parts of South Africa, South America and Australia, and to a lesser extent in the subtropics of these countries. Three cultivars, Australia, Siro Seedmaster and Sirocco, have been registered in Australia (Barnard, 1972). *Phalaris aquatica* var. *stenoptera* (Hack.) Hitchc. (*P. stenoptera* Hack.), a more vigorous form usually provided with short rhizomes (Whyte *et al.*, 1959), is also cultivated although in Australia this form is usually not recognized as a separate taxon which may be of hybrid origin. *P. arundinacea* L. (*Digraphis arundiacea* (L.) Trin.) is a vigorous rhizomatous species of wet situations common in northern temperate areas and now introduced throughout the world and often naturalized. It has been tried with some success in warm temperate areas and sometimes in the subtropics; this species is also used for crosses with *P. aquatica* and several hybrids have been produced. A hybrid known as Ronpha grass is a particularly vigorous tufted or slightly stoloniferous grass tried in the subtropics and in the high-altitude tropical areas where it did not flower and had to be propagated by splits. In Kenya it was one time popular with the farmers but was soon abandoned because of its low palatability except the first growth of the year of establishment which is readily grazed. Palatability of *P. aquatica* is usually good but that of *P. arundinacea* and its hybrids with *P. aquatica* is much lower. The two species of *Phalaris*, including their hybrids, contain toxic substances which may, under certain circumstances, cause 'phalaris staggers' and 'sudden death' metabolic diseases of sheep (Barnard, 1972).

Rhynchelytrum Nees

Contains up to 40 species which occur in tropical Africa, Madagascar, Arabia, India and Indonesia.

Rhynchelytrum repens (Willd.) C. E. Hubbard (*R. roseum* (Nees) Stapf & C. E. Hubbard: *Tricholaena repens* Willd.; *T. rosea* Nees). Natal grass; Redtop Natal grass. $2n = 36$

Short-lived tufted perennial or annual. Stems 30–100 cm high. Leaves 5–30 cm long and 2–10 mm wide. Panicle moderately loose, white or more usually pink or purple, silkily hairy, 5–20 cm long; ultimate branches capillary with long hairs. Spikelets with two florets, ovate,

2.5–6 mm long, hairy, with the hairs adpressed in young panicles and spreading at maturity. Lower glume up to 1.5 mm long, slightly (up to 0.5 mm) distant from the rest of the spikelet; upper glume as long as the spikelet, gibbous in the middle, with a short beak and a fine awn up to 4 mm long; it is densely pilose with the hairs exceeding the glume by 5 mm. Lower floret male, its lemma similar to the lower glume but narrower and less gibbous. Upper floret fertile, 2–2.5 mm long. Grain 1.5 mm long.

Occurs naturally throughout tropical Africa, Madagascar and in South Africa, under moderate rainfall and mostly on light-textured or sandy, often poor soil, as a fallow grass where it can be abundant. It is also found in various other habitats including natural grasslands where it is however not numerous and of little importance for grazing. *Rhynchelytrum repens* also occurs as a weed of arable land but it is not a noxious weed.

In its native Africa *R. repens* is not much valued as a pasture or fodder plant although in South Africa it has been occasionally used for grassing the land after cropping had been discontinued and the fallow period began. *Rhynchelytrum repens* was introduced to USA, South America, Hawaii, India and Australia where it is more valued than in the countries of its origin and where it has naturalized. In India it is established by sowing only 500 g seed/ha; further propagation being effected by self seeding because seed setting is good and reliable. Under experimental cultivation in India moderate rates of NPK are considered to be sufficient for its vigorous growth on sandy soils where it sometimes reached a height of 150 cm. Indian data also show that yields of 40 to 80 t of green fodder/ha were obtained (Singh, 1969) but such high yields must be an exception. The quality of herbage is reputed to be high and it can contain 11.5 per cent CP. Sen & Mabey (1966) observed however that only young, 8-week growth contained 11.5 per cent CP and this was reduced to 9.0 per cent in 16-week-old herbage and to 5.6 per cent after another month of growth; CF content increased from 30.6 to 31.9 per cent and to 35.0 per cent, respectively. Ca content was nearly adequate, 0.15–0.30 per cent, but P content low – 0.09 to 0.11 per cent. A Kenya analysis (Dougall & Bogdan 1960) showed 11.3 per cent of CP, 32 per cent of CF and 45 per cent of NFE.

Some 10–12 kg/ha of spikelets or 7–9 kg of clean seed (fertile florets) can be harvested which is satisfactory bearing in mind the small amounts of seed needed for sowing. Seed germinates well and soon after harvesting, and no seed dormancy was observed. *Rhynchelytrum repens* is also grown for pasture or hay in Brazil and in Florida, USA (Whyte *et al.*, 1959).

Rhynchelytrum villosum (Parl.) Chiov. is a closely allied species with larger spikelets and more distant lower glume; occurs in eastern and north-eastern Africa under more arid conditions than *R. repens*. It is found mostly in sandy soil pastures.

Rottboellia L.f.

There are four species growing in tropical and subtropical regions of Africa and Asia.

Rottboellia exaltata L.f.

Erect but often declining annual 1–3 m high. Leaf blades and sheaths with long stiff hairs. Spikes, terminal and on stem branches, 6–15 cm long, cylindrical, dense, with the rhachis wider than the spikelets. Spikelets awnless, in pairs, one sessile, the other on a pedicel which is firmly pressed to the rhachis. Sessile spikelet 4–7 mm long, with two florets of which the upper is fertile. Mature spikes easily break into cylindrical joints containing one seed each.

Distributed in tropical Africa, India and eastwards to the Philippines often on black, seasonally waterlogged clays in semi-arid areas or in other situations under more humid climates, often on waste land or as an arable weed. It was not infrequently mentioned in the 1940s and 1950s as a grass of value for grazing or hay and Derbal et al. (1959) state that it produces 20–22 t good-quality fresh forage/ha. There are reports that the herbage contains about 10 per cent of CP, and is highly palatable and suitable for ensiling (Eggeling, 1947; Dalziel, 1948). However, in natural stands the animals often avoid R. exaltata because of discomfort caused by stiff, easily breaking hairs. It seems that the interest in this grass faded in 1960s and 1970s. Sexual reproduction, and diploid, tetraploid and hexaploid chromosome numbers, $2n = 20$, 40 and 60, and also $2n = 36$, have been reported (Brown & Emery, 1958; Whyte, 1964).

Rottboellia selloana Hack. $2n = 18$

Tufted perennial 30–65 cm high. Leaves 12–20 cm long. Racemes cylindrical, erect, 7–10 cm long, easily disarticulating at maturity. Joints and sessile spikelets 4–5 mm long.

Occurs in Uruguay, southern Brazil and north-eastern Argentina, in dry grasslands on sandy soil. Well grazed by cattle.

Saccharum L.

Tall robust grasses five species of which occur in tropical and subtropical Asia, and one species reaches Africa. Some species are used for manufacturing sugar.

Saccharum benghalense Retz. (S. munja Roxb.) Bharra grass. $2n = 60$

Similar to S. officinarum but differs in having thinner stems and narrower leaves which are up to 2.5 cm wide. Spikelets 4–6 mm long, larger than in S. officinarum. A native of India, mainly of north-western parts, it occurs at spring banks and forest edges. Cattle and buffaloes eat only young tender leaves and avoid well-advanced plants, especially in the wet season; it is sometimes recommended to cut young growth for hay. *Saccharum benghalense* has been introduced to Sri Lanka.

Saccharum officinarum L. Sugar cane; Noble cane. $2n = 60, 80$ or 90

Tufted perennial. Stems 2.5–6 m high and 3–6 cm in diameter, glabrous below the panicle. Leaves 70–200 cm long and 1.5–10 cm wide, erect or drooping. Inflorescence a large panicle (known as the arrow or the tassel) 25 to over 70 cm long, loose to fairly dense, feathery, of varying colour, much branched, with fine ultimate branches terminating in articulated racemes. Spikelets 3.5–4 mm long, paired, one sessile, the other on a pedicel, each supported by long fine hairs which are longer than the spikelet. Each spikelet has two glabrous boat-shaped glumes and two florets; the lower floret reduced to a hyaline scale, the upper bisexual, awnless.

Sugar cane originated from South Pacific Islands and New Guinea but has now spread throughout the world tropics where it is cultivated for sugar under moist tropical climates. *Saccharum officinarum* has been studied from the point of view of its utilization as fodder in India, Taiwan, Brazil, Panama and perhaps in other countries. In a trial in Brazil (Zúñiga *et al.*, 1967) sugar cane yielded a large bulk of fodder, 236–284 t fresh material/ha during the 13 months of trial or 62 t DM in unfertilized plots and 72–78 t when fertilized with NPK. The quality of sugar cane forage seems to be satisfactory and in Trinidad (Harrison, 1942) immature plants contained 7.4 per cent CP and 4.2 per cent DCP. The same forage contained 35.4 per cent CF and 46.3 per cent NFE and their digestibility in trials with cattle was determined to be 65.3 and 58.2 per cent, respectively. Sugar cane tops as a residue of canes cut for sugar production and constituting up to about 10 per cent of the crop are perhaps of more interest as a source of fodder than the whole plant. The tops were studied in India, Taiwan, Hawaii, Trinidad, Mexico and Brazil and in Mexico they contained 3.9 per cent CP, 29.8 per cent CF, 1.5 per cent fat and 55.4 per cent NFE (Teunissen & Villarreal, 1966), but higher contents of CP, 4.5–9.0 per cent, CF, 30.5–35.4 per cent, and NFE, 53.7 per cent, were recorded earlier in Trinidad (Harrison, 1942). The quality of sugar cane tops as feeding material, although on a low side, especially in regard of DCP content, is not lower than of some other tropical grasses and can be used for feeding cattle, especially if concentrates are added.

Saccharum sinense Roxb. Uba cane. $2n = 116–118–120.$

Similar to *S. officinarum* from which it differs in having narrower leaves, up to 5 cm wide, and thinner stems; this taxon can also be regarded as a form of *S. officinarum*. Cultivated for sugar in southern China and Japan and to a limited extent in subtropical India (Bor, 1960).

Uba cane is grown to a limited extent for fodder in Sri Lanka

(Senaratna, 1956) and is also considered to be one of the principal fodder grasses for the USA Virgin Islands (Oakes, 1969b). Tried in Fiji under the name of Urban cane (possibly distorted Uba cane), it yielded 37 t/ha of leafy fresh forage to which the leaves contributed 93.9 per cent (Roberts, 1970). Uba cane has also been tried as a fodder grass in Surinam where the herbage contained 8.3 per cent CP: the leaves contained 9.5 per cent and the stems 5.3 per cent. CP digestibility was determined to be 77 per cent. The herbage had a high content of CF, 34.6 per cent (Dirven, 1962). High content of CP, 9–10 per cent in the DM, was also determined in Trinidad (Harrison, 1942).

Saccharum spontaneum L.

A wild grass, tough and usually unpalatable to the animals except to domesticated elephants in India but worth mentioning because it is common and widely spread in the Old World tropics, mainly in southern Asia but also in north-east Africa. It is a tall and vigorous perennial with strong long rhizomes, thin stems which are up to 4.5 m in height and long narrow leaves.

Sehima Forsk.

This genus includes seven species which occur in tropical and subtropical Africa, India and Australia. Grasses of dry, open habitats.

Sehima nervosum (Rottb. ex Willd.) Stapf. Sain grass (India)

$2n = 20, 40$ or 34

Tufted perennial with slender stems 50–120 cm high. Leaves linear, glaucous. Stems terminate in a single spike-like raceme 7–12 cm long. Spikelets paired, one sessile the other pedicelled, 8–11 mm long. Florets two per spikelet. Upper floret only of the sessile spikelet is fertile. Spikelets armed with two straight bristles and a kneed and spirally twisted awn.

Occurs in semi-arid areas of East Africa, the Sudan, southern Arabia, India, Burma, Thailand, tropical Australia and Papua. It is numerous and often dominant on slopes and rocky ground in central and northern India from 5°N. northward, and also in northern Australia. In India a plant community with a dominance of *S. nervosum* which is tolerant to grass fires, yielded 4,162 kg/ha, and 7,561 kg when fertilized with 60 kg N/ha (Dabadghao & Shankarnarayan, 1970). In India it is considered a palatable grass especially for sheep although the leaves are harsh and fragile. In Kenya 7.0 per cent of CP was found in herbage harvested at the early flowering stage.

In semi-arid western India (Singh, 1967) CP content was 6.8 per cent in young plants, 4.8 per cent at the early flowering stage and 2.5 per cent when the plants ripened and shed seed. The contents of CF were 19.6, 20.1 and 25.2 per cent, respectively; EE content decreased from 2.4

per cent at the early stage of growth to 1.4 per cent in mature plants. The content of NFE was maintained at about the same level of 56–61 per cent and so was the content of Ca which fluctuated from 0.42 to 0.53 per cent. P content decreased gradually from 0.11 to 0.06 per cent during the plant development and maturation. The nutritive value of *S. nervosum* was favourably compared with other main pasture grasses of the area. Its value was mainly in the high content of energy and low content of CF but the level of CP was low and so was the level of P, inadequate for cattle requirements. The average digestibility was reasonable: 46.5 per cent for DM, 36.9 per cent for CP, 61.9 per cent for EE, 50.6 per cent for CF and 52.6 per cent for NFE, and the balance of N in the animals positive. The herbage was palatable and the animals consumed 1.66–2.13 kg/day/100 kg liveweight. In Australia this species is regarded as a poor quality grass and Tothill & Hacker (1973) state that sparse foliage makes this grass of little value in natural pastures.

Reproduction has been reported to be sexual (Brown & Emery, 1958) and diploids ($2n = 20$), tetraploids ($2n = 40$), and also plants with an unusual number of chromosomes ($2n = 34$), have been found. Seed germination is poor and in Madras it was found to be 16–26 per cent after 1–2 years of storage, and still lower in the third and fourth years (Whyte, 1964).

Setaria Beauv.

Annuals or perennials, mostly tufted but sometimes rhizomatous. Panicle from dense and spikelike to open, but never very loose. Spikelets awnless, usually ovate in shape and supported by one to a few bristles. Lower glume short, the upper longer and can reach the length of the spikelet. Florets two: the lower male or sterile, the upper bisexual, its lemma and palea crustaceous, often transversely rugose or reticulate; the lemma clasping the palea. A large genus of 140 species, mostly tropical, but there is a number of species, mostly annual arable weeds, which occur in temperate areas. The majority of species grow in moderately humid areas and there are also forest species, some of them with broad leaves the vegetation of which resembles that of young palm-tree leaves. Some species are good grazing grasses, the best one is *S. anceps*, better known as *S. sphacelata*. This species has been introduced into cultivation under the names of Kazungula setaria and Nandi setaria.

Setaria anceps Stapf ex Massey. Setaria grass; Napierzinho (Brazil). (Fig. 32)

Tufted perennial. Stems erect, sometimes with ascending bases, 1–2 m high, moderately thick and usually compressed in the lower part. Leaves up to 40 cm long and 8–20 mm wide, glabrous, with tightly compressed and keeled sheaths which, at the stem bases and on short sterile shoots, are arranged fan-fashion. Panicle (spike) dense, cylindrical, mostly 10–30 cm long, but sometimes longer. Spikelets in groups on short branched peduncles and supported by 5–15 bristles varying in length and colour. Spikelets 2.5–3 mm long, elliptic, flat on the lower-glume side and convex on the other side, with two florets. Glumes soft, membranaceous, the lower glume about one-third and the upper one-half to two-thirds the length of the spikelet. Lower floret male or sterile, with soft membranaceous lemma and palea as long as the spikelet. Lemma and palea of the upper, fertile floret hard, crustaceous, finely wrinkled. Grain elliptic, convex on the embryo side and flat on the other side.

This species has been widely known to agriculturists under the name of *S. sphacelata* from which *S. anceps* differs mainly by its vegetative character; folded and sharply keeled leaf sheaths which are arranged fan-fashion at the tuft bases, and glabrous leaves. It seems that *S. anceps*

Fig. 32 *Setaria anceps.*

is the name now universally accepted by the taxonomists for the cultivated setaria grass and its acceptance by the agriculturists is inevitable. *Setaria sphacelata* is closely allied to *S. anceps* and, together with a few other species, they form the so-called *Setaria sphacelata* complex. Similarly with other species of the complex, *S. anceps* occurs naturally mainly in tropical Africa at 600–2,600 m alt, extends to subtropical South Africa and can constitute a considerable portion of the herbage in natural grasslands but seldom dominates in the grass cover.

Environment. *Setaria anceps* grows naturally under an annual rainfall of over 750 mm, although in South Africa some varieties can exist in the areas where the annual rainfall ranges from 500 to 750 mm; it cannot, however, survive long dry seasons. This grass develops better at temperatures slightly lower than those of the hot tropical climate. In southern Australia (Australian Report for 1968–9) it tends to respond to long photoperiods of 12 to 16 hours by increasing its growth, especially under relatively low temperatures of 20–25°C; this does not mean, however, that flowering would also respond to the same long photoperiods. *Setaria anceps* tolerates light frosts but a considerable proportion of plants may die when temperature falls below −4°C, although R. M. Jones (1969) observed 23–91 per cent survival during a winter in which the lowest temperature reached −9°C. Frost tolerance is apparently an inheritable feature; when progenies of plants collected in Kenya at various altitudes were tested in Australia their frost resistance correlated with the altitudes at which the original plants were collected. *Setaria anceps* occurs on a variety of soils, except those of high alkalinity or acidity, and tolerates temporary flooding and waterlogging. It is also resistant to grass fires.

Introduction. *Setaria anceps* is cultivated to an appreciable extent in some countries of its origin, namely in South Africa, Kenya and Rhodesia and on an experimental, and often on a small farm scale, in the majority of other African countries. Outside Africa it has been introduced to the Philippines, New Guinea, India, Malagasi (Madagascar), Fiji, Taiwan, Florida, USA, and possibly to some other areas, but the country where it has now become an important grass is Australia, mainly the medium-moist areas of Queensland and New South Wales, where considerable research and breeding have been done. Outside the tropics *S. anceps* is grown under irrigation in Morocco and Israel.

Establishment. *Setaria anceps* can be established vegetatively or from seed. For vegetative propagation the herbage is cut to about 10–15 cm from ground level, the tufts are split to units containing one to a few shoots and planted immediately or after a week of storage in a cool and moist place until new roots begin to appear. The splits take easily and under suitable weather over 90 per cent take can be expected. Each split should be planted in the right position and scattering planting material on the ground and discing results in poor establishment. Split planting is a laborious process and can be done only when labour is not a problem. In Kenya vegetative propagation has been used only for the establishment of seed production stands. Any large-scale establishment for fodder or grazing is done by seed and broadcasting can be used especially for legume mixtures. In Kenya setaria grass is often undersown to maize; seed is then broadcast after the final weed cleaning, when maize plants reach 70–100 cm in height (Poultney, 1963); establishment under maize is, however, unsuitable for setaria/clover mixtures. For the establishment without cover crop seed is sown to a seedbed free from weeds and well, but not too finely tilled. Seed is drilled in lines 25–50 cm apart and to a depth not exceeding 3 cm. In Kenya maximum emergence was observed from a depth of 5 mm on easily compacted poor soil and to 30 mm on structural red loams (Bogdan, 1964a). Seed rates up to 18 kg/ha have been recommended but recommendations can be meaningless for seed of unknown quality which can vary very widely. In Kenya 1.2 kg/ha (1 lb/ac) of pure germinable seed (PGS), otherwise referred to as pure live seed (PLS), roughly 120 seeds/m^2, is considered adequate but 1.5 kg/ha may be a safer rate. Higher seed rates are used in Australia and Middleton (1970) in south-east Queensland has shown that 6.6 kg seed/ha gave 21–31 seedlings/m^2 and 2.2 kg gave 6–14 seedlings; 10 months after sowing there was no difference between herbage yields for the two seed rates. Hacker & Jones (1969) consider that 1.1–2.2 kg seed/ha (1–2 lb/ac) can be adequate. Superphosphate drilled with seed usually promotes early seedling growth and helps the establishment, but nitrogenous fertilizers are usually only given when the grass has been well established. Seedling emergence and the early growth of seedlings is slow, and seedlings can be overgrown with weeds and early cutting of weedy growth, or grazing if the weeds are palatable, may be necessary. Small seedlings can however develop into large tufts and Hacker & Jones (1969) write that the establishment which may be considered a failure in the first year of growth can produce good stands in the second year. Young growth is particularly slow under low temperatures and this may inhibit the cultivation of *S. anceps* at particularly high altitudes. Seedlings and very young plants are flat and have a red spot (at least in cv. Nandi) on the leaf sheaths, the features which help to recognize them in the field.

Association with legumes and other grasses. A number of pasture legumes have been grown with *S. anceps* with a good to moderate success. Mixtures with *Desmodium intortum, Macroptilium atropurpureum, Glycine wightii, Trifolium repens* and with annual *M. lathyroides* were successful in Australia, with *Desmodium intortum, D. uncinatum, G. wightii, T. repens, T. semipilosum* and annual *T. rueppellianum* in Kenya and with *G. wightii* in Rhodesia and South Africa. Slow establishment of

S. anceps encourages an early start and a good development of the legumes which are easier to establish in setaria mixtures than in mixtures with a number of other grasses, e.g. *Chloris gayana*. In its third and fourth years of growth *S. anceps* becomes more aggressive and there are reports that it is difficult to maintain the legumes on 4-year-old setaria swards. Phosphate fertilizers are needed more for setaria/legume mixtures than for pure setaria not only in the year of establishment but also in subsequent years. In Kenya a *S. anceps* cv. Nandi mixture with *Trifolium repens* yielded 13.45 t DM/ha during the last three years of a trial without applied P and 19.85 t when about 1,000 kg single superphosphate/ha were applied (Suttie, 1970). A similar trend was observed by Blunt & Humphreys (1970) in Queensland for mixtures with *Desmodium intortum, Lotononis bainesii, Stylosanthes guianensis* and *Macroptilium atropurpureum*, but fertilizer N usually weakens and can even eliminate the legume. Legumes grown in mixtures with *S. anceps* seldom increase the yields of grass but they usually increase total DM and especially CP yields of mixed herbage compared with grass grown alone. Lower yields of a *S. anceps* (cv. Nandi)/*Desmodium intortum* mixture than of *S. anceps* alone were reported by Riveros & Wilson (1970) which was due mainly to *D. intortum* being dormant during the winter, whereas the grass was not and maintained its dominance in the sward. Setaria association with other grasses was not particularly successful. Attempts to grow *S. anceps* with *Chloris gayana* did not succeed mainly because *C. gayana* is more aggressive in the first one or two years and suppresses *S. anceps* of which only scattered plants survive and they cannot form sufficiently dense stands when *C. gayana* weakens and begins dying out in the third and fourth years. In trials in south-eastern Queensland, where *S. anceps* was grown with *C. gayana*, three species of *Paspalum* and *Digitaria decumbens* only the last named species survived and formed a satisfactory mixture with *S. anceps* (From Hacker & Jones, 1969).

Management, fertilizing. Information on *S. anceps* productivity has been obtained mainly from the yields of cut rather than grazed grass. The highest yields of herbage were mostly obtained from the grass cut at 3- to 7-week intervals and grazing once in 4 to 8 weeks is usually recommended. Low cutting or grazing is usually not advocated and in Queensland pure stands of *S. anceps* cut to 15 cm from the ground yielded 27.5–28.2 t DM/ha compared with 23.4 t obtained from the grass cut at 7.5 cm (Riveros & Wilson, 1970). *Setaria anceps* is grazed or cut during the rainy season but is unproductive in the dry periods; it is, however, one of the first grasses to produce fresh growth at the onset of the rains. Setaria responds well to fertilizer N and produces over 30 kg DM per kg of applied N and J. G. Boonman even records up to 65 kg DM resulting from 1 kg N. Nitrogenous fertilizers applied to natural grassland containing perennial species of *Setaria* can substantially increase their proportion in the herbage. Split application may not necessarily result in yield increases compared with single applications and sometimes can give lower yields. The effects of P fertilizers, except those applied at the establishment, depend on P content in the soil and the application of P may not always be necessary. Potassium effect has been little explored but pot trials by I. F. Ferguson and A. E. Martin (from Hacker & Jones, 1969) indicate that K application is essential when K content in the plants falls below 1 per cent. It was shown in South Africa (De Bruyn & McIlrath, 1966) that Mn, and to a lesser extent B, were needed for the normal growth of *S. anceps* but Cu and Zn requirements can be low. Symptoms of nutrient deficiencies in *S. anceps* are expressed in leaf reddening under N or S deficincies and dry leaf tops under K deficiency. The deficiency of Fe was occasionally observed and was expressed in leaf chlorosis, which was of a temporary nature (Hacker & Jones, 1969). Lester & Carter (1970) observed herbage yield increases from spraying with gibberellic acid.

Pests and diseases. Leaf diseases are not common although *Pyricularia tirsa* causes some damage in Australia and in South Africa; the disease is expressed in red spots on the leaves. Inflorescence diseases caused by a species of *Sphacelotheca* and *Fusarium nivale* var. *majus* are serious in Zaire, and *Tilletia echinosperma*, known as setaria bunt, can ruin the seed crop in some areas of Kenya, especially in its western parts where the main bulk of seed is grown. It is manifested in inflated spikelets, the content of which is destroyed and the diseased spikelets contain a mixture of spores and pieces of mycelium. This fungus apparently spreads from one flowering head to another and there seems to be no danger of introducing the disease to other countries with seed (Bogdan, 1971). There seems to be no serious pests damaging *S. anceps* except occasional attacks by army worm and some other pests common with other tropical grasses and not specific for *Setaria*.

Herbage production. Setaria anceps develops large numbers of tillers. Boonman (1971b) counted up to 1,900 tillers/m^2 (including short sterile tillers) in *S. anceps* cv. Nandi Mark 3 and 1,400 tillers in the original cultivar. In other grasses under trial the number of tillers/m^2 ranged from 1,280 to 1,800. Yields of *S. anceps* herbage can be high and reach their maximum in the second or third year of growth. In Queensland cv. Nandi grown with two legumes produced in the year of establishment 11.2 t DM/ha and was outyielded only by *Chloris gayana*; in the following 4 years Nandi setaria totalled 32.0 t DM/ha without fertilizer N, 44.0 t when given 112 kg N/ha and 58.6 t when given 336 kg N, considerably outyielding *C. gayana* (R. J. Jones, 1970). In Fiji (Roberts, 1970), two introductions of *S. anceps* yielded 19.3 and 26.9 t DM/ha in 11 months when cut every 8 weeks, and yields of over 40 t fresh herbage/ha were reported from India. In Brazil, 176 t fresh herbage/ha from unfertilized grass were obtained in the 13 months of a small-plot trial (Zúñiga, *et al.*, 1967). In another trial in Brazil (Pereira *et al.*, 1966) *S. anceps* yielded 70 t green material/ha in wet summer but only 5 t during the dry and cool season. When fertilized with moderate amounts of NPK + B and Z the wet season yields increased to 100 t but little or no increase was obtained in the dry season; however, fertilizers plus sprinkler irrigation increased dry-season yields by 30 t/ha.

Conservation. Satisfactory silages were prepared from *S. anceps* herbage, especially when molasses was added; fermentation was of the acetic acid and not lactic acid type. In Australia (Catchpoole, 1968) silage prepared from cv. Nandi had pH values ranging from 4.5 to 5.2 and the concentration of lactic acid was as low as 1 per cent which was explained by the low concentration of water-soluble carbohydrates, 4.5–6.1 per cent. The proportion of ammonium N to total N was moderate, around 18 per cent, and the acetic acid silage was stable, especially when the grass was ensiled in the afternoon when it was slightly wilted. In well-sealed experimental silos losses of DM ranged from 9 to 16 per cent, those of nitrogen from 3.4 to 18.1 per cent, the highest losses being observed for silages prepared late in the season from grass well fertilized with N. Hay was successfully made in South Africa although drying rather thick stems of cv. Kazungula presented some difficulties. During hay curing curing losses of organic matter and of N did not exceed 10 per cent of the original weight and Catchpoole (1969) also states that hay of cv. Nandi was successfully prepared in Queensland. The curing period lasted 50–70 hours, depending on the weather, and DM losses were usually below 5 per cent. No slashing or crushing of the herbage was necessary.

Chemical composition, nutritive value. The contents of main nutrients in *S. anceps* herbage vary and its nutritive value is assessed differently by different authors. Bredon & Horrell (1961) consider the grass to be of low nutritive value. Data given by Hacker & Jones (1969) show 1.39, 1.27 and 1.08 per cent N content in the DM of 5-week-old herbages of the Nandi, Kazungula and Bua River cultivars, respectively, which correspond to 8.7, 7.9 and 6.7 per cent of CP, all at a level low even for tropical grasses. Kenya data for 16 samples of cv. Nandi analyzed at various stages of growth ranged from 5.1 to 19.4 per cent CP content and from 1.8 to 14.3 per cent DCP. The content of CF ranged from 25.5 to 37.4 per cent and the quality of herbage from very high to low (Dougall, 1960). A comparable variation in herbage quality is given by Butterworth (1967) in his review on the digestibility of tropical grasses in which CP content in *S. anceps* ranged from 4.8 to 18.4 per cent, DCP content from 2.1 to 13.2 per cent and CF content from 24.0 to 34.4 per cent. The coefficients of digestibility for CP are given as varying from 44 to 72, a relatively high level, and for CF from 65 to 77. NFE contents were within the limits of 40 to 50 per cent and EE contents of 2.4 to 4.7 per cent, a high level for a non-ariomatic grass. The content of P usually varies from 0.10 to 0.23 per cent and is in most cases just adequate or slightly short of animal requirements, and of Ca from 0.25 to 0.50 per cent, although very low content, 0.15–0.18 per cent, was reported from Queensland (Hacker & Jones, 1969) for the Nandi and Bua River cultivars. A low content of soluble carbohydrates was noted from time to time. An important feature of species of *Setaria*, including *S. anceps*, is a high acidity of its fresh herbage reported by Birch *et al.*, (1964). The high acidity is connected with the presence of free ammonium which accumulates in grass tissues, especially when nitrogenous fertilizers are applied. This is also linked with high contents of organic acids, including oxalic acid, in *Setaria* herbage, which are much higher than in *Brachiaria ruziziensis* and a number of other grasses grown under comparable conditions.

Animal production. Animal production from *S. anceps* is not particularly outstanding and is about similar to production obtained from *Chloris gayana*. Hacker & Jones (1969) quote a few known direct trials assessing animal production from grazing. In Kenya 332 kg/ha liveweight gains by cattle were obtained from *S. anceps* cv. Nandi in the first year of utilization, 40–50 kg lower than from *Chloris gayana* or *Melinis minutiflora*. In the second year the liveweight gains decreased to 190 kg/ha and were about equal to those from *C. gayana* and much higher than from *M. minutiflora*. In Rhodesia, cv. Kazungula did well in the dry season and cattle gained 22 kg/head, about the same as from *Panicum maximum* cv. Sabi, whereas cattle lost weight on two other grasses in the trial. Also in Rhodesia, according to H. Weinmann, fertilizer N increased the animal production from *S. anceps* and 135 kg N/ha increased the liveweight gains of cattle from 174 to 290 kg/ha, but there were little or no increases in gains per head. In Kenya 2,000 l milk/ha were obtained from cows grazed on cv. Nandi given no fertilizer N but grown in a mixture with efficiently nodulated *Trifolium repens*. Hamilton *et al.* (1970) report from Queensland, Australia, that cows grazed on cv. Kazungula gave 6–7 kg milk/animal/day which was about 70 per cent of their capacity under optimum feeding; *C. gayana* gave similar yields.

As it has already been noted, setaria herbage is acid. Its pH value reaches 4.8 and the herbage has a high concentration of anhydrous oxalic acid in the DM Kazungula and Bua River cultivars have higher concentrations of oxalic acid than cv. Nandi. High concentrations of oxalic acid, rare in grasses, can be dangerous to the grazing animals; e.g. in Queensland nine cows died after a long period of grazing on the Bua River setaria and hypocalcaemia was diagnosed. In the *Setaria* introductions tested in a special trial, the content of oxalic acid ranged from 2.78 to 7.13 per cent and selection for its low content is considered to be feasible (R. J. Jones *et al.*, 1970).

Flowering, reproduction. Flowering in cv. Nandi begins in the morning, the anthesis taking place between 7 and 9 a.m. The florets open first in the middle part of the spike but the regularity of floret opening is obscured by the complex nature of the inflorescence. Flowering of a single panicle lasts from 1 to 7 weeks, and is usually the longest in the early emerged panicles (Boonman, 1972b). Species of *S. spacelata* complex are cross-fertilized plants and pollen is transferred by wind although insect pollination cannot be excluded; the common honey bee is the main insect visiting in large numbers the flowering panicles and collecting pollen. Self-pollination can occur in isolated plants but self-pollinated progenies are much weaker than those resulting from cross-pollination and they often die before flowering.

The basic chromosome number (x) of *Setaria* is 9 and diploids ($2n = 18$) and tetraploids ($2n = 36$) are common amongst the cultivated types of *S. anceps* although hexaploids ($2n = 54$) and occasionally octoploids ($2n = 72$) and even decaploids ($2n = 90$) have been found. *Setaria anceps* inter-pollinates in its own populations and a great variety of forms result from recombinations of genes, some of the combinations being of considerable practical value and improved varieties and cultivars can be developed by recurrent mass selection as the plants have usually a good combining ability, meaning that the progenies retain to a high degree the characters of mother plants.

Not all species of *S. sphacelata* complex are sufficiently understood and it is not excluded that some ploidy groups within *S. anceps* may belong to different species or botanical varieties. Cross-pollination and hybridization between different but allied species of the complex is possible but it is not so easy as cross-pollination within the same species. Hybridization between *S. anceps* and *S. trinervia* and between *S. anceps* and *S. sphacelata* in the narrow sense was achieved by intermixing plants of a single clone of one species with plants of a single clone of the other and the resulting F_1 plants often have intermediate characters and show some hybrid vigour. Some features, such as the presence of 'fibre' (split old basal leaf sheaths) or hairy leaves were found to be of the dominant type of inheritability (Bogdan, 1961a). Hybridization taking place in natural grasslands where the species can grow together may create considerable difficulties in sorting out herbarium species and the use of living plants and their progenies for solving taxonomic problems of the *S. sphacelata* complex is of real importance.

Seed production. Commercial seed is grown mainly in Australia and Kenya. It was estimated that about 43 t of setaria seed were harvested in Australia in 1968 (Hacker & Jones, 1969). In Kenya cv. Nandi was in 1970 grown for seed on more than 400 ha producing some 30–35 t of seed. In Kenya seed is harvested twice a year and in experimental plots seed yields of fertilized grass ranged from 30 to 150 kg/ha per harvest or double these amounts per year. The yields of pure germinating seeds (PGS) were estimated to be 10 to 70 kg. In trials in Australia (Hacker & Jones, 1971) seed yields averaged from 20 to 90 kg/ha/year. Low yields can be explained by the low percentage of seed recovered by harvesting, and Hacker and Jones estimate it to be 5–7 per cent of the amounts of seed formed on the plants, i.e. the potential yields. The low yields are also a reflection of prolonged emergence of panicles and at the time when some panicles can have ripe seeds some others only emerge; prolonged flowering of the same panicles have a similar but less expressed effect. Early shedding of spikelets and bird damage decrease still more the amounts of recovered seed. Seed yields also depend on the number of panicles per plant and per unit area and at the time of maximum appearance of panicles their numbers in Kenya ranged from 90 to 120/m^2 in the year of establishment and from 170 to 350 in the following year. Widely spaced rows give less uniform panicle emergence than denser stands and in Kenya the optimum results from cv. Nandi were obtained by growing in rows 30 cm apart which yielded 33 per cent more seed than 90 cm rows. The optimum fertilizer is 100–170 kg N/ha per harvest and it was found in Australia that various setaria introductions react differently to different rates of fertilizer N. Early application of N usually gives the best results and split application of N is often unnecessary. Seed is normally harvested when about 10 per cent of seed had already been shed. Boonman (1972b) in Kenya has found that seed ripens, and should be harvested, about 6 weeks after some five panicles/m^2 had emerged, and that there is a safe margin of about 2 weeks during which no significant difference between seed yields has been observed. Within this margin the loss of spikelets by shedding is compensated by the ripening of about the same number of younger spikelets. It has also been observed that early shed spikelets are usually those which contain no seed. Combine harvesting was used in Kenya some years ago but was discontinued because of heavy seed losses, and the grass is now cut with a reaper-mower or by hand, stooked, and the stooks are left in the field for drying for 1–2 weeks before they are threshed either by hand with sticks or by a stationary combine. Seed is preliminarily cleaned at the farm and handed over to seed companies for final cleaning, blending and packing. In Kenya, seed companies blend the seed to bring the percentage of spikelets containing caryopses to certain standard levels in order that a certain standard amount of seed per hectare can be sown. When artificially dried, seed should not be subjected to too high temperatures. Silcock (1971) has shown that freshly harvested seed containing 62 per cent water can be dried safely at temperatures of up to 61°C but drying wet seed at 70/80°C reduces germination percentage. Seed as sold is in the form of a fertile floret containing a caryopsis and there are 1,000,000–1,500,000 such seeds per kg, depending on cultivar.

Cultivars. (Barnard, 1972; Bogdan 1965).

Kazungula. Tall and vigorous plants with slightly bluish leaves, moderately drought resistant. In Kenya it had 14–15 stem internodes in contrast to cv. Nandi which had seven to nine internodes. Seedlings with a faint red spot. Oxalate content high. Tetraploid with $2n = 36$. Originates from Zambia, developed in South Africa and is cultivated mainly in South Africa, Rhodesia and Australia.
Nandi. Vigorous plants of poor drought resistance; leaves dark green. Seedlings with a pronounced red spot. Relatively low oxalate content. Diploid with $2n = 18$. Originates from Nandi highlands in Kenya. Cultivated mainly in Kenya, Australia and Rhodesia. **Nandi mark** 2 has been developed by single-plant mass selection; it is more uniform and more leafy but poorer seed producer than the original cultivar. **Nandi mark** 3 is a further selection in which the features of Mark 2 are still more accentuated.

Fig. 33 *Setaria anceps,* cv. Nandi.

Narok. Resembles cv. Nandi but is more robust and vigorous. Seedlings with or without a red spot. Relatively tolerant to frost, a feature for which this cultivar has been selected. Tetraploid with $2n = 36$. Originates from the Aberdere mountains of Kenya, 2,400 m alt, and developed in Australia.

Other better known cultivars are **Bua river**, originating from Malawi and **Toittskraal** from Rhodesia.

Setaria geniculata (Lam.) Beauv. $2n = 36$ or 72

Perennial, tufted or spreading with short rhizomes which can be up to 40 cm long. Stems erect or geniculately ascending and prostrate in the basal parts. Leaves flat, 10–15 cm long and 2–8 mm wide. Panicles dense, cylindrical, 2–8 cm long but occasionally longer, greenish-yellow, green or purplish. Spikelets 2–3 mm long supported by five to eight yellow or purplish bristles. Lower glume one third and upper glume one half the length of the spikelet.

Distributed in North America, the West Indies, Central America, and South America to Argentina and Chile. Occurs at roadsides and sometimes in grasslands, grazed by stock and in Argentina its CP content was determined as 10.8–14.4 per cent, EE 1.3–1.9 per cent, CF 26.9–32.7 per cent, NFE 44–45 per cent, indicating a reasonable to good nutritive value (Vonesch & Riverós, 1967–8).

Setaria longiseta Beauv. $2n = 18$

Tufted perennial 70–150 cm high with numerous leaves 12–30 cm long. Panicle moderately loose (not spikelike), 8–25 cm long, often reddish. Spikelets 2 mm long supported by one bristle which is 6–12 mm long; in some spikelets the bristle can be absent.

Occurs in tropical Africa, in grasslands and in bush. Cultivated experimentally in Kenya on a small scale for several years and found to be deserving further trials. Well liked by stock. Seed is abundantly produced but bunt on spikelets, similar to that on *S. anceps* and caused by *Tilletia echinosperma*, can seriously damage the seed crop.

Setaria splendida Stapf. Giant setaria.

Tall tufted perennial similar to *S. anceps* but taller, 1.5 to 3.5 m in height, and with larger leaves which are up to 70 cm long and 12–20 mm wide. It usually has longer panicles, and longer and more numerous bristles surrounding spikelets, than *S. anceps*. Spikelets are similar to those of *S. anceps* but can be slightly shorter.

Occurs naturally in tropical Africa but is less common and less widely spread than *S. anceps*; the centre of this species distribution is perhaps south–central Africa. The specific status of this form is not quite clear and it is not excluded that at least some types of *S. splendida* can be F_1 hybrids possessing hybrid vigour (Hacker, 1966). It grows in grasslands in the areas of reasonably high annual rainfall, usually over 1,000 mm, and also on stream banks.

Setaria splendida has been tried in cultivation, mostly experimental, in several African countries with reasonable success, but little if any seed is normally produced although there are reports on its establishment or natural regeneration by seed. It is grown mainly from tuft divisions, the splits taking easily, and vegetative propagation is feasible because the plants are large and can be widely spaced. *Setaria splendida* can be used for grazing and is palatable to stock, and also for soilage or ensiling. Dry matter content of leaves is rather low and the contents of CP not high, usually about 10 per cent in the first flush of growth. The herbage regenerates rapidly after grazing or cutting and in western Kenya the grass can produce some 4 to 9 t DM/ha/year (Bogdan, 1965a). It is not drought resistant and not productive during the dry season, and Bumpus (1958) recorded for Kenya leaf yields of 3.1 t DM/ha during the dry season, the yields being lower than those of most of the other grasses under trial. It is, however, one of the first grasses to produce fresh growth with the onset of rain or when it is cut or closely grazed at the end of the dry season. Successful mixtures with species of *Desmodium* have been established.

Setaria splendida is a cross-pollinating plant and different 'varieties', most of which are actually clones, do not breed true to type. Tetraploids, pentaploids, hexaploids and septaploids, with respective $2n$ chromosome numbers 36, 45, 54 and 63, were found.

Snowdenia C. E. Hubbard

Contains four annual or short-lived perennial species which occur in tropical East Africa.

Snowdenia polystachya (Fresen.) Pilger (*Beckera polystachya* Fresen.). Abyssinian grass.

Annual or short-lived tufted perennial with ascending, much branched stems and numerous soft leaves. Spikes numerous, on fine peduncles, gathered in loose inflorescences on stem branches. Spikelets subsessile, 2–3 mm long. Glumes in the form of minute scales. Florets two, the lower reduced to a lemma which is as long as the spikelet, truncate and terminating in a fine awn 2–10 mm long. Upper floret slightly smaller, fertile, bisexual.

Occurs naturally in Ethiopia, on roadsides, waste land, forest edges, etc. The size of the plant is extremely variable and depends mainly on soil fertility and moisture; under cultivation it can easily reach a height of 70–100 cm. Introduced to Kenya Highlands in 1940s and was cultivated for a few years as the 'Abyssinian grass' but was later abandoned mainly because of its predominantly annual habit; it remained in Kenya as a weed of cultivation, rather serious in some areas. The herbage is readily grazed by cattle and seed setting is outstanding.

Sorghastrum Nash

Differs from *Sorghum* mainly by solitary, not paired, spikelets at the end of ultimate branches; pedicelled spikelets absent.

Sorghastrum nutans (L.) Nash (*Andropogon nutans* L., *Sorghum nutans* (L.) A. Gray). Indian grass.

Tufted perennial 30–200 cm high with narrow, linear leaves. Panicle 15–30 cm long. Spikelets solitary, lanceolate, hairy, brown or rufous in colour, 5–8 mm long, with a geniculate awn 1–2 cm long.

Subspecies *nutans* occurs in North America, from Canada in the north to Arizona and Texas in the south and also in northern Mexico. Common mainly in midwestern states in lowland prairies. Considered to be a satisfactory grazing grass and has been introduced into cultivation mainly for the improvement of natural grasslands. A selected cv. Holt was developed by the Nebraska Agricultural Research Station and released in 1960. Other areas of *S. nutans* distribution include southern Brazil, Uruguay and northern Argentina where two subspecies occur: ssp. *albescens* (Hack.) Burk., a tall grass up to 120 cm high with flat leaves 3–6 mm wide, and ssp. *pellitum* (Hack.) Burk., a smaller type up to 80 cm high and very narrow leaves 0.5–1 mm wide. Subspecies *albescens* produces satisfactory to mediocre grazing.

Closely allied, *S. rigidifolium* (Stapf) Pole Evans (*Sorghum rigidifolium* Stapf), rather similar in appearance and morphology to *S. nutans*, occurs in East and South Africa where it is common and often numerous in seasonal swamps or in waterlogged depressions; it is usually avoided by grazing animals apparently because of its fibrous and hard leaves and stems. Analysed in Kenya (Dougall & Bogdan, 1958), the herbage contained 41.6 per cent CF and 6.8 per cent CP; P content was low.

Sorghum Moench

Tall annuals or perennials. Inflorescence a panicle, loose or contracted; its branches terminate in short racemes, raceme joints bearing a pair of spikelets each, sessile and pedicelled; except the top pair which bears three spikelets: one sessile and two pedicelled. Sessile spikelet usually has hard glumes which can be woolly or shiny at maturity, and two florets; the lower floret is reduced to a lemma and the upper floret bisexual; the lemma of the bisexual floret is often provided with an awn which is kneed and spirally twisted below the knee. Pedicelled spikelet with two florets of which the lower is reduced to a lemma and the upper is male or empty and has no awn; the pedicelled spikelet can be much reduced.

According to Purseglove (1972), Doggett (1970) and some earlier authors, whose concepts of sorghum taxonomy are accepted in this book, the cultivated grain and fodder sorghums, and wild pasture and fodder sorghums belong to subgenus *Eu-Sorghum* (*Sorghum*) which can be divided into:

1. *Halepensia* which includes diploid and tetraploid ($2n = 20$ or 40) perennial species with elongated rhizomes, and
2. *Arundinacea*, a diploid group ($2n = 20$) of tufted perennial or annual species.

Subgenus *Parasorghum* differs from *Sorghum* (*Eusorghum*) in having a simple panicle branched to the first degree only, its branches terminating in racemes; raceme joints and spikelets margins bearded with long and dense brown or white hairs. Here belong grasses of relatively little importance as source of fodder.

Subgenus *Sorghum* (*Eu-Sorghum*) occurs naturally in a wide area of Africa–India–South East Asia. In Africa it was differentiated into species of the *S. arundinaceum* complex and in south-eastern and southern Asia into species of the *Halepensia* group, the main initial form being possibly *S. propinquum*. These two main types hybridized, possibly in the central part of the area, and eventually produced *S. halepense*. On the other hand, *S. arundinaceum* in a wide sense (*sensu lato*) gave rise to cultivated grain sorghum in the north-eastern quarter of Africa which then spread to the west and the south of Africa, and eastward to India and as far east as China. Grass sorghums, in both cultivated and wild forms, have also been developed from *S. arundinaceum s.l.* and produced several types considered by some authors to be distinct species and by the others as subspecies, varieties or even races of *S. arundinaceum*. Of these species *S. arundinaceum* in the narrow sense (*sensu stricto*) is distributed in West Africa, *S. verticilliflorum* is confined to East Africa, *S. sudanense* to north-east Africa and *S. aethiopicum* occurs in the Sudan. Cultivated grain and fodder species of the *S. arundinaceum* group which arose from natural or intentional hybridization and further segregations, were introduced to the areas outside their origin, to North, South and Central America, the West Indies, Australia, Russia, and a number of other countries, mainly to semi-arid areas, and numerous cultivars have been developed.

Grain and sirup sorghums exist in a number of forms to which some authors give a species rank, and Snowden (1936) enumerated 31 such species. Other authors bulk cultivated grain sorghums in one species and Doggett (1970), following Clayton (1961), considers *S. bicolor* to be the correct name for the group.

Harland & de Wet (1972) have recently suggested a simplified classification of sorghums in which all forms of subgenus *Sorghum* (*Eu-Sorghum*) are included in one species – *S. bicolor*. Cultivated grain and fodder sorghums are grouped into ssp. *bicolor* and spontaneous (wild) sorghums into ssp. *arundinaceum*. Subspecies *bicolor* includes five basic races: bicolor, guinea, caudatum, kafir and durra which can be relatively easily distinguished, mainly by the structure of their spikelets, and 10 intermediate races some of which can be hybrids. The race bicolor includes sorgos which have juicy stems and are used as fodder plants, directly or as hybrids between sorghum and wild types belonging to ssp. *arundinaceum*, although juicy-stem sorghums can also be found amongst other races of ssp. *bicolor*. Subspecies *arundinaceum*, or 'grass sorghums', are provisionally classified into six races: arundinaceum, aethiopicum, virgatum, verticilliflorum, propinquum and shattercane. The simplified classification of Harlan & de Wet has not so far been sufficiently tested and the more familiar division of *Sorghum* accepted by Doggett is retained here.

Sorghum aethiopicum (Hack.) Rupr. ex Stapf. Group *Arundinacea*.

$2n = 20$

Tall annual with a narrow panicle 3–10 cm wide. Sessile spikelet usually tomentose, with an awn 20–30 mm long. Occurs mainly in the Sudan where it is widespread and common on river banks, near cultivations and is also grown on a small scale for fodder.

Sorghum almum Parodi. Group *Halepensia*. Columbus grass; Pasto colon; Sorgo negro.

Densely tufted perennial 1–3 m high forming short ascending terminal rhizomes (not from side buds as in *S. helepense*). Stems erect, numerous. Leaves flat, 30–100 cm long and 1.5–4 cm wide. Panicle 20–60 cm long, loose. Sessile spikelets ovate-lanceolate, 4.5–7 mm long and 2.5 mm wide, awnless, or more often with an awn about 1 cm long. Glumes brown or black, hard, ovate, completely covering the caryopsis at maturity. Caryopsis brownish in colour, ovate, 3.3–4 mm long and 2–2.3 mm wide. Spikelets persistent on the racemes. Pedicelled spikelet male, similar to that of *S. halepense*.

Parodi recognized two varieties: var. *almum* (*typicum*), now widely cultivated, which has larger spikelets, 5.5–7 mm long, and an awn up to 1 cm long, and var. *parvispiculum* Parodi with smaller spikelets, 4.5–6 mm long, awnless or with a short awn; the plants are smaller than those of var. *almum*.

First found in Argentina and described and named by Parodi (1943, 1946) *Sorghum almum* was considered to be a hybrid between *S. halepense*, which grew in Argentina as a weed, and a cultivated forage sorghum. Its hybrid origin has been universally accepted and it was later suggested that *S. almum* had more likely originated from crosses between *S. halepense* and grain sorghum. Another hybrid, between *S. halepense* and *S. sudanense*, was designated by Parodi (1946) as × *S. randolphianum* after L. R. Randolph, who produced the hybrid. *Sorghum almum* is a stable and well-balanced hybrid which does not normally segregate into parental types.

Soon after its recognition as a species, *S. almum* was introduced into cultivation in Argentina where it has widely spread and was grown on a large farm scale only a few years later, in late 1940s. In Argentina, and later in South Africa, it was often referred to as the 'five year sorghum' because of its perennial habit and indeed in many cases it lasted up to 5 years and even longer. However, more often than not, its productive life is shorter and Edwards & Visser (1967) reported from South Africa that *S. almum* yielded 9.4 t DM/ha in the first year, 5.8–7.8 t in the second year and very little in the third. From Argentina Columbus grass spread to South Africa where it became one of the important fodder plants, and then to the USA, Australia and a number of other countries with tropical, subtropical and even temperate climates. It has been successfully tried in southern areas of Russia in Asia and in south-western

Canada, from where very high yields, over 15 t DM/ha were reported (Hubbard, 1960). Cultivation of *S. almum* was for some time prohibited in Australia and in some states of the USA because it was feared that this rhizomatous grass could become a serious weed; it was proved, however, to be non-aggressive, the short rhizomes did not spread far from the mother plant and the ban on seed production was lifted. *Sorghum almum* is valued for its good seed production, the ease of establishment, drought and salt resistance, reasonably high yields, good quality of herbage and its perennial habit.

Environment. *Sorghum almum* is grown with good success in warm and dry climates but can withstand a certain degree of frost. In Georgia, USA, it survived a winter temperature to $-15°C$ (Davies & Edye, 1959), and although this may be an exceptional case, the resistance to lighter frosts raises no doubt. *Sorghum almum* is a drought-resistant species and is recommended in South Africa for areas with an annual rainfall from 400 up to 600 mm and in Australia up to 750 mm. In more humid areas it becomes more susceptible to leaf diseases similar to those of *S. bicolor*, and can also become a weed. *Sorghum almum* can grow on a variety of soils; the best seem to be black soils of flats, as e.g. soils of Darling Downs in Australia. In Australia, grey and brown, slightly saline soils of the Brigalow country of eastern Australia are suitable, as well as alluvial soils, including fresh alluvium, although the established plants do not tolerate flooding. *Sorghum almum* can withstand moderate salinity of Argentinian soils.

Establishment. *Sorghum almum* is grown from seed which is only slightly smaller than those of Sudan grass, the seedlings are strong and can force their way through the crust which is formed on black heavy soil. Seed is broadcast or, better, drilled with an adapted cereal drill in rows 80–100 cm apart; closer spacing is also used but in South African trials 90 cm rows usually gave better establishment. Seed rates range from 5 to 7 kg/ha for wide-spaced rows and up to 20–25 kg for broadcasting or closer row planting. Deficiencies of P in the soil should be corrected not only for better early growth but also for reducing the content of HCN in young herbage. *Sorghum almum* was successfully established in Australia in burnt brigalow bush by sowing into ash. Sowing into grassland in the marginal areas of Karroo desert, South Africa, gave good establishment but the plants eventually died from drought.

Management, fertilizing. In Argentina, *S. almum* is used almost exclusively for grazing; in other countries of its large-scale cultivation *S. almum* is grown for grazing, hay or sometimes for making silage. Grazing can begin some 3–4 months after sowing or when the plants reach a height of at least 30 cm or more. Earlier grazing is avoided because of possible high contents of HCN in young herbage. After each grazing the herbage is topped, if possible, to a height of 30–40 cm (Gangstad, 1963). For hay or silage the plants are cut at the early flowering stage and in some trials low cutting resulted in higher yields than cutting at a higher level. Fertilizer effects vary and, depending on soil fertility, the responses to fertilizer N can be from negligible to quite substantial, but rates higher than 200–250 kg N/ha are usually unnecessary. In a Queensland trial (Henzell, 1963) *S. almum* responded well to fertilizer N by increasing its content in stands invaded by *Digitaria didactyla* weed and at 200 kg N/ha or over yielded over 11 t DM/ha/year. High rates of fertilizer N can increase the content of HCN in the herbage to a dangerous level, especially if not balanced by sufficiently high application of fertilizer P.

Association with legumes. Sorghum almum is grown mostly in pure stands, but on well-drained soils mixtures with lucerne can produce good pastures, lucerne being grazed more in the cool season and *S. almum* in summer. Such mixtures can be established by sowing 5–6 kg/ha of *S. almum* seed and 1 kg of lucerne. Other legumes tried with some success in mixture with *S. almum* are *Glycine wightii* and twining species of *Vigna*. In northern Australia, mixtures with *Stylosanthes humilis* were tried and did well in the first season but not so well in subsequent years. *Pueraria phaseoloides* was also tried but found to be too strong for *S. almum* and smothered it during the establishment period.

Conservation. Laboratory ensiling has shown that good silage can be prepared of *S. almum*, provided that the ensiled herbage is at least 9 to 11 weeks old and not moistened by rain (Catchpoole, 1972).

Herbage yields. Sorghum almum is a good herbage producer; it rapidly recovers after grazing or cutting and can give two to three harvests per season. Yields of 13.5 t DM/ha were recorded in Australia (Davies & Edye, 1959) from a single cut of 12-week-old growth which can be quite tall, and in a trial by Yates *et al.* (1964) 10-week-old plants reached over 2 m in height. Smaller yields were obtained in Texas, USA (Gangstad, 1967), where plants cut at the flowering stage gave in a single cut 6.1–7.7 t oven-dry herbage/ha and 8.3–19.4 t DM/ha in two cuts, outyielding fodder sorghum and *S. halepense* and equalling the yields of Sudan grass. Yields of up to 14.5 t DM/ha/year were also recorded in the USA and 4–5 t in South Africa, and in farm practice 4 to 10 t DM/ha/year can be expected. In Kenya, on poor soil and at 1,800 m alt (Bumpus, 1958), *S. almum*, managed for dry-season production yielded, apart from two wet-season grazings, 2.2 t DM/ha in two cuts, which was less than dry-season yields of other grasses in the trial.

Chemical composition. Herbage of *S. almum* is of high quality and CP content is reasonably high, its average content can be accepted as being about 12–14 per cent. In the leaves of fresh green herbage it can occasionally reach over 20 per cent, whereas in dry or frosted plants CP content was recorded as ranging from 2.5 to 9 per cent CP (Yates *et al.*, 1964). The application of 500 kg N/ha increased CP content in the leaves from 15.2 to 21.5 per cent in spring and from 7.6 to 11.0 per cent in summer; the stems contained 2–3 per cent CP.

Animal production. Sorghum almum is readily grazed by all kinds of stock and gives reasonable or high liveweight gains. In trials with cattle in Texas, USA, 280 kg liveweight gains/ha/year were obtained (Gangstad, 1963) and in Australia (Coaldrake & Smith, 1967) *S. almum/Panicum maximum* mixture, grazed at one head/ha, gave yearly liveweight gains of 440 kg/ha. In a 2-year trial with cattle in subtropical Queensland, Yates *et al.* (1964) obtained 367 kg liveweight increase/ha year when the animals were grazed at a rate of one head/0.4 ha and 286 kg at a rate of one head to 0.6 ha; *Sorghum almum* was however unable to withstand heavier grazing of one head/0.2 ha. Liveweight gains per head were however higher under lighter grazing, 175 kg and 143 kg/year, or 0.50 and 0.41 kg/day, respectively, at 0.6 and 0.4 ha per head. The application of N at 500 kg urea/ha/year increased the liveweight gains very slightly at the heavier grazing and none at all at lighter grazing. In *S. almum*, similarly to other species of *Sorghum*, high contents of HCN have been reported. Herbage cut at the flowering stage usually contains low to moderate amounts of HCN and Gangstad (1967) gives them as ranging from 97 to 318 p.p.m. but in younger herbage HCN content can be considerably higher and reach the dangerous level of 750 p.p.m. or even exceed it, and cases of cattle death from HCN poisoning have been reported in Australia by Davies & Edye (1959). It is therefore recommended to begin grazing when the plants reach a height of at least 30–40 cm, and some authors consider that only the plants that have reached the heading stage can be safely grazed. Davies and Edye indicate that allowing hungry cattle a free access to a young flush of growth presents the greatest danger to the animals. Also the plants damaged by frost, diseases, or those mechanically injured, contain increased amounts of HCN and are not safe for grazing.

Seed production. In *S. almum* seed setting may vary considerably from plant to plant, it is an inherited character and lines were found in which seed setting closely approached that of tetraploid grain sorghums (Doggett, 1970). Although lines and cultivars may differ in seed setting it is in general satisfactory. Seed yields up to 1,500 kg/ha have been quoted (Davies & Edye, 1959); this yield level is however seldom achieved and most of the other authors indicate seed yields ranging from 250 to 750 kg/year in two harvests. Seed is in the form of spikelets with an attached pedicel(s) and often also a pedicelled empty spikelet. Although the spikelets are less firmly attached to the panicle than in Sudan grass, relatively little seed is lost at harvest. A standard header-harvester can be used and harvesting presents no difficulties except those caused by a considerable grass height. Seeds of *S. almum* are slightly smaller than those of Sudan grass and Gangstad (1967) in his breeding work recorded 115,000–170,000 and 80,000–100,000 seeds/kg, respectively, for the two grasses, although according to Wheeler (1950) there are about 120,000 sudan grass seeds/kg (55,000/lb) and Barnard gives about the same figures for cv. Crooble of *S. almum*. Sorghum almum seeds can be distinguished from those of Sudan grass by their darker colour. As in Sudan grass, the seeds have no disarticulation scar at the spikelet bottom, but the piece of broken raceme attached to the spikelet is usually shorter than in Sudan grass seed. Seeds of *S. halepense* can be distinguished by the presence of a disarticulation scar and they are smaller (280,000/kg or 130,000/lb according to Wheeler) and lighter in colour than those of *S. almum*. In spite of these differences the presence of seed of *S. halepense*, a rhizomatous weed, is not always easy to detect in seed consignments of *S. almum*.

Reproduction, hybridization. Sorghum almum is a predominantly cross-pollinated plant which shows a certain degree of variability. It is a tetraploid with $2n = 40$ and can easily hybridize with *S. halepense* which has a similar number of chromosomes. *Sorghum almum* can also cross with diploid fodder and grain sorghums and with Sudan grass, also a diploid, but such crosses occur naturally much less frequently. Progenies of these crosses can be either tetraploids or triploids. The former can contain male–sterile plants and be of value for breeding of *S. almum* and of other types of sorghum. The triploids are sterile and they can have longer and stronger rhizomes than it is typical for tetraploid *S. almum*, are more difficult to eradicate, and triploid seeds present an undesirable admixture to *S. almum* seed. *Sorghum almum* has been used in breeding of both grain and fodder sorghums for developing tetraploid forms or for creating perennial types in annual fodder sorghums. When breeding for the improvement of *S. almum*, apart from common general objectives, high yields, high herbage quality, drought resistance etc, a special objective can be the uniformity of seed colour distinctive from that of *S. halepense* seed. Obtaining cultivars with low contents of HCN is another important objective feasible because of the inherited nature of HCN content in the herbage.

Cultivars

Crooble. Developed in New South Wales, Australia, in 1952–3 from material introduced from South Africa, and distributed in 1959. Adapted to summer rainfall areas of Queensland. Susceptible to leaf blight caused by *Helminthosporium turcicum* and leaf rust caused by

Puccinia spp. A prolific seeder; seed is not shed. Drought and salt tolerant (Barnard, 1972).
Nunbank. A South African cultivar which is coarser, more palatable and less persistent than Crooble.
De Soto. Developed in Texas, USA. Slightly shorter (120–150 cm in height) than average for *S. almum*. The rhizomes are persistent and the cultivar is not recommended for arable crop rotations especially in more humid areas. Seed dark mahogany in colour. Low content of HCN (Gangstad, 1967).
Rietondale is another cultivar sometimes mentioned in the literature.

Sorghum arundinaceum (Desv.) Stapf. Group *Arundinacea* $2n = 20$

Annual 2–4 m high with medium-thick stems. Leaves up to 75 cm long and 2–6 cm wide. Panicle large, loose, much branched. Sessile spikelet elliptic-oblong or lanceolate, acuminate, awnless or with an awn 5–10 mm long. Occurs naturally in West Africa reaching Angola in the south, and also in Zaire. Common and fairly widespread. One of the species from which cultivated sorghums originated. When accepted in a wide sense it can include all or nearly all grass-sorghum species of Eu-Sorghum and is known as *Sorghum arundinaceum* complex.

Sorghum bicolor (L.) Moench. Group *Arundinacea* (*S. bicolor* ssp. *bicolor*; *S. vulgare* Pers., *Holcus sorghum* L., *Andropogon sorghum* (L.) Brot.) Sorghum; Sorgum; Sorgo; Jowar; Jonna; Cholam; Chari India), Pyaung (Burma). $2n = 20$

Annuals or short-lived perennials 0.5–6 m in height; the stems in the basal part are 0.5–3 cm in diameter. Leaves 30–100 cm long and up to 12 cm wide. Panicle loose to very dense and of varying shape. Sessile spikelet persistent, 3–10 mm long; glumes about equal, up to two thirds, the length of the grain at maturity. Pedicelled spikelet narrower and smaller than the sessile (fertile) spikelet, persistent or deciduous.

Sorghum has been cultivated for grain in the area of its origin from prehistoric times. During the last century it was introduced to the USA and then to the Central and South Americas, Australia and southern and south-eastern Europe. *Sorghum bicolor* is cultivated mainly for grain or for fodder and to a lesser extent for manufacturing sirup. The sirup type of sorghum, known in the USA as *sorgo*, has sweet and juicy pith and this type or its derivatives are also used as forage sorghums in preference to those with dry pith. When grown for grain the straw of the types with finer stems is often used for fodder after the grain has been harvested. Types with finer, more numerous and more leafy stems are also grown specially for fodder but they are being replaced by hybrids developed from crosses with wild or semi-wild species of the *S. arundinaceum* complex and bred for fodder production in the USA, India, Australia and other countries. These hybrids are dealt with in a separate chapter.

Sorghum bicolor is a drought-resistant species and types with rapid development, and early flowering and seed ripening are particularly adapted to dry conditions and can be grown under an annual rainfall of 300–350 mm. It is a crop which requires high air temperatures, 25–35°C, for its best development. Sorghum is a short-day plant and a 10-hour photoperiod usually results in the earliest flowering; increases in day length by 3 hours can delay flowering by 2 to 4 months although DM production may benefit from the increase in day length. Sorghum can grow on a variety of soils, including black heavy soils of flats and depressions, preferring, naturally, more fertile ground, but soil salinity or alkalinity can reduce yields very considerably; soil alkalinity also decreases the effect of fertilizer N. In arable rotations sorghum is grown after various crops but responds to preceding leguminous crops by yield increases.

Establishment. Propagation is by seed which is sown in 25–90 cm rows or sometimes broadcast. Seeds are placed to a depth of 2 to 5 cm and the best seedling emergence has been observed from a depth of 1 to 3 cm. Sowing rates vary from 3 to 9 kg/ha in dry areas (Whyte *et al.*, 1959) depending on seed size which is different in different varieties, and 15–35 kg/ha in more humid areas. In India, seed rate increases from 25 to 50 and 75 kg/ha resulted only in small yield increases (Srivastava, 1969). Although the seedlings are strong, weeds can reduce herbage yields and in Puerto Rico propazine applied at 1 kg/ha gave good results in controlling weeds, especially when sorghum was sown in the dry season to irrigated land.

Fertilizing, management. In India, the application of f.y.m. at 10 t/ha for dry-land sorghum and 25 t for irrigated crops, plus 100 kg ammonium sulphate is recommended (Relwani & Kumar, 1970). Nitrogenous fertilizers usually increase forage yields but the increases, especially in dry areas, are not always great although Srivastava (1969) obtained DM yields of 12.4 t/ha by applying 25 kg N/ha, and 16.5, 18.6 and 19.8 t when fertilizer rates were increased to 50, 75 and 100 kg N/ha, respectively. CP content in the herbage also increases under the application of fertilizer N. Sometimes, however, N produces no effect as it was observed in Uruguay. The application of P, K and often also Zn may produce good results.

Utilization can begin some 40–50 days after sowing and Srivastava obtained 18.7 t DM/ha when the herbage was cut twice 45 and 90 days after sowing, whereas a single cut 90 days after sowing gave 14.9 t of herbage with lower content of CP. Nevertheless a frequent practice is to cut the grass once, at the milk- or dough-stages of grain maturity, and in Hawaii (Sherrod *et al.* 1968) there were no significant differences in DM yields of herbage cut at these two stages and at heading, but the content of DCP was considerably higher in earlier than later cuts. Early grazing presents some danger of animal poisoning and some authors recommend to graze only when the grass reaches 50–60 cm in height. Sorghum is cut for hay, soilage or ensiling and also grazed.

Ensiling. Silage prepared from sorghum can contain 6–9 per cent CP (Butterworth's review, 1967) and its digestibility can be 40–75 per cent, although there are also reports that CP digestibility in silage can be considerably lower than in green fodder. Ensiling greatly reduces the content of HCN and can be generally considered as a suitable but not always reliable method of sorghum herbage conservation. Sorghum silage is usually well accepted by the animals, better than silage prepared from *Pennisetum purpureum* but not as well as that of maize.

Legumes. Sorghum is usually sown pure although in India it is sometimes grown, mostly experimentally, in mixtures with *Cyamopsis tetragonoloba*, soyabean, *Vigna unguiculata, V. aconitifolia* or pigeon pea. Mixtures usually yield less DM than sorghum grown alone but CP contents and yields are higher.

Parasitic weeds (see in Doggett, 1970). *Striga* weed parasites can exhaust sorghum plants and reduce the yields of grain and fodder, often very considerably. *Striga asiatica* and *S. hermonthica* are the most important and widespread species but there are also more localized species of lesser economic importance, e.g. *S. curviflora* and *S. hirsuta* in Australia. *Striga* can be controlled to a certain degree by crop rotation or by using 'trap crops' susceptible to *Striga* on which seeds germinate and *Striga* plants grow and are then ploughed in before the parasite reaches the flowering stage. These methods can help, especially when combined with high rates of fertilizer N, but in general *Striga* control is difficult and often expensive. Relatively resistant cultivars have been developed, the resistance being affected either by thick-walled root endoderma or by a decrease of the stimulating effect of sorghum roots on *Striga* seed germination. It should be noted that cultivars resistant to one species of *Striga* may not necessarily be resistant to the other species.

Pests and diseases (see in Doggett, 1970). The most important and widespread pest is perhaps sorghum shoot fly (*Antherigona varia soccata*) which damages the growing points and the bases of top leaves; the top leaf dies out and the growth of the shoot stops. The plant can recover by forming new tillers which, under heavy infection, can also be damaged. Some cultivars including the specially bred ones, can recover more rapidly than the others. Correct timing of sowing and insecticide application can reduce the damage. Endosulfan and Thimet (phoret) (highly poisonous) sprayed or applied in rows at sowing, respectively, can be effective. Other important pests can be various stemborers, species of *Chilo, Busseola* and *Sesamia*.

Of leaf diseases the more important are the fungi leaf spot diseases caused by *Helminthosporium turcicum* and manifested in red spots on leaves and consequent leaf dying. The use of resistant cultivars is the main form of control and there are a number of resistant forms in India and East Africa. Infected by downy mildew, a disease caused by *Sclerospora sorghi*, the whole plant becomes grey and downy, and eventually dies. The removal of diseased plants, crop rotation and the use of resistant varieties are the control measures. Anthracnose caused by *Colletotrichum graminicolum* is another widespread disease. Of the panicle and spikelet diseases, grain smut caused by *Sphacelotheca sorghi* is the most damaging. It is seed-borne; the fungus germinates at the time of seed germination, grows inside the plant and destroys all spikelets in the panicles of the infected plant. Seed dressing with copper sulphate or thiram thinned to 1 : 400 is an effective treatment and there are other formulations.

Herbage yields and quality. Herbage yields vary within very wide limits, from under 6 to 75 t green fodder or 3 to 25 t DM/ha, depending on the variety, soil moisture regime, fertilizers used, the stage of growth at which the plants are harvested and some other less important factors. Yields which can be expected in farm practice on dry land may range from 10 to 45 t fresh material or 3 to 12 t DM/ha. Doggett, quoting J. R. Quinby and P. T. Marion, gives average air-dry yields of irrigated and dry-land crops obtained in Texas, USA, as ranging from 14 to 17 t/ha, depending on the variety, or 42–45 t ensiling material/ha, and 12.9–14.7 t DM were obtained in Brazil (Carneiro *et al.*, 1972).

Herbage quality depends mainly on the stage of growth at which the crop is cut and, to a lesser extent, on the amounts of fertilizer N applied. CP content in straw has been quoted by Butterworth (1968) for India and Tanzania as ranging from 2.3 to 4.9 per cent, its digestibility from nil to 43 per cent and DCP content from nil to 2.1 per cent; in late-harvested hay the contents of CP and DCP were not much higher. In Hawaii, the content of DCP at the dough-ripe stage was low, 1.2 per cent, but much higher, 6.1 per cent, in herbage cut at the heading stage (Sherrod *et at.*, 1968). Much higher CP contents were obtained in Argentina, 16.2 per cent in the herbage cut when it reached a height of 40 cm (Hernández & Abiusso, 1969). Doggett (1970) also quotes the work of J. S. Lakke Gowda who analysed sorghum herbage at different stages of growth and found that CP content was reduced from 29.9 per cent at an age of 3 weeks to 11.4 per cent in 13-week-old herbage, whereas the content of CF increased from 24.4 to 33.4 per cent, respectively. Ca content decreased from 1.06 to 0.78 per cent and P content from 1.29 to 0.78 per cent, a very high content by any standard.

Sorghum plants can contain large amounts of HCN, in the form of dhurrin, which can be dangerous to animals when its content exceeds 750 p.p.m. Doggett (1970), quoting C. E. Nelson, gives HCN content in the leaves of three varieties as ranging from 170 to 1,390 p.p.m. and in two varieties it exceeded the safety level. In irrigated aftermath the contents of HCN were still higher, 876–2,006 p.p.m. The contents of HCN were also higher in plants fertilized with N than in unfertilized plants, but in dried or ensiled sorghum the contents of HCN decreased sharply and silage or hay present no danger for animals.

Seed. Sorghums are cross-pollinated plants, but also self-compatible and they produce seed in large quantities. Seed is firmly attached to the panicles and present no difficulties for harvesting although bird damage may be considerable.

Cultivars. Sorgos or sweet, sirup-producing sorghums suit better for fodder production than grain sorghums and although sorghum × Sudan grass and other hybrids are replacing sorgo as forage plants, a number of sorgo cultivars are still grown for the purpose. Of the old cultivars the better known ones are perhaps **Sumac**, **Early amber**, **Early orange**, **Sapling**, **Sugardrip**, **Red orange**, **White African**, and **Tracy**, a hybrid cultivar resulted from crosses between Sumac and White African (Barnard, 1972). Other cultivars worth mentioning are **Lavrense**, **Fartura** and **Santa Eliza** which had some success in Brazil (Carneiro *et al.*, 1972), and **M. P. Chari**, **Dudhia**, **Gwalior** and a few others recommended by Relwani & Kumar (1970) for India. A number of other cultivars suitable for fodder production and developed in the USA, Australia, South America, India and Russia appeared in the last two to three decades, some of them being hybrids between well known old or new cultivars.

Sorghum halepense (L.) Pers. Group *Halepensia.* Johnson grass; Sorgho d'Alep; Pasto Johnson; Massambara; Gumai. (Fig. 34).

A rhizomatous perennial up to 2 m high, similar to *S. sudanense* in other characteristics except the slightly smaller sessile spikelets which are 4–5.5 mm long. The sessile spikelet, which contains seed, disarticulates easily from the raceme joint leaving a round disarticulation scar at the spikelet bottom, whereas Sudan grass spikelets have no disarticulation scar but a pointed broken joint.

Sorghum halepense is believed to be of Mediterranean origin, the areas of its natural distribution extending eastwards to India. Because of its long and strong rhizomes it has become a noxious weed of arable land, especially on alluvial soil and under irrigation. It occurs also in natural and semi-natural lowland grasslands, on river banks, mainly on alluvial soil.

Sorghum halepense was introduced to USA in 1830 as a fodder grass and was first grown by W. Johnson, hence the name of the grass first accepted in the USA and then almost everywhere. Introduced to the southern states, first for grazing and fodder, this grass has spread naturally, by seed, and become a serious weed of arable land in the tropical and warm areas throughout the world.

Sorghum halepense is a short-day species which flowers best under a 12-hour photoperiod, but vegetative growth under this photoperiod does not increase with the rise of temperature above 27°C, whereas the maximum growth observed at higher temperatures was obtained only under longer photoperiods (Ingle & Rogers, 1961). This species is less drought resistant than *S. sudanense* and responds well to irrigation, especially on alluvial soil or other soils of light texture in which the rhizomes can spread easily; hard soils are unsuitable for *S. halepense*. In natural or seminatural grasslands the rhizomes do not penetrate much below 15–20 cm from the ground level but can reach a depth of up to 40–60 cm on arable land. When the ground is free from other plants, a rapid spread of rhizomes is usually observed and Horowitz (1973) records from Israel that single plants spread in $2\frac{1}{2}$ years' time to 3.4 m in each direction and occupied 17 m² producing 190 tillers/m².

For pasture or hay *S. halepense* can be established from seed which is drilled in at about 20–35 kg/ha. When the grass is present in large quantities as a weed in cotton or other crops, seed sowing can be unnecessary, and land simply left fallow can produce a good crop of hay and yield up to 10–15 t/ha (Whyte *et al.*, 1959); the yields can however be much lower, of the order of 2–4 t hay/ha. Hay is usually of good quality and even when cut at an advanced stage of seed development it can contain 4 to 13 per cent CP and 2–9 per cent DCP, although the content of CF in late harvested hay can be high, 40–45 per cent, and the content of NFE low, 32–34 per cent (Larin, 1950). *Sorghum halepense* can also be grazed and strip- and rotational-grazing were found to be superior to continuous grazing or to feeding chopped fresh grass to the animals; it should be kept in mind, however, that, similarly with the other species of *Sorghum*, *S. halepense* can, at a young stage, contain appreciable amounts of HCN and should be grazed with care. In trials by Hawkins *et al.*, (1969) grazing resulted in higher herbage digestibility, 62–63 per cent, than that of chopped fodder, 56 per cent. Milk production was also higher from grazed grass than from cut and chopped fodder, 12.9–13.6 kg and 11.4 kg milk/cow/day, respectively. Grazing can suppress Johnson grass to a considerable extent if it is continuous, heavy and lasts for two to three seasons.

Seed is produced freely but sheds easily and is, therefore, rather difficult to harvest. There is, however, not much demand for *S. halepense* seed because farmers avoid establishing it on land where it is absent in order not to spread further this noxious weed.

Sorghum halepense is a tetraploid with $2n = 40$ but diploids with $2n = 20$ have also been found. Hybrids with *S. sudanense* and with grain sorghum have been obtained.

Sorghum halepense × S. roxburghii Stapf.

This combination is usually known as **Krish**, a diploid ($2n = 20$) hybrid cultivar developed in Australia from F_1 progenies obtained in India from crosses between *S. halepense* and *S. bicolor* var. *roxburghii*. This is a perennial with practically no rhizomes. Stems up to 4 m high, thick, not sweet. The early growth is slow but the plants are productive late in the season and produce leafy herbage palatable to stock. Krish is highly resistant to leaf diseases (Barnard, 1972).

Sorghum plumosum (R. Br.) Beauv. Subgenus *Parasorghum*. (*Holcus plumosus* R. Br.) $2n = 20, 30$

Tufted perennial 60–120 cm tall. Leaves flat or convolute, 2–8 mm wide. Panicle 15–30 cm long, loose to moderately dense. Racemes hairy, 2.5–4 cm long, with five to eight joints. Sessile spikelets 6–8 mm long, brownish or reddish, bearded with rufous, reddish or occasionally white hairs on margins and the callus and with a geniculate and spirally twisted awn. Pedicelled spikelet subequal to sessile spikelet but hairless.

Common and often co-dominant in dry grasslands throughout northern Australia. The tillers develop early in the season but in the dry period the plants may become dry. After seed setting the plants can develop fresh tillers and contribute to grazing. Burning produces little effect on the grass (Lazarides *et al.*, 1965). The stems are hard and the herbage is of moderate quality but is well grazed when not too old.

Sorghum timorense (*S. australiense* Garber & Snyder) is an annual species of the same section *Parasorghum* common and locally abundant in northern Australia and important as a source of fodder.

Sorghum propinquum (Kunth) Hitchc. Group *Halepensia*. $2n = 20$

Tall, stout perennial forming small, loose tufts and sending short (15–30 cm) stout rhizomes. Leaves up to 1 m long and 3–5 cm wide. Panicle large, loose, oblong, 20–60 cm long. Racemes very fragile. Sessile spikelet elliptic-lanceolate 4–5 mm long and 1.2–2 mm wide; lemma with a short mucro or occasionally with a short awn. Grain small, 1.5–1.8 mm long.

Occurs in humid areas of South and South-East Asia and also in the Philippines and is considered to be an ancient type from which other species of the *Halepense* group have developed. Grows in forests, forest edges and on stream banks. Although leafy it is used as fodder to a very limited extent.

Sorghum sudanense (Piper) Stapf. Group *Arundinacea*. Sudan grass; Pasto Sudan. (Fig. 34).

Annual with numerous erect stems up to 3 m high and 3–9 mm thick. Leaves, narrow, 8–15 mm wide and 30–60 cm long. Panicle 15–30 cm long, open when mature. Sessile spikelet 6–7 cm long; glumes loosely hairy, glossy and almost glabrous at maturity. Upper lemma with an awn up to 16 mm long. Pedicelled spikelet as long as the fertile spikelet but narrower. Racemes not easily breaking at maturity and 'seed' (spikelets) fall off by breaking the raceme joint and a piece of joint makes a short point at the base of the spikelet but no disarticulation scar is present at the seed bottom as, e.g. in *S. halepense* or other wild sorghums.

According to Harlan & de Wet (1972) this grass is a non-seed-shattering segregate from a hybrid or hybrids involving *S. virgatum* (known also as *S. arundinaceum* var. *virgatum* or, by Harlan & de Wet, as *S. bicolor* ssp. *arundinaceum* race *virgatum*) and a sorgo type of cultivated sorghums (*S. bicolor* ssp. *bicolor* race *bicolor*).

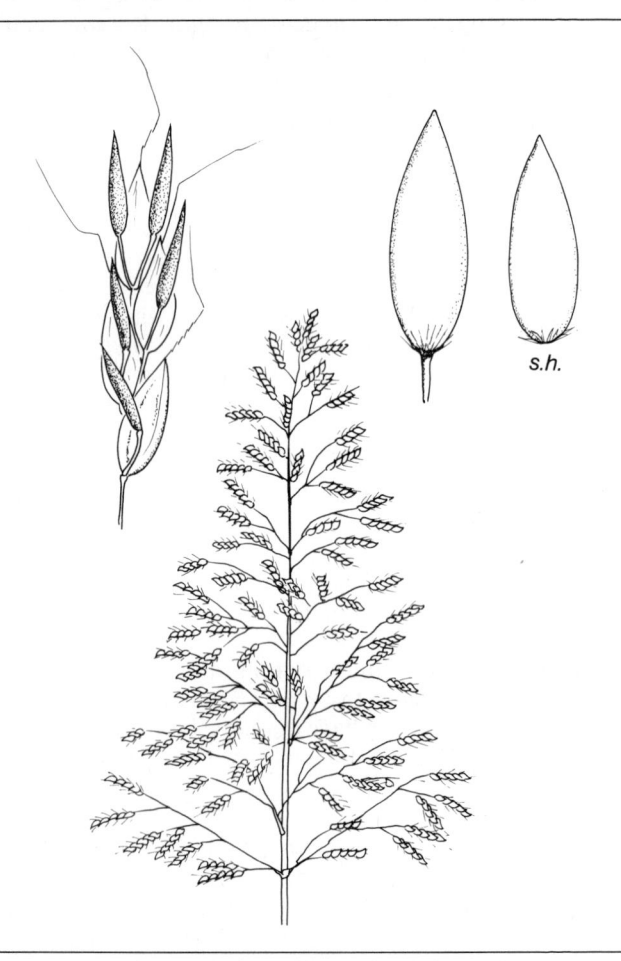

Fig. 34   *Sorghum sudanense* (*S.h.* – spikelet of *S. halepense*).

Originates from Sudan and southern Egypt where in a wild state it occurs only sporadically and only a few specimens have been collected. Occurs on roadsides, in old cultivations, as an occasional weed of arable land and at stream banks. *Sorghum sudanense* was introduced to the USA in 1909 and was almost immediately recognized as a useful fodder plant. It rapidly spread as a cultivated grass, first in the United States and then in other countries. In 15–20 years after its introduction to the USA it was widely tested in a number of countries and soon became a recognized fodder plant grown on a large scale. It is however suitable only for the areas with warm or hot dry summers and has had little success in humid tropics. Sudan grass is valued for its reasonably high seed yields, ease of establishment, the ability to recover after grazing or cutting better than most of the annual grasses, high herbage yields and good quality of herbage. Although it contains cyanogenetic glucosides which, at high concentration, can be poisonous to cattle and other animals, the contents of these glucosides are lower in Sudan grass than in other species or types of Sorghum.

Sudan grass suits continental climates under which it thrives on dry land and responds well to irrigation. Seed germinates well and the seedlings emerge rapidly under high soil and air temperatures, whereas both germination and emergence are slow in cool surroundings, especially when coupled with high humidity. Sudan grass cannot tolerate frosts and dies when the temperature drops to 3–5°C below the freezing point. Fertile and warm soils produce the highest yields although Sudan grass can also grow on poor soil; saline, alkaline or solonets-type soils are unsuitable.

Establishment. A well prepared and firm seedbed is essential and seed is either broadcast or, more usually, drilled through an ordinary cereal drill. The distance between the rows varies widely, from 15 to 100 cm, and most acceptable distances range from 25 to 40 cm in medium-rainfall areas and wider in particularly dry areas. Sowing rates also vary and range from 15 to 75 kg/ha, the higher rates being used for more humid areas and under irrigation. Seed is drilled to a depth of 1 to 3 cm. In India the application of about 20 t f.y.m. is recommended at ploughing (Whyte, 1964) but in most other countries superphosphate at a rate of 100–200 kg/ha is often used, with or without nitrogenous fertilizers. In warm and sufficiently moist soil, seedling emergence can be expected on the fifth to sixth day after sowing.

Management, fertilizing. Sudan grass is used for grazing, chopping and direct feeding to the animals as soilage, or haymaking. Ensilage is seldom practised mainly because better silage can be prepared from fodder sorghum; moreover hay is preferred for conservation because the grass is grown in the areas with hot and dry summers where the herbage together with the relatively thin stems can be easily cured and dried. The grass is usually grazed when it reaches a height of some 60–80 cm and is at the heading stage; for hay it is cut slightly later, at about the flowering time, or still later, often when the seed reaches the milk-ripe stage. In the majority of recorded trials, low cutting, some 8–15 cm from the ground level, gave higher yields than cutting at higher levels. Sudan grass recovers well after cutting and three to four cuts per season are often taken and up to seven cuts per year have been recorded under irrigation. Apart from the application of fertilizers at the establishment, fertilizing during the growing season is usually restricted to about 50 kg N/ha after each or every second cut. Complete fertilizers containing NPK are also occasionally used, especially on soil deficient in K. The grass responds well to irrigation and especially (in India) to sewage irrigation, and it is recommended to irrigate the grass every 2 weeks.

Legume mixtures. Sudan grass is usually grown in pure stands although in the USA Sudan grass/lucerne mixtures are used as an irrigated crop. Other legumes sometimes grown in Sudan grass mixtures are soyabean (*Glycine max*), *Vigna unguiculata* and *Mucuna* (*Stizolobium*), spp. (Relwani, 1968b).

Herbage production. Sudan grass can produce up to over 88 t/green fodder/ha when fertilized with 280 kg N + 140 kg P_2O_5/ha which is about 60 per cent more than for the unfertilized control (Whyte, 1964), but usually yields are lower and average from 10 to 40 t/ha, depending on the stage of growth at which the plants are cut, and on soil fertility and moisture content. Hay yields from 2 to about 10 t/ha have been recorded, and Gill *et al.* (1967) report from subtropical Punjab (India) linear yield increases under the application of up to 280 kg N/ha.

Chemical composition, nutritive value. Sudan grass herbage is rich in CP and according to Russian authors (Larin, 1950) its average content in the herbage reaches about 12 per cent, whereas young plants often contain 16 per cent CP; CP digestibility is usually high and can reach 65–70 per cent. The content of NFE is relatively low, mostly about 40–45 per cent, although their digestibility is high and can be of the order of 70–75 per cent. CF content is not high and seldom exceeds 30 per cent. Sudan grass is highly palatable to all kind of stock, especially to cattle, even at relatively advanced stages of growth, and little herbage is wasted during the grazing. It has been estimated that cattle consume about 80 per cent of grass before the panicles emergence and up to 45–50 per cent at grain ripening (Larin, 1950). Animal production from Sudan grass is usually higher than from fodder sorghum.

Similarly with *S. bicolor* and other species of *Eusorghum* Sudan grass contains appreciable amounts of HCN which can be dangerous when its content reaches 750 p.p.m.; in Sudan grass the dangerous level is, however, seldom reached. The content of HCN is higher when the grass is fertilized with N than when no N is given. Applied N also increases the content of nitrate N in the herbage and a metabolic relationship between nitrate N and HCN has been suggested (Murphy, 1966). Applied K decreases the content of nitrate N in the herbage. In Sudan grass wilted in the field before grazing, the content of HCN is high and can reach the dangerous level as it can in plants weakened by diseases, pests or other adverse factors, and grazing such weakened plants can result in animal poisoning. Normally Sudan grass has, however, a lower content of HCN than fodder sorghum or sorghum/Sudan grass hybrids and grazing healthy and non-wilted plants presents practically no danger. The highest content of HCN has been observed in the leaf blade, excluding the midrib; it was lower in the leaf sheath and still lower in the midrib.

Flowering and seed production. Sudan grass flowering is to a certain extent controlled by photoperiods although under the day lengths normal for he tropics the plants flower freely. Different varieties can however behave differently and Blondon & Lenoble (1973) give an example of a very early cultivar flowering even under continuous light and also of the influence of temperature on the effect of photoperiods.

When grown for seed Sudan grass is often sown in wide, 60–90 cm rows, but stands established for grazing or hay can also be used for seed production. In its tillering and panicle emergence habits Sudan grass resembles perennial grasses; the panicles emerging gradually, at different times, although to a lesser degree than in perennial tropical species. Seed is not shed easily, and harvesting is not difficult. The best time for harvesting is usually considered to be the stage when the early-flowered panicles are fully ripe. Harvesting can be done by a reaper-binder or a windrow cutter, the cut plants dried for 2–4 days in the field and then threshed by a cereal thresher or a stationary combine. Yields of 500–1,500 kg of clean seed can be expected and there are about 100,000 seeds per kg.

Cultivars. *Sorghum sudanense* occurs in diploid and tetraploid forms with $2n$ chromosome numbers of 20 and 40, respectively. It can easily hybridize with *S. bicolor* and with wild sorghums (*S. arundinaceum, S. halepense*), and its bred varieties or cultivars originate from crosses between Sudan grass and other species of *Sorghum*, although some of them have only a relatively small admixture of germ plasm acquired from species other than *S. sudanense*. Cultivars Piper, Tift and Greenleaf belong to this last category and although of hybrid origin they are usually considered to be Sudan grass cultivars and certain information given in the Sudan grass chapter may, perhaps, refer to these cultivars; they are, therefore, considered here. These cultivars have somewhat thicker stems and wider leaves than the ordinary common Sudan grass, features apparently due to generic links with fodder sorghum.

Cv. **Tift**. Bred at Tifton, Georgia, USA; originates from a cross between

Sudan grass and Leoti sorghum backcrossed to Sudan grass. Comparable with Sudan grass but has more juicy and thicker stems, and gives higher yields of herbage but seed setting is poorer. High HCN content. Late maturing and resistant to leaf diseases.
Cv. **Piper**. Bred in Wisconsin, USA, from crosses between cv. Tift and several lines of Sudan grass low in HCN content. Early maturing, vigorous and high yielding. Vigorous tillering and rapid recovery after grazing or cutting. Low HCN content and good resistance to leaf diseases.
Cv. **Greenleaf**. Bred in Texas, USA, from two crosses of Leoti sorghum with Sudan grass lines. Free-tillering cultivar resistant to leaf blight. Medium–low content of HCN.

Sudan grass hybrids

Sorghum sudanense × S. arundinaceum

Suhi-1. An F_1 hybrid between male sterile *S. arundinaceum* and cv. Tift of hybrid Sudan grass; released in the USA in 1961. A high-yielding hybrid which in a trial in Georgia, USA, outyielded both parental types and all the other cultivars under trial. Stems almost dry or slightly juicy. HCN content is high and the grass should be grazed with caution.

Sorghum sudenense × S. bicolor

For obtaining *S. sudanense* × *S. bicolor* hybrids, *S. bicolor* parental plants are usually selected from sorgo (sirup sorghum) which transfer to the hybrids the juicy and sweet nature of the stem, a recessive character and in early crosses Leoti sorghum was particularly favoured as a sorghum parent; parents for final crosses can themselves be of hybrid nature. After a discovery of cytoplasmic male sterility in *Sorghum*, male-sterile types have often been used to facilitate hybridization and for the production of hybrid seed. At present a number of hybrids are F_1 progenies which do not breed true to type but segregate in the following generation into various types and the hybrid cannot be reproduced from F_1 seed. Special hybrid seed needs to be produced in the same way as hybrid maize seed except that no detasseling can of course be applied. Seed of immediate parents of the hybrid is usually a property of the institution which developed the hybrid and cannot be reproduced without its permission. For seed production seed of the male-sterile plants is interplanted with those of the 'restorer' parent ('R line'), usually one row of restorer to three rows of male-sterile plants and seed of the latter is sold as hybrid seed (Doggett, 1970). Isolation required for seed production is the same as for growing seed of ordinary cultivars for seed production, i.e. 200 m from any cultivated sorghum and 400 m from Sudan grass. Hybrid cultivars can differ in the content of HCN which is an inheritable feature, its high content being a recessive or partly recessive character.

Agricultural features of sorgo and Sudan grass hybrids are usually intermediate between those of the two parental types. The stems are not so thick as in sorgo but thicker than in Sudan grass; the leaves are of intermediate width; the ability of producing tillers and of recovery after cutting or grazing and producing sufficient growth for several cuts are also intermediate. Stems of the hybrids are usually sweet and juicy, a character inherited from sirup sorgo. Hybrids closer to the sorgo type are more suitable for chopped soilage or for ensiling and those closer to Sudan grass for grazing or hay. Hybrid cultivars Tift, Piper and Greenleaf, closely approaching the Sudan grass type, have already been mentioned. Other better known hybrids and hybrid cultivars are briefly described below; numbers of hybrids are produced and released almost every year, often based on old cultivars and it is hardly possible or necessary to describe or even mention them all.

Sweet Sudan Grass. Bred in Texas, USA, by crossing Sudan grass with Leoti sorghum. A late type which retaining grazing habit of Sudan grass, remains palatable and productive for a longer period than ordinary Sudan grass and has a good resistance to leaf diseases. Stems sweet and juicy. Seed can be distinguished by its sienna-red colour compared with beige or pale-yellow colour of Sudan grass seed. Other hybrids based on Sweet Sudan had been produced, e.g. cv. SS6 developed in Australia.
Lahoma, bred in Oklahoma, USA, is another hybrid cultivar based on breeding material resulting from crosses between Sudan grass and Leoti sorghum. It tillers freely, has juicy stems and gives high yields of herbage and seed. Leaves are broad and somewhat yellowish.
Sudax, bred in Texas; can last sometimes two seasons if irrigated during critical periods. A tall type which recovers well after grazing.
Zulu, bred in Australia, is an F_1 hybrid between 'Redlan' *S. bicolor* and Greenleaf Sudan grass. Taller than Sudan grass and with coarse but sweet and juicy stems.
Bantu is also an F_1 hybrid developed in Australia and resulting from crosses between 'Redlan' sorghum and Piper Sudan grass. Stems dry, not juicy.

Other hybrids are NK **Sordan**, NK **Trudan**, **Pioneer**, **Horizon** and their derivatives and also other new hybrids, named or designated by letters and digits.

Sorghum sudanense × S. halepense

A hybrid between Sudan grass and *S. halepense* is a tufted perennial without rhizomes which forms broad tufts with numerous tillers and lasts for two to three years. Under the names of **Sorgrass**, **Sweet sorgrass** or **Perennial sweet sorgrass** it has some success in South Africa and the southern United States and has been tried in Kenya. It seems, however, that in the last decade the interest in this grass has subsided.

Sorghum verticilliflorum (Steud.) Stapf. Group *Arundinacea* $2n = 20$
Erect annual up to 2.5 m high. Leaves 10–25 mm wide and 40–60 cm long. Panicle large, up to 40 cm long, loose and usually with pendent low branches at advanced stages of flowering. Sessile spikelet 5–7 cm long with an awn up to 20 mm long. Pedicelled spikelet as long as the fertile spikelet but narrower. Racemes easily disarticulating at maturity.

Common in southern Ethiopia, Somalia and East Africa, extending southwards to South Africa and to the Indian Ocean islands on the African side. Occurs on roadsides, wasteland, stream banks and as a weed of arable land, often on black clay, seasonally waterlogged soils. Introduced, mostly accidentally, to Australia, India, Polynesia and the West Indies where it grows mainly as a weed of cultivation. It was tried without much success as a fodder grass in East Africa. Readily grazed by the animals although it contains relatively high amounts of HCN.

Perennial kavirondo sorghum is supposed to be a natural hybrid between *S. bicolor* and *S. verticihiflorum*. The plants have been brought from the Kavirondo area of western Kenya (Edwards, 1948) and experimentally grown as a fodder plant in Kenya and some other countries but is now almost abandoned. This sorghum can last 2 years and sometimes even longer. When found, it was perhaps an F_1 hybrid as it segregated into various types; those with thick stems resembling grain sorghum and with fine stems and more numerous tillers were rather similar to hybrid fodder types. Snowden considered this form to be *S. arundinaceum* var. *kavirondense* Snowden.

Sorghum virgatum Snowden. Group *Arundinacea*. $2n = 20$
Annual with long and narrow panicles 15–60 cm long and 1–5 cm wide; sessile spikelet 6.5–7 mm long and 2–2.5 mm wide, acute with a slender awn. Occurs in North East Africa, tropical and subtropical. Usually considered as only a variety of *S. arundinaceum* and by Harlan & de Wet (1972) as a race of *S. bicolor* ssp. *arundinaceum*. It is a possible parent of spontaneous hybrids with cultivated forms of sorghum which, when segregated, might have produced Sudan grass.

Sporobolus R. Br.

Annuals or perennials, tufted, stoloniferous or rhizomatous. Panicle open, to contracted, sometimes spikelike. Spikelets small, with one floret. Glumes persistent, herbaceous and so are the lemma and palea. Grain is enclosed in a free pericarp which swells when moistened. A large, predominantly tropical genus of about 150 species. Mostly medium quality grasses but some are of good grazing value but not of outstanding importance, and there are also fibrous species of low grazing quality and hardly palatable as e.g. is *S. pyramidalis* Beauv. which can invade overgrazed pastures. A number of species grow well on saline or alkaline soils.

Sporobolus cordofanus (Steud.) Coss.
Erect annual 8–60 cm high with the stems often branched. Leaves 4–12 cm long and 3–6 mm wide. Panicle ovate, 2–10 cm long, with slender, often reddish branches arranged in whorls. Spikelets 1.6–2 mm long, usually olive-green or dark grey. Lower glume up to 0.6 mm long, upper as long as the spikelet. Grain 0.6–0.8 mm long.

Occurs in tropical East Africa, from Sudan to Rhodesia and also in Senegal, on dry soil or damp sands, in bushland or in open and often overgrazed dry areas where it can be abundant and provide a considerable amount of grazing, especially in the more favourable years. The herbage, together with stems, is relatively soft and is well grazed.

Sporobolus helvolus (Trin.) Th.Dur. & Schinz
Stoloniferous perennial forming small tufts. Stems thin, wiry, 15–60 cm high. Leaves flat, 2–10 cm long and 2–4 mm wide. Panicle small, narrow, 4–12 cm long and 0.5–2 cm wide. Spikelets 1.5–2 mm long, often greenish-brown in colour. Glumes equal or subequal, as long as the floret. Grain ellipsoid, 0.5 mm long.

The stolons of this grass arise as ordinary shoots some of which elongate, bend down, root from the nodes when the stems touch the ground and proceed to grow as long, fast-growing stolons, rooting at some distance from the mother tuft and this can be regarded as an adaptation reducing the competition with the main tuft.

Distributed in East Africa and Sudan extending to Mauritania in the west and Somalia in the east; also in Arabia and India. *Sporobolus helvolus* forms large colonies on black, heavy, often waterlogged soil and also on volcanic-ash soils. In spite of the wiry stems it is well grazed and is one of the best grasses on black clay soils of flats and depressions in semi-arid areas. Cut at the early flowering stage the whole plant contained 12.9 per cent CP, only 0.8 per cent EE, 30.6 per cent CF and 45.4 per cent NFE (Dougall & Bogdan, 1960). The contents of P and Ca were 0.16 per cent and 0.47 per cent, respectively, not particularly high but sufficient for the animal maintenance.

Sporobolus isoclados (Trin.) Nees (*Vilfa isoclados* Trin.; *S. marginatus* A. Rich.)
Tufted perennial often also spreading by stolons. Stems 15–60 cm high. Leaves flat, 2–30 cm long and 2–5 mm wide, usually with stiff hairs on margins. Panicle pyramidal, 3–20 cm long; side branches in four to eight whorls, fine, often reddish. Spikelets 1.5–2.2 mm long, greyish-green to pallid, rarely purplish. Lower glume one-quarter to one-third the length of the spikelet, upper as long as the spikelet. Grain ellipsoid, 0.6–1 mm long.

Occurs throughout tropical Africa and in India from sea level to about 2,000 m alt, on dry soil or on seasonally waterlogged flats and depressions, sometimes on slightly saline or alkaline soil. It occurs in a number of forms, some of them short-lived and approaching annual *S. cordofanus* and some spreading to a various degree by means of stolons. An extereme stoloniferous form develops dense leafy mats on strongly alkaline soil and retained this habit when grown in small plots on nonalkaline soil in Kenya. This small creeping grass analysed in Kenya (Dougall & Bogdan, 1965) showed a very high content of CP, 23.3 per cent, only 25.7 per cent CF and a relatively low content of NFE, 36.8 per cent. This type is avidly grazed by sheep, and, if it is not too low, also by cattle, whereas the larger, tufted types are less leafy and slightly less palatable.

Sporobolus nervosus Hochst. (*S. longibrachiatus* Stapf)

Densely tufted perennial up to 40 cm high. Leaf sheaths yellowish, hard; blades (leaves) 4–10 cm long and 1–3 mm wide. Panicle ovate, 7–20 cm long with solitary primary branches and the spikelets crowded in their upper parts. Spikelets 1.7–2.1 mm long, greyish-green or dark green in colour. Lower glume 0.8–1 mm long, acute, upper glume 1.3–1.9 mm long, shorter than the spikelet.

Distributed in the northern parts of tropical Africa from Mauritania in the west to Sudan, Ethiopia and Somalia in the east and also in tropical Arabia; it also occurs in Namibia. *Sporobolus nervosus* grows in dry areas from 300 to 1,700 m alt, in bushland and in open grassland often on dry shallow soil and can form extensive colonies. The stems are usually not numerous and the grass can be leafy, readily grazed and of high nutritive value. Analysed in Kenya (Dougall & Bogdan, 1960), it contained over 16 per cent CP and only 22 per cent CF.

Stenotaphrum Trin.

Includes seven tropical and subtropical species; mostly perennial creepers.

Stenotaphrum dimidiatum Brongn. Pemba grass. $2n = 36$

Stoloniferous perennial similar in habit to *S. secundatum* from which it differs in having slightly wider, bright green leaves; the axis of the spike is wavy and the spikelets are gathered in short racemes of three to several spikelets in the hollows of the spike axis. Its distribution is more restricted and is confined to coastal areas of East Africa, Madagascar, India and the islands of the Indian Ocean. It usually grows on coastal sands, often in shade. It is well grazed and gave encouraging results as a pasture grass in Madagascar and Uganda.

Stenotaphrum secundatum O. Kuntze. St. Augustine grass; Pasto San Augustin.

Creeping stoloniferous perennial. Stems upright or ascending, often much branched, 10–50 cm high. Leaves flat, blunt, glabrous, slightly bluish, 3–15 cm long and 4–10 mm wide, with tightly compressed and keeled sheaths. Spikes (panicles) 4–10 cm long with a spongy axis, flat on the back and 3–7 mm wide. Spikelets sunk in the hollows of the spike axis, solitary or in twos or threes, lanceolate, acute, 5 mm long, pale. Glumes dissimilar; the lower 1–2 mm long, the upper as long as the spikelet. Florets two: lower floret male, upper bisexual.

A pantropical species which occurs in coastal areas of Africa, Central and South America, India, Australia and in Pacific islands. It grows in relatively humid areas and prefers fertile soil although it can form colonies on poorer soils. It can grow on soil with a high groundwater table, up to 30 cm from soil surface, although in places with particularly high groundwater table the content of CP in the plant can be reduced. It can tolerate flooding for a considerable period of time and a relatively high content of salt (NaCl) in the soil but highly saline sea shores are avoided.

Stenotaphrum secundatum is well grazed by cattle, can withstand high grazing pressure and close grazing and has been cultivated on experimental and farm scale for a few decades, mainly in Florida, USA, as an introduced grass. None or very little seed is produced and the grass is propagated vegetatively by pieces of stolons which can be planted in 60–75 cm rows and about 30 cm apart in the rows (Whyte *et al.*, 1959), or scattered on soil surface and then disced in. Five to six months are required for the establishment of a continuous sward. *Stenotaphrum secundatum* responds well to fertilizer N by increasing herbage yields. Close grazing inflicts no harm and the highest yields of herbage have been obtained when the grass was cut or grazed to a height of about 5 cm from the ground level. In a trial in Florida, the herbage contained 16.7 per cent CP in spring and 12.8 per cent in autumn (Haines *et al.*, 1965) and, in another trial, the digestibility of CP was determined as 51.6 per cent, DM digestibility 59.6 per cent, CF 44.4 per cent, EE 60.4 per cent and NFE 69.1 per cent (Chapman *et al.*, 1960). It was estimated (Haines *et al.* 1965) that young cattle used in a trial consumed 56.5 t fresh herbage/ha/year and gained 1,280 kg liveweight/ha/year; other Florida trials gave comparable or slightly lower gains.

Stenotaphrum secundatum is also much used for soil conservation and as a lawn grass in several countries, including south-west Europe. Several cultivars have been selected, the best known one is Roselawn.

Diploids, tertraploids, hexaploids and octaploids with $2n$ chromosome numbers 18, 36, 54 and 72, respectively, have been found and an irregular number $2n = 20$ was also recorded.

Themeda Forsk.

Up to ten species in tropical and subtropical Africa, Asia and Australia.

Themeda australis (R. Br.) Stapf (*Anthistiria australis* R. Br.). Kangaroo grass.

This species is essentially similar to *T. triandra* (see below) and the difference that exists between the two species is not quite clear; it is therefore preferable to treat *T. australis* as a geographical form of widely distributed *T. triandra* pending a taxonomic revision of the group. The chromosome numbers of the Australian material, $2n = 20$ or 40, do not differ from those of typical *T. triandra*. Reproduction is predominantly apomictic but sexual diploids have also been found, mainly in New Guinea (Evans & Knox, 1969).

Themeda australis is common and often numerous in the whole continent of Australia, in Tasmania and New Guinea. In Australia Kangaroo grass is considered to be palatable to cattle; its CP content is however on the low side and according to Siebert *et al.* (1968) ranges from 6.3 per cent in green, growing plants to 2.5 per cent in dry mature plants; CF content is given as 25–28 per cent which is low for any tropical grass at an advanced stage of growth.

Themeda triandra Forsk. (*Anthistiria glauca* Desf., *A. imberbis* Retz.) Red oatgrass; Rooigras (S. Africa). (Fig. 35). (The panicles create some resemblance to oats and maturing plants are well tinged with red or orange colouration, hence the common name).

Tufted perennial 40–150 cm high. Leaves 5–25 cm long and 2–6 mm wide, flat or folded. Panicle loose, nodding or erect, 10–40 cm long, with clusters of racemes pendent (sometimes erect in young plants) on filiform branches or sessile on the stem. Each raceme, which is supported by a spatheole, gives the impression of a single spikelet but actually consists of seven spikelets: the four basal green or reddish, sterile, awnless spikelets are 8–12 mm long; the single fertile, dark-brown, spindle-shaped spikelet is 5–6 mm long and terminates in a stout awn 4–6 cm long which is kneed and spirally twisted below the knee; the two uppermost spikelets are male. The dark and glossy fertile spikelet, which contains a single caryopsis, is easily detached at maturity and functions as seed. There are about 280,000 such 'seeds' per kg.
Themeda triandra is predominantly an aposporous apomict and there are diploids, tetraploids, hexaploids and octoploids with $2n$ chromosome numbers 20, 40, 60 or 80 (Brown & Emery, 1957), although chromosome numbers $2n = 30$ and 50 have also been reported. It occurs in a great variety of forms throughout the Old World tropics and subtropics and reaches temperate areas, e.g. in South Australia. Plant height, stem thickness, the shape of the panicle and minor details of the floral parts vary in different geographical areas and habitats, and several botanical varieties have been recognized and described which are however of little practical significance, and the *T. triandra* complex needs a revision. Stapf, in *Flora of Tropical Africa* considers true *T. triandra* to be an African species which also occurs in Arabia, but he

does not include in *T. triandra* the types grown in India, the Philippines and other Asiatic areas. However, in the pastoral and agricultural literature plants of these areas are usually referred to as *T. triandra*, and for the time being it is perhaps wise to use this well known name for the whole complex of varieties, including the Asiatic ones. The same treatment could perhaps be applied to *T. australis*.

Fig. 35   *Themeda triandra* (from Edwards & Bogdan, 1951).

Distribution. In Africa *T. triandra* is one of the most common grasses which covers large areas and often dominates in natural or semi-natural grasslands. It occurs from sea level to 3,500 m alt, in the equatorial belt, is particularly numerous above 1,000 m, and at somewhat lower altitudes in the rest of the African tropics and subtropics and in South Africa. *Themeda triandra* is also common in India, mainly in its western part, in Pakistan, Burma and the Philippines. It is utilized for grazing in natural grasslands but not in sown pastures, and a few attempts at cultivation were not successful because of its slow establishment and the lack of seed which is difficult to harvest and to clean. Moreover, quite a number of cultivated tropical grasses are more productive and more nutritious than *T. triandra*.

Environment. *Themeda triandra* is moderately drought resistant but cannot grow under an annual rainfall of much less than 500 mm and is distributed in areas where the rainfall ranges from 500 to about 1,200 mm. It occurs in open grasslands, often as a dominant grass, and can form almost pure stands, or in grassland with scattered bush or trees. Grassland dominated by *T. triandra* is considered to be the fire climax, the final stage of grassland development under regular grass fires. In the absence of fire it is gradually ousted by other grasses or by bush, and beneficial effects of grass burning on *T. triandra* were reported from Kenya, Uganda, Rhodesia and elsewhere. In Kenya, at 2,300 m alt, yearly burning of completely protected grassland with the dominance of unpalatable *Pennisetum schimperi* resulted in almost pure stands of *T. triandra*. However, in some areas, too frequent burning suppressed *T. triandra* in favour of *Heteropogen controtus*, whereas less frequent burning maintained its dominance. In other areas only late season burning was reported to encourage *T. triandra*. The reasons for *T. triandra* resistance to grass fires is not fully understood, although the ability of its seeds to bury themselves in the soil to a depth of 2–4 cm is a contributing factor. Seed of *T. triandra* has a sharp, barbed callus and under changing air humidity the spirally-twisted awn screws the seed into the soil. The development of a large number of secondary basal tillers due to the effect of fire, as reported by Tainton & Booysen (1965) can also contribute to the fire resistance of *T. triandra*. *Themeda triandra* can grow on a variety of soils, but prefers either red soil or black clays and usually avoids too sandy habitats. However, when grown on sandy soil it develops a mucilaginous layer on the surface of larger roots to which sand particles stick and form a protective cover (Taerum, 1970). It can withstand waterlogging but not prolonged flooding. It tolerates relatively infertile soils but responds well to fertilizers as was shown by Rethman & Malherbe (1970); in their trials in South Africa, grassland dominated by *T. triandra* produced increased yields of organic matter and of N by 163 and 303 per cent, respectively, when NPK fertilizer was applied, whereas N digestibility was increased from 58.3 to 70.8 per cent in the wet season and from 39.3 to 63.1 per cent under drier conditions. Wild (1965) reports that *T. triandra* can grow well on toxic serpentine soil with high contents of Ni and Cr.

Management. *Themeda triandra* thrives under lenient grazing but heavy grazing suppresses it and eventually excludes it from the sward. Under rotational grazing *T. triandra* can usually withstand heavier grazing pressure than under continuous grazing and Acocks (1966) observed that 2 weeks of sheep grazing followed by 6 weeks of rest resulted in rapid recovery of *T. triandra*, previously suppressed by heavy continuous grazing, and increased the grazing capacity of the land. In South Africa low tolerance of *T. triandra* to grazing has been linked with the position of the growing points. In contrast to the majority of South African and tropical grasses, the growing points of sterile shoots are gradually elevated by the growth of basal internodes and reach some 8 cm above ground level (Booysen *et al.*, 1963). The rising of the growing points in ungrazed grass can take up to 9 months, before they reach the reproductive phase and initiate panicles and spikelets. Animals grazing in the middle or the late parts of the season destroy the growing points which at that time rise above the level of grazing. If, however, the grass is grazed early in the season when the growing points are not raised above 1–2 cm, they are not damaged and can develop panicles and set seed in the following season.

Palatability, chemical composition, nutritive value. In a number of published reports *T. triandra* has been regarded to be palatable or highly palatable to cattle and sheep. In South African trials with sheep on the palatability of local grasses (Theron & Booysen, 1966) it was the most palatable of eight species under trial. *Themeda triandra* is also well grazed by wild ungulates (Stewart & Stewart, 1970). Engels *et al.* (1969) has shown in trials with sheep that natural grassland dominated by *T. triandra* has sufficient nutritive value for maintenance and production of the animals throughout the year. CP content of *T. triandra* was reported by Dougall & Glover (1964) to be 17.23 per cent in young growth regenerating after burning, but reduced to 12–13 per cent 2 months later, and to 6.11 per cent 3–4 months after burning; lower CP values were obtained after cutting when *T. triandra* contained 6.68 per cent CP 2–3 weeks after cutting, and about 3–4 per cent 3–4 months later, all CP values of plant regrowth after cutting being below the maintenance level. Similar or still lower CP values were reported from Uganda, and Marshall & Bredon (1967) determined 2.9–3.5 per cent CP in *T. triandra* herbage with CP digestibility ranging from negative values to 33 per cent and that of DM from 53 to 64 per cent, although higher digestibility coefficients were also recorded. CF content in young growth of *T. triandra* can be as low as 24 per cent, but in advanced or mature herbage it is high and ranges between 35 and 42 per cent. P content is medium to low, usually 0.15-0.25 per cent, but can be below 0.10 or even 0.05 per cent. Herbage intake by the animals is low to medium and varies from 0.5 to 1.8 kg DM per 100 kg animal weight.

On the basis of low CP content and digestibility, Marshall & Bredon (1967) re-examined the traditionally accepted grazing potential of *T. triandra*, and disproved its positive value in natural grasslands. Further research is however needed, especially bearing in mind that animal production from grasslands dominated by *T. triandra* can be reasonable. In trials in Uganda steers grazed on *T. triandra*-dominated pastures gained 150-190 kg liveweight per animal during the 2 years of trial when grazed at stocking rates of 2.4-3.6 ha per animal, and 118-140 kg at 1.2 ha per animal; gains per ha were higher at 1.2 than at 3.6 ha per animal. Particularly high gains/ha, up to 240 kg at a stocking rate of 0.6 ha/animal were obtained in paddocks cleared from unpalatable *Cymbopogon afronardus* (Harrington, 1969). In practical range management in East Africa, the stocking rates for *T. triandra*-dominated grasslands usually range from 1.2 to 5 ha per head of adult cattle.

Trachypogon Nees

Tufted perennials of dry, open habitats. Up to ten species in tropical America and tropical and South Africa.

Trachypogon spicatus (L.f.) O.Ktze (*T. plumosus* Nees, *T. polymorphus* Hack., *Andropogon plumosus* H.B.K.). Horo (Malagash) $2n = 20$
Densely tufted perennial 40-140 cm high; stems bearded at the nodes and surrounded at the base with firmly attached, old leaf-sheaths. Leaves convolute or flat, glabrous or hairy, up to 30 cm long and 2-5 mm wide. Racemes solitary or in two to three on stem apex, erect, 7-20 cm long, with several joints and pairs of spikelets, one of the pair subsessile the other on a longer pedicel, both of about the same length, 3.5-4.5 mm. Subsessile spikelet male or sterile, awnless. Pediceled spikelet with a sharp callus, bisexual and fertile, with a silvery-hairy felxuous awn 3-8 cm long.

Distributed in tropical Africa, Madagascar, the Philippines and in America from Texas and Arizona in the north to Argentina in the south. *Trachypogon spicatus* is often numerous and sometimes dominates in African savannas and is very common and abundant in the Madagascar plateau, in slightly degraded pastures (Cabanis *et al.*, 1970). In South America it is particularly common in Venezuela where the *Trachypogon* savanna covers large areas free from flooding, mainly on somewhat sandy soils. Ramia (1967) indicates that in South America four more species of *Trachypogon*, including common *T. montufari* (H.B.K.) Nees, form the main bulk of herbage, but as these species are difficult to distinguish and are of a similar growth habit, they are usually treated by South American pastoralists as a *T. spicatus* complex. *Trachypogon* herbage is coarse, practically uneaten by cattle towards the end of the dry season and mature herbage is usually burnt before the rains begin; the animals then graze on the young growth during the wet season. Herbage quality is in general mediocre or poor and the average CP content was determined by Cunha *et al.* (1971) as 5.9–6.3 per cent, CF 33.2–36.1 per cent, lignin 11.7–12.5 per cent and EE 1.9–2.1 per cent. Grass analysed 15 days after the previous cut contained 8.1 per cent CP, 45 days 6.9 per cent, and 105 days 4.7 per cent; CF contents were 25.6, 33.6 and 39.6 per cent, respectively. Low CP content, 5.7 per cent, and high CF content, 40.2 per cent, were also demonstrated in Kenya by an analysis at the early flowering stage, which also showed very low contents of Ca and P, 0.15 and 0.10 per cent, respectively (Dougall & Bogdan, 1958).

Tripsacum L.

Includes seven species native of tropical, mostly Central America. Tall tufted or spreading grasses; related to maize.

Tripsacum dactyloides L. Eastern gama grass (USA)

A strong perennial resembling *T. laxum*; it forms large tufts or clumps up to more than 1 m in diameter which spread by means of short rhizomes. Stems 1–3 m high. Leaves 40–60 cm long and 1 to 3.5 mm wide. Panicle with one to three racemes of the structure rather similar to that of *T. laxum* panicles.

Occurs in a wider area than the other species of *Tripsacum*, from southern USA to Central America, West Indies and the northern parts of South America where it forms dense or clumpy stands in moist situations, mainly on stream and river banks and is used as fodder or for grazing, but close grazing is not recommended because it can weaken or even kill the grass. The herbage is palatable to all stock and reputed to be of high nutritive value. Cultivated as fodder in countries of its origin and is usually propagated vegetatively by pieces of rhizomes. Introduced to India where it is regarded as a nutritious grass. A ploidy sequence of $2n = 18$–36–54–72–90–108 has been observed and occasional irregular chromosome numbers reported.

Tripsacum latifolium Hitchc. Prodigioso (León & Sgaravatti).

Tall perennial resembling *T. laxum* from which it differs in having slightly narrower (up to 7 cm wide), more attentuate and more hairy leaves, drooping upper, staminate parts of the raceme and smaller staminate spikelets which are 3–4 mm long; both spikelets in a pair are almost sessile (Randolph, 1970). It also has a somewhat more procumbent growth. Occurs naturally in Guatemala and other parts of Central America, in Mexico and in some Caribbean islands and is used for fodder as a local grass in its natural habitats and in cultivation. It is also cultivated in neighbouring countries where it has naturalized. As a fodder grass it has much in common with *T. laxum* but the stems develop earlier and the grass is therefore somewhat less leafy than *T. laxum*. *Tripsacum latifolium* is a fertile diploid with $2n = 36$. The pollen is 70–100 per cent fertile and seed is usually well formed. Atypical tetraploid plants ($2n = 72$) have also been found (Randolph, 1970).

Tripsacum laxum Nash (*T. fasciculatum* Trin.). Guatemala grass; Pasto guatemala. (Fig. 36).

Large tufted or spreading perennial. Stems glabrous, up to 3.5–4.5 m high and 1 cm thick, erect or ascending from a prostrate base arising from thick short rhizomes. Leaves usually covered with short scattered hairs, up to 80 cm long and 9 cm wide, narrowed at the base into a keeled leaf sheath. Inflorescence subdigitate, with three to eight racemes, which are up to 20 cm long, with flattened joints in the lower part. Each joint bears a pair of pistillate (female) spikelets which are 5 mm long; one of the spikelets can occasionally be fertile. The upper, longer part of the raceme bears numerous pairs of staminate (male) spikelets which are 3–4 mm long; one of the pair is almost sessile, the other on a fine pedicel.

Distributed in Mexico and Central America, almost exclusively as a cultivated fodder plant; reports on the occurrence of wild plants need confirmation. As a cultivated grass it has spread to South America where it is much grown, especially in Brazil, and also to the Caribbean islands. It has also been introduced to Ghana, Ivory Coast, Kenya, Rwanda, Madagascar, US Virgin Islands, Fiji, Sri Lanka and probably to other tropical countries, where it is grown experimentally or on a limited farm scale and is much valued for its vigour, almost complete absence of stems in the herbage for a considerable period of time and for retaining its nutritive value at advanced stages of growth.

Being almost completely sterile, *T. laxum* is propagated by tuft divisions or stem cuttings and takes a longer time to establish than *Pennisetum purpureum* or *Panicum maximum*. It forms semi-spreading large stools and should be planted at about the same spacing as the larger types of *P. maximum*. Whyte et al. (1959) consider 90–100 cm by 18–24 cm as suitable but wider spacing can often give better results. The plants are tall and robust but the stems develop very late and Tardin et al. (1971) in Brazil observed that after 196 days of growth the plants produced no stems and were still at the vegetative stage. In Kenya the tall flowering stems only seldom developed. *Tripsacum laxum* grows relatively slowly and in Brazil it produced only 7 g DM/m^2/day (Tardin et al., 1971), less than the other large grasses under trial. In spite of the slow growth the grass can give satisfactory to high yields which are, however, lower than those of *Pennisetum purpureum*. In Brazil (Zúñiga et al., 1965) a single (second) cut gave 68 t fresh herbage or 12 t DM/ha and similar yields were obtained in Thailand (Holm, 1972) from a well-fertilized crop. Higher yields were obtained in another trial by Zúñiga et *al.* (1967) in which 195 t fresh herbage or over 40 t of dry forage/ha were harvested in the 13 months of the trial. In a trial by Pereira *et al.* (1966), also in Brazil, *T. laxum* yielded 70 t of green material/ha in the wet season and 4 t in the dry season. Moderate amounts of NPK plus B and Zn increased the wet season yield to 100 t/ha but decreased slightly that of the dry season; however, in combination with sprinkler irrigation the fertilizers increased dry season yields to 16 t/ha, but yearly yields of *Pennisetum purpureum* tested in the same trial exceeded about twice those of *T. laxum*. In Kenya (Kenya Report 1970) the yields ranged from 11.2 to 26.2 t DM/ha/year, depending on the frequency of cutting, the lowest yield being obtained at 2-month intervals between the cuts and the

Fig. 36 *Tripsacum laxum.*

highest at 8-month intervals. In the same trial *P. purpureum* yielded 22.4 and 47.8 t DM/ha, respectively; leaf yields were, however, higher in *T. laxum*, 17.0 and 17.1 t/ha, at 4- and 8-month intervals, respectively, whereas those of *P. purpureum* were 13.7 and 10.5 t. The contents of stem in herbage cut over 2, 4 or 8 months were 0, 9 and 35 per cent for *T. laxum* and 32, 57 and 78 per cent for *P. purpureum*. CP contents in the leaves of the two grasses were about equal and in *T. laxum* they ranged from 8.3 to 13.7 per cent, depending on the frequency of cutting. However, Dirven (1962) in Surinam determined the content of CP as 6.1 per cent in the leaves and 4.6 per cent in the stems, and the content of CF in leaves was surprisingly high, 40.4 per cent. In another trial in Surinam

(Appelman & Dirven, 1962) CP content reached 15.9 per cent in grass cut every three weeks, decreased with the decrease in cutting frequency and was 7.5 per cent in grass cut every 8 weeks. The digestibility of CP remained, however, at about the same level of 58–62 per cent. *Tripsacum laxum* can persist for a number of years but its drought resistance is poor, possibly because of a relatively shallow root system, and the plants do not produce much herbage in dry seasons if not irrigated.

Because of a considerable vigour of growth, only a few legumes can mix well with *T. laxum* but *Demodium intortum* and *D. uncinatum* formed good, balanced mixtures, and DM yields of the mixtures were over double those of pure grass. Suttie & Moore (1966) found that the presence of *D. uncinatum* increased CP content in the grass from 6.6 per cent to 10.0 per cent and CP content of the mixture was 14 per cent.

Tripsacum laxum is usually cut for soilage and it is recommended to cut the herbage at a height of 10–15 cm from the ground level when the grass reaches 120–150 cm in height (Tardin *et al.*, 1971). In Kenya *T. laxum* was also used for grazing with reasonable success.

The apical points of tillers are situated near the ground level and usually escape damage by cutting or grazing. They can, however, reach 1–4 m from the ground level in the grown-up stems. The panicles and the spikelets are well formed but the plants are almost completely sterile which can partly be explained by the low fertility of pollen: only 1 to 5 per cent of good pollen was observed (Randolph, 1970). There are distinct cultivated types which could have risen only sexually and this confirms the view that fertile seed can be occasionally formed. *Tripsacum laxum* can apparently hybridize with *T. pilosum* Scribn. & Merr., an assumption based mainly on the presence of seedlings under *T. laxum* plants grown in the vicinity of *T. pilosum* (Randolph, 1970). *Tripsacum laxum* is a tetraploid, the chromosome number being $2n = 72$, but, according to Randolph, triploids ($2n = 54$) are also known as well as plants with irregular chromosome numbers ranging from over 54 to under 72. Some of triploid plants might have resulted from crosses with *T. latifolium*.

Urochloa Beauv.

Includes some 25 species of tropical Asia and Africa.

Urochloa bolbodes (Steud.) Stapf (*U. oligotricha* (Fig. & De Not.) Henrard) $2n = 36$

Tufted perennial 30–60 cm high with three-noded stems. Leaves 15–20 cm long and 6–9 mm wide. Panicle usually with many (up to 20) racemes, distant below and crowded on the panicle top. Racemes dense, 1–6 cm long, with crowded spikelets which are lanceolate or lanceolate-ovate, 3–4 mm long. Florets two, the upper, bisexual and fertile, 2–2.5 mm long.

Urochloa bolbodes occurs naturally in Ethiopia, and in tropical West and East Africa reaching Rhodesia in the south; it grows in semi-arid areas up to 1,800 m alt. In experimental cultivation it performed well in Rhodesia, but gave low yields and did not persist under grazing in Kenya trials at 1,800 m alt. It is not common in natural grasslands, but where it occurs, is considered to be palatable and of good nutritive value. Its CP content at the early flowering stage was determined as 13.7 per cent and CF as 30.0 per cent; Ca and P contents were low (Dougall & Bogdan, 1960). Reproduction is apomictic.

Urochloa mosambicensis (Hack.) Dandy. Little para (Fiji).

U. pullulans (Steud.) Stapf is botanically rather similar and is not treated here as a separate species. Although unlikely, it is possible, however, that in a future revision of the genus *U. pullulans* would be regarded as a distinct species. In agricultural and pastoral literature *U. mosambicensis* is mentioned much more frequently than *U. pullulans*.

Perennial, often short-lived, up to 100 cm high. Stems erect or ascending from a geniculate base rooting from nodes. Leaves 15–25 cm long and 10–15 mm wide. Panicle consists of 5–10 erect or suberect dense racemes, which are 5–9 cm long and arranged on a common axis 5–7 cm long. Spikelets 4–4.5 mm long with the two lower glume three quarters the length of the spikelet. Florets two, the upper, bisexual, fertile, elliptic 2.5 mm long.

Urochloa mosambicensis occurs naturally in tropical and South Africa, from Tanzania in the north to Transvaal in the south, in grassland, fallows, at roadsides, etc., in semi-arid areas with 400–800 mm of annual rainfall, at relatively low, frost-free altitudes. It grows on well-drained soils and its content in the sward increases from the application of N and P fertilizers. *Urochloa mosambicensis* has been tried in cultivation in Africa and introduced to India, Hawaii, Fiji and Australia.

For cultivation of *U. mosambicensis*, 4 kg seed/ha is a recommended rate, and when grown in mixture with *Stylosanthes humilis*, 2 kg suffice. Sowing fresh seed is not recommended as it requires post-harvest maturation, seed dormancy can, however, be broken by hammer-milling which destroys the hard lemma clasping the seed. Quick establishment and moderate to good yields of herbage were observed at altitudes up to 700 m, and yields of about 4 t DM/ha in India, and 18 t fresh material in Fiji (Roberts, 1970) were obtained. Low yields and poor persistence were however observed in Kenya at 1,800 m alt. In Australia, where *U. mosambicensis* is grown mainly with *Stylosanthes humilis*, grass yields in the mixtures ranged from 830 to 6,520 kg DM/ha and those of the legume from 25 to 3,438 kg. (from Whiteman & Gillard, 1971).

Urochloa mosambicensis is an apomict and both tetraploids, with 28

$2n$ chromosomes, and hexaploids, with $2n = 42$, were found. Seed yields of about 100 kg/ha were recorded from Australia.

Urochloa stolonifera is another perennial species tried in cultivation in India. *Urochloa panicoides* is an annual common in tropical Africa which has now spread to a number of other tropical areas. The protein content of the herbage is high, it is well grazed and, although an annual, this grass is occasionally favoured in natural grassland (Long *et al.*, 1969).

Basic information from Whiteman & Gillard (1971).

Vossia Wall. & Griff.

A monotypic genus with one species only.

Vossia cuspidata (Roxb.) Griff.

Perennial with floating or submerged stems which can also creep on the ground when the water recedes. Stems root freely from numerous nodes and their erect parts reach up to 1 m high. Leaves up to 1 m long and 15–20 mm wide, somewhat fleshy. Panicle digitate with up to six racemes 15–20 cm long, articulate, with joints about 8 mm long. Each joint has a pair of spikelets, one sessile the other on a stout pedicel, both alike in shape and sex; 7–10 mm long; their lower glumes are continued into a flat awn 2–3 cm long. Florets two, the lower male, the upper bisexual, fertile.

Occurs on swampy ground, lake margins and in streams and rivers of northern Nigeria, Sudan, Uganda, Kenya, Zaire, Rhodesia, Malawi and India and is common in the upper Nile River area. *Vossia cuspidata* is eaten by buffaloes and cattle if the animals can reach it. On flat low land of the southern shores of Lake Baringo in Kenya it provides, together with *Cynodon nlemfuensis*, a considerable amount of grazing. It may, however, become a nuisance in irrigation canals and ditches. This species can be considered a counterpart of South American grasses of similar ecology – *Hymenachne amplexicaule* or *Panicum elephantipes*.

Zea L.

Maize. One polymorphic species of Mexican origin now widely cultivated throughout the world in tropical, subtropical and warm regions.

Zea mays L. Maize; Corn; Indian corn.

Tall, robust annual 2–3 (1.5–6) m high with overlapping leaf sheaths. Leaves (blades) 30–150 cm long and 5–15 cm wide. Male (staminate) and female (pistillate) spikelets in separate inflorescences, although the plant is monoecious and the male and female inflorescences develop on the same plant. The male inflorescence (tassel) is a terminal panicle of several racemes beset with paired spikelets which are 8–13 mm long, each containing two florets. The female inflorescence (cob) arises from the leaf axil and there can be one to three cobs per plant. The cob is a spike or an ear with the greatly thickened axis densely packed with spikelets in 8 to over 16 dense rows. The cob is hidden within an involucre of several bracts or husks. The ovary has a long, thread-like stigma and stigmas of all the spikelets protrude from the involucre and form a long tuft of 'hair' known as silks. The naked caryopsis have only minute floral scales, except in 'pod corn' in which the floral scales are well developed. The caryopsis (grain) is large and of different shape and texture in different varieties or types of maize.

Maize is unknown in a wild state although there is evidence that it existed as a wild plant in the remote past. Its nearest wild relation is *Euchlaena mexicana* which differs mainly in having pistillate spikelets arranged on a flat, articulate axis. Hybrids between maize and *Euchlaena mexicana* have been obtained. Another genus related to maize is *Tripsacum* and hybrids between some species of this genus and maize have also been produced.

Maize has been cultivated in Mexico from prehistoric times and later also in adjacent American countries. In the sixteenth century it was introduced to Europe and then to other parts of both Old and New Worlds, to the tropical, subtropical and warm temperate areas, but particularly wide distribution of maize occurred in the United States of America. For a long time it was grown almost exclusively for grain but in the last 70–80 years also as a forage crop. *Zea mays* is an extremely variable species, six forms of which are usually recognized. Of these forms, only dent maize (var. *indentata* Montg.), which has somewhat angular grain concave on top, and, to a lesser extent, flint maize (var. *indurata* Montg.), which has round grain with hard endosperm, are grown for forage.

Maize is a cross-pollinating plant, a diploid usually with the chromosome number of $2n = 20$, although irregular numbers of $2n = 21$, 22 or 24 have also been found, and is genetically complex and variable. 'An open pollinated cv. is a mixture of many complex hybrids and few seeds, even on the same ear, will have exactly the same genotype' (Purseglove, 1972). Innumerable cultivars and hybrids, some of them bred for the tropics, have been developed by cross-pollination between different forms or local cultivars combined with single plant selection, the breeding work being based on extensive studies of maize plant genetics. Maize hybrids require special techniques for seed production and the nature of the hybrids used on farms and the principles of seed production require brief explanatory notes for those not familiar with the problem.

Hybrids between forms not closely related genetically show heterosis or hybrid vigour and F_1 generations are also relatively uniform. They have been extensively used for grain crops and fodder production in the countries with more advanced agriculture, especially in the USA.

For obtaining hybrid seed, two selfed lines, selected for similar desirable characteristics connected with dominant genes, are grown in alternate rows and to secure cross-pollination between the lines all plants of one of the lines are detasseled and seed is harvested only from the line with detasseled plants; one row of the pollinator line with intact tassels is usually sufficient to pollinate two to four rows of detasseled plants. Seed obtained in this way is hybrid seed which produces uniform, vigorous and high yielding farm crops. The resulting crop cannot, however, be used for sowing but only for commercial use and fresh hybrid seed should be acquired each year from hybrid-seed growers. This single-cross seed is expensive because the selfed plants are weak and produce little seed. To make seed cheaper, double-crossing is used; the hybrid is crossed with another hybrid similarly produced and seed of the resulting double-cross hybrids, which produce more seed than single-cross hybrids, is used as commercial seed. The crop grown from such seed is less uniform and may be slightly inferior to that grown from single-cross hybrids. Still less expensive seed, which can be reproduced by the farmer, is obtained from synthetic cultivars and presents an advanced generation of multiple hybrids. For the development of synthetics selfing is also used and the inbreds are produced from selected plants tested for their quality and for the combining ability; these are intercrossed and the crosses used for the production of synthetics which can then be sown for obtaining both commercial grain and seed for further planting. The multiple synthetic cultivars produce less uniform progeny and are not so vigorous and productive as the crops grown from hybrid seed but they are better adapted for a variety of conditions, whereas the hybrids are suitable only for the areas for which they had been intended. The multiple synthetic hybrids (cultivars) are, however, superior to unimproved local cultivars. The development of composite hybrid cultivars is similar but includes crosses between remote hybrids. The breeding procedure and the production of hybrid seed are infinitely more complex than is described here, the description being based mainly on the relevant chapter in Purseglove's manual (1972).

The development of maize hybrids and hybrid seed production have been confined mainly to grain maize although some forage maize hybrids have also been produced, e.g. Stewart's Multi-T (multi-tiller) hybrid which develops side tillers, undesirable in grain maize but useful in forage maize; tried in southern Canada it was markedly superior to non-tillering hybrids in the production of digestible nutrients/ha. It seems, however, that so far the introduction of forage-maize hybrids is still at its initial stage.

Environment. Maize is widely cultivated from 50°N. to 40°S. and from sea level to 3,300 m alt in the Andes of the northern parts of South America and in Mexico. Maize for forage can be grown within still wider limits, in the areas with shorter growing periods as the plants are not expected to reach maturity. In the tropics, maize requires some 600–900 mm of rain during the growing season, but less rain is needed for growing forage maize. Maize can be grown on a variety of soils, the best results being obtained on fertile and well-drained soil, although it can be grown, as in Kenya, on seasonally waterlogged black soils waterlogged only for a short period.

Establishment. Maize can be easily established. the seedlings are strong and grow fast but a certain temperature minimum, about 10°C; is required before germination can begin and rapid germination and seedling growth are usually observed only when soil temperature reaches 15°C. Under a true tropical climate, soil temperature seldom drops below 15°C, but at higher altitudes night temperatures can be low. In Rhodesia, under low night and high day temperatures in September–October, a disproportion of growth of different seedling parts, resulting in uncurling of the first leaves below the soil surface, was observed (Buckle, 1972); shallow planting, not deeper than 6 cm, is recommended for obtaining normal seedlings. Seed should be sown into a seedbed free from weeds but fine soil tillage should be avoided and seed can even be planted into ploughed soil without any harrowing or discing. The depth of sowing varies from 5 to 12 cm, depending on soil moisture and other soil characteristics. Sowing rates usually range from 9 to 16 kg/ha to obtain maize stands of about 50,000 plants/ha or 5 plants/m^2, but for forage maize denser stands are sometimes preferred, especially on good and well fertilized soil. In Nigeria very dense stands, 430,370 plants/ha, planted at 15 x 15 cm produced high yields of DM (Haggar & Couper, 1972). Seed is drilled in rows 100 or 90 cm wide but sowing in hills, three to four seeds per hill, can also be used. Denser stands of forage maize can produce more herbage but the proportion of cobs in them is reduced compared with the standard plant density. When maize is grown for grazing, much closer stands are normally used. On weedy land inter-row cultivation is essential and it can be combined with the application of herbicides, the most commonly used being 2,4-D or simazin. In India, for instance, 1.5 kg simazin/ha combined with hand weeding increased DM yields from 5.25 t/ha for untreated maize to 15.69 t (Panday *et al.*, 1969). Simazin should be used with care as it may retard the early growth of maize. Mulching with straw can produce good results and in India it reduced water requirement of maize plants by 25–30 per cent and increased fresh herbage yields from 39.7 t/ha to 43.2 t (Pal & Pandey, 1969).

Legumes. Fodder maize can be grown in mixtures with vigorous annual legumes and cowpea or soyabean, the latter mainly in the subtropics, are the more common components in the mixtures. The legumes usually increase CP content of the mixture but seldom the yields of forage.

Fertilizing. The application of fertilizer nitrogen is essential and in nearly all trials considerable yield increases from applied N have been obtained. The highest rates usually applied are in the order of 150 kg N/ha which give economic responses mainly from highly productive hybrid maize; the low rates range from 30 to 50 kg/ha. In a trial in India (Gill *et al.*, 1972) 50 kg N/ha increased the yields of fresh forage from 9.66 t/ha unfertilized, to 18.38 t but a further increase of N to 100 kg/ha produced little effect and resulted in a yield of 18.66 t, increasing however CP content on the forage from 4.4 to 6.2 per cent, and a similar trend was observed in Nigeria. Phosphorus fertilizers are less effective but are necessary when relatively high rates of fertilizer N are applied. K fertilizers are seldom effective although it has been noted that K is essential on heavy black soils. On soils deficient in Zn the application of this micronutrient is essential although overdoses can reduce the intake of P and may be detrimental to maize yields.

Yields. Fodder yields can be high, 100–200 t fresh fodder/ha and even higher, but those recorded from the tropics usually range from 8 to 70 t fresh fodder/ha or from 2 to 20 t DM. Low yields are reported from semi-arid areas and 3.8 t DM/ha harvested in Kenya were obtained from a short-time maize composite hybrid cultivar bred specially for dry areas with short growing periods (Semb & Garberg, 1969). Low yields were also reported from Surinam where 1.7 to 5.9 t DM/ha were harvested depending on the soil and the age of plants at which they were harvested; the maize was outyielded by *Brachiaria mutica* which produced 16 t DM/ha (Dirven, 1963). The quoted yields, up to 70 t fresh material/ha, were obtained from one crop/year, but in Trinidad 60 t DM/ha/year were harvested from four crops grown in the same year one after another and harvested 3 months after sowing.

Harvesting. Harvesting is done at an advanced stage of growth, sometimes at tasseling, but mostly considerably later, at the milk-ripe or dough-ripe stages of grain maturity, or even still later – at the waxripe-stage. The advantage of late harvesting is the accumulation of more DM in cobs in relation to the total yield, although the content of CP in total forage may decrease. The plants are cut by machine or by hand and the field can then be occupied with a catch crop.

Utilization, silage. The stover of maize grown for grain left in the field after the cobs have been removed is often used as rough grazing of low-protein material. Some legumes undersown to maize can be used to improve the quality of stover grazing and in Kenya encouraging results were obtained in trials with species of *Mucuna* (*Stizolobium*). In Rhodesia (Topps & Manson, 1967) early harvesting of cobs with subsequent utilization of younger, still partly green stover has been suggested. The content of CP in the younger stover was high enough to allow the preparation of satisfactory silage or direct feeding.

Forage maize can be grazed, especially if planted dense, or used for feeding green, as soilage. However, the main use of forage maize is ensiling and maize is perhaps the most useful crop for making silage. The ensiled material is usually chopped to pieces 6–20 cm long and well compressed to prevent the penetration of air. Good silage of lactic or acetic acid type, with low pH, can then result almost without fail. Wilting the cut maize before ensiling may result in a better silage than that prepared from freshly harvested maize and in trials in Egypt (Nowar *et al.*, 1973) pH decreased and lactic and acetic acid contents increased with the decrease in water content and the best silage was obtained from maize containing about 26 per cent DM, whereas maize as harvested (15.4 per cent DM) or wilted to a lesser degree gave silage of somewhat lower quality. In maize silage the contents of fat and CF are usually higher than in fresh material before ensiling and those of CP and NFE lower. The digestibility of nutrients in silage remains usually at the same level as in fresh material or are slightly lower. Butterworth (1967), summarizing published data obtained in a few tropical countries, gives 57–61 per cent digestibility for DM, 22–67 per cent for CP, 45–83 per cent for NFE and 39–72 per cent for CF.

The Legumes

Introduction

The legumes, which belong to the botanical family Leguminosae, are another important group of plants comparable to the grasses in their significance for agriculture and especially for grassland productivity and animal nutrition.

Classification and distribution

The Leguminosae (Fabaceae) family is divided into three distinct groups or subfamilies: Mimosoideae, Caesalpinioideae and Papilionoideae which some botanists regard as families – Mimosaceae, Caesalpiniaceae and Papilionaceae – of the botanical order Leguminosae. Only two species of Mimosoideae, namely *Leucaena leucocephala* and *Desmanthus virgatus*, are included in this book; the other species of this subfamily are of insufficient importance for agriculture or pastoral husbandry to warrant their inclusion although a number of shrubs and trees, notably species of *Acacia*, are browse plants of some significance in natural environments. *Leucaena leucocephala*, although a woody plant, a small tree, is often grown as an arable legume, sometimes even in mixture with larger grasses, and its management and the utilization for fodder are not in any great contrast with those of conventional legumes and so is *D. virgatus*. No species of Caesalpinioideae, predominantly woody plants, are cultivated for fodder and only very few are used as natural browse plants. On the other hand, species of Papilionoideae are widely grown as pasture or fodder crops and are of considerable importance for natural grazing or browsing. Papilionoideae is a very large subfamily which includes up to 200 genera and some 12,000 species distributed throughout the world.

Leguminosae in general are of tropical origin and of them the Caesalpiniaceae is considered as the most primitive type of the three subfamilies (Tutin, 1958), and the morphology of their flowers and other plant parts are close to the original ancient types of Leguminosae; species of Caesalpinioideae have remained almost exclusively tropical dwellers represented mainly by trees, woody lianes and shrubs and by a relatively few herbs. The Mimosoideae have also retained a number of primitive features and are distributed mainly in the tropics and subtropics although a number of species has penetrated into warm temperate areas. On the other hand, the Papilionoideae is the most advanced subfamily with a great morphological variability and adaptation and its numerous species penetrated all parts of the world, reaching arctic areas and high mountains in the tropics. The great majority are either small shrubs or, more commonly, herbs of which there is a considerable number of annuals, and trees are in minority.

A comprehensive classification of Papilionoideae was presented by Hutchinson (1964). Since then some alterations have been suggested and in the *Flora of Tropical East Africa* (1971) J. B. Gillett recognized 17 tribes of Papilionoideae of which Indigoferae. Aeschynomeneae, Sesbanieae, Genisteae, Psoraleae and Trifolieae are of some importance in the tropics as fodder or pasture plants, but the majority of useful tropical legumes now in cultivation belong to the tribes Stylosantheae, Desmodieae and Phaseoleae. The indicated nine tribes, and especially the three last named, are distributed throughout the world tropics although the main wealth of species, and particularly of Desmodieae and Phaseoleae tribes, is concentrated predominantly in Latin America.

The structure of the leguminous plant

The Papilionoideae subfamily includes trees, shrubs, woody lianes and herbs. Herbaceous legumes can be erect, suberect or creeping and also climbing by means of twining stems often supported by the leaves which rest on surrounding plants; plants climbing by means of tendrils developed instead of terminal leaflets, such as species of *Vicia* or *Lathyrus*, are relatively rare in the tropics. Some species vary in habit and there can be bushy, prostrate or climbing cultivars of the same species, e.g. in some species of *Mucuna*. In some species the leaves can be simple but they are usually compound and a large number of species, including most of the cultivated legumes, have trifoliolate leaves (Fig. 37c). In a number of other species the leaves are paripinnate (terminating in a pair of leaflets) (Fig. 37e) or imparipinnate (terminating in a single leaflet) (Fig. 37d). There is often a pair of stipules (Fig. 37x) at the base of the leaf petiole which in some species are fused with the low portion of petiole (Fig. 37b). Leaflets can be sessile or have petiolules which can have stipels, usually small and very narrow, at their bases. Young seedlings have two cotyledons; the first true leaf to appear is usually simple, with one leaflet, but further leaves are trifoliolate or pinnate. The plants usually have a well developed tap root penetrating deep into the soil and in some species side roots can be thickened and serve as storage organs. In prostrate forms the creeping stems can root at the nodes but these roots are relatively shallow.

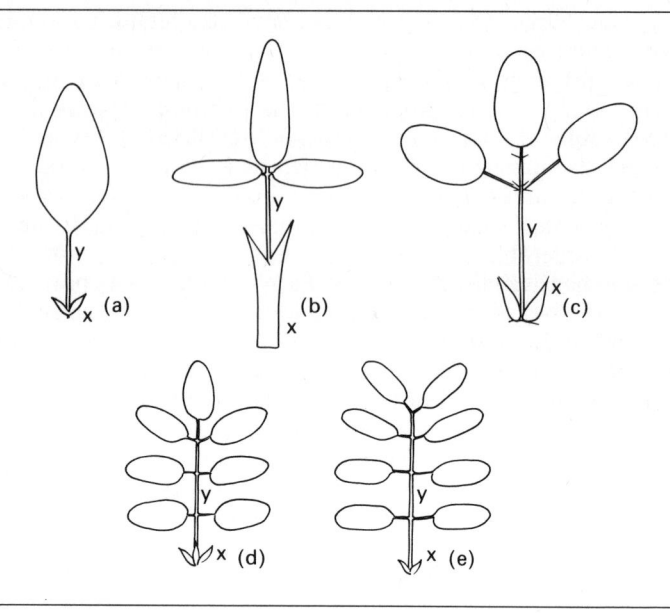

Fig. 37 Leaves of leguminous plants: (a) simple, (b) and (c) trifoliolate, (d) imparipinnate, (e) paripinnate; (x) – stipules, (y) – petioles.

Nitrogen fixation by the legumes

The importance of legumes in agriculture as arable fodder and pasture crops and as components of natural grasslands, and perhaps also their wide spread throughout the world, depend, as it is universally known, on their ability to fix the atmospheric nitrogen in symbiosis with *Rhizobium* bacteria found in the legume root nodules. Being thus rich in protein the legume may greatly increase the nutritive value of fodder and consequently the animal production. After the leguminous plants die and decompose, or even during their lifespan, they, or their roots, enrich the soil with nitrogen and thus benefit the crops grown after legumes. This allows economy in costly nitrogenous fertilizers and saving, at least in temperate countries, of millions of tons of these fertilizers. In the tropics the ability of leguminous plants to fix nitrogen has so far been used to a lesser extent than in temperate areas. Symbiosis of tropical legumes with active and productive *Rhizobium* is difficult and often hardly possible to achieve and it was thought that no practical solution of the problem was forthcoming. However, investigations by D. O. Norris, C. S. Andrew, M. F. Robyns, E. F. Henzell, M. Obaton and numerous other workers on the legume/*Rhizobium* symbiosis, its genetic background and the physiology of both participants of the symbiosis carried out during the last two decades made it possible to achieve reliable practical results in symbiotic fixation by tropical legumes. The cultivation of fodder and pasture leguminous crops, mostly as grass/legume mixtures, is now rapidly spreading and it is hoped that fodder and pasture legumes will soon become a common feature of the tropics and subtropics and that nitrogen fixation by tropical legumes on a farm scale would approach the level of nitrogen fixation by temperate legumes.

Root nodules of leguminous plants, the *loci* of *Rhizobium* activity, are of various shape and size. In a few species, e.g. *Lotononis bainesii*, the nodules are formed on the tap root only, in some other species on the tap root and its branches but in most species they occur mainly on fine root branches, not far below the ground level. *Rhizobium* bacteria can exist in the soil not being in symbiosis with leguminous plants but when a suitable species of legume is found in the vicinity, they move to the plant, penetrate the roots and form nodules. Where there are no suitable *Rhizobium* bacteria in the soil, the legume forms no nodules and newly introduced leguminous plants should be inoculated with the right type of bacteria. Some *Rhizobium* bacteria are highly specialized and can enter into symbiotic relation only with a certain species of legume, some are less selective and can live and work actively in a few species, usually closely philogenetically related, and there are also 'promiscuous' or indiscriminating bacteria which can inoculate a large number of species. These 'promiscuous' *Rhizobium* bacteria belong to the so called cowpea type; they can be in an active symbiosis with a number of tropical legumes, are widely spread in poor and acid soils of the tropics and, conversely, most of the tropical legumes can be inoculated by one and the same cowpea *Rhizobium* although some of these legumes can fix nitrogen better with certain strains than with the others.

Obaton (1974), quoting several authors, lists *Cajanus cajan, Macroptilium atropurpureum, M. lathyroides, Stylosanthes guianensis* and species of *Vigna* as species which enter in symbiosis with almost any strain of cowpea type found in the soil and for which seed inoculation seldom produces any beneficial effect. The plants moderately specific and for which inoculation with certain strains of cowpea *Rhizobium* is desirable are *Arachis hypogaea, Canavalia ensiformis, Clitoria ternatea*, species of *Desmodium, Indigofera hirsuta, Lablab purpureus, Glycine wightii* and *Pueraria phaseoloides*. The third group mentioned by Obaton includes leguminous species highly specific in regard of *Rhizobium* and for which inoculation with the right strain is necessary; these are *Centrosema pubescens, Glycine max, Leucaena leucocephala* and *Lotononis bainesii*. Here also belongs *Trifolium semipilosum*.

Temperate legumes, especially those well known in cultivation, are, in general, much more specific in their *Rhizobium* requirements than tropical species. That tropical legumes are less discriminative in relation to *Rhizobium* bacteria than temperate species can be linked with their origin (see in Norris, 1964). As it has already been mentioned, the legumes have originated in the tropics and genera and species which did not migrate far but remained in tropical environments in the course of their evolution have not developed specific relations with *Rhizobium* or these relations are rather loose; on the other hand, genera and species which penetrated to remote parts of the world, and particularly into temperate countries where they found calcareous soils better provided with calcium than the acid soils of the tropics, developed symbiosis with specific *Rhizobium* bacteria which lost their ability to exist on an acid substrate and formed highly specific strains. However some tropical, or rather subtropical, legumes also require specific *Rhizobium*; but these species are in a minority in the tropics.

Two types of *Rhizobium* bacteria have been distinguished which differ in their reaction to Ca and to pH of the soil: alkali-producing, which decrease the acidity of the substrate, and acid-producing, which increase it. The first named type is characteristic of the cowpea group found in the nodules of tropical legumes adapted to acid soils poor in calcium, whereas the second type is characteristic of a number of temperate legumes grown in their natural habitats on non-acid soils rich in calcium. Transferred to soils poor in calcium and with low pH these temperate legumes (lucerne, white clover) require liming at a rate of 1 to several tons/ha. Most of the tropical legumes growing in symbiosis with the alkali-producing bacteria do not usually require liming, which can even produce a detrimental effect. Although in field trials lime can sometimes produce a positive effect on tropical legumes, this effect can usually be explained by an indirect effect of lime through other nutrients, but when all deficiencies in nutrients have been corrected, the beneficial effect of lime can disappear and Norris (1964) advises to apply hundredweights rather than tons of lime/acre in farm practice with tropical leguminous plants new to the area, at the same time experimenting with liming in order to prove or disprove the need for its application. It has to be kept in mind that *Rhizobium* bacteria, and not necessarily the leguminous plant itself, need for their activity only very small amounts of calcium, so small that impurities in almost any substance coming in contact with *Rhizobium* can provide sufficient amounts of calcium.

For productive symbiotic fixation of nitrogen the plants should be supplied with adequate amounts of phosphorus, potassium, sulphur (unless single superphosphate, which contains S, is used as P fertilizer) and also micronutrients of which the most important one is molybdenum and sometimes also copper. The presence of nitrates in the soil can retard or inhibit symbiotic N fixation and heavy rains under which the nitrates are leached to deep soil layers encourage N fixation. Nitrogen fixation can only occur in moist soil and even the survival of *Rhizobium* bacteria in the soil can be impaired in dry soils and it may become *Rhizobium* sterile after severe and prolonged droughts (Obaton, 1974). One of the most important preconditions of effective N fixation is a correct combination of the host plant and the strain of *Rhizobium* and in a trial by Norris (1964) with *Stylosanthes guianensis* the amounts of fixed·N ranged from 15.4 to 70.1 mg/plant depending on the *Rhizobium* strain. The ability of a *Rhizobium* strain to fix nitrogen is not necessarily related to its vigour and a vigorous strain may not be the most productive one. It often happens that vigorous *Rhizobium* of low N-fixing productivity suppressed productive strains of low vigour and this stresses the importance of selection of productive strains and of isolating them from unproductive but vigorous ones.

Although a considerable improvement in nitrogen fixation by the legume/*Rhizobium* system has been achieved in the tropics in the last two decades and E. F. Henzell (Hutton, 1970), e.g., estimates that N fixation by tropical legumes in Australia ranges from 20 to almost 300 kg N/ha/year, tropical legumes in general still remain less effective than the best temperate cultivated species such as lucerne or white clover and this can be explained not only by inherent abilities of the host plant but also by the relatively slow N fixation by *Rhizobium* bacteria common in the tropics.

A number of 'promiscuous' tropical species can usually do without any outside *Rhizobium* as it is likely to be present in the soil, but to secure more efficient N fixation *Rhizobium* bacteria should preferably be introduced by seed inoculation with commercial cowpea inoculants. Still better results can often be achieved by the use of pure cultures of strains especially suitable for the species or cultivar to be sown and this is particularly important for 'moderately selective' tropical legumes as defined by Obaton. The legumes highly selective in regard of *Rhizobium* require inoculation with special bacterial strains otherwise no N fixation would be effected. Commercial inoculants are normally distributed in peat culture and are mixed with seed before sowing. To secure the adherence of *Rhizobium* to the seed and its better survival if sowing is delayed, and also to provide some P nutrient for immediate use, the inoculated seed can be pelleted, i.e. coated with phosphates, lime or both with an addition of a sticker such as skimmed milk or methyl cellulose. Pelleting with lime is important for acid-producing *Rhizobium* used for the inoculation of white clover or lucerne but for the alkali-producing *Rhizobium* of the cowpea group pelleting with lime is unnecessary and can be done by coating with rock phosphate.

In mixture with the grasses

The legumes are seldom grown in pure stands although certain erect or sub-erect species, such as lucerne, guar (*Cyamopsis tetragonoloba*), *Lablab purpureus* or *Trifolium alexandrinum* are often or even predominantly grown alone. The main importance of fodder and grazing legumes, temperate or tropical, is however in their mixtures with grass, in which the grass normally forms the main bulk of herbage, whereas the legume increases the bulk and improves the quality of herbage. The increase in bulk is due partly to a different degree of utilization of different soil nutrients by the two components and partly to the nitrogen fixation by the legumes. The quality of herbage is increased because of the high content of proteins in the legumes, whereas the content of this nutrient in the grass is low, especially at later stages of growth. Sometimes, however, the bulk of mixed herbage is not increased by the presence of a legume, especially if the grass is vigorous and high-yielding, as e.g. sorghum or maize, but the yields of CP and the quality of herbage increase invariably. Yields of the grass as a component of grass/legume mixtures only seldom increases, compared with the grass grown in pure stands, and often even decreases and the increases in total yields of mixed herbage are usually contributed by the legume. This indicates that during the growth of mixed herbage the nitrogen fixed by the legume is mainly and often almost exclusively used by the legume itself and is transferred to the grass to a very limited extent. The transfer of nitrogen to the companion grass is not necessarily correlated with the amounts of nitrogen fixed by the legume. In a sand-culture trial by Henzell (1962) lucerne fixed 25.8 g N per pot, white clover 21.1 g and tropical legumes, *Indigofera spicata* and *Desmodium uncinatum* 18.4 and 11.1 g, respectively. When the same legumes were grown with a companion grass (*Paspalum commersonii*) *D. uncinatum*, which fixed the smallest amount of nitrogen, transferred to the grass the largest portion of it, 1.66 per cent, white clover transferred 1.12 per cent, lucerne 0.87 per cent and *I. spicata* 0.59 per cent. It was noted that *D. uncinatum*, from which the highest proportion of fixed N was transferred to the grass, formed a kind of natural mulch of its dead leaves which could have been the source of transferred nitrogen. It was also observed in Kenya that perennial grass grown in mixture with annual legumes, which died in the first dry season, developed in the next season as vigorously as the grass receiving fertilizer N; drought-susceptible perennial legumes partly dying during the dry season produced a much smaller effect, whereas drought-resistant perennial species persisting well during the dry season had no effect (Bogdan, 1964b); it seems feasible that the legumes release nitrogen to the soil only when at least parts of the plant die out and decompose. Henzell (1962) writes that there are two ways of transferring nitrogen from the legume to the grass: (a) direct from the legume through the soil, and (b) through the animal and soil with dung and urine. The second way is considered to be much less important than the first because dung and urine affect only a small portion of land surface.

Cultivation

The legumes are established almost exclusively from seed although attempts have been made to grow some creeping species from rooted cuttings and in perennial species of *Arachis* this is a normal method of establishment. Stem cuttings of a number of perennial leguminous species can root and produce vegetative progenies which are used in plant breeding for the evaluation of progenies of cross-pollinated plants, for examining the effects of environment on genetically identical material, etc.

Seedbed for farm sowing is prepared in the usual way and seed is sown preferably in rows, especially in grass/legume mixtures in which the two components are often sown in alternate rows. Seed rates differ widely depending mainly on seed size: the number of seeds/kg ranging from 2–2.5 thousand in species of *Mucuna* to over 3 million in *Lotononis bainesii*. Similarly with a number of species of other botanical families, leguminous plants often have a considerable proportion of hard seeds, i.e. seeds which remain viable in the soil for a few months to several years without germination. The presence of hard seed is normally an inheritable character, it can be reduced by selection and in species or cultivars grown for a number of generations the percentage of hard seed is usually negligible; on the other hand it can be very considerable, especially in recently introduced species. The presence of hard seeds is apparently an adaptative feature which prevents all seeds from germination at the first opportunity and then die if a sudden drought occurs. This character may be useful in annual legumes grown in mixtures with perennial grasses. In hard seeds water cannot penetrate through the seed skin or testa and seeds remain unimbibed. Soaking in water for some 24 hours or longer can reduce the proportion of hard seeds but the reduction is relatively small. Mechanical scarification (seed scratching) done by hammermill operated at experimentally determined speed, or rubbing gently with fine sandpaper if the amounts of seed to be treated are small, can considerably reduce the proportion of hard seed; scarification should be done with great care in order not to destroy the easily breaking seeds. Soaking in concentrated sulphuric acid is perhaps the more reliable and efficient method for reducing the percentage of hard seed, and increases of germination percentage from 20 to 80 have been recorded. The optimum duration of soaking varies from 2 to 20–25 min and should be determined experimentally for each species because keeping in sulphuric acid for too long a time can destroy the seeds. After soaking in sulphuric acid seed should be thoroughly washed with water. Hard seeds should of course be treated before the inoculation with *Rhizobium*.

If seed is inoculated it should be sown as soon as possible after the inoculation but delays in sowing are much less harmful if seed has been pelleted. If during or after sowing, when the soil is bare and its temperature reaches 35°C at the depth of sowing, *Rhizobium* bacteria of the cowpea group lose their ability to grow and can die at a temperature of 40°C or over. Hot weather can be particularly harmful to *Rhizobium*-inoculated small seeds which require shallow sowing.

Fertilizers used for the legumes include mainly phosphorus in the form of double or single superphosphate; the latter is preferred as it contains sulphur, the deficiency of which can reduce the vigour and productivity of the legume. Potassium is another nutrient commonly used as fertilizer. The legumes are sensitive to deficiencies in micronutrients and molybdenum is particularly important for *Rhizobium* activity. On soils deficient in boron and copper these nutrients have also to be added. The tolerance of legumes to aluminium and manganese ions varies and some species, e.g. *Stylosanthes humilis* or *Lotononis bainesii*, are more tolerant than many others. Nitrogenous fertilizers are usually not needed but small amounts of N normally produce no harmful effect on legumes and can even help during the early growth of the grasses and of the legumes before the *Rhizobium* has gathered its full strength. Moderate or high rates of N, over 50–70 kg/ha, usually reduce the content of legumes in the sward, and even eliminate them, partly because the fertilizer N increases the competitive vigour of the grass and partly, probably, because of the harmful effect of nitrates on *Rhizobium*.

The management of grass/legume mixtures for the benefit of the legume, or for maintaining a desirable grass–legume balance, include rotational grazing. Close and frequent grazing may be more harmful to the legume than to the grass and in trials with Siratro (*Macroptilium atropurpureum*) grown with *Setaria anceps* it resulted in a strong dominance of the grass (R. J. Jones, 1973a). Frequent and low cutting usually produces a similar effect.

Pests and diseases

A number of legumes is attacked by various root-knot nematodes and some species or even cultivars are more resistant than others, and some can even be fully resistant. The small-leaf virus disease is common but not often serious and it seldom produces such a devastating effect as it does to the grasses. Other pests and diseases may affect certain species and they are mentioned in the main part of this book dealing with individual species.

Yields, utilization

Yields of DM are usually lower than those of the grasses except of some high yielding perennial legumes such as lucerne which, under irrigation, can produce in several cuts up to 12 or more t DM/ha/year. In pure stands well-managed legumes can be expected to yield anything from 1 to 15 t DM/ha/year and similar, or somewhat lower yields can be obtained from legumes as components of grass/legume mixtures. CP yields are higher than those of grasses because of its high content in the herbage. Grass/legume mixtures are used for grazing and also for haymaking or ensiling. The legumes grown in pure stands are grazed less often and hay is usually prepared, e.g. of lucerne or *Trifolium alexandrinum*. When drying and curing pure legume for hay a great care should be taken in order to retain the leaf: the succulent stems dry much slower than the leaves which fall off easily and the resulting hay may lose its most vluable parts. Haymaking of mixed grass/legume swards is easier: the green mass, as cut, is less solid than pure legume, the legume then dries faster and loses less leaf. Ensiling pure legume herbage is difficult because it contains less sugar than the grasses and to achieve lactic acid fermentation is not easy. Much better results can be obtained from ensiling grass/legume herbage, especially if the grass is sorghum or maize, or when molasses is added to secure a sufficiently high level of sugar in the ensiled material.

Chemical composition, nutritive value

Tropical legumes contain more crude protein (CP) than tropical grasses and its content usually varies from 10 to 25 per cent in the dry matter but may occasionally be even higher. The content of nitrogen-free extractives (NFE) is normally lower than in the grass, the average being about 40 per cent. Crude fibre (CF) content varies widely and cultivated legumes usually contain 20–25 per cent CF, which is lower than in the grasses, although some stemmy and fibrous species of *Crotalaria* and *Tephrosia* can contain up to 35 per cent and even 40 per cent CF (Bartha, 1970). The content of ether extract (EE) is rather similar to that of the grasses, 1–3 per cent, and occasionally up to 5 per cent. The content of Ca is relatively high and usually ranges from 0.5 to 2.0 per cent, and that of P is also higher than that of the grasses and normally ranges from 0.20 per cent to well over 0.30 per cent, which satisfy the animal requirements; some legumes may however contain very little P and Bartha (1970) has determined the content of P below 0.20 per cent in six species out of 16 wild legumes examined in the Sahel zone of West Africa. The digestibility of the main nutrients is usually slightly higher than those of the grasses and the digestibility of CP considerably higher. The stems of leguminous plants become more fibrous with the age and contain more CF and less CP than the younger stems although this loss of nutritive value is less evident than for the stems of grasses. Little change has however been observed in the leaves of legumes and they remain nutritious and palatable at advanced stages of growth, often throughout the dry season, and the legumes are of particular value for late season grazing.

Palatability, toxicity

Palatability of the legumes, and especially of wild legumes, varies much more than the palatability of grasses. Whereas grass palatability can be very low because they can be unduly fibrous, or a few are aromatic, the legumes can be unpalatable if they are bitter, aromatic, have some other unpleasant taste to cattle, are prickly or exceedingly hairy. If a legume, even normally well grazed, is new to the animals they require some time to get used to it, even to lucerne, but when accustomed to a certain leguminous species they would eat it well. Young plants are often eaten less willingly than at more advanced stages of growth as it was e.g. observed for *Trifolium rueppellianum* or *Stylosanthes humilis*. A large number of legumes, mainly wild, are toxic; adult local animals usually recognize toxic species in natural grasslands and avoid them but young calves or lambs can occasionally be poisoned. Some legumes have toxic properties which cattle or sheep would not recognize; the harmful effect in these cases can sometimes be delayed and the herbage become toxic only after having been fermented in the digestive tract. Some legumes may contain estrogenous substances and affect the fertility of female animals and even cause abortions; fortunately, species with estrogenous properties are rare or practically absent amongst the main tropical and subtropical cultivated legumes. Seeds are often more toxic than the herbage although a reverse position has also been observed. A number of species of *Tephrosia* and *Crotalaria* are toxic but they are not used in cultivation. Species of *Indigofera* can also be harmful and this is especially worrying in *I. spicata*, otherwise a good and productive legume extensively tried and well suited for cultivation. Toxic properties of *I. spicata* were first observed in 1941 and it was later found that it causes liver degradation in grazing animals. *Leucaena leucocephala* can cause the loss of hair in cattle and sheep if fed pure and in large quantity. Tannins are often found in legumes in considerable quantities and may affect plant palatability and digestibility; of the important cultivated legumes only species of *Desmodium* contain tannins in quantities that can affect the animals. Tropical legumes only seldom cause bloat and *Trifolium alexandrinum* and *Lablab purpureus* are so far the only cultivated species known to harm the animals in this way (Hutton, 1970), but some temperate clovers introduced to the tropics may cause bloat and should be grazed with care.

Reproduction

The flowers

The flowers of Papilionoideae are irregular or 'zygomorphic', i.e. two-sided, mostly symmetrical, except the keel in the Phaseoleae tribe in species of which it is often spirally-twisted. The flowers are seldom single but gathered into inflorescences: racemes, spikes or heads, in which the flowers are sessile or on short stalks and are often supported by bracts or bracteoles. The flower consists of a calyx, corolla, stamens and a pistil

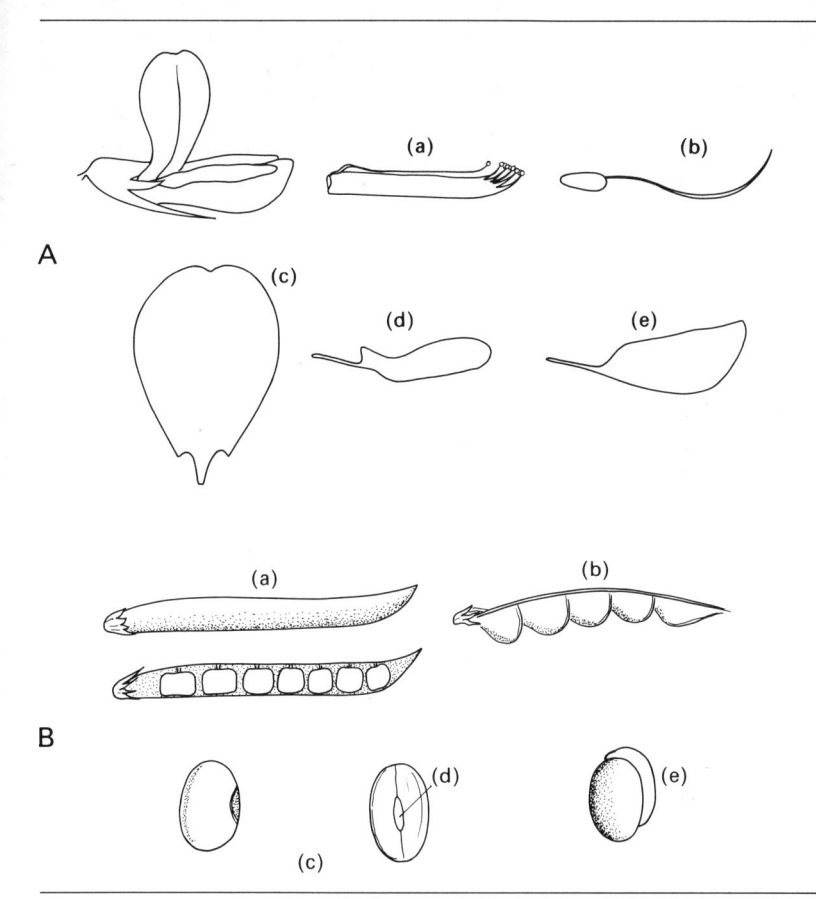

Fig. 38 Flowers and pods of leguminous plants. (A) Flower of *Rhynchosia minima* (a) stamens, (b) pistil, (c) standard, (d) wing, (e) keel (from Edwards & Bogdan, 1951). (B) Pods and seeds: (a) dehiscent pod, (b) articulate pod, (c) seed, (d) hilum, (e) aril.

(Fig. 38). The calyx has five sepals united in the lower parts and is usually two-lipped; the upper lip has two and the lower lip three teeth; the calyx is herbaceous and usually green or variously tinged. The corolla consists of five free petals. The uppermost petal or **standard** (Fig. 38Ac) is usually larger than the others, it often has an upright orientation and is the most showing; its coloration is normally less bright on the back which can be greenish or brownish. Two side petals or **wings** (Fig. 38Ad) vary in shape but are mostly elongated, with a lobe in the upper part, and narrowed below to a strip. Two lowermost petals form together a boat-shaped structure or **keel** (Fig. 38Ae); they can be fused at the apex and usually embrace the stamens and the pistil. The petals are variously coloured or white and the standard is usually the brightest. The ten stamens are united by their filaments (Fig. 38Aa), they surround the pistil and are usually enclosed in the keel; the top stamen is usually free in most species and in some species the lowermost stamen is also free and the filament tube is then divided and only four stamens on each side are fused. In certain genera and species some of the ten anthers can be reduced, usually in alternate stamens, to a sterile state. The **pistil** has an **ovary**, the future pod, a **style** and a **stigma** which receives pollen. The shape of the stigma is often characteristic for the species and can then be used for plant identification. The ovary develops into the **pod** (Fig. 38Ba–e), a typical example of which can be the pod of the ordinary pea. The pod has dry papery, or sometimes hard, walls and there is no dividing wall inside the pod. The seeds are situated on the edge of the pod wall, where it forms a rib, and are arranged in two rows. The pod usually opens longitudinally and releases the seeds. In a number of species of the Desmodieae and Aeschynomeneae tribes the pod consists of two to several one-seeded joints or segments which break off and separate at maturity. Mature seeds have no endosperm but only the embryo, which includes two large cotyledons containing stored nutrients, and skin or testa. Seeds vary in shape from globose to cubical, elongated, oval, cylindrical, etc. When the seed breaks from its stalk, a scar or **hilum** (Fig. 38Bd) of different shape and size is left on the seed. Some species have an **aril** (Fig. 38Be), an outgrowth, mostly near the hilum.

Flowering, pollination

Leguminous plants of the Papilionoideae subfamily with their zygomorphic flowers seem to be well adapted to pollination by insects and have honey usually concealed at the base of the standard and of the upper, usually free stamen, so that only the insects with sufficiently long tongues, usually bees, can reach the honey. The bee sits down on the wings of the flower which are connected with the keel; under the insect's weight the keel bends down freeing the anthers and the stigma which touch the insects; the anthers leave pollen on the insect, whereas the stigma receives pollen from it. When the insect flies off, the keel, the anthers and the stigma may return to their original position making thus possible new insect visits. In other species of legumes, such as lucerne or species of *Desmodium*, the stamen tube, together with the stigma, are under a strong upward tension and are kept inside the keel only by its clasping end; the tension is released by the weight of the insect which sits down on the keel and the stamens and the pistil jump upwards, pressing themselves firmly to the standard. This process, known as tripping, is fairly common in leguminous plants; apart from the insects tripping can occur accidentally when the flower touches the surrounding plants and it can also be done by hand. In spite of these often elaborate adaptations, which suggest cross-pollination, the majority of cultivated tropical pasture legumes are predominantly self-pollinated and pollination often takes place before the flower opens. In temperate cultivated legumes, at least those of the Trifolieae tribe, perennial species are mostly crosspollinated and the annuals self-pollinated, whereas in the tropics both annuals and perennials are mostly self-pollinated, although crosspollination has also been observed. Of the better known tropical legumes *Indigofera spicata, Lotononis bainesii, Centrosema pubescens* are self-pollinated species in which the pollination occurs before the flowers open. *Macroptilium atropurpureum* and *M. lathyroides* are selfpollinated, pollinating in open flowers. *Desmodium uncinatum* and *D. sandwicense* are self compatible species in which cross-pollination can occur and they both require flower tripping (Hutton, 1960). Pollination in the *Trifolium rueppellianum* complex, which is a series of a few closely allied species, presents an interesting case. In this complex there are cleistogamous forms, in which pollination and fertilization occur in closed flowers, forms which require flower tripping for pollination but are predominantly self-pollinated under field conditions, whereas still another species requires tripping but does not form any seed if foreign pollen is not introduced and is entirely cross-pollinated (Bogdan, 1966b).

Seed production and harvesting

Seed formation of the legumes is normally good and seed yields satisfactory to good but seed harvesting in species recently introduced into cultivation present the obstacles similar to those of grass harvesting: the flowers appear gradually during a long period of time, especially in perennial species, and while some flowers only begin flowering the others can bear ripe pods or even have shed their seed. The pods of some species burst vigorously, sending their seed to a considerable distance from the plants on which they have been formed. As a result of uneven ripening and seed shedding only a relatively small proportion of seed that were formed and ripened can be harvested and Hopkinson & Loch (1973) have recommended to pay more attention to the development of techniques of picking up Siratro (*Macroptilium atropurpureum*) seed from the ground rather than concentrating all attention on seed yield increases and on the techniques of harvesting direct from the plant; these considerations can also refer to some other legumes. Seed harvesting in plants with articulated, breaking-off pods have also their difficulties. Those pulses which are also used as fodder plants, e.g. *Lablab purpureus* or *Cyamopsis tetragonoloba*, and which have been in cultivation for a long period of time, and have been bred or selected, mostly unintentionally, for non-shattering of their seed, present usually no difficulties in harvesting.

Breeding

The methods and the results of breeding depend to a high degree on the mode of reproduction which in tropical legumes differ in general from that of the grasses. It has already been indicated that most of the tropical cultivated legumes are predominantly or almost exclusively self-pollinated plants although the reproduction of species of *Desmodium*, such as *D. uncinatum* or *D. sandwicense*, can be partly or entirely of cross-pollinated nature (Hutton, 1960), and so is *Trifolium semipilosum*, and some degree of cross-pollination was also observed in *Glycine wightii*. Self-pollinated plants vary little and populations of cultivated legumes so far introduced and grown on any appreciable scale are mostly uniform and there is little scope for simple selection of superior types. For finding heterogenous or uniform but different populations as sources of genetic variability for breeding, new material has to be introduced from the areas of the species origin, mainly from Latin America. This work is in progress, it receives an increasing attention and a number of plant types distinct from those already in cultivation have been introduced. Crossing self-pollinated plants of different populations using emasculation and artificial pollination has already been done with good results and used by Hutton (1962) for the development of Siratro, a well known cultivar of *Macroptylium atropurpureum*. Another source of variability can be found in induced mutations resulting from irradiation with ^{60}Co radioactive rays. This method can preferably be used for obtaining characters not found in wild plants but which can be useful in cultivated legumes. The transfer of genes from plant to plant and from species to species can be achieved by the effect of colchicine but such transfer is even more difficult to achieve than in the grasses. In spite of the outlined difficulties fair progress in breeding pasture legumes has been achieved.

The main objectives of the improvement of legume cultivars are: (*a*) herbage and (*b*) protein production, (*c*) nitrogen-fixing ability, (*d*) drought and cold resistance, (*e*) balanced mixtures with grasses and increasing legume longevity under grazing, (*f*) palatability and freedom from toxic and repellent substances, (*g*) resistance to pests and diseases, and (*h*) seed production. The first three objectives – herbage and protein production and good nitrogen fixation – present a single complex and can only be achieved by combined breeding of the leguminous plant and of *Rhizobium*, to make them fit each other and develop highly compatible and productive combinations, and considerable positive results in this direction have been obtained, mainly in Australia. Freedom from toxic substances is of importance, in the first instance for *Indigofera spicata*, and a big step forward has already been achieved by the establishment of the identity of its toxic substance (Hutton, 1970), but other species, e.g. those of *Crotalaria*, are still awaiting to be approached. Breeding for resistance to pests and diseases concerns mainly the resistance to root-knot nematodes and to small-leaf virus disease. Most of the important cultivated legumes are to a various degree resistant to nematodes, but some are not, and considerable work has yet to be done. A number of cultivated legumes are also resistant to the small-leaf virus and breeding to induce similar resistance into *Lotononis bainesii*, *Desmodium intortum* and some of the African clovers is in progress but is far from being completed. Good balance of the legume and the grass in grass/legume mixtures during the first 3 years of growth have been achieved but most of the legumes used in mixtures tend to weaken and disappear in 4 to 5-year-old mixtures, especially if grazing density is on the high side and more persistent and competitive cultivars are being bred. Finally, breeding cultivars with non-shattering pods is perhaps one of the most needed and most difficult objectives to achieve but obtaining mutants by ^{60}Co irradiation can perhaps be a promising line of approach.

The more important species

Aeschynomene L. Joint vetch

Annual or perennial, erect or prostrate herbs or shrubs and a few species are climbers. Stems often thickened and pithy and there are some species with the stems floating in water. Leaves paripinnate with two to many pairs of leaflets. Flowers yellow, the standard often lined with purple. Pods linear to elliptic, compressed, articulated to one-seeded joints which disintegrate at maturity. A large genus of over 150 species, mostly South American or African, with a few species in tropical Asia and northern America. Occurs mostly in wet or seasonally wet situation and can withstand waterlogging and flooding. Some species constitute a considerable proportion of herbage and are reasonably well grazed or browsed but may also be toxic. The genus is of little importance in cultivation although a few species have been under trial.

Aeschynomene americana L.

Annual herb (Pittier (1944) considers *A. americana* to be an annual, whereas León & Sgaravatti (1971) and some other authors regard it as a perennial; it can be that some forms of this species are perennials although a possibility of identity confusion cannot be excluded). Stems and leaves glabrous or with scattered glandular hairs. Leaves with 10–30 pairs of leaflets. Flowers yellow, in axillary racemes. Pods with four to eight joints, straight at the upper margin and markedly and deeply constricted between the joints at the lower margin.

Occurs naturally in all warm areas of America in permanently or seasonally wet habitats. In a trial in Florida (Moore & Hilmon, 1969) *A. americana* was established in seasonally waterlogged plains where NPK fertilizer, applied in the year following the establishment, increased its proportion in the sward and it yielded 1.40 t DM/ha after the first application of fertilizers and up to 4.76 t after the second application; this legume was palatable to cattle. For good germination seed should be scarified and in Florida germination increased from 2–6 per cent to 74–78 per cent when the pods were shelled and seed surface scratched (Hanna, 1973).

Aeschynomene indica L. $2n = 40$

Annual or perennial herb resembling *A. americana* from which it differs by more numerous, 5–13, joints of the pod which is only slightly constricted between the joints at the lower edge. Widespread in tropical Africa and in tropical and subtropical Asia and Australia and also in North America. Occurs mostly in wet places, swamp margins, seasonally flooded or waterlogged grass plains, sometimes on black saline soils, up to 1,500 m alt and is usually eaten by stock although it may sometimes be toxic and death of animals ascribed to the effect of this plant has been reported in Malawi.

Alysicarpus Desv. Alyce clover

Annual or perennial, erect or prostrate herbs. Pods of a few segments (one-seeded joints) breaking up at maturity. A relatively small genus of up to 30 species distributed throughout the tropics, and to a lesser extent subtropics, of the Old World. The plants are well eaten by stock and two to three species have been tried with some success in cultivation and grown on a moderate farm scale, but lost their popularity in the last one to two decades.

Alysicarpus glumaceus (Vahl) DC. $2n = 20$

Annual herb 15–150 cm high, not unlike perennial *A. rugosus* but with narrower leaflets, larger pods, pod joints and seeds. It is very variable and a few botanical subspecies and varieties have been distinguished. Occurs in grasslands, and other habitats, mostly in moderately wet or semi-arid areas throughout tropical Africa and is often common but seldom produces any large bulk of herbage. Palatable to cattle. In the early research in Kenya the name *A. glumaceus* was erroneously applied to *A. rugosus*.

Alysicarpus rugosus (Willd.) DC. ssp. *perennirufus* J. Léon. Alyce clover
$2n = 16$

Erect, prostrate or ascending perennial herb up to 150 cm high but usually much lower. Stems thin, somewhat wiry. Leaves with one or sometimes three leaflets 1.5–10 cm long and 0.2–2 cm wide, oblong to linear-lanceolate, rounded at the base, almost glabrous, on petiole up to 2 cm long. Stipules prominent, brownish. Flowers in terminal and axillary racemes. Calyx 6–10 mm long. Standard whitish to reddish-purple or bluish, 6–7 mm long; keel often greenish. Pods 5–10 mm long, with three to six joints, constricted between the joints and scarcely exceeding the calyx. Joints 1–1.5 mm long and 2–2.5 mm across, reticulate or strongly transversely ridged. Seeds reddish-brown or olive in colour, compressed, 1.5 mm long.

This is the perennial form of the species extensively tried in cultivation; in East Africa it was in the past misnamed as *A. glumaceus*. It differs from the annual subspecies *rugosus* and *reticultaus* Verdc. by having brown calyx hairs. *A. rugosus* as a species is widely distributed in tropical Africa and Asia.

Alysicarpus rugosus ssp. *perennirufus* has been introduced in Kenya from local flora and has shown a great variety of forms differing in growth habit, leafiness, leaflet shape, etc., and grown with some success in mixtures with perennial grasses. Grass mixtures also gave good results in India where mixtures with *Chloris gayana* and *Panicum antidotale* were grown (Bondale, 1969). In India *A. rugosus* yielded 16 t green fodder/ha, more than the other monsoon species tested in the same trial (Tiwari, 1966) but in another Indian trial yields of DM ranged from 1.88 to 2.96 t/ha only, depending on the number of cuts, the highest yield being for a single cut at the ripe stage (Whyte, 1964). In India young herbage contained 26 per cent CP and only 14 per cent CF, but the content of CP decreased and that of CF increased with plant age (Bondale, 1969). In Kenya this species was one of the most palatable legumes out of ten species tried in stall feeding. Seed develops and ripens unevenly and sheds easily and seed production is difficult. In spite of some success obtained in Indian and East African trials, *A. rugosus* is now considered less important than 10-20 years ago and does not receive much attention by the agriculturists.

Alysicarpus vaginalis (L.) DC. Alyce clover; One-leaf clover; Trebol Alicia $2n = 16$ or 20

Erect or spreading perennial or annual herb resembling *A. rugosus* from which it differs by the leaflets reticulate on both sides and by flowers and pods. The standard is orange to pinkish or purple in colour, 4-6 mm long. Pods 1.2-2.5 cm long, *not constricted between the joints*; the four to seven joints are 2.5-3 mm long and 1.5-3 mm across, subcylindrical and with raised reticulate ridges. Seeds 1.7 mm long, yellowish to brown, often speckled.

It is the best known cultivated species of *Alysicarpus* grown in India and Sri Lanka but mainly in Latin American countries and also in Florida, USA. In India it is regarded to be 'a perennial but is often grown as a summer annual' (Whyte *et al.*, 1953) and is also grown as an annual in Bolivia and Florida (Kretschmer, 1970). *Alysicarpus vaginalis* originates from the Old World tropics and is grown as a pasture or fodder crop, mostly experimentally but also on a moderate farm scale in the countries mentioned. In Fiji (Roberts, 1970b) it was found unsuitable for cultivation but has naturalized and is used in natural grasslands and comparable results were obtained in Mali (Derbal *et al.*, 1959) where this legume has been recommended for the improvement of natural pastures but not for arable cultivation. Better results were obtained in Sri Lanka (Fernando, 1961) where it was tried with *Brachiaria brizantha*, the mixture yielding about 12.5 t DM/ha when fertilized with N and P, and also with *B. distachya* and *Paspalum dilatatum*. In Florida *A. vaginalis* has been grown for several years on a moderate farm scale but in trials by Kretschmer (1970) with Pangola grass (*Digitaria decumbens*) mixtures it was one of the low yielding legumes although it increased the total yield of herbage to 8,470 kg/ha compared with 5,240 kg for grass alone. In Bolivia (O. Braun) it is considered to be an important cultivated legume which can regenerate itself easily from seed fallen to the ground and has an advantage over introduced temperate legumes in not causing bloat to cows; it is well liked by cattle and horses. Ten to twenty kg seed/ha is required when sown broadcast and there are about 650,000 seeds/kg. The plants are highly susceptible to root-knot nematodes.

Except perhaps in some South American countries *A. vaginalis*, similarly with *A. rugosus*, is now losing its importance.

Arachis L.

Annual or perennial, erect, suberect, prostrate or creeping herbs. Leaves paripinnate, usually with two pairs of leaflets. Flowers yellow, single or in two to seven axillary spikes. The lower parts of calyx, corolla and stamens form a long, narrow receptacle tube (often mistaken for pedicel) on top of which the free parts of calyx, corolla and stamens are borne. The ovary is inside the receptacle tube and has a long style. The pod develops on a 'gynophore' or 'carpophore' which bends down and buries the pod in soil. Pods indehiscent, mostly oblong, with one to six seeds.

A small genus of some ten species all distributed in South America. *Arachis hypogaea* or groundnut is widely cultivated; wild species are mostly perennials, some of them are grazed in their natural habitats or are under trial as forage plants, *A. glabrata* being the best known and showing good performance in trials in Florida. The other two promising species, which, similarly with *A. glabrata*, grow in northern Argentina, Uruguay, Paraguay and south-eastern Brazil are *A. marginata* Gardner, a species rather similar to *A. glabrata* but of more xerophytic nature and with leaflets of different texture, and *A. villosa* Benth., an erect, non-creeping plant.

Arachis glabrata Benth. (*A. prostrata* Benth.)   2 = 40

Perennial with underground creeping, much-branched rhizomes (rootstocks) producing short, suberect, above-ground shoots. Leaves with four leaflets; leaflets broadly elliptic, 6–20 mm long and 5–14 mm wide, subglabrous or glabrous underneath. Flowers axillary, in the lower part of the stem. Receptacle tube filiform, 2.5–10 cm long; calyx 6–7 mm long; standard yellow to orange, orbicular, 10–12 mm in diameter. Pods small, 10 mm long and 5–6 mm thick, ovoid, acute, longitudinally striate. Seeds ovoid, pale. *Arachis glabrata* var. *hagenbeckii* (Harms)

Hermann (*A. hagenbeckii* Harms, *A. diogoi* Hoehne) differs from the main variety by narrow, acute leaflets (Hermann, 1954).

Indigenous in northern Argentina, Paraguay, Uruguay and extending to north-eastern Brazil, it occurs on well-drained light soil but can also be grown on 'black rich earth' (Prine, 1964). The above-ground parts can be killed by frost but underground rhizomes usually survive. The plants are relatively low and form dense mats. *Arachis glabrata* is grazed in its natural habitats and the main variety has been introduced to cultivation, mainly to Florida, USA, where several introductions have been tried and cvs **Arb**, **Arblick** and **CSl** selected as the most promising (Prine, 1973). Propagation is effected by cuttings or pieces of rhizomes; the establishment is slow and two or more years are required for the development of vigorous stands. The uptake of nutrients by the plants is considerable and a high-yielding crop removes from the soil about 160 kg of Ca, 45 kg Mg, 270 kg K, 34 kg P and 240 kg N/ha. High requirements for soil N indicate that the amounts of N fixed by *Rhizobium* bacteria of root nodules is not great and insufficient for vigorous plant development. Yields of well-fertilized *A. glabrata* can be high and Prine (1964) experimenting in 1961-4 with cv. Arb reported them to range from 4.48 to 5.54 t DM/ha/year in two cuts, and yields were even higher in his further trials and reached up to 9 t/ha (Prine, 1973). CP content of herbage ranged in Prine's trials from 10.7 to 17.9 per cent, CF from 26.9 to 30.2 per cent, NFE from 44.0 to 47.7 per cent, P from 0.34 to 0.57 per cent and Ca from 0.53 to 2.76 per cent. Organic matter digestibility was determined as 60-62 per cent and the plants were well eaten by farm animals.

Arachis glabrata is an aggressive plant and can, to a certain degree, suppress the companion grasses when grown in mixtures, although total yields of mixed herbage increased in comparison with those of grass alone from 900 kg DM/ha to 1,910 kg for Pangola grass (*Digitaria decumbens*) mixture, from 1,050 kg to 1,860 kg/ha for Pensacola Bahia grass (*Paspalum notatum*) mixture, and from 1,400 kg to 2,520 kg DM/ha for Bermuda grass (*Cynodon dactylon*) mixture (Prine, 1964).

Arachis hypogaea L. Groundnut; Peanut; Cacahuete; Mani

Prostrate or bunch-forming annual. Leaf with four leaflets, petiole 3-7 cm long, stipules 2.5-3.5 cm long, connate with petiole for less than half their length. Leaflets 3-7 cm long and 2-3 cm wide. Flowers are mainly on the lower parts of the plant, single to several in axillary spikes. Receptacle tube 4-6 cm long. Standard yellow, often lined with red, orbicular, about 1.5 cm in diameter. Pods on long gynophores (carpophores) developing after flowering and known as 'pegs'; they bury pods in soil to a depth of 2 to 7 cm. Pods oblong, but can be almost globose, reticulate, with one to six seeds. Seeds 1-2 cm long and 0.5-1 cm thick.

Arachis hypogaea is of South American origin but has now spread throughout the world's tropics, subtropics and also warm temperate areas up to 40°S. and 45°N. It is an important crop grown for seeds which are rich in oil and protein and used for the production of oil, peanut butter, industrial proteins, confectories and for eating as nuts. It has also been tried as a fodder crop for making hay and occasionally for grazing or ensiling, but as such it has not received any wide application. *Arachis hypogaea* is however of considerable importance for a combined utilization for seed and hay production under which the herbage is prepared to hay after seed has been detached. In the USA about 20 per cent of the acreage under peanut, or well over 100,000 ha, are used for preparing peanut/vine hay and the extent of such utilization was even greater in the 1950s before combine harvesting, which does not allow the use of the tops for hay, was introduced on a large scale. In USA about 3 per cent of peanuts are also used in the field for fattening pigs (Leffel, 1973).

Arachis hypogaea requires a warm climate, is easily killed by frosts, and in temperate countries can be cultivated only in the areas with long summers. It requires not less than 400–500 mm of rain during the growing period but prefers dry weather during seed maturation and harvesting. Light textured or loose soils, such as sandy loams, facilitate pod burying and harvesting.

Establishment is effected from seed and seed rates vary from 20 to 90 kg shelled seed/ha, depending on the type of plant – bunch cultivars require less space than prostrate types – and on methods of sowing. Responses to fertilizers are erratic and usually small, and if the preceding crop was well fertilized there is usually no need to apply fertilizers to peanut. The best responses were obtained from fertilizer P, and S as fertilizer can improve root nodulation. Nodulation is usually observed but the amounts of atmospheric N fixed by *Rhizobium* is apparently restricted and responses to fertilizer N have been observed. Seed ripens in 90 to over 140 days after sowing, depending mainly on the cultivar. On loose, sandy soil harvesting can be, and is, done, especially in the less technologically developed countries, by pulling out the whole plants and picking up the pods. On heavier, compact soils ploughing out the roots and collecting pods is also used and in the USA special peanut combiners have been developed and are now in general use in the southern states of USA.

When *A. hypogaea* is used as a duel purpose crop, harvesting should be done at the earliest possible date at which the crop reaches its full production of seed, and unnecessary delay in harvesting decreases the yields and quality of vines or tops. In trials in Venezuela (Combellas *et al.*, 1971) harvesting 87, 94, 101 and 108 days after sowing resulted in herbage yields of 4.5, 5.1, 3.9 and 2.8 t DM/ha, respectively; CP contents were 15.3, 13.5, 14.8 and 10.3 per cent and DM digestibility 60, 57, 56 and 52 per cent; however, taking in account seed yields the best time of harvesting was at 101 days after sowing. Mazzani *et al.* (1972), also in

Venezuela, obtained the highest yields of seed and of herbage, 6.21 t DM/ha, 104 days after sowing. Yields of *A. hypogaea* herbage can be expected to be anything from 2 to 7 t DM/ha. The content of CP can sometimes reach almost 20 per cent but Prine (1964) gives CP content of groundnut hay as ranging from 7.5 to 15.6 per cent, CF from 19.1 to 43.8 per cent and NFE from 36.6 to 54.7 per cent, the averages being 11.9, 27.0 and 45.9 per cent, respectively. He also gives average contents of P as 0.18 per cent and Ca 1.26 per cent. In *in vivo* trials with sheep (Velásquez & Gonzáles, 1972), DM of well-prepared groundnut hay was digested to 53.1 per cent, CP to 62.1 per cent, CF to 55.2 per cent and TDN content was 56.4 per cent. The uptake of herbage by sheep was high, 80 g/kg $W^{0.75}$.

Main diseases which affect herbage yields are the leaf spot caused by species of *Cercospora* and rosette virus. *Cercospora* disease is widespread and heavy infestation can substantially reduce herbage yields. In trials by Cummins & Smith (1973) an infested crop yielded 1.03 t DM/ha but the yields increased to 2.91 or 3.15 t/ha when the plots were treated with 420 g benomyl/ha or 1.7 kg chlorothalonil applied at 2-week intervals. Rosette-virus infested plants are dwarfed, the leaves become small and distorted, and the disease can inflict considerable damage, mainly in Africa. Increasing plant density helps to keep down the level of infection (Purseglove, 1968).

Flowering of *A. hypogaea* begins 4–6 weeks after sowing and reaches its maximum in a further 4 to 6 weeks. The flowers open and pollination takes place at sunrise, the anthers bursting and pollination taking place within the closed keel. Bees and other insects visit the flowers but, judging on the uniformity of progenies, cross-pollination is rare (Purseglove, 1968). *A. hypogaea* is a tetraploid with $2n$ chromosomes = 40.

There is a number of groundnut cultivars in cultivation and A. H. Bunting (see in Purseglove, 1968) recognizes two main groups of cultivars: Virginia and Spanish-Valencia.

In the Virginia group the plants can grow for a long period, sometimes surviving until the next season, and they are mostly late maturing, requiring 120 to over 140 days from sowing to ripening. The foliage is dark green in colour. The plants are of prostrate or spreading-bunch types.

In the Spanish-Valencia group the plants are strictly annual and mature early, 90–110 days after sowing. The herbage is relatively light green.

There is a number of cultivars developed or selected mainly or exclusively for seed production; they can differ in the bulk and vigour of herbage and some cultivars are more suitable for the combine seed + hay production than the others. For example, in India (Rao *et al.*, 1969) cv. Asiriya Mwitunde produced 3.0 t hay/ha (after the pods have been harvested) compared with 1.4 and 1.7 t yielded by the two other cultivars in the trial, in which the herbage also contained less CP and less P in the DM than that of cv. Asiriya Mwitunde. The development of new cultivars is not easy because the plants are almost 100 per cent self-pollinated and artificial hybridization is time consuming and slow.

Atylosia Wight & Arn.

Erect or climbing herbs or shrubs. Leaves with three leaflets. Flowers small or of medium size, yellow. Over 30 species distributed in the tropics of the Old World.

Atylosia scarabaeoides (L.) Benth. $2n = 22$

Perennial climbing or trailing herb. Stems slender, covered, together with petioles, with short ferruginous hairs. Leaflets elliptic, up to 7 cm long and up to 3 cm wide, pubescent and gland-dotted on both surfaces. Inflorescences short, axilliary, with few flowers. Standard yellow, tinged with red, 9–10 mm long and 5 mm wide. Pods oblong, compressed, 2.5 cm long and 5–7 mm wide, with three to six seeds, grooved between seeds, densely pubescent. Seeds reddish, speckled with brown or black, 4–5 mm long and 1.5–2 mm thick.

Distributed in tropical Asia and in adjacent areas of Africa. Common in tropical India up to 24°N., and considered to be useful and well grazed in natural pastures. It has been tried, with good success, for oversowing natural pastures with the dominance of *Heteropogon contortus* by drilling in just before the monsoon rains begin or during the early monsoon season. Sowing rates of 8 kg seed and application of 20 kg P_2O_5/ha resulted in good mixed pastures which at the ripe stage contained up to 7 per cent CP compared with about 2 per cent for unimproved *H. contortus* stands analysed at the same stage of growth (Dabadghao & Shankarnarayan, 1970).

Cajanus DC.

A small genus of two or three species.

Cajanus cajan (L.) Millsp. (*C. indicus* Spreng.). Pigeon pea; Red gram; Pois d'Angola; Arhar; Tur (India). (Fig. 39).

Erect shrub or perennial herb 1 to 4 m high, often grown as an annual crop. Leaves with three leaflets; leaflets elliptic to lanceolate, green and velvety pubescent above, silvery greyish-green and with longer hairs below, 2.5–10 cm long and up to 3.5 cm wide. Flowers in dense groups or heads on common penducles 2–7 cm long. Calyx velvety-fulvous. Standard yellow, often with reddish-brown or red lines, or red outside, almost orbicular, 1.2–1.7 cm in diameter. Wings yellow. Pods almost straight, 5–10 cm long and 0.5–1.5 cm wide, glabrous or hairy and glandular. Seeds cream-coloured to brown, subglobose of various size.

Fig. 39 *Cajanus cajan* (from Purseglove, 1968).

Two varieties are usually distinguished: var. *fulvus* DC, early maturing, usually annual, short type with yellow standard and glabrous, light-coloured pods containing about three seeds, known as **tur** in India where it is grown predominantly in the south; var. *bicolor* DC, is a late-maturing, tall perennial type with the standard red on its back; hairy pods with four to five seeds, often with maroon-coloured blotches. This variety is known in India as **arhar** and is grown mostly in northern India.

Apparently of African or Indian origin, *C. cajan* is now widely cultivated throughout the world tropics as a pulse crop mainly for grain but also for green pods, as a cover or green-manure crop and as a fodder plant. Fodder production is often or predominantly combined with the production of grain and the forage is fed green or less often grazed (grazing may damage the plants), but is usually made to hay or occasionally to silage.

Environment. *Cajanus cajan* is a short-day or photoperiodically neutral plant which can flower at 10–12 h photoperiods but a number of cultivars can flower throughout the year under all photoperiods observed in the subtropics or tropics. It requires a warm climate, does not produce any growth at temperatures below 10°C and most cultivars are susceptible to frost. Rainfall requirements are moderate, the plants are relatively drought resistant and can be grown under an annual rainfall of 600–800 mm although it grows and yields better under higher rainfall, up to 1,200–1,500 mm. Almost any soils are suitable except waterlogged clays but it responds to deep fertile soils on which it is supposed to last for a longer period.

Cultivation. Establishment is from seed sown in rows about 1 m apart or broadcast, especially when grown as a green cover or for fodder only. Seed rates vary from 4–6 kg/ha to 10–20 kg broadcast. Little or no hard seeds are usually present and no scarification is necessary. Inoculation is hardly needed as *C. cajan* is almost non-specific in its *Rhizobium* requirements (Obaton, 1974) and is normally naturally inoculated by ordinary cowpea type *Rhizobium* usually present in the soil. Fertilizer P produces usually good responses but smaller and less frequent responses have been obtained from K; moreover different cultivars react differently to applied nutrients. Fertilizer N, even when applied at low rates of the order of 20 kg/ha, usually reduces the yields.

Utilization. The crop is cut for forage at the stage when the first pods begin to ripen or, less often, at the preflowering stage. To obtain satisfactory regeneration the plants are cut high, 50–75 cm from the ground level when allowed to grow to 150 cm or over, or at some 15 cm when shorter cultivars are used or when younger herbage is cut. When cut close to the ground the plants may die or regeneration would be slow and poor. *Cajanus cajan* can also be harvested by plucking vegetative shoots after the pods have been harvested for grain. Yields of green fodder when cut at the seed-maturing stage as estimated in Colombia (Crowder, 1960) range from 20 to 60 t/ha; in Cuba 11.9 t DM/ha were obtained by Febles & Padilla (1970), which seem to be within reasonable limits. The nutritive value of herbage is high and CP content ranges usually between 10 and 18 per cent, although higher contents of CP were also reported. The digestibility of CP can range from 60 per cent to as high as 88 per cent, the last figure being reported by A. J. Oakes and O. Skov (see in Akinola *et al.*, 1975). CP digestibility of the older plants can equal or even exceed that of young plants mainly, perhaps, because the plants cut at the more advanced stages of growth may contain a certain proportion of flowers and young seeds and pods. High animal production was obtained by F. G. Krauss (Akinola *et al.*, 1975), who reported daily liveweight gains of 0.68–1.25 kg/head of cattle fed on *C. cajan* forage and considered that one hectare of *C. cajan* can support from 0.8 to 3.6 head/ha. Akinola *et al.* (1975) state in their review that *C. cajan* can be used for cattle fattening and can increase milk yields of cows and goats.

Pests and diseases. Root nematodes – species of *Meloidogyne* and other genera – often attack *C. cajan* and there are a number of insect pests which damage the plants to various degrees; they attack mainly the young plants, and the flowering parts and developing and ripe seeds and pods. Of the diseases, the most common and damaging, especially in India, is wilt caused by *Fusarium udum*, a soil-borne pathogen. Other important diseases are those of roots and stem bases such as root rot caused by *Phaeolus manihotis* (Akinola *et al.*, 1975) and also canker of stems caused by *Physalospora cajanae* (Purseglove, 1968).

Reproduction. Most of the flowers open between 11 a.m. and 3 p.m. and can remain open for up to 6 hours (Purseglove, 1968) or even longer, depending on weather conditions; rain during the flowering reduces fertilization and the formation of seeds. Pollination can occur before the flowers have been opened and the plants are essentially self-pollinated and self-fertile but after the flowers have been opened they are visited by various insects, mainly bees, and some other insects, such as thrips (*Taneothrips distilis*) which can apparently pollinate the flowers when they are still closed. Cross pollination can be considerable and the authors quoted by Akinola *et al.* (1975) report from under 1 to 65 per cent cross-pollination. The plants are normally diploids with $2n$ chromosomes $= 22$ and although tetraploids and hexaploids with the corresponding numbers of chromosomes of 44 and 66 have also been reported; they have been associated with abnormal plants only.

Seed ripens in 5–6 months from sowing in early cultivars and in 8–9 months in late cultivars, and in this essentially pulse crop seed production is good and yields of 500 to 1,000 kg seed/ha are normal. Seeds retain their viability for a considerable period of time and 80–100 per cent of seed can be expected to germinate after having been stored for 4 years.

The species is very variable and numerous varieties and cultivars exist, especially in India, where this crop has been grown for a considerable number of years and received much attention. The cultivars are mostly classified and selected for grain and it seems that little attention has been given to forage yields and quality and other features important for fodder plants.

Calopogonium Desv.

A small genus of American tropics and subtropics.

Calopogonium mucunoides Desv. Calopo; Rabo de iguana. $2n = 36$

Trailing or twining perennial herb densely covered with ferruginous hairs. Leaves with three leaflets which are elliptic to rhomboid-ovate, 4–10 cm long and 2–5 cm wide. Flowers small, the blue corolla is 7–10 mm long. Pods with five to eight seeds, linear, 2–4 cm long and 3.5–5 mm wide, straight or curved, densely covered with brown hairs. Seeds dark brown or yellow-brown, 2.5–4 mm long and 2.5–3 mm wide.

Indigenous in tropical South America, Central America and the West Indies, and introduced early in this century to tropical Africa and Asia, first as a cover crop for tree plantations and for green manure but later used in numerous trials as a forage plant and often naturalized.

Calopogonium mucunoides requires a moist and warm climate and is grown in the humid tropics at altitudes from sea level to about 2,000 m but is more suited to altitudes from 300 to 1,800 m (Crowder, 1960). It cannot withstand any long periods of drought and dies out but can regenerate itself from seed, and is adapted to a wide range of soils. Shade tolerance is poor.

Establishment is effected by seed although Whyte *et al.* (1953) indicate that propagation by cuttings is used in Malaya. Seed can be sown broadcast at 6–10 kg/ha or in 1 m rows at 3–6 kg. A large percentage of hard seeds has been noted and scarification is advised. Seed yields are good and there are 70,000–75,000 seeds/kg. Inoculation is hardly necessary as the plants can be naturally inoculated by soil *Rhizobium*; *C. mucunoides* accepts *Rhizobium* strain CB756 developed in Australia (Obaton, 1974) suitable for a number of legumes which respond to the ordinary cowpea inoculant.

Calopogonium mucunoides grows fast and in 4–5 months after sowing can form a dense growth of entagled herbage 30–60 cm high. The ability to fix atmospheric nitrogen has been demonstrated by Oke (1967a) but the amount of fixed N was much smaller than that produced by *Pueraria phaseoloides*; young plants were more effective in fixing N than the old ones. *Calopogonium mucunoides* has been tried with moderate to good success in mixtures with *Melinis minutiflora* and *Chloris gayana*, and Riveros (1960) obtained good results from mixtures with *Digitaria decumbens* in which the legume was oversown into harrowed stands of the grass. Yields of herbage at the peak of growth can be very high and Herrera *et al.* (1966) obtained 13.55 t DM/ha from a single cut at the stage of mature pods, but lower yields are commonly reported. The main negative feature of *C. mucunoides* is its low palatability usually ascribed to the abundance of hairs on the leaves and stems. The herbage is often refused by cattle although they may eat it to a certain extent during the dry season. Low palatability is perhaps the reason why the interest in this legume as a forage plant has faded in the last decade.

Canavalia Adans.

A genus of some 50 species, mostly large woody climbers, distributed throughout the world tropics and subtropics, but mainly in America. *Canavalia ensiformis* and *C. gladiata* are cultivated for edible seed (mainly *C. ensiformis*), as green manure or for fodder. Some other species are also used for fodder but to a much lesser extent, one of them is

C. bonariensis Lindl. which was grown by Warmke *et al.* (1952) in Puerto Rico. This plant was the least palatable of the legumes they tried and only 29 per cent of herbage given to cattle was eaten compared with 64 per cent for *Stizolobium deeringianum*.

Canavalia ensiformis (L.) DC. Jack bean $\qquad 2n = 22$

Large bushy perennial or annual herb. Leaves with three leaflets; leaflets large, 6–20 cm long and 3–11 cm wide, ovate or elliptic. Flowers in racemes 5–12 cm long on long common peduncles, pink or purple, large, standard 2–7 cm long. Pods linear oblong, 15–35 cm long and about 3 cm wide, with two ribs near the margins of both surfaces. Seeds 1.5–2 cm long and up to 1 cm thick, white with a brownish spot near the hilum; hilum 6–9 mm long.

Originates from South and Central America [perhaps from wild *C. plagiosperma* Piper (Sauer, 1964)] where it has been grown from prehistoric times for its edible seeds and is now cultivated throughout the tropics for seed, as a cover or green manure crop or for fodder. In the tropics it can be grown from sea level to 1,800 m alt, under humid or semi-humid climates or in the drier areas under irrigation. If accurately placed 60–100 cm apart, about 10 kg seed/ha can suffice but in Central America 25–30 kg/ha is sown for seed production and 65 kg for green manure (Whyte *et al.*, 1953). The plants produce a large bulk of green fodder, 20–60 t/ha. In analyses by Elliott & Fokkema (1960) the herbage contained 13.8–16.0 per cent CP, 2.1–2.9 per cent EE, 26.5–35.7 per cent CF and 41.2–43.5 per cent NFE. The digestibility of these nutrients was determined as 56–59, 57–69, 38–61 and 70–72 per cent, respectively, and of DM 54–60 per cent, and it was almost invariably higher in hay cut late than early. In pure stands the palatability is mediocre but improves in grass mixtures and is better for dry than fresh green herbage (Derbal *et al.*, 1959). Seed, though edible, can be toxic when eaten by humans or fed to cattle in large quantities. The flowers are visited by insects and over 20 per cent cross pollination has been observed (Purseglove, 1968).

Canavalia gladiata (Jacq.) DC. Sword bean $\qquad 2n = 22$

Rather similar to *C. ensiformis* from which it differs by a somewhat twining habit when grown in shade, more acute leaflets, usually black or variously coloured (cinnamon, brown) seeds and longer seed hilum which is 10–25 mm long. This cultivated species has probably originated from wild *C. virosa* (Roxb.) Wight & Arn., and Sauer (1964) even considers *C. gladiata* only as a cultivated form of the latter. This species is of Asian origin and is used as green manure or as a forage plant.

Centrosema (DC.) Benth.

Climbing or prostrate herbs or subshrubs. Leaves mostly with three leaflets (in some species with one to seven leaflets). Inflorescences axillary. Pods linear, with raised nerves or wings at margins. There are 40–50 species, all indigenous to American tropics. *Centrosema pubescens* is now cultivated throughout the world tropics as a pasture plant; another pasture or fodder species of lesser importance is *C. plumieri* and there are a few more species grown as cover crops.

Centrosema brasilianum (L.) Benth.

Rather similar to *C. pubescens* (see below) from which it differs in having finer stems, smaller, usually glabrous leaflets, and narrower pods with a longer and finer beak. Common in Brazil and neighbouring South American countries. In a trial in northern Queensland, Australia, *C. brasilianum* yielded 15.0 t DM/ha/year, more than any other *Centrosema* introductions, the main part of the yields being produced in the post-wet period (Grof & Harding, 1970).

Centrosema plumieri (Pers.) Benth. Butterfly pea $2n = 20$

As *C. pubescens* (see below) but the leaflets are larger and glabrous, the stems less fine, the pods broader, 12–16 mm wide, with wings instead of hard ribs on margins; seeds about twice as large as in *C. pubescens*. Originates from tropical South America and has been used as a cover crop in rubber-tree plantations in southern Asia and elsewhere. Also tried as a fodder plant in Australia and a few other countries but with only moderate success.

Centrosema pubescens Benth. Centro; Jitirana (Brazil) (Fig. 40) $2n = 20$

Climbing and trailing perennial with slender pubescent stems. Leaves with three leaflets; leaflets oblong to ovate-laceolate, 1.5–7 cm long and 0.6–4.5 cm wide, finely pubescent on the lower side, seldom glabrous. Flowers in dense and short axillary racemes, on a peduncle up to 6 cm long. Calyx 6–10 mm long. Standard whitish, mauve, purple or yellow, 1–3.5 cm long and 2–3 cm wide. Pods linear, 4–17 cm long and 6–7 mm wide, with hard raised margins and a narrow straight beak at the apex, dehiscent at maturity, with up to 20 seeds. Seeds red-brown, streaked with black, 4–5 mm long, 3 mm wide and 2 mm thick.

Occurs naturally in South America, where it is widespread and common in grasslands, on river banks, roadsides, etc. Introduced first to southern Asia, apparently in the last century (Barnard, 1969), where it was outstandingly successful as a cover crop in tree plantations, but not in coffee where it climbs to the top of small coffee trees, and as green manure. It spread then to other areas and is now widely grown throughout the world tropics and subtropics, mainly as a forage plant, on experimental and often on a farm scale. *Centrosema pubescens* has become one of the most popular tropical fodder legumes highly valued for its vigorous and productive growth, good quality of herbage and the ability to form balanced mixtures with grasses; its vulnerable side seems to be seed production affected by difficult harvesting.

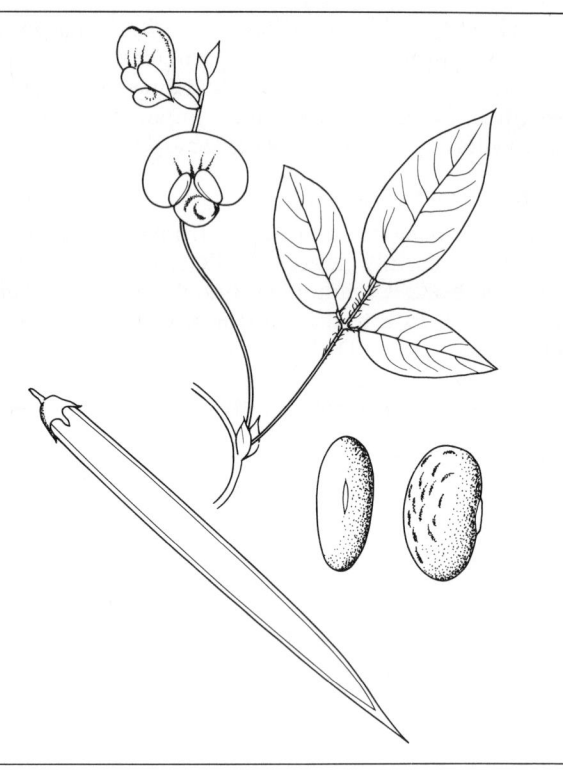

Fig. 40 *Centrosema pubescens.*

Environment. Centrosema pubescens is adapted to humid or moderately humid tropics and subtropics with an annual rainfall of about 1,500 mm although it is also reputed to be drought resistant and Toutain (1973) states that it grows well in the areas receiving 800 mm rain p.a. but prefers those of 1,200–1,500 mm. Its drought resistance is probably understood in the sense that it can survive droughts; during the droughts the leaves can however be lost and the tops dry out but the plants remain alive and can sprout again with the onset of rains. Under irrigation dry season production can however be good. Not particularly high temperatures are required for the optimum growth but the plants stop growing under low temperatures and are easily killed by frost unless there is a dense cover of herbage which protects the crowns. *Centrosema pubescens* can grow on relatively poor soils but requires well-drained soils and cannot withstand waterlogging or flooding. High concentrations of Mn in the substrate can be toxic although *C. pubescens* is slightly less sensitive to Mn than the majority of other legumes.

Establishment, inoculation. Establishment is effected by seed which is sown at 4 to 6 kg/ha in about 1 m wide rows. The proportion of hard seeds is usually high, about 60 per cent, and Whyte *et al.* (1953) recommend soaking seed in hot water prior to sowing, whereas Phipps (1973) has found that the best results can be obtained by dipping seed in boiling-hot water for 1 second or by keeping the seed under deep freeze for 16 days; these treatments increased seed germinability from 30 to 55 per cent. Concentrated sulphuric acid has also been recommended for reducing the percentages of hard seeds.

Centrosema pubescens is highly selective in its *Rhizobium* requirements, special strains of suitable *Rhizobium* have been selected and a standard strain is in commercial supply. After treatments seed can be pelleted with either rock phosphate or lime, or with peat cultures of *Rhizobium* in which form commercial inoculants are issued and adding methyl cellulose as a sticker (Norris, 1973b). In Nigeria (Odu *et al.*, 1971) nodulation and nitrogen fixation were the best at soil pH of 6.0 and decreased with the increase of pH value. Nitrogen fixation of inoculated plants is usually satisfactory to good although it seems to be less effective than that of the Siratro cultivar of *Macroptilium atropurpureum* and it has been estimated that in grass mixtures the effect of nitrogen fixed by *C. pubescens* can be compared with that of up to 75 kg of applied N in terms of DM production and over 100 kg in terms of CP yields (R. D. Singh *et al.*, 1968). It has also been estimated that from 75 to 280 kg N/ha/year is accumulated in the soil as a result of *C. pubescens/Rhizobium* activity (Toutain, 1973).

Centrosema pubescens seedlings develop slowly but soon gather momentum and in 4 to 6 months after sowing form vigorous stands, strong enough to compete with *Imperata cylindrica*, a very persistent and vigorous grass weed, and Whyte *et al.* (1953) state that in India *C. pubescens* has been used for the suppression of this weed.

Fertilizers. Responses to fertilizer P vary and strong effects of 100–200 kg P_2O_5/ha have been observed in Brazil, but smaller responses have been reported from Tanzania where 80 kg P_2O_5/ha increased DM yields only by 126 kg DM/ha, whereas gypsum or lime gave higher yield increases. Boron and Mo can increase the response to P. Erratic responses to P can perhaps be partly explained by the presence of endotrophic mycorrhiza of *Endogene* fungus in the roots which can compensate to a certain extent for the scarcity of available forms of phosphorus in the soil. This was shown in trials by Crush (1974) in which young plants of *C. pubescens* weighed 1.67 g when grown in soil deficient in available P, 3.88 g if grown in the same soil but with the mycorrhiza present, and 4.95 g when fertilizer P was applied but mycorrhiza absent. Responses to fertilizer K are still more erratic than those to P.

Herbage yields. When grown in pure stands herbage yields vary, and in Brazil (Buller *et al.*, 1970) *C. pubescens* yielded 3 to 5.5 t DM/ha, depending on P and K rates; the yields being below the average yields of other legumes in the trial. Some 2 to 5 t DM/ha can be expected in farm practice, although much higher yields have also been obtained: 7.6 t DM/ha in Ghana (Tetteh, 1972) and over 12 t in Australia from cv. Belalto (Harding & Cameron, 1972).

Grass mixtures. Except as a cover crop or green manure *C. pubescens* is seldom grown alone but mostly in mixture with grasses and can form well-balanced mixtures even with strong grasses if helped by suitable management. It has been tried and often grown on a farm scale with *Andropogon gayanus* and *Chloris gayana* in Nigeria, *Hyparrhenia rufa* in Uganda, *Heteropogon contortus* and *Pennisetum polystachion* in India, *Digitaria decumbens* in Brazil and USA, *Brachiaria mutica, Melinis minutiflora, Panicum maximum, Panicum coloratum* and *Pennisetum clandestinum* in Australia, *Pennisetum purpureum, Ischaemum indicum* and *Brachiaria humidicola* in Fiji, *Cenchrus ciliaris* in Tanzania, and with *Paspalum notatum* and *Setaria anceps* in USA. In grass mixtures *Centrosema pubescens* usually increases total yields of herbage compared with grass grown alone; in trials in India (R. D. Singh *et al.*, 1968) it increased the total yields of unfertilized *Pennisetum polystachion*, 5.27 t/ha, to 9.23 t for the mixture and in Florida (Kretschmer, 1970) where *Digitaria decumbens* yielded 5.24 t/ha the inclusion of *C. pubescens* raised the yield to 9.39 t, and increased the content of CP in the herbage from 6.0 to 9.9 per cent. Yields of the grass component usually either decrease or remain at the initial level and only seldom the legume improves its growth.

Nutritive value. *Centrosema pubescens* herbage contains from 11 to 23–24 per cent CP, its stems are fine and not lignified to any considerable extent at advanced stages of growth, and its nutritive value is high. Although palatable, it can be less palatable than the companion grass, especially in the wet season; moreover the cattle have to be trained or get used to eating *C. pubescens* herbage before any large quantities are consumed. The herbage contains some oxalates, their content was reported to be 2.22 per cent, but only 0.10 per cent of water-soluble oxalates was found (Ndyanabo, 1974).

Animal production. Admixture of *C. pubescens* to grass usually increases the animal liveweight gains and in trials in Brazil (Aronovich *et al.*, 1970) liveweight gains of calves increased from 349 kg/ha when grazed on pure *Digitaria decumbens* to 410 kg on a *D. decumbens/C. pubescens* mixture.

Pests and diseases. In Australia red spider or mite (*Tetranychus* sp.) can damage the leaves which are also susceptible to attacks by *Cercospora* sp. causing leaf-spot disease.

Reproduction. Pollination occurs in expanded flower-buds before the flowers open; the plants are almost exclusively self pollinated and progenies of commercial seed are uniform and true to type (Hutton, 1960; Barnard, 1969). Differences between various introductions have however been observed and selection of superior cultivars was undertaken in Brazil and Australia resulting in the development of two outstanding cultivars – **Deodoro** in Brazil and **Belalto** in Australia. Belalto is based on an introduction from Costa Rica; it out-yielded commercial varieties and differed from the ordinary commercial type by being of less twining and more creeping habit, with the runners freely rooting from the nodes.

It has been found that the degree of nodulation and nitrogen fixation of *C. pubescens* is an inheritable feature which can be utilized for the development of cultivars with an increased ability of nitrogen fixation.

Seed production. Seed ripens 6 to 12 months after sowing and is well produced but harvesting can be difficult because of prolonged flowering and easy seed shedding. Nevertheless some 150 kg seed/ha has been harvested in Australia and hand-picking may result in 300–600 kg/ha (Toutain, 1973). Seed germinates poorly in the first months after harvesting and in a trial by Sepra (1971) germination was improved from 12 per cent soon after harvesting to 64 per cent after 1 year of storage. There are 33,000–40,000 seeds/kg.

Clitoria L.

Mainly climbing or erect herbs 30–40 species of which occur throughout the tropics and subtropics.

Clitoria ternatea L. Cordofan pea $2n = 16$

Twining perennial with fine stems 0.5–3 m long. Leaves pinnate with five to seven leaflets; leaflets oblong-lanceolate to almost orbicular, 1.5–7 cm long and 0.3–4 cm wide, glabrous above and pubescent below. Flowers single or paired, with the pedicel twisted to 180° so that the funnel-shaped standard is turned down, white or bluish, 2.5–5 cm long. Pods linear, flat, 6–12 cm long and 0.7–1.2 cm wide, with up to 10 seeds. Seeds olive, brown or almost black in colour, often mottled, 4.5–7 mm long and 3–4 mm wide.

Occurs naturally in tropical Africa, Madagascar, Arabia, India, China, Malaya, Indonesia, Pacific Islands, and North, Central and South America; in South America there are also a number of other species of *Clitoria*. *Clitoria ternatea* grows in bush, grasslands, often on seasonally-waterlogged black clays, old cultivations, wasteland, etc., up to 1,500 m alt. It is cultivated in some countries for fodder and as a pulse crop. In the Sudan *C. ternatea* has been grown for some time for fodder or grazing, often in mixture with *Chloris gayana*; this mixture has also produced good results in trials in semi-arid Kenya. Encouraging results were also obtained in Australia, the Philippines, Senegal and Zambia.

Fresh seeds do not germinate or imbibe water, but when stored for 6 months 15–20 per cent germination can be obtained. Germination can be improved by grinding the seed with sand and also by hot water, sulphuric acid or KOH treatments (Mullick & Chatterji, 1967). Soaking in a 100 p.p.m. solution of NaCN also improved germination and early plant growth. Initial growth is fast, and in India 24 t fresh material/ha was obtained after 2 months of growth (Katiyar *et al.*, 1970); the herbage contained over 21 per cent CP, 33.3 per cent CF, 34.7 per cent NFE, 0.8 per cent Ca and 0.28 per cent P. Digestibility, as tested on sheep, was 74.2 per cent for DM, 85.2 per cent CP, 61.6 per cent CF and 72.9 per cent for NFE and the herbage contained 18.2 per cent DCP and 68.6 per cent TDN. CP content can be even higher than in the trial by Katiyar *et al.* and reach 24–30 per cent of the DM, and in pot trials by M. B. Jones *et al.* (1970) the content of N in the fodder, average for various treatments, was 5.03 per cent which corresponds to 31.4 per cent CP, and was higher than in the other plants under trial; the yields of herbage and of N were, however, the lowest.

Clitoria ternatea varies considerably in its morphology and in the agriculturally important characters. In Cuba, where five cultivars were compared (De Matos & De la Torre, 1971), the Mexican cv. **Conchita clara** produced the highest yields of herbage and it was also the most drought resistant: its yields under dry land conditions, 81.8 t fresh fodder/ha, were almost as high as under irrigation – 84.0 t. The four other cultivars under trial, **Indio Hatuey** and **Oriente** from Cuba, and **Negra** and **Jaspeada** from Mexico, gave much lower yields, which ranged from 40 to 64 t/ha without irrigation, and from 55 to 78 t when irrigated.

Crotalaria L.

Plants of various habit, herbaceous or shrubby. Leaves with one to seven, but mostly three leaflets. Flowers medium in size to large, usually yellow but sometimes orange-coloured or blue. Keel rounded or bent in a right angle, often with a prominent beak. Anthers dimorphic: five large alternating with five small. Seed usually heart-shaped or oblong.

A very large genus of some 550 species distributed throughout the tropics, and to a lesser extent in subtropics, and particularly in tropical Africa where about 400 species have been recorded. Most of the species are avoided by grazing or browsing farm animals apparently because of their bitter or otherwise unpleasant taste and the presence of toxic glucosides, and there are no outstanding fodder or pasture plants amongst numerous crotalarias. Herbage of some species is however eaten, especially when made to hay as e.g. is the case with *C. juncea*. A few species are used as cover crops or for green manure but apart from the three species described below information on non-toxic plants is scanty. *Crotalaria lanceolata* was tried in Australia (Milford, 1967) but found unsuitable for cultivation: the content of CP was low and of CF high, and the intake by the animals inadequate at any advanced stage of growth. *Crotalaria anagyroides* of South American origin and also grown in southern Asia, Indonesia and Sri Lanka is another non-toxic species; analysed in Kenya at the pre-flowering stage it contained 23 per cent CP. Some of the common features of species of *Crotalaria* are the resistance of most species to root-knot nematodes, a considerable proportion of hard seeds and self-pollination in the majority of species.

Crotalaria incana L.

Erect, usually much branched annual or short-lived perennial up to 1.5 m high, rarely taller. Stems with long, usually spreading hairs or shortly pubescent. Leaves trifoliolate; leaflets elliptic or obovate, glabrous above and often hairy below. Flowers 12–40 in lax racemes, yellow or purplish; beak of the keel, straight, 8–11 mm long. Pods 3–5 cm long and 8–12 mm across, with 40–50 seeds. Seeds heart-shaped, 2–3 mm long, brownish or olive-green, often mottled. Distributed in tropical America and Africa. According to Whyte *et al.* (1953) *C. incana* is non-toxic and palatable, used to some extent as fodder and is considered promising. However, in palatability tests in Kenya, both cattle and sheep refused to eat the herbage. The nutritive value can be good and the herbage analysed in Kenya (Dougall & Bogdan, 1966) at the preflowering stage contained 27.8 per cent CP, 2.9 per cent EE, 27.1 per cent CF, 36.4 per cent NFE, 0.81 per cent Ca and 0.26 per cent P. There are about 200,000 seed/kg.

Crotalaria intermedia Kotschy $2n = 16$

Erect or sometimes spreading annual or short-lived perennial herb 0.5–2 m high. Leaves trifoliolate; leaflets linear to elliptic, 4–13 cm long and 3–25 mm wide. Flowers in terminal and lateral racemes, from various shades of yellow to almost white, of various size; keel with untwisted, long-projecting beak up to 17 mm long. Pods deflected, narrow-cylindrical, often slightly curved at the apex, 3–6 cm long and 5–20 mm wide. Seeds heart-shaped, pale yellow to orange in colour, 2–3 mm long. The name *C. intermedia* is well known in agricultural literature, but Polhill (1971) considers *C. intermedia*, as it was usually understood, to include two distinct taxa: *C. ochroleuca* G. Don which has wider pods, 15–20 mm across, and pale flowers, and *C. brevidens* Benth. with pods 5–7 mm wide and bright-yellow flowers, both species distributed in tropical Africa, the former mainly in its southern parts and the latter in more northern parts of Africa.

Occurs mainly in grasslands, sometimes on seasonally waterlogged ground. *Crotalaria intermedia* has been tried in some African territories and introduced to other areas, notably to the USA where it was grown on a fairly wide scale but is now losing its popularity. It is non-toxic and reasonably palatable and in trials in Rhodesia yielded 2–3.5 t hay/ha. Analysed in Kenya (Dougall & Bogdan, 1966) at the early flowering stage *C. intermedia* contained 28.8 per cent CP, 3.3 per cent EE, 22.1 per cent CF, 35.9 per cent NFE, 0.72 per cent Ca and 0.33 per cent P; at late flowering the content of CP was decreased to 21.4 per cent, Ca to 0.44 per cent and P to 0.27 per cent, whereas the content of CF was increased to 35.7 per cent. Seed setting is reasonably good and there are about 220,000 seeds/kg (Whyte *et al.*, 1953).

Crotalaria juncea L. Sunn hemp $2n = 16$

Erect annual up to 1.5 m high. Leaves simple, oblong lanceolate, 7–13 cm long and 1.5–2.5 cm wide, almost sessile. Flowers 8–20 in long racemes, yellow; keel with spirally-twisted beak. Pods cylindrical, 3–6 cm long and 1–2 cm across, with about six seeds. Seeds heart-shaped, up to 6 mm long, brown to black.

Originates from India and is now widely grown throughout the tropics and subtropics for green manure, as a cover crop, for fibre and, to some extent, as a fodder plant. It is mildly toxic, mainly to horses. Cattle can eat it safely provided that no more than 10 per cent of sunn hemp hay is added to other feeds (Whyte *et al.*, 1953). In trials in India (Reddy, 1968) it yielded 5.9 t hay (5.19 t DM), 749 kg DCP and 3,835 kg TDN/ha, considerably out-yielding lucerne; the herbage contained 18.1 per cent CP, 1.1 per cent EE, 38.1 per cent CF and 34.1 per cent NFE. *Crotalaria juncea* responds well to fertilizer P and the application of 45–360 kg P_2O_5/ha increased the yields of herbage and its nitrogen content; it was estimated that 213–296 kg N/ha would be fixed by plants given P_2O_5 compared with 132 kg for unfertilized plants (Rao & Sadasivaiah, 1968). There are about 30,000–35,000 seeds/kg.

Cyamopsis DC.

A small genus of the Old World tropics. Annual herbs.

Cyamopsis tetragonoloba (L.) Taub. (*C. psoralioides* DC.). Cluster bean; Guar (Fig. 41) $2n = 14$

Erect annual 60 cm to 3 m high with stout stems. Leaves trifoliolate with ovate leaflets serrate on margins. Flowers 8 mm long, with white standard and pink or mauve wings, gathered in dense axillary racemes. Pods linear, compressed, 4–10 cm long, with a double ridge on the dorsal edge and with 5–12 seeds. Seeds oval or angular, 5 mm long, white, grey or black.

Unknown as a wild plant and possibly originates from India where it is widely grown for young pods eaten as a vegetable, for seed which are highly viscose when milled and are used in culinary or industry (Purseglove, 1968), and also as a fodder or green-manure crop. *Cyamopsis tetragonoloba* is also grown in Pakistan, Indonesia and other parts of southern and south-eastern Asia and has been tried in tropical and subtropical Africa, Australia, South and Central America (Whyte *et al.*, 1953) and in some warm temperate areas such as the USA or Middle-Asian republics of the USSR. It has met with good success in south-western USA and is grown on a farm scale, mainly for seed.

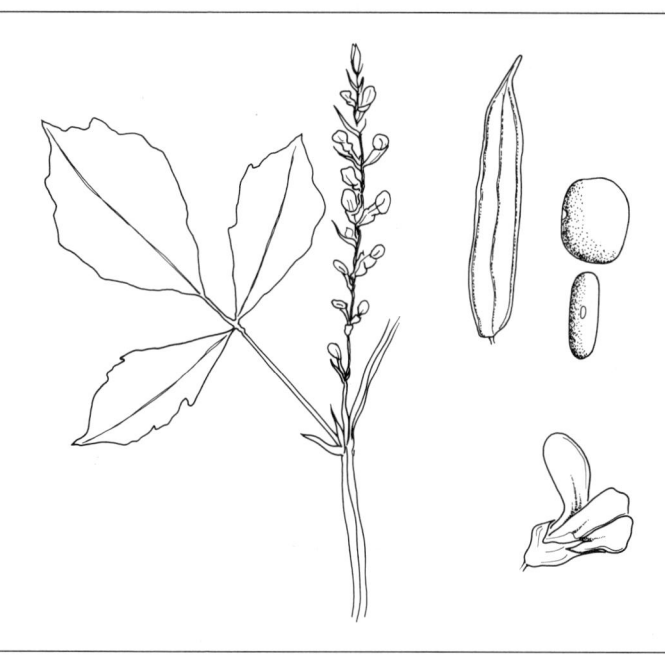

Fig. 41 *Cyamopsis tetragonoloba.*

The plants are fairly drought resistant and are grown in semi-arid parts of India, mainly on light- or medium-textured soils but cannot tolerate waterlogging or excessively moist soils (Narayanan & Dabadghao, 1972). When grown for seed or pods *C. tetragonoloba* is sown in rows at 10–15 kg seed/ha, or more in the USA, but 40 to 60 kg/ha are sown, in rows or broadcast, when intended for fodder, and in India an increase of sowing rate from 20 to 60 kg seed/ha increased fresh fodder yields by over 50 per cent. In India *C. tetragonoloba* is grown for fodder mainly in mixtures with sorghum and in a trial in a subtropical environment a 2 : 1 mixture of sorghum : *C. tetragonoloba* was the most successful and gave 8.8 t DM/ha, only a slightly lower yield than for sorghum grown alone, and 407 kg CP/ha; the mixed herbage contained 4.86 per cent CP compared with 3.44 per cent for pure sorghum; pure *C. tetragonoloba* yielded 1.54 t DM/ha and 251 kg CP/ha but its CP content was 16.3 per cent (Bhan, 1967). In semi-arid areas of India *C. tetragonoloba* has been sown, mainly experimentally, to established

Cenchrus ciliaris, *C. setigerus* or *Lasiurus hirsutus* and the presence of the legume increased fodder yields by 20–30 per cent, and also improved herbage quality by enriching it with proteins.

Symbiotic N fixation has been confirmed in north-western Australia where *C. tetragonoloba*, grown experimentally, added 220 kg N/ha to the plant-soil system during the three growing seasons of trial. *Cyamopsis tetragonoloba* is palatable to cattle and Narayanan & Dabadghao (1972), quoting P. E. Lander, give the content of CP 3.1 per cent, EE 0.4 per cent, CF 4.4 per cent, NFE 8.0 per cent, Ca 0.61 per cent and P 0.07 per cent for green herbage containing 80.8 per cent water which correspond to about 16.1 per cent CP, 2.1 per cent EE, 22.9 per cent CF, 41.6 per cent NFE, 3.2 per cent Ca and 0.35 per cent P in the DM, and comparable figures are given for hay. The digestibility was determined as 76–77 per cent for CP, 16–39 per cent for EE, 26–45 per cent for CF and 70–73 per cent for NFE; the content of TDN ranged from 57 to 64 per cent and of DCP from 12 to 14 per cent.

The plants develop rapidly and fruiting begins in about 8–9 weeks after sowing and even in 6–7 weeks in early cultivars. Seed yields of varieties grown for seed are estimated to range from 650 to 900 kg/ha and yields of similar or somewhat lower order can be expected from cultivars grown for fodder. There are about 45,000 seeds per kg, but this figure can vary greatly in different cultivars. There are numerous cultivars grown in India and those more vigorous and leafy, as e.g. selected cvs. **Pusa Sadabahar**, **Sirsa 1** and **Sirsa 2** are grown as fodder plants. In Texas cvs. **Brooks**, **Halls** and **Mills** are used (Leffel, 1973).

Desmanthus Willd. Subfamily Mimosoideae.

Trees, shrubs or perennial herbs with bipinnate leaves; leaflets small. About 40, almost exclusively American, species.

Desmanthus virgatus (L.) Willd. (*Mimosa virgata* L.) Dwarf koa

$$2n = 28$$

Shrubby or herbaceous plant with erect stems up to 2 m high. Leaves bipinnate, 4–8 cm long, with three pairs of pinnae, each with six to eight pairs of leaflets; leaflets 4–9 mm long. Flowers white, a few to several in small dense heads. Pods three to five per head, erect or suberect, dehiscent, 6–9 cm long and 3–4 mm wide. Distributed naturally in Central and South America and southern Asia and introduced to other tropical and subtropical areas. It is palatable to cattle, tolerates grazing or cutting and used as fodder or for grazing in Hawaii and Mauritius (Whyte *et al.*, 1953). In India (Sundararaj & Nagarajan, 1963) it was tried as a hedge plant, the prunings used for forage, and about 30 kg of green fodder was obtained in a single cut from one meter length of hedge. The contents of CP and P were comparable to those of lucerne. In Cuba *D. virgatus* is however considered to be potentially toxic.

Desmodium Desv.

Perennial, rarely annual herbs, subshrubs or shrubs of erect to rambling or prostrate habit. Leaves with one or three, rarely five, leaflets. Flowers in terminal or axillary racemes. Corolla variously coloured, mostly pink or mauve. Stamens connate and form a tube but the uppermost stamen usually free; anthers uniform. Pods articulate, with two to several joints (segments), usually indehiscent but breaking at maturity into one-seeded segments. Seeds oblong, orbicular or rectangular.

A large genus of over 300 species, mainly tropical and subtropical, but there are some temperate species, mostly American or Asian. The largest concentration of species occurs in Mexico and Brazil, and also in eastern Asia. A few species, mainly of South American origin have been introduced to a number of tropical countries for trial as fodder or pasture plants, the most successful being *D. intortum* and *D. uncinatum* which are now becoming important pasture legumes in Australia and Africa. A number of species contain appreciable amounts of tannins which may reduce their palatability to farm animals, and some species are regarded as pasture weeds.

Desmodium adscendens (Sw.) DC. $2n = 22$

Perennial herb up to 1 m high, rambling or erect. Leaves trifoliolate; leaflets elliptic-obovate, almost glabrous, terminal leaflet 2–4 cm long and 1–3 cm wide, larger than lateral leaflets. Flowers in terminal and axillary racemes, usually pink. Standard 4–5 mm long. Pods with one to five joints, straight on the upper edge and curved below. Seeds elliptic, 2.5–5 mm long.

Occurs throughout tropical Africa and in South America, mostly at forest edges and stream banks. In Brazil it provides forage for all stock and especially horses, and three to four cuts per year can be taken. In Brazilian analyses the herbage contained 10.5 per cent CP, 3.4 per cent EE, 49.8 per cent NFE and 31.4 per cent CF and the contents of digestible nutrients were determined as 7.7, 2.1, 37.8 and 17.3 per cent, respectively. The digestibility of organic matter was 64.9 per cent (Pio Corrêa, 1931).

Desmodium barbatum (L.) Benth. $2n = 22$

Spreading to erect perennial of varying habit, 10–100 cm high. Leaves with one or three leaflets which are orbicular to ovate or obovate, glabrous or hairy. Terminal leaflets 1–5 cm long and 0.5–2 cm wide; lateral leaflets slightly smaller. Flowers in terminal and axillary, short and dense, sometimes capitate racemes. Corolla blue, pink or red, often tinged with white. Standard 4–5 mm long. Pods with one to six joints which are 2–3.5 mm long and 2–2.5 mm wide. Seeds purple to greenish-tawny, 1.5–2 mm long.

Distributed in tropical America and almost throughout tropical

Africa. In Central America it is a valuable and important constituent of grasslands, particularly above 1,000 m alt. A considerable intraspecific variation has been observed and Central American ecotypes which differ in agronomically important characters such as the proportion of leaves and fine stems in total herbage or seed setting are a promising feature for selection and breeding. Some ecotypes were noted for their abundant seed setting, the weight of seed reaching 27 per cent of the total weight of plant. Only a relatively small degree of seed shattering has been observed, suggesting a possibility of mechanical harvesting (Semple, 1964).

Desmodium canum (J. F. Gmel.) Schinz & Thell. (*D. incanum* (Sw.) DC.) Kaimi; Kaimi clover $2n = 22$

Perennial herb of varying habit, from ascending to erect, sometimes a subshrub, 0.3–3 m high. Leaves trifoliolate with the leaflets varying in size and shape from almost orbicular to lanceolate, dark green on the upper surface often with a paler streak along midrib, and pale beneath. Flowers in terminal and axillary racemes. Corolla blue, red or purple; standard up to 6 mm long. Pods with up to eight joints which are 3.5–5 mm long and 2–3.5 mm wide. Seed oblong, small, 1 mm long.

Distributed naturally in America from Florida and Texas in the north to Uruguay and Argentina in the south, and now spread throughout the world tropics, mostly on roadsides, wasteland and other disturbed ground. *Desmodium canum* has been extensively tried in Hawaii, in pure stands and in mixtures with *Digitaria decumbens* or *Pennisetum purpureum*. In trials by Whitney *et al.* (1967), it fixed 94 kg N/ha, much less than *D. intortum*, which fixed 406 kg, and no nitrogen was transferred to the companion grass from *D. canum* plants. In another trial in Hawaii (Whitney & Green, 1969) a mixture of *Digitaria decumbens*/*D. canum* yielded 6.84 t DM/ha/year without inoculation and 7.51 t when *D. canum* was inoculated, whereas *D. decumbens* alone yielded 3.78 t. The effect of the legume was equal to that of 210–240 kg applied N/ha. *Desmodium intortum* mixtures out-yielded, however, those of *D. canum*. Under high rates of P and K fertilizers potential yields of *D. canum* are usually lower than those of *D. intortum* but the former is preferred under the conditions of poor grazing management or under low rates of fertilizers (Younge *et al.*, 1964). *Desmodium canum* is now under trial in other areas, including Australia. Hybrids between *D. canum* and *D. uncinatum* have been obtained but they seem to be sterile.

Desmodium discolor Vog. Marmelada de caballo; Capparicho (Brazil)
$2n = 22$

Perennial herb up to 2 m high but usually much smaller, woody at the base and much branched higher up. Leaves trifoliolate or sometimes simple. Leaflets ovate-oblong, large, up to 15 cm long and 7 cm wide.

Flowers violet or mauve, in large paniculate inflorescences. Pods with four to seven joints; joints almost orbicular, 3 mm in diameter. Distributed in South America and introduced to Africa. An important forage plant in Brazil of which three to four cuts of hay per year can be taken. In Brazilian analyses the herbage contained 12.1 per cent CP, 2.7 per cent EE, 43.4 per cent NFE and 35.1 per cent CF and 8.8, 1.8, 33.0 and 19.3 per cent, respectively, of digestible nutrients. The digestibility of organic matter was 62.8 per cent (Pio Corrêa, 1931). Good results have also been obtained in Rhodesia and in Swaziland where it yielded about 7 t hay/ha, out-yielding the other legumes under trial.

Desmodium heterocarpon (L.) DC. $2n = 22$

Shrubby or procumbent perennial up to 1–2 m high. Leaves with three leaflets. Leaflets almost glabrous above and pale and pilose below, terminal leaflet ovate or obovate, 4–6 cm long, larger than the lateral leaflets. Flowers in dense terminal and lateral racemes, pink or dark purple. Standard about 5 mm long. Pods with two to six joints 3–4 mm long and 2.5–3 mm wide, not adherent to cloth. Seeds somewhat rectangular, 2 mm long and 1.5 mm wide, cream or orange in colour. Distributed in eastern and south-eastern Asia, Malaysia and Polynesia and also on the east coast of Africa. Occurs mainly in wet situations. Introduced to America, *D. heterocarpon* was tried in Florida (Kretschmer, 1970) where it out-yielded all the other legumes in the trial except *D. intortum*. Its mixtures with *Digitaria decumbens*, *Paspalum notatum* and *Setaria anceps* yielded 10.0, 9.6 and 9.2 t oven-dry matter/ha, respectively, whereas the grasses grown alone yielded 3.3, 3.0 and 2.2 t/ha.

Desmodium heterophyllum DC. Hetero; Spanish clover $2n = 22$

Perennial prostrate and ascending herb forming swards up to 15–20 cm high and rooting strongly from the nodes of creeping stems. Stems and leaf-petioles covered with brown hairs. Terminal leaflets 15–20 mm long and 10–15 mm wide but lateral leaflets are smaller. Flowers 3–5 mm long, pink. Pod of three to six segments (joints), 12–25 mm long and 4–5 mm wide, finely pubescent. Seeds kidney-shaped, 2–2.5 mm long, yellowish brown, becoming darker with age.

Distributed naturally in Mauritius and in south-east Asia: Indochina, Malaysia and Indonesia, and introduced to several tropical areas. Trials in Sri Lanka (Whyte *et al.*, 1953) and in Fiji (Roberts, 1970b) were met with only a moderate success, but in Australia *D. heterophyllum* was found suitable for mixtures with dense-growing grasses, *Digitaria decumbens* or *Brachiaria decumbens*, in those areas of north-eastern Queensland where the annual rainfall reaches 1,500 mm or over. It formed well-balanced mixtures with the grasses, especially under heavy grazing, was palatable to cattle, contained some 17–18 per cent CP and cattle grazed on *D. decumbens* or *B. decumbens* mixtures with *D. heterophyllum* gained over 750 kg liveweight/ha/year (Mackay, 1973). In Australia this legume is established by seed using 0.3–0.5 kg seed/ha for grass mixtures and once established it can regenerate itself from seed. *Desmodium heterophyllum* requires a specific *Rhizobium* strain and strain QA 982 selected in Australia was found suitable for the purpose, and the inoculated plants showed a satisfactory fixation of atmospheric nitrogen. Up to 40 kg seed/ha can be harvested but harvesting is difficult because of the gradual appearance of flowers during a long period of time, easy breaking of ripe pods and because the flowers and pods are not much raised above the herbage (Mackay, 1973). *Desmodium heterophyllum* is susceptible to small-leaf virus which does not however reduce the legume productivity to any great extent. Cv. **Johnstone**, based on an introduction from New Guinea, was registered in Australia in 1973.

Desmodium intortum (Mill.) Urb. Greenleaf desmodium; Kuru vine (Rhodesia) $2n = 22$

Large perennial herb with erect, ascending or scandent, much branched, often reddish-brown stems. Leaves trifoliolate with ovate leaflets which are 2–7 cm long and 1.5–5 cm wide, usually brown- or red-speckled on the upper side. The numerous pink or mauve flowers are in terminal and axillary racemes. Pods falcate, with 8–12 one-seeded joints, easily separating. Seed small.

Desmodium intortum occurs naturally in Central and South America from southern Mexico in the north, to Colombia, Venezuela, Equador, Peru, Bolivia and reaching southern Brazil, in cool climate of high ground at about 1,500–2,500 m alt in the equatorial zone, and down to 800 m in subtropical Brazil. In controlled environment, the optimum growth of *D. intortum* was observed at day/night temperatures of 30/25°C (Whiteman, 1968) but was much reduced at temperatures below 15°C. It tolerates moderately-low temperatures better than a number of other tropical legumes but less so than some grasses and in south-eastern Queensland it was dormant in winter, whereas the companion *Setaria anceps* was not, and the mixture moved towards grass dominance (Riveros & Wilson, 1970). *Desmodium intortum* occurs naturally and is cultivated in areas with an annual rainfall of 1,000 mm or over. It is not drought resistant and often sheds its leaves during the dry season. *Desmodium intortum* can grow on a variety of soils, usually preferring well-drained ground but it can also tolerate seasonal or occasional short-term waterlogging. Tolerance to soil acidity of pH 5 is good; nevertheless it often, but not always, responds favourably to lime. Saline soils are unsuitable.

Desmodium intortum was apparently first introduced to Australia from where, similarly with *D. uncinatum*, it rapidly spread to a number of tropical and subtropical countries within the limits of about 30°N.

and 30°S. A large bulk of palatable herbage and reasonable seed yields made *D. intortum* popular; in preliminary and also in extensive trials it produced encouraging results in many areas and is now being introduced into farm practice, especially in Australia where seed is commercially produced.

Establishment. Seed is sown into a well-prepared seedbed, broadcast, or drilled in rows, and 45 cm-wide rows result in higher herbage yields than rows at twice this width. After sowing the field is rolled, but not harrowed to avoid burying the small seeds too deep. Seed rate is 1–2 kg/ha. In a trial with a *Setaria anceps* mixture, seed rates of 1.1, 3.3 and 9.9 kg/ha resulted in about equal herbage yields, although the number of seedlings was greater at the higher rates (Middleton, 1970). For establishment on rough ground aerial sowing has been used at reduced seed rates when a few legumes were sown together. *Desmodium intortum* is highly specific for *Rhizobium* inoculant, and strain GB627, which suits this species well, was developed in Australia (Diatloff, 1968). For better survival of *Rhizobium* bacteria seed is often pelleted, and pelleting with cellofas sticker was more successful than with lime; the latter resulted in a low percentage of effectively nodulated plants. In Tanzania successful establishment on unploughed land was obtained by Northwood & Macartney (1969), when the natural grass was killed with paraquat. Vegetative establishment from stem cuttings is also possible, but can hardly be done on any large scale.

Association with grasses. When grown alone *D. intortum* can, under favourable conditions, develop tall dense stands which are however used more for seed production than for herbage. In pure stands the intake of the herbage by the animals can be restricted by insufficient palatability and Stobbs (1971) reports that on pure *D. intortum* cows produced only 7.7 kg milk/day, the low intake of digestible energy being the main factor limiting milk yields. For herbage utilization *D. intortum* is normally grown with grasses; a number of grass species have been tried, mostly with good success, these species include: *Digitaria decumbens, Brachiaria mutica, B. brizantha, Paspalum dilatatum* and other species of *Paspalum, Pennisetum clandestinum, Eragrostis curvula* and some other grasses. *Desmodium intortum* proved to be strong enough to grow with *Tripsacum laxum, Pennisetum purpureum* and vigorous types of *Panicum maximum.* The presence of *D. intortum* increases the CP content of grass mixtures compared with pure grass, and usually also their yields. In trials in Hawaii, pure *Digitaria decumbens* yielded 3.78 t DM/ha and its mixtures with *D. intortum* 10.84–11.96 t. Crude protein yields of the mixtures were equal to those of pure grass fertilized with 440–525 kg N/ha (Whitney & Green, 1969). It was estimated in the same trial that *D. intortum* fixed 213–264 kg of atmospheric nitrogen/ha and Whitney (1970) reports over 300 kg of fixed nitrogen/ha.

Fertilizing, management Phosphorus and potassium fertilizers are essential for obtaining high yields of herbage, and in Swaziland a *Paspalum dilatatum*/*D. intortum* mixture yielded 3.86 t DM/ha, unfertilized, and 4.72 t when 25 kg P were applied (I'Ons, 1969). In Uganda the application of 67 kg P/ha increased herbage yields of *D. intortum* from 11.5 t DM/ha to 20.3 t (Wendt, 1971). In pot trials 1,350 kg superphosphate/ha were needed to achieve the maximum herbage production, only 35 per cent of which was obtained without P fertilizer (C. S. Andrew & M. F. Robyns, from Bryan, 1969). Lesser yield increases from K were usually obtained. The critical points of P and K concentration in the herbage were determined as 0.22 per cent for P and 0.78 per cent for K; concentrations below these levels indicated that fertilizing becomes necessary. Responses to Ca are erratic, but in Uganda (Wendt, 1971) 2.4 t lime/ha increased DM yields from 14.78 to 16.8 t/ha although there are also reports of no yield increases from applied lime and the same erratic responses refer to sulphur. Responses to molybdenum, especially in the establishment period, were reported, and Mo incorporation into pellets produced about the same effect as its soil application.

Heavy grazing and low or frequent cutting reduce the yields of *D. intortum* and its percentage in mixed herbage, and reductions of the *D. intortum* proportion in grass mixtures resulting from heavy grazing were reported in southern Queensland and from frequent cutting in Hawaii. In Uganda, cutting at a height of 20 cm and at intervals of 6–9 weeks was necessary for maintaining an adequate proportion of *D. intortum* grown in mixture with *Setaria anceps*; decreases in the intervals between cutting also decreased total yields of the mixture, whereas the reduction of the height of cutting to 8 cm decreased the proportion of *D. intortum* but did not affect total yields of herbage (Olsen, 1973), R. J. Jones (1973b) has however observed in Australia that low cutting decreases the yields of *D. intortum* only if combined with high cutting frequency and high yields were obtained when the herbage was cut low every 12 weeks.

Chemical composition, nutritive value. CP content of *D. intortum* herbage varies in different areas, under different conditions and at different stages of growth, and ranges from 9.9 per cent to over 24 per cent (from Bryan, 1969) or from 1.44 to 3.90 per cent N. Two samples of *D. intortum* herbage analysed in Kenya at full flowering contains 18.2 and 20.0 per cent CP, 2.5 and 2.3 per cent EE, 30.7 and 31.3 per cent CF, 39.5 and 38.0 per cent NFE, 1.01 and 0.86 per cent Ca and 0.28 per cent P (Dougall & Bogdan, 1966). Minute quantities of oestrogens in the herbage have also been reported. The content of tannin was high in comparison with other cultivated pasture legumes, and ranged from 1.5 to 3.7 per cent in the stem, and from 3.2 to 8.8 per cent in the leaves (from Bryan, 1969), reaching the level which can affect the rumen cellulose digestion; incidentally, slow decomposition of the leaves, compared with those of *Macroptilium atropurpureum*, noted by Vallis & Jones (1973), can possibly be ascribed to high contents of tannin. *Desmodium intortum* palatability is usually estimated as good or fair, but when grown alone its consumption may not be high enough to secure high animal productivity.

Herbage productivity and conservation. Herbage yields vary widely from some 3 t DM/ha to as much as 20 t in pure stands and yields of the same order were recorded for *D. intortum* fraction in grass mixtures, except that maximum recorded yields were about 3 t lower. Granier & Chatillon (1972) in Madagascar obtained 61.8 and 83.3 t fresh fodder/ha in the first and second years after the establishment, and in South Africa *D. intortum* yielded 8.96 and 11.84 t DM/ha in two localities (Mappledoram & Theron, 1972). Reasonable quality hay has been prepared and in experiments in Queensland it retained the protein content of the fresh material. Hay cured in the sun can lose a considerable proportion of carotene it contains, and Granier & Chatillon (1972) report that hay dried in the sun contained 99 mg carotene/kg DM, whereas hay cut at the same time but dried in shade contained 532 mg carotene/kg. Silage can be prepared, but ensiling is not easy as the ensiled material tends to be light and fluffy, and the concentration of soluble carbohydrates is not sufficiently high to secure lactic acid fermentation. In the laboratory, ensilage with the addition of 8 per cent molasses resulted however in lactic acid silage. Stable silage containing 2.4 per cent N in the DM was also made in Rhodesia (Catchpoole, 1970; Catchpoole & Henzell, 1971).

Diseases and pests. *Desmodium intortum* is reported to be only occasionally and mildly affected by little-leaf virus, although in Brazil low yields of herbage in small-plot trials were attributed to the little-leaf disease. High resistance to root-knot nematodes has been noted. In East Africa Meloid beetles eat the flowers and can reduce seed yields.

Animal production. In trials with young cattle grazed on *D. intortum* in Queensland (R. J. Jones, 1974) the highest annual liveweight gains reached 256 kg/ha at a stocking rate of 2.12 head/ha, or 120 kg/head. Lower stocking rates are however normally beneficial to this species, light grazing is normally recommended and one head/ha is perhaps a more reasonable stocking rate than that mentioned above; little experimental data are however available.

Reproduction. *Desmodium intortum* is a short-day plant sensitive to photoperiods, and in the southern hemisphere subtropics it flowers in May, June or July (from Bryan, 1969); plants grown in the tropics, and especially in the equatorian zone are, naturally, little affected. The plants are self compatible but cross-pollination also occurs, often to a high degree. Crosses between *D. intortum* and *D. uncinatum* have been obtained, the resulting hybrids showing leaf features characteristic of both species, and *D. intortum* has also been crossed with *D. sandwicense*. In a preliminary trial when *D. intortum* was crossed with allied species promising hybrids were obtained in which flowering lasted for a shorter priod of time than in parental plants of *D. intortum*, in which flowering can last 3–4 months; a short period of flowering is a useful feature from the point of view of seed harvesting and production. The only known cultivar is **Greenleaf** developed in Australia from three introductions. It was first released as cv. Beerwah but was later renamed and registered under the name of Greenleaf in 1964 (Barnard, 1972); this name is also used as the common name for the species.

Seed production. *Desmodium intortum* produces abundance of flowers but seed yields are moderate and can be estimated at about 100–120 kg/ha; the moderate yields can partly be explained by prolonged flowering resulting in a relatively small proportion of ripe seeds at any one time. There are about 600,000 seeds/kg.

Desmodium leiocarpum (Spreng.) G. Don. $2n = 22$

Erect or suberect perennial with somewhat woody stem bases. Leaves trifoliolate; leaflets ovate to oblong-elliptic. Flowers purple, in large paniculate inflorescences. Pods pubescent, with two to five joints which are ovate-oblong. Distributed naturally in south-eastern Brazil and Uruguay. In a 1-year trial in Ghana *D. leiocarpum* grown alone yielded 5.0 t DM/ha but was out-yielded by *Centrosema pubescens*. A mixture with *Andropogon gayanus* yielded 34.3 t/ha, less than pure *A. gayanus*, but a mixture of *Digitaria decumbens*/*D. leiocarpum* yielded 20.0 t DM/ha, outyielding *D. decumbens* grown alone (Tetteh, 1972).

Desmodium pabulare Hoehne. Feijao de boi (Brazil)

Shrubby perennial up to 3 m high but usually shorter, woody at the base. Leaves with one or three leaflets; leaflets elliptic or ovate about 2 cm long and 1–1.3 cm wide. Flowers in terminal racemes which are up to 50 cm long. Pods of five to seven joints. Distributed mainly in Brazil and can be abundant in relatively cool areas, mostly in grass-bush habitats. A valuable forage plant which gives good-quality hay, mainly in the Minas Gerais and the St Paulo areas of Brazil. Analysed in Brazil, it contained 20.2 per cent CP and 22.6 per cent CF, and the contents of digestible nutrients were 14.7 per cent CP, 31.6 per cent NFE, 12.4 per cent CF, 3.6 per cent EE and of digestible organic matter 62.4 per cent, indicating high nutritive value of herbage. Seed is abundantly produced and the species is considered promising for introduction into cultivation (Pio Corrêa & Pena, 1952). In pot trials by M. B. Jones *et al.* (1970) the plants contained 2.62 per cent N, which corresponds to 16.4 per cent CP, which was lower than in the other tropical legumes under trial.

Desmodium sandwicense E. Mey. $2n = 22$

Perennial suberect or trailing herb up to 60 cm in height. Leaves trifoliotate with ovate, slightly succulent leaflets almost glabrous and dark green on the upper surface but with a pale, somewhat obscure strip. Flowers in relatively dense racemes, cream and pink in colour. Pods with three to six joints covered with minute hooked hairs and adhering to clothing and animal hairs.

The name *D. sandwicense* has been used by agronomists and some botanists in the last two decades. It is nevertheless controversial and *D. pilosiusculum* DC. should perhaps be the valid name given by De Candolle in 1825. Moreover the species status of this taxon is not recognized by some taxonomists who consider it as a variety of *D. uncinatum*, *D. intortum* or *D. limense* Hook.

Desmodium sandwicense has been known in Hawaii for over 100 years but is apparently an old South American introduction (Fosberg, 1968) which has now spread to several tropical countries, being introduced for trials as a forage plant. Little seems to be known about the environmental requirements of *D. sandwicense* although Rotar *et al.* (1967) found it photoperiodically neutral, i.e. not affected by day length, and Whiteman (1968) determined the optimum day/night temperatures for its growth to be, similarly with some other species of *Desmodium*, 30/25°C. Hutton (1971) observed its poor performance on saline ground.

Desmodium sandwicense has been tried in a few countries only. In Australia, in a mixture with *Paspalum plicatulum* it lasted only 2 years (R. J. Jones *et al.*, 1967). In Florida, USA, *D. sandwicense* grown with *Digitaria decumbens* gave the lowest yields of the seven legumes under trial, yields of oven-dry matter being 7.38 t/ha with 5.24 t for grass alone (Kretschmer, 1970). It was more successful in East Africa, and in Tanzania, in small plots, a mixture with *Panicum coloratum* in which the grass : legume ratio was 1 : 1, yielded about 17 t DM/ha during the 15 months of growth, and good results were also obtained from a mixture with *Setaria anceps* (Naveh & Anderson, 1965). Satisfactory but inconsistent yields of herbage were also obtained in Kenya. *Desmodium sandwicense* can hybridize with *D. intortum*. Seed is well produced.

Desmodium tortuosum (Sw.) DC. (*D. purpureum* (Mill.) Fawcett & Rendle). Florida beggar weed $2n = 22$

Erect perennial or annual herb 1–2 m high with woody stem bases. Leaves trifoliolate; leaflets elliptic to ovate, 2–8 cm long and 1–3 cm wide, resticulate on the upper surface. Flowers variously coloured: white, yellow, mauve or purple, sometimes tinged with green, in large paniculate inflorescences. Standard 2.5–3.5 mm long. Pods with five to seven joints; joints almost orbicular, 3–5 mm long and 2.5–4 mm wide.

Seeds 1.5 mm in diameter. Occurs in the tropics and subtropics of America and has now spread throughout the Old World's tropics and naturalized in some areas. A short-day plant which responds to short photoperiods by reducing the period from sowing to flowering (Wang, 1961). It was grown, mostly experimentally, in several areas, sown at about 9–11 kg seed/ha in 90 cm (3 ft) rows. *Desmodium tortuosum* grows fast, covers the ground rapidly and competes well with weeds. It is recommended for hay rather than for grazing (Whyte *et al.*, 1953). It seems that at present its popularity is waning.

Desmodium triflorum (L.) DC. $2n = 22$

Prostrate creeping perennial or annual herb often rooting from nodes and forming dense mats. Leaves trifoliolate or occasionally simple. Leaflets small, obovate, 4–12 mm long and of almost the same width. Flowers blue, pink or purple in three to four pairs on short axillary racemes. Standard up to 5 mm long and 2–3 mm wide. Pods with up to five joints; joints almost square, 2.5–4 mm long and wide. Seed square to orbicular, 1.5–2 mm in diameter. Distributed throughout the world tropics, in grassland, at roadsides and in arable land as a weed, from sea level to 1,800 m alt. It forms attractive grazeable cover and pot trials in India indicate that *D. triflorum* is worth a further trial, under field conditions.

Desmodium uncinatum (Jacq.) DC. Spanish clover; Tick clover; Silverleaf desmodium (Fig. 42). $2n = 22$

Large perennial herb with trailing stems up to 5 m long densely covered with short, brownish, hooked hairs. Leaves trifoliolate with ovate leaflets which are 3–10 cm long and 2–5 cm wide, slightly hairy and dark green with a well-defined silvery spot or a strip along the midrib on the upper surface, and uniformly light-green on the lower surface. Flowers, in racemes, short at early flowering but extending later up to 50 cm. Corolla mauve to pink, 8–9 mm long. Pods falcate, articulated, breaking into four to eight one-seeded segments, densely covered with minute hooked hairs and adhering to clothing and animal hairs. Seed glossy, lens-shaped, light brown or olive green in colour.

Distributed naturally in Central and South America, from Mexico in the north to northern Argentina and Uruguay in the south, in the warm climate of lowlands up to 1,000 m alt. In controlled environment its optimum growth was observed at day/night temperatures of 30/25°C (Whiteman, 1968), but it can also tolerate frosts which may kill the tops, whereas the plant can recover rapidly and produce fresh growth from the basal parts. In Queensland, 56 per cent of plant survived a winter with frosts up to $-10°C$ (R. M. Jones, 1969). In its natural environment *D. uncinatum* grows in humid or sub-humid areas with an annual rainfall $>1,000$ mm and comparable, or slightly lower, rainfall is required for cultivation. It grows on a variety of soils including moderately acid soils

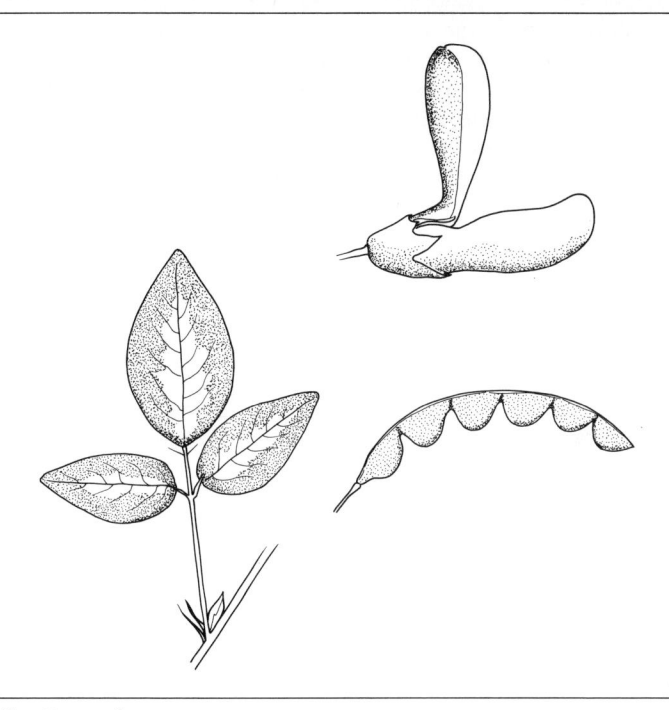

Fig. 42 *Desmodium uncinatum.*

with pH values of 5–5.5 but cannot tolerate soil salinity; well drained sandy soils or loams are preferred but the plants can withstand seasonal short-term waterlogging.

Desmodium uncinatum was first introduced to Australia from Brazil and then tried under cultivation in some other countries in the early 1950s, spread rapidly in the following years and is now grown on experimental or a farm scale in the majority of tropical and subtropical countries in a broad belt extending from 30°N. to 30°S. (Bryan, 1969). It was particularly well received in East Africa and Australia where it was extensively studied and often grown on a farm scale. *Desmodium uncinatum* is valued for its vigour, persistence, high herbage and seed yields and a relative ease of establishment.

Establishment. *Desmodium uncinatum* is usually grown from seed although vegative propagation from slips is also possible and has been done on a small scale when seed was difficult to obtain. Seed is drilled or broadcast into a well-prepared seedbed at a rate of 1 to 4, but mostly 2 or 3 kg/ha; the field is then rolled or harrowed. On rough ground seed can be sown from the air. In trials in Kenya *D. uncinatum* was successfully established in *Hyparrhenia* grassland variously treated or untouched and 15 months after the establishment the resulting sward yielded 10.9–11.5 t DM/ha, the herbage containing 58–70 per cent of *D. uncinatum*; high rates of applied P were necessary for this establishment. (Keya *et al.*, 1972). Hand-harvested seed contains a large proportion of hard seed, but even slight scarification by harvesting machinery induces good germination. Inoculation with *Rhizobium* culture is essential; *D. uncinatum* is not particularly specific in regard to inoculants, and can be inoculated even with *Rhozobium* cultures prepared from *D. intortum* or *Glycine wightii* nodules; a strain of *Rhizobium* culture, CB627, particularly suitable for *D. uncinatum*, was however developed in Australia (Diatloff, 1968). Seed coating or pelleting with neutral rock phosphate or bauxite is recommended; lime pelleting also produces good results if seed is sown soon after pelleting; the delay in sowing of lime-pelleted seed for 4 weeks resulted in severe decline of nodulation (Norris, 1972). P fertilizer is applied at sowing, the amounts used depend on the content of available P in the soil. The critical content of P in the herbage is about 0.23 per cent and when P content approaches or drops below this value, fertilizer P needs to be applied. The critical content of K is about 0.80 per cent. Ca requirements are intermediate compared with the majority of tropical legumes. *Desmodium uncinatum* is tolerant to low Cu content of the soil and moderately tolerant to Mn excess. It tolerates the presence of Al, small concentrations of which, 0.5 p.p.m., can even improve plant growth and 1 p.p.m. increases P content in the plant (Andrew *et al.*, 1973).

Association with grasses. *Desmodium uncinatum* grown in pure stands, the almost exclusive practice for seed production, will, under favourable conditions, form a dense mass of entangled stems supporting each other. For herbage it is normally grown in grass/legume mixtures. There seems to be no information of *D. uncinatum* herbage used for silage or hay, but it has been successfully established and utilized for grazing in mixtures with *Chloris gayana*, Nandi and Kazungula cultivars of *Setaria anceps*, *Panicum maximum*, *P. coloratum*, *Digitaria decumbens*, *Paspalum commersonii*, *P. dilatatum*, *P. notatum*, *P. plicatulum* and *Pennisetum clandestinum*. In Kenya, Uganda and Hawaii it was experimentally grown with tall and aggressive grasses such as *Pennisetum purpureum* and *Tripsacum laxum*. When left ungrazed, *D. uncinatum* often overgrows the grass and can suppress it to a considerable extent. It cannot however withstand close or intensive grazing, and grazing management can be an instrument for maintaining the desirable proportion of *D. uncinatum* in grass mixtures. *Desmodium uncinatum* is one of the most persistent tropical pasture legumes but sometimes survives only for a relatively short period of time as e.g. in trials with *Setaria anceps* mixtures by Young & Chippendale (1970).

Nitrogen fixation. The nitrogen-fixing ability of *D. uncinatum* inoculated with an effective strain of *Rhizobium* is well established, the amounts of fixed atmospheric N contributing to plant nutrition in grass mixtures ranging from 90 to 160 kg/ha, which is however less than the highest recorded or estimated amount of N fixed by *D. intortum*. Increases in N content due to the presence of *D. uncinatum* have been observed not only in total mixed herbage but in some cases also in grass components of mixtures, and Suttie & Moore (1966) have shown that the contents of CP in *Pennisetum purpureum* and *Tripsacum laxum* grown with *D. uncinatum* were increased by 18 and 51 per cent, respectively, compared with its contents in the two grasses grown alone. The ability to fix nitrogen was observed also in sand culture in which *D. uncinatum* fixed relatively small amounts of N; however, it transferred 1.66 per cent of the fixed N to the companion grass, and this percentage was higher than for the other three legumes under trial (Henzell, 1962).

Chemical composition. Crude protein content in the herbage usually ranges from 12 to 24 per cent of the DM and the plants analysed in Kenya at full flowering contained 23.0 per cent CP, 2.4 per cent EE, 28.0 per cent CF, 38.5 per cent NFE, 0.74 per cent Ca and 0.33 per cent P. Similarly with some other species of *Desmodium* the content of tannins is high, especially in the leaves where it was determined to be 6-7 per cent of the DM approaching the level above which plant palatability is reduced. Oestrogens have been found in *D. uncinatum* herbage but in such minute quantities that they can hardly affect the animals (Bryan, 1969). The herbage is usually evalued as fairly palatable to very palatable, but Barnard (1972) writes that 'palatability is not high and stock take a little time to get used to it'.

Herbage yields. Barnard (1972) estimates DM yields of Silver Leaf cultivar to be about 5 t/ha, and in Tanzania, in small plots and under a cutting regime, yields ranged from 2 to 21 t DM/ha for pure *D. uncinatum*, and 10-12 t for grass mixtures (Bryan, 1969). Febles & Padilla (1972) in Cuba obtained 14.7 t DM/ha from a *D. uncinatum*/*Panicum maximum* mixture and Mappledoram & Theron (1972) 8.63 t DM from pure stands of the legume grown in subtropical South Africa. Yields can however be much lower and under grazing utilization they were found to range from as little as 90 kg DM/ha to as high as the yields already mentioned, the average yields approaching the figure given by Barnard. Liveweight gains from grazed *D. uncinatum*/grass mixtures average some 340 kg/ha (from Bryan, 1969).

Pests and diseases. In Australia *D. uncinatum* was found susceptible to little-leaf virus disease if the plants were undernourished. In the southeastern United States anthracnose caused by *Colletotrichum dumatium* damaged the plants and they are also highly susceptible to root-knot nematodes. In East Africa the flowers are often damaged by Meloid beetles and this can reduce seed yields.

Reproduction. Desmodium uncinatum flowers profusely but only under short photoperiods, and in the southern hemisphere subtropics flowering usually takes place in April or May when the day length is reduced to less than 12 hours, but there is no definite time of flowering in equatorial Africa where the day length only slightly fluctuates around 12 hours. It is essentially a self-pollinated and self-fertilized plant but the flowers require tripping by insects which may bring pollen from other flowers and cross-pollination is fairly common (Hutton, 1960). *Desmodium uncinatum* is a diploid with the chromosome number $2n = 22$. Crosses between *D. uncinatum* and *D. intortum* and also *D. canum* have been obtained. There is only one recognized cultivar – **Silver leaf** – selected in Australia from a Brazilian introduction and registered in 1962 (Barnard, 1972). The name silver leaf is also often used to denote *D. uncinatum* as a species.

Seed production. Desmodium uncinatum is a prolific seeder and seed is harvested by hand or machinery. When hand-harvested the seed is stripped off the racemes and pressed into balls. The hooked hairs on pod segments make them stick together, but they can easily be separated by rubbing. Suttie & Ogada (1967) in Kenya harvested seed mechanically by mowing the ripening stands, leaving the cut plants in windrows for a week to dry and then threshing them with a combine harvester fitted with a pick-up arrangement. This mechanical harvesting gave 200–300 kg seed/ha, higher yields than hand-harvesting, and mechanically-harvested seed germinated better than that harvested by hand. For smooth flowing the pod-segments should be shelled and shelling does not present any great difficulty. In Australia seed yields are reported to be around 340 kg/ha and Strickland (1970), reviewing the position with the commercial seed production of tropical herbage plants, estimates that the average seed yields of *D. uncinatum* range from 200 to 350 kg/ha and states that yields up to 550 kg/ha have been recorded. In trials in Australia, Gibson & Humphreys (1973) have found that fertilizer nitrogen can increase seed yields by 21–31 per cent if about 17 kg N/ha/week is applied for a few weeks after the beginning of the flower-initiation stage. There are 200,000–240,000 seeds/kg. Basic information from Bryan (1969).

Glycine Willd.

Perennial twining or procumbent herbs (except *G. max* which is an annual pulse crop). Leaves with three leaflets. Flowers small. Pods dehiscent, linear or oblong. A small genus of about ten species of predominantly tropical and subtropical areas of the Old World. *Glycine wightii* is a pasture legume which has already been cultivated on a farm scale and its importance is increasing. *Glycine max*, although essentially a pulse crop, is also grown for fodder, usually in mixtures with large annual grasses.

Glycine max (L.) Merrill (*G. soja* (L.) Sieb. & Zucc.; *G. hispida* (Moench Maxim.; *Soja max* (L.) Piper). Soyabean; Soja

Erect, semi-prostrate or twining annual 20–180 cm high, grey or brownish pubescent. Leaves with three, or occasionally five, leaflets which are 3–10 cm long and 2–6 cm wide. Flowers three to over 15 in short axillary racemes, small, with the standard about 5 mm long, white or purplish. All stamens usually united or sometimes the upper stamen free. Pods hairy, of various colour, slightly curved and somewhat flattened, 3–7 cm long and about 1 cm wide, with one to five, but mostly two to four, seeds. Seed pale-yellow, green, brown or black, often mottled and varying in size.

Glycine max or cultivated soyabean originates from temperate and subtropical areas of the Far East and has been cultivated in China and neighbouring countries from times immemorial. It was introduced to Europe and America in the last century and from the early years of the present century it spread rapidly throughout the world tropics, subtropics and temperate areas and was first grown predominantly for fodder but later mainly as a pulse crop for its seed rich in oil and proteins. In 1966 it was grown on over 25 million ha, practically in every country between 50°N. and 40°S.

Environment. *Glycine max* is a short-day plant. It requires warm but not excessively hot summers for its optimum growth and can withstand light frosts, and in the tropics it can grow from sea level to about 1,800 m alt. The optimum growth has been observed under substantial annual rainfall of 1,200–1,500 mm although some cultivars can be of medium drought resistance. The majority of soil types are suitable but *G. max* thrives best on moist but well-drained and fertile loams and responds well to P and K fertilizers, and, in drier areas, to irrigation.

Cultivation and uses. Being an annual, *G. max* is grown in crop rotations, sometimes in rice fields for soil improvement and fodder, between two crops of rice. It is often grown in pure stands but more usual utilization is in mixtures with maize, sorghum or Sudan grass. The herbage is seldom grazed but cut and fed green or used for making hay or silage. The ripe plants can also be used as roughage after the pods have been removed.

The establishment is by seed which is fairly easily produced. Seed is sown broadcast or in rows some 50 cm apart (or at wider spacing when grown as pulse crop) and sowing rates range from 30 to 65 kg/ha when grown for forage. An application of 35–70 kg P_2O_5 and 70–90 kg K_2O is usually recommended (Purseglove, 1968) and in India f.y.m. is applied at about 25 t/ha together with a small dressing of fertilizer N, about 20 kg/ha (Narayanan & Dabadghao, 1972). Seed should be inoculated unless previously inoculated soyabean has been recently grown in the same field. Soyabean requires a specific inoculant, *Rhizobium japonicum*, which has a number of strains and work is in progress on the selection of *Rhizobium* strains and cultivars of soyabean symbiotically compatible. Inoculated seed can be pelleted either with lime or with neutral rock phosphate. However, Herridge & Roughley (1974) have found that pelleting may not be necessary if a sticking agent, e.g. gum arabic slurry, is used. In a trial in India (Yadav & Vyas, 1971) soyabean *Rhizobium* tolerated the presence of NaCl and $MgSO_4$ in the soil in concentrations of up to 3 per cent but was less tolerant to $MgCl_2$. Similarly with other legumes under trial, pH 3.0–4.0 and below inhibited *Rhizobium* growth which, however, tolerated pH as high as 10.

Seedlings emerge 5–7 days after sowing and they require some protection from weeds at early stages of growth; later they grow fast and the plants can be harvested for fodder in 45 to 100 days after sowing depending on the cultivar and the climate. The crop is cut for hay or other uses usually at the stage of developing pods. Herbage is then dried in windrows and stacked. Yields of fresh herbage vary and Crowder (1960) gives 30 t/ha as an average yield for Colombia; in India the best selected cultivars yield 30–38 t and some 20–25 t/ha have been reported from Brazil. In the USA green fodder yields average 20–25 t/ha (9–10 t/ac) and yields of hay range from 2.5 to 12.5 t (Purseglove, 1968) but lower yields have also been reported.

Pests and diseases. Soyabean plants are relatively free from pests and diseases; those of some importance are a few species of *Meloidogyne* – root-knot nematodes, and of diseases, bacterial blight (*Pseudomonas glycinae*), bacterial pustule of leaves (*Xanthomonas phaseoli* var. *sojensis*) and pod and stem blight (*Diaporthe phaeolorum*, var. *batatis*) (Purseglove, 1968).

Nutritive value. In India (Johri *et al.*, 1971), *G. max* as fed to animals contained 15.7 per cent CP, 2.5 per cent EE, 29.5 per cent CF, 40.5 per cent NFE, 1.46 per cent Ca and 0.34 per cent P, which shows its high nutritive value and sheep consumed 2.6 kg DM/100 kg body weight. Straw contained 6.8 per cent CP, 1.2 per cent EE, 41.2 per cent CF, 39.4 per cent NFE, 1.69 per cent Ca and only 0.16 per cent P and was of much lower feeding value than fresh herbage; its consumption by sheep was however almost equal to that of fresh herbage. In trials with sheep in Brazil (Melotti & Velloso, 1970–1) DCP content of soyabean plants harvested 95 days after sowing was 7.0 per cent and the digestibility of DM 56.5 per cent, of CP 69.1 per cent, NFE 63.3 per cent and EE 40.3 per cent.

Reproduction, variability. Soyabean flowers open in the morning and pollen can easily reach the stigma just before the flowers open. After pollination, the open flowers are visited by insects, mostly small bees, and cross-pollination is not excluded but is of the order of 1 per cent or less (Purseglove, 1968). Artificial cross-pollination is not easy because effective emasculation of small flowers is difficult to achieve; nevertheless artificial hybridization has been used, mainly in the USA, for the development of superior cultivars. The number of somatic ($2n$) chromosomes is 40 and the plants are diploids. Tetraploids, with $2n = 80$, have been obtained under the effect of colchicine; these tetraploids were of no immediate practical value but can perhaps be utilized in further breeding work.

Seed production presents no particular difficulties, especially in the tropics and subtropics where the days are short. Fodder varieties usually give lower yields of seed than grain cultivars, but from the point of view of establishment this is compensated by smaller seeds, i.e. the larger numbers of seeds per kg. The number of seeds per kg, as given by Metcalfe (1973), is about 11,000, but it varies a great deal depending on the cultivar.

Glycine max is a very variable species and innumerable cultivars have been recognized in the Far East, most of them grown for human consumption. There is, however, a number of fodder cultivars and these are fine-stemmed and often twining. Seed of fodder varieties is usually black and smaller than in grain types.

The plants are sensitive to photoperiods and some cultivars do not flower at all if the day length exceeds 14 hours (or the length of the dark period is less than 10 hours) and in the USA the cultivars are zoned to fit the length of the day. This feature is less important in the tropics, and partly in subtropics, where the days are short; moreover the photoperiods affect more grain than fodder utilization. Fodder cultivars grown in the south of the USA have shown good results in other warm countries, including the tropics, and those grown in the gulf states of USA include **Avoyelles**, **Gatan** and **Yelnando** (Whyte *et al.*, 1953). Cv. **Santa Maria** is a good Brazilian fodder and hay cultivar.

Glycine wightii (R. Grah. ex Wight & Arn.) Verdc. (*G. javanica* auctt., *G. petitiana* Hermann pro parte). Glycine; Soja perene. (Fig. 43).

This species had been known by the agriculturists and pastoralists, and also by botanists, as *G. javanica* until Verdcourt (1966) found that the type specimen of Linnaeus represents actually not *Glycine* but a species of *Pueraria* and suggested the name *G. wightii*.

Perennial climbing, trailing or procumbent herb. Stems 60–450 cm long. Leaflets 1.5–15 cm long and 1–12 cm wide, ovate to elliptic; lateral leaflets somewhat oblique, sometimes lobed, glabrous or hairy on both sides. Inflorescence usually racemose of 20 to 150 flowers. Flowers 4.5–11 mm long; standard white, mauve-blue or white with a mauve

Fig. 43 *Glycine wightii* (from Edwards & Bogdan, 1951).

spot. Pods linear-oblong, straight or slightly curved at the apex, 1.5–3.5 cm long and 2.5–5 mm wide, glabrous to densely hairy, often with rusty or brown hairs. Seeds 2–4 mm long.

Distributed throughout tropical and subtropical Africa, mainly in its eastern parts, Arabia and India. A polymorphic species and a number of intraspecific forms have been distinguished and described. Hermann (1962) revised the genus and suggested a system of *G. wightii* (as *G. javanica*) classification. Verdcourt (1966) remodelled and simplified the system, and he recognizes the following intraspecific taxa:

Ssp. *wightii* with small flowers, 4.5–7.5 mm long, and hairy pods; var. *longicauda* (Schweinf.) Verdc. of this subspecies has lax inflorescences and occurs in north-east tropical Africa, whereas var. *wightii* has congested inflorescences and occurs in India.

Ssp. *pseudojavanica* (Taub.) Verdc. has small flowers and glabrous pods.

Ssp. *petitiana* (A. Rich.) Verdc. Flowers 7.5–11 mm long; pods densely covered with brown hairs. In this subspecies var. *petitiana* has slightly larger flowers, 8–11 mm long, with the corolla blue to mauve in colour, whereas var. *maernsii* (De Wild.) Verdc. has flowers 7.5–8.5

mm long, the corolla is white or tinged with blue or mauve, and the standard has a blue or mauve blot.

The taxa mentioned have different geographical distribution. The cultivated types belong mostly to ssp. *wightii* var. *longicauda* and subsp. *petitiana* var. *maernsii*, and perhaps also to crosses between them.

In the areas of its natural distribution *G. wightii* occurs in grasslands and in bush from sea level to over 2,000 m alt and is sometimes cultivated, mostly experimentally, as a pasture plant; at the early stages of introduction it was used mainly as a green-cover crop for soil conservation. It received some attention in Kenya and Rhodesia although in Kenya the attention was limited to experimental studies. Introduced to a number of tropical and subtropical countries *G. wightii* has been met with good success in Brazil, and to a lesser extent in other Latin America countries, but especially in Australia where considerable research and practical work has been done and the cultivars selected in Queensland have now spread to America and Africa. *Glycine wightii* is valued for its fair drought resistance, good adaptation to various soil conditions, reasonable yields of herbage, an ability to mixing well with grasses and a reasonable longevity; the main negative feature is slow initial growth and slow nodulation in the establishment period.

Environment. *Glycine wightii* is reasonably drought resistant and can grow under an annual rainfall from 700 mm upwards and withstand not very long seasonal droughts. The optimum rainfall is of the order of 1,100–1,200 mm (Toutain, 1973) but the plants do not thrive under a wetter climate and Grof & Harding (1970) report that in parts of north-eastern Australia receiving over 3,000 mm p.a. the performance of *G. wightii* was much poorer than of other perennial legumes. The optimum growth was observed at day/night temperatures around 30°/25°C but the plants can also survive some frost and R. M. Jones (1969) recorded about 50 per cent survival in a winter with the lowest temperature reaching $-10°C$. Under cool conditions of subtropical winters the plants cease to grow when the air temperature drops below 10°–15°C. For satisfactory growth *G. wightii* requires fertile soil and prefers heavy rather than light soils; it can grow on black seasonally waterlogged clays if the soil is waterlogged for not very long periods. The level of tolerance to manganese and aluminium is low and the plants require the presence of molybdenum perhaps even more than other legumes, especially during the establishment period. Soil salinity tolerance is moderate and fertilizer N increases the tolerance and a gradual increase in soil salinity is less harmful than a rapid increase in the content of NaCl.

Establishment. Establishment is effected by seed only and seed rates vary from 5 to 10 kg/ha (Delgadillo & Rossiter, 1972). Seed can be sown in rows, to a depth of about 1–3 cm at the distance between the rows of 40–50 cm or 90–100 cm as indicated by Toutain (1973). Sowing in furrows is sometimes recommended and in a trial in Brazil Lovadini (1971) obtained 27.2 t green fodder/ha from furrow planting, 23.7 t from flat row planting and 17.7 t from broadcast. *Glycine wightii* can be sown on unploughed land when paraquat has been applied (Northwood & Macartney, 1969) and it has also been oversown to natural grasslands, either drilled in or broadcast which in some cases produced better results than drilling, indicating a possibility of sowing from the air. Seed scattering on burnt Brigalow bush in eastern Australia resulted however in poor establishment. A large proportion of hard seed has often been observed and treatment with concentrated sulphuric acid, hot water or mechanical scarification are recommended.

Nodulation. *Glycine wightii* is moderately selective in its *Rhizobium* requirements although data on its nodulation and inoculation are controversial. Some authors obtained satisfactory results without inoculation, natural nodulation being effected from local *Rhizobium* in the soil; some obtained little nodulation from reputed *Rhizobium* strains and it appears that a combination of suitable variety or cultivar of the host plant and of certain *Rhizobium* type of the cowpea group is essential. Selection of suitable combinations is complicated by the fact discovered by Diatloff & Ferguson (1970) that different combinations can be optimal for (a) the speed of initial nodulation, (b) nodule numbers and (c) symbiotic efficiency. The work on selecting compatible strains is however in progress and encouraging results have been achieved, although effective inoculation seems to remain more difficult than in the majority of other cultivated legumes. The nodules appear 24–32 days after sowing compared with 9–12 days for *Macroptilium atropurpureum*, they are small and globose in shape and initially only a few, two to five per plant (Diatloff & Ferguson, 1970). Nodulation is often uneven and Whiteman (1972) observed that only 42 per cent of inoculated plants formed nodules. It was also noted that tetraploids nodulate in general better than diploids.

Nitrogen fixation is slow to begin and the seedlings can be weak and discoloured because of N deficiency, and it is sometimes recommended to apply small amounts of fertilizer N to encourage early development of young plants. At later stages, however, *G. wightii* fixes N well, often at the same rate as the best N-fixing legumes. High soil temperature soon after sowing, when the seeds absorb water can be harmful, but when the radical has appeared high temperatures become less harmful. Pelleting usually improves nodulation and pelleting with phosphate gives more consistent, positive results than pelleting with lime and there is conflicting evidence about the necessity of lime pelleting; addition of Mo usually improves nodulation. Nodulation and N fixation are affected by soil salinity and in sand culture trials (Wilson & Norris, 1970) the nodules which developed before NaCl was applied were not affected but no new nodules were formed; after NaCl has been removed from the substrate the formation of nodules was resumed. The evidence given indicates that the establishment can be difficult and slow and is the most vulnerable period in *G. wightii* cultivation; however, once established the plants become strong and competitive.

Grass mixtures. *Glycine wightii* forms well-balanced and productive mixtures with grasses. It has been grown on a fairly large scale with *Panicum maximum* cv. Petri and *Chloris gayana*, and also tried with a number of other species: *Cenchrus ciliaris, Setaria anceps, Melinis minutiflora, Digitaria decumbens* and *Pennisetum clandestinum*. Because of the slow initial growth it mixes better with the slow, rather than fast established grasses, as, e.g., with cv. Petri of *Panicum maximum*. When grown in mixtures with rapidly established *Chloris gayana* some authors recommend sowing the legume 2 weeks earlier than the grass. After *G. wightii* has been well established it is more persistent in a grass mixture than most of the other legumes even when subjected to frequent low grazing or cutting to 5–7 cm above the ground level. It is also more vigorous than other legumes in uncut herbage swards. Similarly with other legumes, yields of herbage increase with the decrease in cutting or grazing frequency from 4 to 12 weeks although Delgadillo & Rossiter (1972) recommend rotational grazing with the intervals of 5 to 6 weeks between grazings.

Fertilizing. It is normally recommended to apply 150 to 250 kg superphosphate/ha and usually no other fertilizers are needed. Responses to fertilizer P are nearly always high but responses to K erratic and have been recorded mainly on soils particulaly deficient in this nutrient. Responses to lime have also been erratic although in Brazil the application of about 1,000 kg/ha are recommended, whereas further applications may result in yield decreases. Small rates of N are sometimes recommended for young swards but in the years following the establishment fertilizer N can reduce the proportion of legume in the sward.

Utilization. Mixtures with grasses are normally used for grazing but hay is also prepared. In hay, the protein content and the nutritive value are usually reduced compared with fresh herbage mainly perhaps because of the loss of leaves, and in trials by Melotti *et al.* (1969) the percentage of leaf was reduced from 41.3 per cent in fresh herbage to 33.1 per cent in hay, CP content from 15.7 to 13.3 per cent and its digestibility from 70.5 to 66.9 per cent; TDN content was also lower in hay: 55.3 per cent compared with 57.3 per cent in fresh herbage. Silage is seldom prepared but Barker & Kyneur (1962) report that good silage was obtained from a mixture with *Panicum maximum*. Fermentation was good, mainly of lactic acid type. pH was of the order of 4–4.5 and the silage was readily accepted by cattle.

Pests and diseases. In Kenya seeds are eaten by the larvae of *Bruchus* beetles which reduce seed yields.

Nutritive value. Delgadillo & Rossiter (1972) give CP content of *G. wightii* herbage as 12.10 per cent, EE 1.48 per cent, CF 42.60 per cent and NFE 33.47 per cent. However, in the majority of other analyses CP content is usually higher and ranges from 11 to 20 per cent and occasionally even to 30 per cent; its content can decrease with plant age as was shown by Lima & Souto (1972), who reported that in hay cut at the preflowering stage, at early flowering and full flowering CP content decreased from 19.6 to 14.21 per cent and to 11.26 per cent, respectively. The decrease however does not always occur and little if any reduction in CP content with the plant age was also reported. CF content is usually lower than that given by Delgadillo & Rossiter and NFE content higher and approaches 40 per cent. The contents of Ca and P, as determined in Kenya, were 1.50 and 0.29 per cent, respectively (Dougall *et al.*, 1964).

The digestibility of CP is usually high, of the order of 65–80 per cent, and no significant differences between the digestibility of CP and other nutrients in old and young plants have usually been observed, and the percentage of TDN can sometimes be even higher in older than younger plants. *Glycine wightii* is highly palatable and is readily grazed.

Herbage yields. Yields of *G. wightii* herbage do not differ much from those of other high-grade pasture legumes except in the year of establishment and especially during the first 3–4 months of growth, and range from 3 to 8 t DM/ha in well-managed stands. In Brazil (Buller *et al.*, 1970) DM yields of pure *G. wightii* ranged from 1.30 to 5.88 t/ha depending on the cultivar and fertilizer rates.

Animal production. Admixture of *G. wightii* to grass increases the animal production and in Brazil (Buller *et al.*, 1970) its mixture with Pangola grass (*Digitaria decumbens*) increased the liveweight gains of steers from 231 g/day/head when grazed on unfertilized grass to 316 g when grazed on the mixture; the gains on grass given 100 kg N/ha were 294 g/day. A mixture with *Stylosanthes guianensis* gave, however, a slightly greater liveweight increase than that with *G. wightii*. Increases in liveweight gains were also obtained at higher levels of pasture and animal productivity, and good intake of glycine herbage by the animals was reported. In a trial in Australia by Park & Minson (1972) meat of lambs grazed for 6 weeks on pure *G. wightii* acquired occasionally a characteristic and objectionable flavour; its intensity was reduced by further grazing up to 18 weeks.

Reproduction, flowering. Flowering in *G. wightii* begins 46 to 130 days from seedling emergence depending on cultivar but mainly on the day length (Wutoh et al., 1968b). It is a short-day species and at the daylength limits of 8 to 11 hours flowering begins at 45–58 days depending on cultivar. At 12-, 13- and 14-hour days the preflowering period increases gradually and at 16- to 18-hour days reaches its maximum of 130 days and some varieties or cultivars do not flower at all under long days. Temperature can also have some effect on the time of flowering. *Glycine wightii* is in general a cleistogamous plant, the pollination taking place in closed flowers before they open and before the insects, mostly small wild bees, visit the flowers (Hutton, 1960). Some variation in seed colouration was, however, observed in progenies of single plants of certain varieties which showed that some cross-pollination may occur. Hybridization was obtained by Wutoh et al. (1968a) in emasculated flowers and heterosis was observed in F_1 (the first hybrid generation) expressed in average herbage yield increases of 65 per cent over those of parental plants, but was reduced to 33 per cent in F_2. The progenies resulted from hybridization differed from parental plants in a number of characters.

Two ploidy levels occur: diploids with $2n$ chromosome numbers $= 22$ and tetraploids with $2n = 44$, both distributed throughout the whole area of *G. wightii* natural distribution although the diploids occur predominantly in true tropical areas of Tanzania, Kenya and Rhodesia, whereas the tetraploids are found mainly in subtropical areas or at high altitudes in the tropics (Edye et al., 1970). Chromosome number $2n = 20$ has also been reported.

Seed production. Seed is usually well produced and Verhoeven (1958) in Australia developed a useful method of seed harvesting with a 5-foot mower equipped with a protective cover under the drive-shaft for preventing the vines from wrapping around the shaft. The cut herbage is then dried in windrows and threshed, and some 250 kg seed/ha were obtained. Neme (1958) in Brazil quoted still higher seed yields, up to 10 kg/100 m² (1,000 kg/ha) but such high yields are uncommon and nothing approaching them was recorded in other trials. Delgadillo & Rossiter (1972) in Bolivia give seed yields of some 115 kg/ha in a single cut, but in practice seed yields can be still lower. In the main cultivated varieties the number of seeds/kg ranges from 130,000 to 170,000. A large proportion of hard seeds, sometimes up to 80–90 per cent, has often been observed.

Variability. Although essentially self-pollinated, *G. wightii* is very variable and its main subspecies and varieties have been mentioned earlier. Apart from the purely taxonomic classification based on the morphology and geography, Edye et al. (1970) have divided *G. wightii* into six groups on the basis of a combination of morphological, geographical, cytological, physiological and agricultural characteristics:

1. Vigorous, early maturing, white-pubescent plants. Good seeders. Poor frost resistance. $2n = 22$.
2. Early maturing plants of medium to poor vigour. Leaflets small, pods rounded. Good frost resistance. $2n = 44$ and occasionally 22.
3. Plants of medium vigour and medium frost resistance and late maturing. Brown hairs present. Flowers large. $2n = 44$ and occasionally 22.
4. Late maturing vigorous plants of poor frost resistance. Leaflets small, cordate, pointed. Brown hairs present. $2n = 22$.
5. Very vigorous, medium-late maturing plants of poor frost resistance. Leaflets large, pointed. $2n = 22$.
6. Vigorous, very early maturing plants of medium frost resistance. Seed yields poor. Leaflets and stems with brown hairs. $2n = 22$.

There are however some introductions which do not fit any of the six groups.

Cultivars. A number of selected introductions, some of them of the cultivar status, have been tried in Australia, Brazil, Rhodesia, Kenya and other countries. The best-known cultivars extensively tried in a number of areas are those selected in Australia and released for general use in 1962:

Tinaroo. Originates from Kenya. Diploid of group 5. A medium-late maturing type, very productive, much spreading and stoloniferous but with slow initial growth. Poor frost resistance.

Cooper. Originates from Tanzania. Diploid of group 1. An early-maturing type, less stoloniferous than Tinaroo. Good seeder and relatively drought resistant but frost resistance poor.

Clarence. Originates from South Africa. Tetraploid of group 2. Less stoloniferous than the two other cultivars. Relatively fast initial growth; good frost resistance.

Indigofera L. Indigo

Annual or perennial herbs or shrubs. Leaves usually imparipinnate although some species have only three or even one leaflet. Flowers usually in axillary racemes which in some species may branch and form panicles. Corolla red or pink, mostly small. Pods narrow, straight or curved, usually dehiscent, with two to many seeds. A very large genus of some 700 species distributed throughout the world tropics and subtropics. A number of species are grazed or browsed but some are toxic or suspected to be toxic to the animals, especially to horses, and none are of any great importance as sources of fodder although *I. spicata* and *I. hirsuta* are cultivated, mostly in mixtures with fodder or pasture grasses. A number of other species, e.g. *I. tirta* L.f. (also known as *I. subulata* Poir, or *I. retroflexa* Baill) and *I. schimperi* Jaub. & Spach (*I. tettensis*

Klotzsch) in East Africa, and *I. cordifolia* Heyne ex Roth. and *I. medicaginea* Bak. in India, and many others in various countries are regarded as useful components of natural grazing and some are suggested for introductory trials.

Indigofera cordifolia Heyne ex Roth. $2n = 16$

Prostrate hairy annual. Leaves simple, ovate. Flowers red in sessile heads. Pods oblong. Distributed in India and other parts of southern Asia. In India it is frequent and often abundant and contributes to natural grazing. In digestibility trials with sheep it contained 7.8 per cent DCP and 57.5 per cent TDN (Nath *et al.*, 1971).

Indigofera enneaphylla L. $2n = 16$

Prostrate perennial herb forming dense low growth. Leaves with seven to nine leaflets. Flowers small, red, in dense heads. Pods small, about 5 mm long and 2 mm wide. Occurs in India and also in other parts of southern Asia and in tropical Australia, and can be frequent in grasslands. In digestibility trials with sheep it contained 10.7 per cent DCP and 55.5 per cent TDN (Nath *et al.*, 1971). No toxic effect on sheep has been observed although it was reported to be toxic to horses.

Indigofera hirsuta L. Hairy indigo. $2n = 16$

Erect or spreading annual up to 1.5 m high but mostly much smaller. Leaves imparipinnate, up to 12 cm long, with five to nine leaflets; leaflets 2.5–4 cm long, pilose on both sides. Flowers numerous, in dense long racemes; corolla red, 5 mm long, the standard with white hairs on the back. Pods straight, 15–20 mm long and 2 mm wide, covered with mixed white and brown hairs. Seeds six to nine per pod, angular, pitted.

Distributed in tropical Africa, Madagascar, southern Asia and northern Australia, in moderately humid areas, in thin bush, wasteland and in arable land as a weed. Introduced to South America, and to Florida and Texas in the USA where it is cultivated to a limited extent, pure or in mixture with pasture grasses. In trials by Kretschmer (1970) in Florida, a Pangola grass (*Digitaria decumbens*) mixture yielded 4.88 t oven-dry material/ha, whereas pure Pangola grass yielded 3.31 t. Mixtures with *Setaria anceps* yielded 4.52 t as against 2.24 t for pure grass, but there were no yield increases for mixed stands of bahia grass (*Paspalum notatum*); grass mixtures with other legumes in the same trials greatly outyielded those of *I. hirsuta*. An analysis of *I. hirsuta* herbage (Dougall & Bogdan, 1966) gave 23.8 per cent CP, 2.0 per cent EE, 15.2 per cent CF, 46.8 per cent NFE, 1.88 per cent Ca and 0.37 per cent P, showing its excellent quality; however, silage prepared of *I. hirsuta* satisfied only the maintenance requirements of stock and was not well eaten (Catchpoole & Henzell, 1971). *Indigofera hirsuta* is a prolific seeder. There are 440,000 seeds/kg.

Indigofera spicata Forsk. (*I. hendecaphylla* Jacq.; *I. endecaphylla* Jacq.). Trailing indigo. (Fig. 44)

Prostrate or ascending perennial herb with flattened stems. Leaves imparipinnate up to 5 cm long, with 5–11 leaflets. Leaflets up to 3 cm long, obovate or oblong, acute. Flowers in dense racemes; corolla 4 mm long, red or pink. Pods reflected, straight or slightly curved, 1–2 cm long and 1.5–2 mm wide, five to eight seeded. Seeds suborbicular, yellow, smooth. It varies greatly and several forms have been recognized, but Gillett (1971) is of the opinion that 'all forms grade into one another and it seems advisable to await genetical and transplanting data before recognizing infraspecific taxa'.

Distributed in tropical Africa, South Africa, Madagascar, India, Sri Lanka, southern and south-eastern Asia and tropical America. Occurs in grasslands, rocky situations, etc., but mostly on roadsides, wasteland and other disturbed habitats. *Indigofera spicata* is relatively drought resistant and can grow under moderate annual rainfall and on relatively poor soil. The plants are moderately specific in their *Rhizobium* requirements and can be inoculated by a few strains of cowpea-type

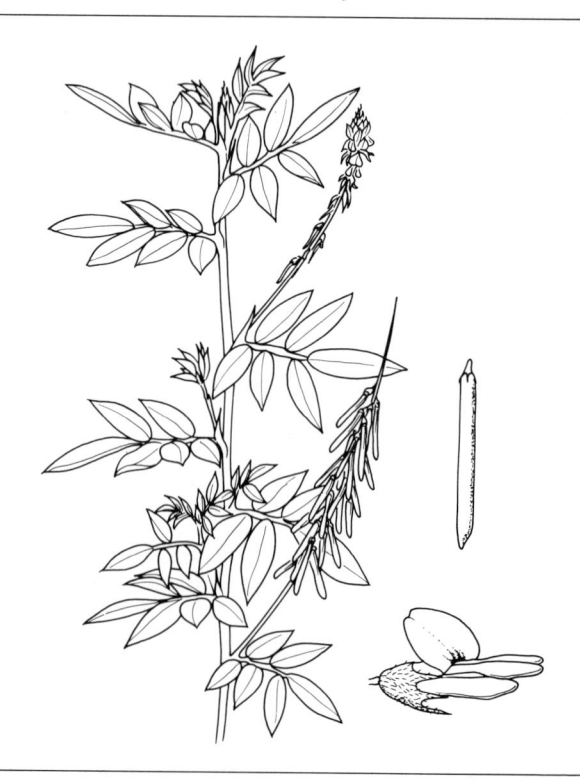

Fig. 44 *Indigofera spicata* (from Edwards & Bogdan, 1951).

Rhizobium (Obaton, 1974). In pot trials the rate of nitrogen fixation was lower than those of temperate species but higher than those of the other five tropical legumes tested, although the transfer of fixed nitrogen to the companion grass was low, 0.59 per cent of the amount of nitrogen fixed (Henzell, 1962).

Introduced into cultivation, *I. spicata* yielded well, formed balanced mixtures with grasses and was reasonably well grazed; the use in cultivation is however restricted by its toxicity to the animals expressed in liver degeneration in cows and sheep, and especially in horses, and pregnant animals can abhort. It was found that the toxic agent is 1–2–amino–6–amidinohexanoic acid which was named indospicine (Hutton, 1970). This discovery made it possible to breed for the absence of the specific amino acid, i.e. for non-toxic herbage. If the presence of indospicine is not taken into the account, the quality of herbage is usually high.

Flowers of *I. spicata* open in the morning and flowering is completed before 11 a.m.; a considerable number of flowers are tripped by insects. Pollination takes place about 4 hours before the flowers open; the plants can thus be considered cleistogamous and no intra-variety variation has been observed (Hutton, 1960). Diploids with the chromosome number $2n = 16$ and tetraploids with $2n = 32$ have been recorded. Seed production is moderate to poor and the establishment is sometimes effected by cuttings which root fairly easily (Whyte *et al.*, 1953). There are about 450,000 seeds/kg.

Lablab Adans.

Only one species; it is allied to *Dolichos*.

Lablab purpureus (L.) Sweet (*Dolichos lablab* L., *D. purpureus* L., *Lablab niger* Medic., *L. vulgaris* Savi). Lablab bean; Hyacinth bean; Bonavista bean; Lubia (Sudan). (Fig. 45). $2n = 20, 22, 24$

Climbing or erect annual or short-lived perennial up to 1 m high and with longer stems in climbing types. Leaves trifoliolate; leaflets large, 3–15 cm long and 1.5–14 cm wide. Inflorescences axillary, 4–20 cm long on peduncles 2–40 cm long. Standard purple or white, almost orbicular in shape and 1.2–1.6 cm in diameter; keel bent in a right angle. Pods of different shape depending on subspecies. Seeds large, white, red, brown or black with a long white aril around a third of the seed.

Verdcourt (1971) distinguishes three subspecies:

1. Ssp. *purpureus*. Pods scimitar-shaped, 4–10 cm long and 2–4 cm wide, with two to five seeds. Cultivated throughout the tropics as a pulse crop.
2. Ssp. *benghalensis* (Jacq.) Verdc. Pods linear-oblong to oblong, falcate, 3.5–14 cm long and 1.2–4 cm wide. A pulse crop of Asiatic

origin but also grown in Africa; it is also used as a forage crop and a very vigorous form which contained 33.5 per cent CP in the leaves was under observation as a forage plant in Kenya.
3. Ssp. *uncinatus* (A. Rich.) Verdc. (*Lablab uncinatus* A. Rich.). Grows wild throughout tropical Africa. Pods and seeds as in ssp. *purpureus* but smaller. The plants are readily eaten by stock.

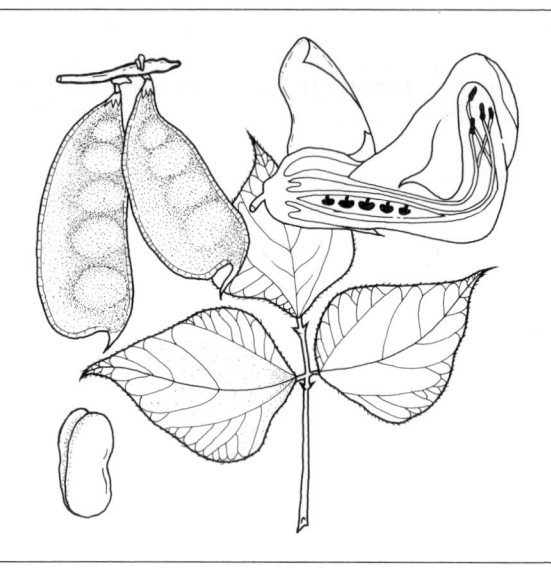

Fig. 45   *Lablab purpureus* (from Purseglove, 1968).

Several types were tried as fodder plants in Kenya, Rhodesia and the Sudan where it is one of the main fodder crops, and some were introduced to Brazil, the Philippines and Australia. A white-flowering variety of ssp. *purpureus* was grown on a farm scale in the Rongai area of Kenya and the herbage was mixed with Sudan grass and ensiled. Introduced to Australian subtropics, this late-flowering cultivar as more vigorous and productive than the majority of other introductions and was registered as cv. **Rongai** (Barnard, 1972). It was later spread to tropical parts of Australia. The data given below are mostly based on trials with the Rongai cultivar, but there can also be other types under experimental and farm-scale cultivation for fodder and grazing.

In Australia cv. Rongai grows well under humid and warm conditions but it also shows good drought resistance and can be grown under 500 mm of annual rainfall, the drought-resistance being attributed mainly to deep root penetration. In the subtropics it can withstand cold weather better than cowpea (*Vigna unguiculata*) or velvet bean (*Stizolobium*). It can grow on a variety of soils but does not tolerate waterlogging.

For fodder or grazing *L. purpureus* is grown in rows, at 10–18 kg seed/ha, although higher rates have also been used. In the countries of its origin *D. purpureus* grows well without inoculation but in Australia the plants do not nodulate naturally and seed inoculation with a suitable cowpea-group strain of *Rhizobium* is required. Seed pelleting with *Rhizobium* inoculant mixed with rock phosphate, lime or Cellofas A produced a positive effect (Norris, 1971). *Lablab purpureus* can be grazed, fed as soilage or conserved as silage or hay.

Lablab purpureus is a fast-growing legume and grazing or cutting can begin 7–10 weeks after sowing. It can withstand severe grazing and Philpotts (1969) reports that after three close grazings 50 per cent of plant survived, compared with 16 per cent for cowpea. Herbage yields of *L. purpureus* vary widely even in the same locality and in subtropical New South Wales Murtagh & Dougherty (1968) report leaf yields ranging from 0.40 to 2.21 t DM/ha and total herbage yields from 0.64 to 4.62 t, which compares favourably with the yields of velvet bean grown in the same trial. Yields of over 12 t fresh material/ha were obtained in the Sudan, and Toutain (1973) states that 15–20 t fresh material/ha can be expected under dry-land cultivation and up to 40 t under irrigation.

Hay has been prepared but haymaking is not easy because the thick stems dry slowly and this results in a considerable loss of leaves. Crude protein content in the herbage can reach over 25 per cent in the DM but in trials in Queensland its content in hay ranged from 12.7 to 14.1 per cent and little changed with plant age, but organic matter digestibility was reduced from 61.3 per cent in younger plants to 48.3 per cent in more mature material (Thurbon *et al.*, 1970). Silage made of *L. purpureus* provided only for maintenance requirements of stock although DM digestibility in trials with ensiling was 56 per cent and of CP 67 per cent (Catchpoole & Henzell, 1971). Milk production from *L. purpureus* is normally higher than from grasses. In an Australian trial (Hamilton *et al.*, 1969) milk from the cows fed on *D. purpureus* was tainted but became palatable and acceptable after pasteurization.

Leucaena Benth. Subfamily Mimosoideae.

This genus includes about 50 species which occur almost exclusively in Tropical America.

Leucaena leucocephala (Lam.) De Wit (*Mimosa leucocephala* Lam.) formerly known as *L. glauca* (L.) Benth. W. T. Gillis (Gillis & Stearn, 1974) suggested another combination – *Leucaena latisiliqua* (L.) Gillis; the name *L. leucocephala* is, however, retained here pending general recognition of the new combination by the taxonomists. Ipil-ipil (Philippines); Koa haole (Hawaii); White popinac (Fig. 46).

Shrub or tree up to 20 m high. Leaves bipinnate, 15–20 cm long, with 4–10 pairs of pinnae, each with 5–20, but mostly 10–15 pairs of leaflets; leaflets 7–15 mm long and 3–4 mm wide. The globose flowering heads have numerous white flowers of which one quarter to one third develop into pods. Pods linear, flat, 12–18 cm long and 1.5–2 cm wide, with 15–30 seeds. Seeds elliptic, flat, shiny, brown in colour, 6–8 mm long and 3–4 mm wide.

Originates from Mexico but spread by intentional or accidental introduction first to the Caribbean islands and then to other areas and has now pan-tropical distribution. The early introductions were used mainly as shade trees in plantation crops but later *L. leucocephala* became an important fodder crop in several countries, especially in

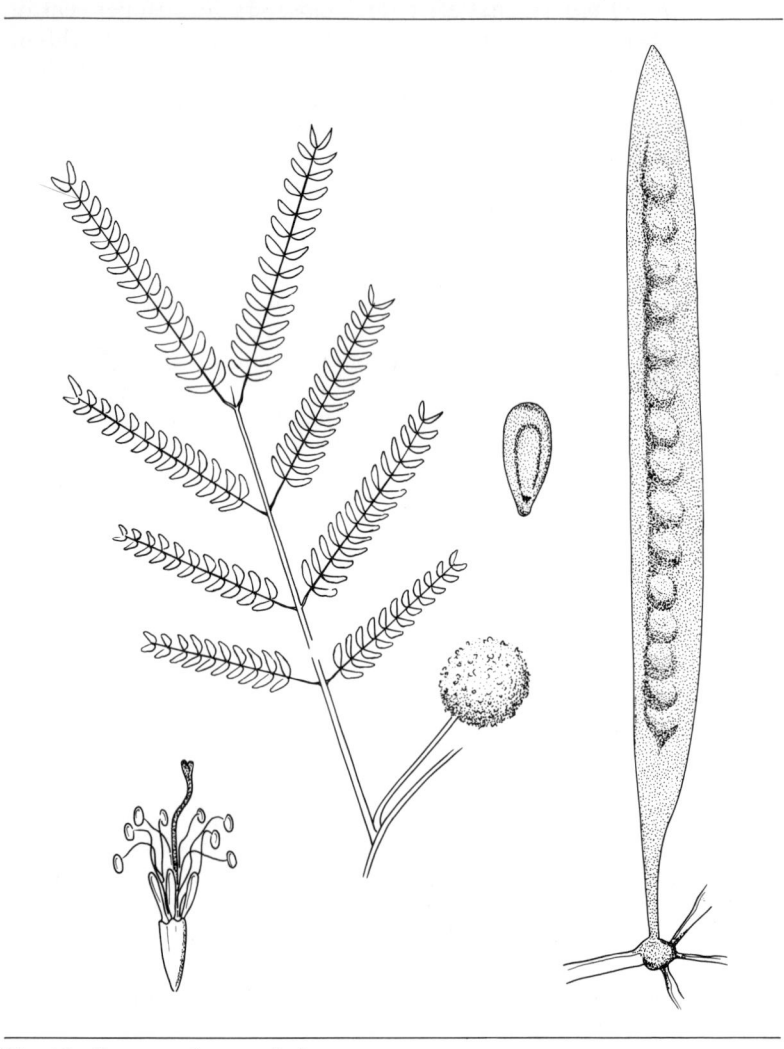

Fig. 46 *Leucaena leucocephala.*

Hawaii and Australia. This plant is valued for its ability to withstand repeated defoliation, high yields of foliage and its tolerance to low soil fertility and relatively low rainfall. Slow early growth and a risk of animal poisoning are its weak points. *Leucaena leucocephala* is usually grown in pure stands but in Hawaii it was experimentally grown with *Tripsacum laxum* and *Panicum maximum* (Plucknett, 1970); there are also reports on its cultivation in mixture with *Pennisetum purpureum*.

Environment. *Leucaena leucocephala* is a purely tropical plant although on the eastern Australian coast it can be grown as far south as Brisbane. The altitudes at which it thrives do not usually exceed 700 m and in Hawaii it is mainly grown at elevations not exceeding 300 m; there are, however, reports of *L. leucocephala* cultivation at up to 1,500 m altitude. *Leucaena leucocephala* is remarkably tolerant to adverse moisture conditions, apparently because of its deep roots, and can be grown at an annual rainfall ranging from 500 to 5,000 mm; at the low rainfall range it responds well to irrigation. Well-drained soils are required for good growth and high yields, and waterlogging or flooding are not tolerated. It can withstand slight soil acidity of up to pH 5 but grows much better in neutral or slightly alkaline soils. *Leucaena leucocephala* is more tolerant to low phosphorus status of the soil than a number of other tropical legumes and this may be due to the presence of endotrophic mycorrhiza which Possingham *et al.* (1971) have found in the roots. Tolerance to grass fires is poor.

Establishment. In the countries where *L. leucocephala* has been naturalized self-established stands are often used for fodder. When sown, the plants are grown in 60–120 cm rows or as hedges between other crops. Seed can be sown into rough, untilled ground, but a well-prepared seedbed results in more rapid establishment. Recommended seed rates range from 15 to 40 kg/ha. There is usually a high percentage of hard seed and germination of 2 to 12 per cent can be expected from untreated seeds. Mechanical scarification by shaking seed + sand mixture, submerging in sulphuric acid for 10 min. or in hot water heated to 60°–80°C can increase germination percentage to 76–97. In Hawaii (Takahashi & Ripperton, 1949) local ranchers feed *L. leucocephala* seed to cattle and seed recovered from dung germinated to 87 per cent. The plants can also be established vegetatively, from seedlings grown in the nursery or from stem cuttings, but this is seldom practical.

Leucaena leucocephala is specific in its *Rhizobium* requirements and is said to be effectively nodulated only with fast-growing, acid-producing strains. However Norris (1973a) in a study of two contrasting types of *Rhizobium*, acid-producing NGR8 and alkali-producing cowpea-type CB81, has found that on acid soils the latter should be used and seeds should be lime-pelleted, whereas on soils of high pH NGR8 can be applied and then seed can be pelleted with the methyl cellulose sticker only, although Oakes in his review (1968a) wrote that 'it would appear that its inoculation requirements are either rather nonspecific or nonexistent'. Diatloff (1973) has recently confirmed that inoculation of *L. leucocephala* is necessary.

The growth of seedlings and young plants is slow and newly sown stands require repeated weeding for obtaining high yields. Satisfactory results have also been obtained with chemical weeding.

Management, fertilizing. The herbage is grazed or cut and fed fresh but satisfactory hay, silage and Leucaena meal can also be prepared. Leucaena herbage is cut first 2 to 8 months after establishment, and repeatedly cut when it reaches a height of 90 to 150 cm. In trials in Hawaii (Takahashi & Ripperton, 1949), cutting close to the ground gave higher yields, over 50 t fresh material/ha/year, than cutting at 76 cm from the ground which gave 40 t; cutting at 38 cm resulted in intermediate yields. Four cuts per year gave higher yields than three or six cuts. In New Guinea (from Hill, 1971) cutting at 12-week intervals, or four cuts/year, gave higher yields of herbage than cutting at 6-week intervals but losses in bulk were partly compensated by higher herbage quality which contained 22 per cent CP in the DM compared with 19 per cent for herbage cut at 12-week intervals.

On acid soils liming is essential and in Papua (Hill, 1970) *L. leucocephala* grown on acid soil increased herbage yields about $2\frac{1}{2}$-fold when 25 t lime/ha were applied. Responses to P, especially when applied with lime, can be considerable but there are hardly any responses to K. Nitrogen is sometimes applied in moderate quantities at establishment and some authors consider that it acts as a 'starter' for the initial growth before the root nodules have been formed and N fixation begun. Molybden and B are also effective in regard of nitrogen fixation and responses to S were noted in New Guinea (Hill, 1971).

Chemical composition. Crude protein contents in the majority of references quoted by Hill (1971) range from 14 to 19 per cent in the DM for the whole herbage as fed to the animals but Oakes (1968) gives a wider range, from 15 to 25 per cent. The content of CF usually fluctuates from 33 to 38 per cent, of NFE from 35 to 44 per cent, and CP and CF contents in the leaves are given as 28.8 and 12.8 per cent, respectively. CP content varies with plant age which in its turn depends on the frequency of cutting. Deficiencies in the contents of tryptophane and in sulphur-containing amino acids have been noted. The contents of vitamins A and C are normally high.

Herbage productivity. Yields of DM vary very widely. In Hill's review (1971) they range in Virgin Islands from 3.05 to 20.52 t/ha/year, depending on rainfall and from 13.63 to 18.56 t depending on the cultivar. In south-eastern Queensland, Australia, DM yields ranged from 1.50 to 12.59 t over a 9-month period and under dry conditions of the Australian Northern Territory from 2.05 to 4.31 t over a period of 20 months. In New Guinea *L. leucocephala* grown as hedges yielded 13.33 t in a 9-month period. Anslow (1957) reported a very high yield of 34.6 t DM/ha when *L. leucocephala* was grown under irrigation. A yield of 3.5 t CP/ha was obtained in Australia.

Animal production. Information on animal production is somewhat erratic. Hill in his review (1971) reports that in Australia steers grazed *L. leucocephala* gained 200 to 522 g/animal/day. Plucknett (1970) in Hawaii observed direct correlation between the rainfall and liveweight gains in steers grazed at one animal to 0.8 ha; the animals gained 233, 171 and 90 kg/ha in the years with 1,800, 860 and 510 mm of rain, respectively. He also reports that an irrigated *L. leucocephala/Panicum maximum* pasture gave 400 kg annual liveweight increase/ha and cows grazed on the same mixture produced 9,770 kg milk/ha/year, the maximum annual production reaching 4,900 kg/milk/cow, 12 kg being the average daily milk yield/cow. Higher milk yields from *L. leucocephala* than from *Pennisetum purpureum* were also obtained in Hawaii.

Ill effect on the animals. When eaten as the sole feed *L. leucocephala* can adversely affect animal health, apparently because of the high content of mimosine, 2 to 5 per cent of the DM. The mimosine effect is manifested in some metabolic disorders and in the loss of hair observed mainly in young cattle, and also in the disruption of reproductive activity in cows: in most cases the reproductive cycle of the cow is little affected but the newly borne calves are often smaller than those borne by cows receiving the diet not overloaded with *L. leucocephala*. These ill effects of mimosine on cattle are erratic and no such effects were recorded from Hawaii where *L. leucocephala* is grown on a farm scale. A stronger effect has been observed in sheep who can lose part or all their wool. However, sheep would not normally eat *L. leucocephala*, as the taste of it is repulsive to these animals. A strong ill effect on horses was also observed.

Reproduction, variability. Flowers of *L. leucocephala* open during the night (Hutton & Gray, 1959) but pollen is shed between 7 and 8 a.m. At the time of anthesis the anthers are excerted slightly above the stigmas and cross-pollination can occur although self-pollination prevails. Emasculation for cross-pollination between known parental plants can be achieved by submerging the flowers in water contained Gardinol K.

A considerable variability exists and selection and breeding has been practised, mainly in Australia and in Hawaii, where numerous introductions have been accumulated. The aim of breeding is usually the development of high-yielding cultivars with considerable plant vigour and of certain forms of growth. Various forms have been grouped into three main types: (a) low-growing, bushy, early flowering and low yielding plants; (b) tall, late-flowering, high yielding plants sparsely branched at the base; (c) as (b) but abundantly branches at the base. Tall erect growth was found to be a dominant character and crosses between types (a) and (c) were most suitable for developing high-yielding cultivars. The better known geographical cultivars are **Hawaii, Peru, El Salvador, Guatemala** and **Australia**. Cultivar Hawaii, which belongs to type (a), appears to be the least productive; relative yields of the other four cultivars seem to depend on the environment rather than on the cultivar itself. Breeding for low mimosine content is at a less advanced stage because of the difficulties in the mimosine content determination and the scarcity of information on its heritability; moreover it can perhaps be of lesser importance than breeding for high production. Nevertheless Gonzales *et al.* (1967) have selected forms with relatively low mimosine content. *Leucaena leucocephala* is an octaploid with $2n$ chromosome number of 104 although $2n = 36$ has also been reported. Seed production received no special attention apparently because seed is abundantly produced and can easily be harvested by hand. Basic information mainly from Gray (1968), Oakes (1968a) and Hill (1971).

Lotononis (D.C.) Eckl. & Zeyh.

Prostrate or suberect annual or perennial herbs. Leaves mostly with three leaflets. Pods oblong. A genus of some 100 species, predominantly South African although a few species occur in tropical and northern Africa and in South East Asia. *Lotononis bainesii* is a pasture legume grown on experimental and, mainly in Australia, on a farm scale.

Lotononis angolensis Baker $\qquad 2n = 18$

Perennial herb not unlike *L. bainesii* (see below) in habit but less prostrate, forming less dense herbage and with the stems rooting only from the lower nodes. It also differs from *L. bainesii* in having larger pods which are up to 18–20 mm long and larger seeds – about 1 mm in diameter. Distributed naturally in East and Central Africa, including Angola, mostly in upland grassland, up to 2,200 m alt. Attempt to cultivate *L. angolensis* as a pasture legume were not particularly successful.

Lotononis bainesii Baker. Lotononis; Miles lotononis (Australia) (Fig. 47).

Glabrous perennial herb with creeping stems producing upright shoots. Leaves trifoliolate, on petioles 0.5–6 cm long. On the main stem branches the central leaflet is 2.5–6.0 cm long and 0.5–1.5 cm wide, lanceolate, larger than the lateral leaflets; the leaflets of short shoots arising from the axils of large leaves are much smaller. Inflorescence dense, mostly globose or umbellate racemes on peduncles up to 15 cm long, with 8–25 flowers. Flowers yellow, 8–10 mm long. Pod linear-oblong, 8–12 mm long, many-seeded. Seed small, creamy-yellow or pinkish, oblong to asymmetrically heart-shaped, about 0.5 mm in diameter.

Creeping stems of *L. bainesii* root freely from the nodes, they can spread rapidly and form a dense growth of vertical shoots the height of which depends on soil conditions and utilization; ungrazed cover can reach up to 50 cm in height, but heavily grazed plants can form a low carpet 2–8 cm high.

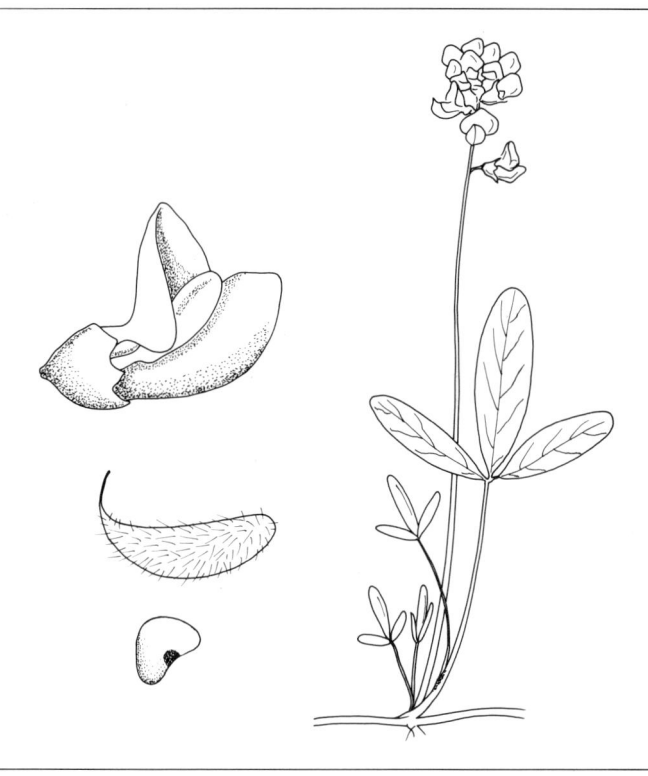

Fig. 47 *Lotononis bainesii.*

Indigenous to South Africa where it occurs in an area southward of the Tropic of Capricorn. Introduced to Rhodesia, Kenya, Malawi, Tanzania, Brazil, Florida in the USA, and Australia. In Australia, where *L. bainesii* was introduced by J. E. Miles in 1952, it has acquired some importance as a pasture plant grown mostly in grass mixtures, mainly in the subtropics of south-eastern Queensland, in New South Wales, but also further north, in the Australian tropics. The outstanding qualities of this plant were little recognized until 1955–60 when Norris

(1958) found a 'red' active and productive *Rhizobium* strain compatible with *L. bainesii*. *Lotononis bainesii* is valued for its ability to grow in mixtures with highly competitive pasture grasses, for the quality of its herbage combined with reasonable herbage yields, resistance to cold, and reasonable seed production. It is however not drought resistant and does not last in grass mixtures for more than 2–3 years.

Environment. For its optimum growth *L. bainesii* requires a moist climate with an annual rainfall of 900 mm or over. The plants also need a warm climate or at least long warm and wet periods and cease growing at temperatures below 9–10°C but can withstand ground-level frosts up to about $-9°C$ and have an ability of frost hardening, i.e. increasing frost tolerance under a gradual decrease of temperature, which is lacking in a number of tropical legumes. Its frost resistance depends on plant age and the supply of nutrient P. *Lotononis bainesii* can grow on various soils including poor soils such as sandy, lateritic or podzolic, which are usually acid with pH not far above 5.0. Well-drained soils are preferred. *Lotononis bainesii* is not particularly demanding in regard of soil fertility and can grow reasonably well on soils with a low level of available phosphorus and has an ability of rapid absorption of P from the soil and of transferring P from the roots to the tops. It tolerates a comparatively high content of aluminium and manganese in the soil and in trials by Andrew *et al.* (1973) tolerated 2 p.p.m. of applied Al and the presence of aluminium increased P absorption from the substrate to a greater extent than in the other legumes used in the trial. Salt tolerance is poor and *L. bainesii* was one of the most sensitive legumes to the presence of salt applied at 3,500 p.p.m. in trials by Hutton (1971).

Establishment. The establishment can be effected by creeping-stem cuttings, not necessarily rooted, which are planted at a distance of about 60 cm, but this legume is normally established by seed. One kg seed/ha is sufficient but the seedbed should be well prepared as seeds are very small, should be sown very shallow and only lightly covered just to secure some protection from dry and hot air for the survival of *Rhizobium*. Seed inoculation is necessary as *L. bainesii* is highly specific in its *Rhizobium* requirements. A suitable *Rhizobium* strain has been brought from the area of its natural distribution and no other *Rhizobium* culture seems to be compatible with *L. bainesii* plants and, *vice versa*, the *L. bainesii Rhizobium* strain does not inoculate effectively any other leguminous species. Commercial inoculant is available and it is based on the CB376 strain of *Rhizobium*. Root nodules are of a specific shape: a small nodule appears first on the tap root and gradually increases in size often completely surrounding the root: only occasional smaller nodules can be found on lateral roots. The nodules are of a characteristic red colour (Norris, 1958).

The seedlings are small and vulnerable at first but soon they begin growing fast and in about 8 weeks the plants become fairly strong and competitive, and 3-month-old plants have been recorded to yield 1.5 t DM/ha and some 35 kg N/ha which corresponds to about 220 kg CP; 4-month-old plants yielded 2.0 t DM and 45 kg N (280 kg CP)/ha (Bryan, 1961). Three- to four-month-old plants can already be utilized.

Diseases. *Lotononis bainesii* seems to be resistant to root-knot nematodes and the main disease which often attacks the plants, especially those grown under poor conditions, is small-leaf virus transmitted by the Jassid (*Orosius argentatus*). In Queensland, Australia, intense grazing in summer to prevent the accumulation of mature and dying plant tops at a later period helps to keep away the virus. Occasional virus attacks do not usually affect to any great extent the ability of *L. bainesii* to fix nitrogen (Bryan, 1961).

In mixtures. *Lotononis bainesii* mixes well with grasses and can compete with strong species such as *Digitaria decumbens, Chloris gayana* or *Paspalum dilatatum* and it has also been tried, mostly with good success if correctly inoculated, with some other grasses: *Paspalum commersonii, P. plicatulum, P. notatum, Cynodon nlemfuensis, Acroceras macrum* and some others. The management of *L. bainesii* swards, in pure stands and in grass mixtures, includes rotational grazing or cutting often used in trials to simulate grazing. Eight-week intervals between grazings seems to be most suitable and in a clipping trial by Bryan *et al.* (1971) with pure *L. bainesii* resulted in a yield of 6.9 t DM/ha; the yield was reduced to about 4.8 t when the plants were cut at 4-week intervals, whereas an increase of the intervals to 12 weeks produced no yield increases and resulted in an increase of weed population. Similar results were obtained in trials with *L. bainesii*/grass mixtures. In grass mixtures the proportion of *L. bainesii* in the sward often decreases under heavy grazing as this legume is very palatable and is grazed preferably but the opposite effect of heavy grazing has also been recorded. In ungrazed mixed swards of *Chloris gayana/L. bainesii* the proportion of legume was also low and rotation grazing at moderate stocking seems to be the best grazing treatment. When grown in mixtures in which other legumes are included, the reaction of *L. bainesii* depended on the companion legume species. It was unable to compete with strong climbing, prostrate or trailing legumes such as *Desmodium uncinatum* or *D. intortum* but was highly compatible with *Trifolium repens*, a plant with rather similar form of growth and even less drought tolerant than *L. bainesii*, and also with annual *Macroptilium lathyroides*. In the majority of trials *L. bainesii* did not last long in grass mixtures and practically disappeared on the third or fourth years of growth.

Management yields. Only moderate rates of fertilizers are usually applied and 125 kg superphosphate/ha is often the rate above which no yield increases were obtained, and the same refers to potassium chloride at about 200 kg/ha.

Yields of pure, well-managed and inoculated with the red *Rhizobium L. bainesii* range mostly from 2 to 8 tonnes DM/ha/year depending on the fertilizers applied, the frequency of cutting and perhaps mainly on the weather conditions and the amount of rain during the period of active growth, but lower yields can be expected from *L. bainesii* as a component of grass mixtures. In Queensland *L. bainesii* is a summer-growing legume and it produces little, often practically nothing, during the winter, at temperatures below 10°C (Bryan *et al.*, 1971).

Nutritive value, animal production. Lotononis bainesii is used mainly for grazing and sometimes also for haymaking when grown in grass mixtures. Laboratory ensilage has also been tried with reasonable success. The herbage contained about 9 per cent of sugars in the DM and the production of lactic acid was slow but continued for 20 days. The percentage of ammonium N was low, about 6 per cent of total N, pH of the silage was 4.9 and the preservation of silage very good (Catchpoole, 1970).

According to an analysis by R. Milford, reported by Bryan (1961), *L. bainesii* herbage 30–45 cm in height, cut at the flowering/seeding stage, contained 19.3 per cent CP, 27.0 per cent CF, 4.0 per cent EE and 41.6 per cent NFE; the digestibility determined in a trial with sheep was 60 per cent for DM, 74 per cent for CP, 47 per cent for CF, 54 per cent for EE, 64 per cent for NFE and the content of TDN was 58 per cent. N balance was positive with +12 g N/day/animal weighing 61 kg. In trials by Bryan *et al.* (1971) the contents of Ca ranged from 0.60 to 0.69 per cent and of P from 0.29 to 0.37 per cent. CP content can vary and in the trials by Bryan *et al.* it ranged from 19.1 to 24.1 per cent (3.05–3.87 per cent N) but 13–17 per cent of CP were reported elsewhere. The data given indicates a very good nutritive value of *L. bainesii*.

Data on the effect of *L. bainesii* on animal performance and production are scarce but there are references on the stocking rates which show that the presence of *L. bainesii* in the sward increases the stocking capacity of pasture to about the same extent as the majority of other good legumes and an inclusion of *L. bainesii* into natural grassland in New South Wales increased the carrying capacity four-fold. No cases of poisoning cattle or sheep grazed on *L. bainesii* have been recorded but as *L. laxum* and also some species of *Crotalaria*, a genus allied to *Lotononis*, are toxic to cattle, a possibility of *L. bainesii* toxicity was examined on rabbits and by chemical analysis and the plants were found to be safe for grazing (Bryan, 1961).

Reproduction. In a study of flowering habits in Australia (Hutton, 1960), *L. bainesii* flowered mainly in spring with another, smaller peak of flowering in late summer. Pollination was observed to take place a few hours before the flowers have opened and the plants were exclusively self-pollinated although, at flowering, they attracted numerous bees. Seed formation is usually good and seed yields can reach about 100 kg/ha under hand picking but it has been reported (Barnard 1972) that seed shattering through a split at the base of the pod may result in considerable seed losses. Using a blade mower equipped with an attached tray for catching pods, seeds, leaves, etc., 25–50 kg seed/ha can be harvested (Bryan, 1961). Seeds are very small, 3 to 3.5 million per kg. Seeds can germinate soon after ripening and 85 per cent germination has been obtained in 3-week-old seeds. Seed viability does not last long and when stored at room temperature and air humidity all seeds can die after a year of storage.

Being a self-pollinated plant, *L. bainesii* forms remarkably uniform populations lacking genetic variability on which plant improvement could be based. Even new introductions from South Africa were found to be identical with those introduced earlier and search for genetic variability is continued. The known plants are tetraploids with the $2n$ chromosome number of 36.

Lotus L.

Predominantly a temperate genus of the Old World, particularly common in the Mediterranean area.

Lotus corniculatus L. Birdsfoot trefoil.

Perennial herb of Euro-Asian and Mediterranean origin now widely cultivated in temperate areas throughout the world and sometimes also at high altitudes in the tropics and subtropics. Var. *eremanthus* Chiov. (*Lotus maernsii* De Wild.), which is glabrous and in general smaller than the main form, occurs in tropical East Africa north of the equator. It has been tried in cultivation in Kenya but without success.

Lotus uliginosus Schkuhr., or Big trefoil, is larger than *L. corniculatus* and grows in more moist situations. Similarly with *L. corniculatus*, it has had some success at high elevations of the tropics.

Macroptilium (Benth.) Urb.

An American genus of two species, formerly known as *Phaseolus*.

Macroptilium atropurpureum (DC.) Urb. (*Phaseolus atropurpureus* DC.). Siratro (Fig. 48). $2n = 22$

Perennial with trailing or creeping stems rooting from the nodes. Leaves with three leaflets. Leaflets dark green and slightly hairy above, and silvery and more hairy below, ovate, 3–8 cm long and 2–5 cm wide; lateral leaflets often with a lobe on the outer side. Flowers 3–12, crowded on axillary peduncles 10–30 cm long. Corolla with protruding wings,

15–17 mm long, deep purple; standard smaller than wings, greenish-purple; the pink keel forms a complete spiral. Pods linear, straight but slightly curved at the apex, about 8 cm long and 5 mm wide, with about 12 seeds. Seeds light brown to black, ovoid in shape and flattened, 4 mm long, 2.5 mm wide and 2 mm thick.

Distributed naturally in northern, Central and South America from southern Texas in the USA to Mexico, Central America, Colombia, Peru and Argentina, and perhaps in other parts of South America, mostly in western areas.

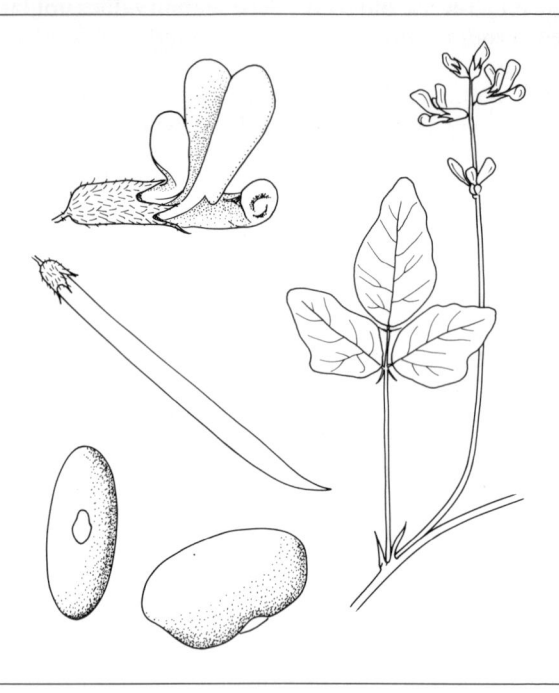

Fig. 48 *Macroptilium atropurpureum.*

Little is known on the performance of wild plants but cv. **Siratro** developed by Hutton (1962) in Australia is now widely spread in the tropics and subtropics and all experimental evidence and other information given below have been obtained from cultivated Siratro. Siratro originates from two Mexican introductions which were crossed and superior recombinations of F_2 populations were selected, intercrossed again and the best ones were tested under grazing in mixtures with highly competitive Rhodes grass (*Chloris gayana*). Outstanding groups of F_4 were then selected and a mixture of their seed formed the original Siratro which combined the more stoloniferous habit of one introduction with the higher yields of herbage and seed of the other. The cultivar was then released for commercial seed production (Hutton, 1962). Slight genetic variations which can be expected from mixed seed may be useful when Siratro is grown under various conditions or when weather conditions vary greatly from year to year. Siratro was first evalued in Queensland and New South Wales, Australia. It has been introduced to other tropical and subtropical countries and tried in Africa (Kenya, Rhodesia), India, Jamaica, Cuba, Brazil, the Philippines, New Guinea, Fiji, Florida, USA and in other areas. Siratro is highly valued for its reasonable drought tolerance, the ability to withstand some frosts, good symbiotic nitrogen fixation, productivity, the ability to form mixed stands with grasses and is now one of the most popular leguminous pasture plants in the tropics and subtropics.

Environment. Under suitable temperatures Siratro yields much better under long-day conditions than under short photoperiods but under photoperiods of 16 hours or over it does not flower and is classified as a short-day plant. Good herbage production was obtained at day/night temperatures ranging from $24/19°$ to $33/28°C$ with a maximum production at $27/22°$ to $30/25°C$ (Hutton, 1970), and R. J. Jones has found that high yields can be obtained at soil temperature above $24°C$. Siratro can also survive frosts although the leaves can easily be damaged. In subtropical Queensland 65–83 per cent survival was observed after a winter with the lowest temperature of $-9°C$ (R. M. Jones, 1969). Rainfall requirements are not particularly high and Siratro can be grown with some success in moderately dry areas with an annual rainfall of 700–800 mm and with the best results under 800–1,600 mm; in humid tropics of Coastal Queensland with over 3,000 mm of annual rainfall its performance was poor (Grof & Harding, 1970). In the monsoon tropics it is a summer-growing plant. Siratro can grow on a variety of soils and does not require particularly good soil; it grows well on volcanic ash and in general prefers light-textured and well-drained soils although it can perform reasonably well on infertile and somewhat waterlogged ground, but not on saline soils.

Establishment. A clean seedbed is required and Siratro, which is nearly always grown in grass mixtures, is usually sown simultaneously with grass. Two to three kg seed/ha are usually sown, the sowing rates depending mainly on seed price, which is high, rather than on agronomic requirements, and Kretschmer (1972) writes that Australian ranchers plant sometimes only 2 oz./acre (about 150 g/ha) and rely on further spreading by natural seeding. Seeding rates for the optimum establishment can however be much higher than those normally used and in trials in Australia (Middleton, 1973), in which 3.3, 9.9 and 29.7 kg seed/ha were sown to establish *Setaria anceps*/Siratro mixtures, the highest sowing rate produced higher DM and CP yields of the mixtures and higher proportions of the legume in the sward than the two lower rates; the effects of high seeding rates lasted, however, 2–3 years only.

Siratro can be oversown to established grass or to natural grassland and successful trials have been reported from India where 8 kg Siratro seed plus 20 kg P_2O_5/ha spread over *Heteropogon contortus*-dominated grassland produced mixed stands in which Siratro constituted about one third of the herbage; when cut for hay at the ripe stage, at which hay is usually cut in the area, the mixture contained 5.5 per cent CP compared with 2.1 per cent for the original *H. contortus* grassland (Dabadghao & Shankarnarayan, 1970). Encouraging results from oversowing were also obtained in Australia. In Queensland, Siratro seed scattered on ash of burnt brigalow bush gave reasonable establishment but the legume did not persist beyond 1 year (Coaldrake & Russell, 1969).

A considerable proportion of hard seeds is often reported and in trials by De Mattos (1970–1) germination percentage was increased from 26.6 per cent for untreated seed to 38.2 per cent for seed soaked in water for 24 hours, 59.8 per cent for mechanically scarified seed, and 65.4 per cent for seed submerged for 3 min in concentrated sulphuric acid; even more spectacular results were obtained by Phipps (1973), who increased seed germination from 18 to 94 per cent by submerging them in sulphuric acid for 15 min.

Seed inoculation is normally advised and the 'cowpea' type of *Rhizobium* culture is suitable; in pot trials by Rotar *et al.* (1967) inoculation increased DM yields by over 40 per cent and the content of N in the plant by 29 per cent. Inoculation by seed pelleting with peat mixed with the inoculant, and with 1 per cent methyl cellulose as a sticker, can be useful (Norris, 1973b). In the field no inoculation may be needed as germinating Siratro seed can be inoculated naturally with the 'cowpea' type of *Rhizobium* often present in soils. Siratro nodulates easily and root nodules are well spread over the roots. Ca and Mo usually improve nodulation.

Grass mixtures. Sitrato is seldom grown alone but nearly always in grass mixtures. A number of grasses have been tried in Siratro mixtures and several have been grown on a farm scale; these include *Digitaria decumbens, Panicum maximum* cv. Petri and cv. Sabi, *Chloris gayana, Paspalum notatum, P. plicatulum, Cenchrus ciliaris, Dichanthium caricosum, Ischaemum indicum, Brachiaria humidicola.* Siratro almost invariably increases the yield of herbage compared with grasses grown singly and in a trial in Florida (Kretschmer, 1972) *Digitaria decumbens* produced 19.0 t DM/ha during the 4 years of trial compared with 42.3 t for its mixture with Siratro; the average content of CP of mixed herbage was increased from 6.1 to 9.5 per cent and yields of CP from 1,140 kg/ha, 4-years total, to 3,900 kg. In another trial in Florida *D. decumbens*/Siratro mixtures yielded as much as *D. decumbens* grown alone and given 125 kg N/ha, and a *Paspalum notatum*/Siratro mixture considerably exceeded the yields of *P. notatum* grown alone and receiving 125 kg N/ha. The contents of CP in Siratro mixtures were even higher than in the first trial. The same trend was observed in a number of trials in different countries and with different grass companions. Siratro has also been tried in mixtures with another legume plus grasses, e.g. with *Macroptilium lathyroides*, a suberect annual which grows fast and increases the proportion of legumes in the year of establishment.

Fertilizers, management. Application of 400 kg/ha of the 0–10–20 (N–P$_2$O$_5$–K$_2$O) fertilizer is recommended by Kretschner (1972) in anticipation of P and K soil deficiency but Siratro is relatively tolerant to comparatively low level of available P and this can perhaps be linked with the presence of endogenous mycorrhiza in its roots as it was shown in Australia by Possingham *et al.* (1971). Norris (1967) indicates that no response can be expected from liming but some authors recommend liming at a rate of 1–2 tons/ha. Nitrogenous fertilizers at low rates, 30–50 kg N/ha, can be applied safely in order to support the companion grass at the early stages of growth but higher rates suppress Siratro plants, possibly by increasing the competitive ability of the grass, and R. J. Jones (1970) reports that 112 kg N/ha reduced by half the yields of Siratro in mixtures with *Setaria anceps* or *Chloris gayana*, and 336 kg N/ha resulted in negligible yields of the legume which disappeared on the fourth year of growth. Some micronutrients can increase yields of DM and the content of CP, and Tang & Lin (1970) obtained 11 per cent yield increase from Mo applied alone and from Mo in the presence of applied Ca; the Ca effect can perhaps be attributed to the increase of pH resulting from its application. Both Mo and Ca improved nodulation and increased N content in Siratro plants.

Low and frequent grazing or cutting produce deterimental effect on Siratro grown in mixture with grasses and in Australia (R. J. Jones, 1973a) yields of a *Setaria anceps*/Siratro mixture were doubled when the intervals between cuttings increased from 4 to 16 weeks and from 3 to 9 weeks in another trial; longer intervals also increased the percentage of the legume in mixed swards (R. J. Jones, 1973a). Low cutting or grazing (4 cm from the ground level) were damaging to Siratro cut at frequent intervals but produced little effect under infrequent cutting. In general, severe defoliation, whether by removing the leaves only or by cutting the plants low and frequently, decreases the proportion of Siratro in mixed swards; heavy grazing also tends to weaken and eventually eliminates the legume and stocking rates should therefore be maintained at a reasonable level and usually not exceed 1–1.5 head of cattle/ha.

Chemical composition. In India, Saxena *et al.* (1971) analysed Siratro herbage at different stages of growth and estimated the average content of CP as 23.0 per cent, CF 30.4 per cent, NFE 29.7 per cent, Ca 1.42 per cent and P 0.21 per cent. Digestibility of the nutrients in trials with sheep was 82.9 per cent for CP, 39.5 per cent for CF, 52.5 per cent for NFE and 55.1 per cent for total DM, and the contents of DCP and of TDN were 19.1 and 53.4 per cent, respectively. Brazilian data given by Lima *et al.* (1972) for hay cut at 7- and 9-week intervals showed less CP, 13.8–20.0 per cent, and more CF, 33.1–40.7 per cent. In trials with sheep the digestibility of the nutrients and of total DM varied very considerably and those of CP and CF increased with the delay in cutting the plants for hay, and so was the uptake of hay by sheep. Siratro is well liked by the grazing animals and is readily eaten by cattle and sheep.

Animal production. When mixed with grass, Siratro can considerably increase liveweight gains of grazing animals in terms of kg per ha, and especially in kg per animal. 'T Mannetje (1972) obtained liveweight gain increases from 96 kg/head for steers grazed on *Cenchrus ciliaris* alone to 173 kg when the animals grazed a *C. ciliaris*/Siratro mixture at the same stocking rate of 0.7 steer/ha; animal performance was linear improved with the increase of Siratro in the pasture, but not beyond about 1 tonne Siratro DM/ha. Animal production from heavily fertilized pure-grass pastures can however exceed that of grass/Siratro mixtures with the same grass, and R. J. Jones (1974) obtained 256 kg/ha liveweight gain in young cattle from grazing Siratro/*Setaria anceps* mixture at 2.42 head/ha, the optimum stocking rate, whereas the same grass grown in pure stands and receiving 336 kg N/ha gave 491 kg liveweight gains in animals grazed at the optimum rate of 5.58 head/ha. Gains per head were, however, higher, 106 kg, than from the fertilized grass which gave 88 kg. Feeding pure Siratro herbage cannot be advocated and milk production in trials by Stobbs (1971) was low, 7.7 kg/cow/day, and low intake of digestible energy was considered to be the main factor limiting milk production.

Pests and diseases. Siratro is reasonably resistant to root-knot nematodes. It is also practically resistant in the field to little-leaf virus often attacking various legumes, and it seems that no other particularly serious diseases have been recorded. In Australia, however, it was found that *Pseudomonas phaseolicola*, causing halo blight parasiting on Siratro and also affecting french beans, can be transferred by siratro seed (Moffett, 1973), a factor which may affect the french-bean industry.

Reproduction. Induced tetraploids produced in India (Singh & Patil, 1970) were superior to diploid plants of Siratro in vigour, stem thickness and persistence, and they contained more CP in the whole plant and in the leaves than the usual diploid plants. The tetraploids had, however, low pollen and seed fertility.

Siratro can produce about 900 kg seed/ha, but commercial yields are normally only in the order of 100–160 kg/ha, mainly because of seed shattering (Hutton, 1970). Hopkinson & Loch (1973) state that flushes of seeding, which can be induced by irrigation, may produce some 1.5 t seed/ha but the amounts of actual seed harvested can hardly exceed half a tonne because of seed shattering and the shattered seeds accumulate in large amounts on the ground. The development of methods and techniques for picking up seed fallen to the ground is recommended by the authors in preference to the efforts devoted to the improvement of seed yields and the methods of harvesting. Attempts to breed varieties with non-shattering pods, possibly by obtaining induced mutations, may perhaps be a promising line in *M. atropurpureum* improvement.

There are about 75,000 seeds/kg.

Macroptilium lathyroides (L.) Urb. (*Phaseolus lathyroides* L.). Phasey bean. $2n = 22$

Annual, biennial or short-lived perennial herb 60–80 cm high. Stems erect, somewhat woody in the lower part, much branched, and when grown mixed with tall grasses the plant can acquire a twining habit and reach up to 150 cm long. Leaflets ovate to lanceolate, 4–8 cm long. Racemes about 15 cm long, on axillary peduncles up to 30 cm long. Flowers red or purple; standard almost orbicular about 12 mm in diameter. Pods linear, 8–10 cm long and 3 mm wide, slightly curved, glabrous, with about 20 seeds. Seeds grey, mottled, slightly compressed.

This plant, better known as *Phaseolus lathyroides*, originates from tropical South America and was introduced to tropical and subtropical India, Australia, Africa and to south-eastern North America. It can grow under moderately dry to humid conditions with an annual rainfall ranging from 600 to 2,000 mm, and on a variety of soils, and tolerate some waterlogging and flooding. *Macroptilium lathyroides* is much grown in Rhodesia and Sudan and is also used in Mali in irrigated cotton rotations in which the legume is utilized for grazing and also cut for hay; in these rotations it is grown for a dual purpose to supply grazing and fodder and to improve the soil and is followed by 1 year under cereal and 2 years under cotton (Toutain, 1973).

Nodulation is easy and is effected either by local *Rhizobium* bacteria present in the soil or by 'cowpea' inoculant. This legume is usually grown in mixture with grasses and in Australia moderately good results were obtained from mixtures with perennial grasses: *Paspalum commersonii, P. dilatatum,* cv. Petri of *Panicum maximum* (Barnard, 1972), *Chloris gayana, Setaria anceps* and *Digitaria decumbens,* often also with an admixture of other legumes. In grass mixtures *M. lathyroides* yields well only in the first year of growth and R. J. Jones (1970) obtained 2.75 t DM/ha in the first year of establishment, of which 2.17 t were harvested in the first cut, 75 days after sowing. Toutain (1973) gives however very high yields of DM, 13 t/ha 90 days after sowing and 60 t/ha/year; these yields were obtained under irrigation and the forage contained 17 per cent CP. *Macroptilium lathyroides* usually dies out towards the end of the first season although some plants may survive to the second year and occasionally persist for a few years. The plants are susceptible to root nematodes. In Queensland, Australia, seeds of *M. lathyroides* scattered in the ash of burnt Brigalow bush established well but did not persist beyond the year of sowing (Coaldrake & Russell, 1969). Under arable conditions this legume can easily regenerate itself from seed. The plants are not toxic and are reasonably palatable to stock.

Macroptilium lathyroides is a self-pollinated plant and seed is well produced but seed shattering makes the harvest difficult. There are about 120,000 seeds/kg. Cv. **Murray** selected in Australia is more vigorous and has larger leaflets than unselected commercial types; it is also less hairy (Barnard, 1972).

Macrotyloma (Wight & Arn.) Verdc.

About 25 species distributed mainly in Africa and Asia; these were formerly considered to belong to the genus *Dolichos*.

Macrotyloma axillare (E. Mey.) Verdc. (*Dolichos axillaris* E. Mey.)
$2n = 20$

Twining perennial with trifoliolate leaves; leaflets ovate, 3–5 cm long and 2.5–3.5 cm wide, slightly pubescent. Flowers yellow in two to five flowered short axillary racemes; standard about 13 mm long. Pods 3–5 cm long and 6–8 mm wide, with seven to eight seeds. Seeds 3–4 mm long, subovoid, mottled.

Macrotyloma axillare is indigenous in tropical Africa, Madagascar, Mauritius and Sri Lanka. It was tried with some success in Kenya, and introduced to the Philippines and Australia, where it showed good results in New South Wales and Queensland, and cv. **Archer** was selected and registered in 1967. *Macrotyloma axillare* can be grown in the tropics and frost-free subtropics, and is reputed to be drought resistant, although Barnard (1972) states that it is best adapted to areas with an annual rainfall exceeding 1,000 mm. *Macrotyloma axillare* is a vigorous plant of good palatability, but animals have to be acustomed to it. The pods shatter easily and flowering lasts for a considerable period of time which makes harvesting difficult; nevertheless over 200 kg seed/ha were harvested (Barnard, 1972). There are some 88,000 seeds/kg.

Macrotyloma uniflorum (Lam.) Verdc. (*Dolichos uniflorus* Lam., *D. biflorus* auct. non L.). Horsegram. $2n = 20, 22$ or 24

Twining annual or perennial herb forming dense growth 30–60 cm high. The whole plant is softly tomentose. Leaves trifoliolate with ovate leaflets 3–7 cm long and 2–4 cm wide. Flowers in two to four flowered axillary racemes, yellow with a violet blot on the standard which is about 10 mm long. Pods 6–8 cm long and 4–8 mm wide with six to seven seeds.

Seeds ovoid, 6–8 mm long and 4–5 mm wide, pale brown, occasionally mottled.

Var. *uniflorum* is a cultivated annual form which differs from other varieties in having wider pods which are 6–8 mm wide as against 4–5.5 mm wide in wild varieties. Of Indian origin it is now cultivated in a number of countries in Asia, Africa, West Indies and also in southern parts of the USA as a pulse crop and is also used as a cover crop and for fodder. In Australia, where it has been naturalized, it can be grown in tropical and subtropical areas with an annual rainfall ranging from 600 to 1,100 mm; at higher levels it can be subjected to severe leaf spot attacks caused by *Ascochyta* sp. (Barnard, 1972). It is sufficiently drought resistant to grow at a seasonal rainfall of 380 mm and its drought resistance was also noted in Mali (Derbal *et al.*, 1959), where a sowing rate of 5–10 kg of seed/ha is recommended. It grows on various soils but cannot withstand waterlogging. In Australia it nodulates naturally and a commercial cowpea type of *Rhizobium* inoculant can also be used. In Mali *M. uniflorum* yielded 18–30 t fresh material/ha, and in Australia mature stands produced over 6.5 t DM/ha, 44 per cent of which was seed; this mature herbage contained 18 per cent CP in the DM. When grown with grass it climbs the grass stems and leaves and forms dense stands up to 90 cm high. The plant is palatable at all stages of growth. Seed is abundantly produced and the pods do not shatter easily. The Australian introduction was tested and registered in 1967 as cv. **Leichhardt**.

Medicago L.

Perennial or annual herbs with trifoliolate leaves; stipules dentate, fused with the petiole. Flowers yellow or blue. Pods usually spirally twisted or coiled, with one to a few turns, or can be sickle-shaped. Species about 50, distributed mostly in the Mediterranean area understood in a wide sense. *Medicago sativa* or lucerne is of primary importance for agriculture; other perennial species or their hybrids with *M. sativa* and also annual medics are of lesser if any importance.

Medicago laciniata (L.) Mill. Burr medic; Cut leaf medic (Barnard, 1969) $2n = 16$

Spreading or erect slender annual up to 50 cm high. Leaflets up to 12 mm long and 5 mm wide. Flowers single or paired, corolla yellow 4–5 mm long. Pods (burrs) in globose coils of about five turns, about 5 mm in diameter, with numerous spines terminating in minute hooks and about five seeds. Distributed in North Africa and extends to Punjab in India and southward to East Africa; also in South Africa; it has now spread to other areas of the world with the climate approaching Mediterranean. Occurs in grasslands, often seasonally waterlogged and in central Kenya can be abundant but is only short-lived. The plants are very palatable and if abundant can contribute to the quality and the bulk of herbage but this medic is a nuisance in sheep pastures because of the spiny pods adhering to wool from which they are very difficult and costly to separate.

There is a number of other annual medics some of which were introduced to temperate Australia. Some of these medics have burr pods and there are also species with smooth or almost smooth pods. The better known species are *M. truncatula* Gaertn., *M. orbicularis* (L.) Bart., *M. polymorpha* L. (*M. denticulata* Eilld.) and *M. scutellata* (L.) Mill.

Medicago sativa L. Lucerne; Alfalfa

An important perennial fodder plant of which innumerable research papers and a large number of books have been published. It is grown on about 5 million hectares predominantly in warm temperate areas and a considerable proportion of the area occupied by lucerne is irrigated. Lucerne is grown for hay, lucerne meal or, to a lesser extent, for grazing and then usually in grass mixtures.

Lucerne is grown in most tropical countries, more in South and Central America than in Africa or Asia (Heinrichs *et al.*, 1972; Bolton *et al.*, 1972). In the countries extending from the tropics to temperate areas, such as Australia or Argentina, in which lucerne is grown on about 1.1 and 1.5 million ha, respectively, it is grown almost exclusively under temperate conditions, much less in the subtropics and still less in the tropics. In other tropical countries lucerne is cultivated mostly at elevations over 500 or 1,000 m alt. In the tropics the most important and highest yielding cultivar is Hairy Peruvian selected in Peru, followed by Hunter River of Australian breeding, Smooth Peruvian, Saladina and some other cultivars. In Mexico locally developed productive and persistent cultivars are in use.

Melilotus Mill.

About 25 species, mainly in the Mediterranean area and Central Asia.

Melilotus indica (L.) All. Indian sweet clover; Senji. $2n = 16$

Erect annual or biennial up to 2 m high with trifoliolate leaves. Flowers small, yellow or white, in many-flowered narrow spike-like racemes. Pods 3–4 mm long, ovate, prominently reticulate, with one to three seeds. Distributed in Ethiopia, subtropical India and further north in the Mediterranean area and Asia, mostly as a weed of arable land. This species is grown on a small scale for fodder in India, mostly in the northern, subtropical areas. Similarly with most other species of *Melilotus*, *M. indica* contains coumarin and is fragrant; its hay or silage, if poorly prepared, can cause internal bleeding. The plants require neutral or slightly alkaline soil and produce poor growth on acid soils unless heavily limed. It is grown pure or in mixtures with annual grasses, e.g. with oats, and in trials in northern India *M. indica* yielded 13.1 t green fodder/ha unfertilized and 17.8 t when given NP. Mixtures with oats yielded 15.9 and 30.3 t/ha, respectively (Tomer & Singh, 1969).

Melilotus alba Desr., a European temperate species with white flowers, has also been tried in the tropics and subtropics, e.g. in Colombia (Crowder, 1960), mostly at high elevations, sometimes with reasonable success, as fodder or green manure plant. This species, and the yellow-flowered *M. officianlis* (L.) Pall., can be found in the tropics and subtropics as naturalized plants.

Mucuna Adans.

About 100 species which occur throughout the world tropics.

Mucuna pruriens (L.) DC. var. *utilis* (Wall) Burck. (*M. utilis* Wall. ex Wight, *Stizolobium pruriens* (L.) Medic. var. *utile* Wall., *S. utile* (Wall.) Piper & Tracey). Velvet bean. $2n = 22$

Twining or bushy annual or short-lived perennial. Leaves trifoliolate, leaflets large, 5–20 cm long and 3–15 cm wide, lateral leaflets oblique. Flowers white, mauve or dark purple, in 2–3 cm long racemes. Wings and keel 3–4 cm long, much longer than the standard which is about 2 cm long. Pods 5–10 cm long and 1–2 cm wide, often S-shaped, black or straw-coloured, glabrous, velvety hairy or with woolly hairs, four to five seeded. Seeds 1–2 cm long and 5–6 mm thick, black or white, shiny; hilum 3–5 mm long with a long aril.

Originates from South Asia, introduced throughout the world tropics, subtropics and warm temperate areas, and is grown in Brazil, Colombia, southern states of the USA, Rhodesia and some other countries. The taxonomy of this variable species is not quite clear. Piper and Tracy (see in Verdcourt, 1971) distinguished eight taxa of species rank within the *Stizolobium* (*Mucuna*) *pruriens* complex but most taxonomists consider them to be no more than varieties or even cultivated forms of one polymorphic species, and to clarify the position further work is needed, including studies on living material. The main forms often mentioned in the literature under *Stizolobium* are *S. deeringianum* Bort. or Florida velvet bean, *S. aterrimum* Piper & Tracy or Bengal velvet bean, usually a larger plant than *S. deeringianum* and distributed in warmer areas, and *S. cochinchinensis* (Lour.) Burk. (*S. niveum* (Roxb.) Kuntze). As to the generic name, *Mucuna* is preferable. *Mucuna* is a conserved name which according to the rules of nomenclature cannot be changed and there seems to be no sufficiently clear distinction between *Mucuna* and *Stizolobium* to consider the latter as a separate genus. *Mucuna* is however the name little known to agriculturists, whereas the name *Stizolobium* is widely used.

Velvet bean can be grown in moderately humid to moderately dry areas, on a variety of soils including poor or sandy soils. This vigorous legume is grown pure or, more often, in mixtures with maize, sometimes with fodder sorghum, or with strong perennial grasses such as *Brachiaria mutica, Panicum maximum, Hyparrhenia rufa*, etc., and mixtures with annual *Eragrostis tef* or *Setaria italica* have also been reported from Brazil (Pio Corrêa & Pena, 1952). Seed is sown either broadcast at 25–35 kg seed/ha or in 90–120 cm (3–4 ft) rows and some 30 cm apart in the row, and then 10–15 kg seed/ha would be required (Crowder, 1960). When grown in mixture with tufted grasses, the plants can regenerate to a certain degree from seed fallen to the ground. Grown with maize as a fodder mixture, velvet bean is usually sown at the time of maize planting and in the same rows. Velvet bean can also be grown in maize for the improvement of stover grazing after the cobs have been harvested and then it is sown after the last maize weeding and between its rows.

Green fodder yields can be of the order of 20–35 t/ha (Crowder, 1960) but also considerably lower, especially in subtropical and warm temperate areas, e.g. in New South Wales, Australia. In Colombia the herbage can be utilized some 4–5 months after sowing but this depends on the variety, the conditions of growth and the locality. Good results have been recorded from Brazil where *S. deeringianum* was productive and had good quality herbage which contained 18 per cent CP, 3.7 per cent EE, 36.7 per cent NFE and 32.0 per cent CF; the contents of digestible nutrients were 10.4–10.8, 1.7–2.9, 23.9–26.6 and 10.3–11.2 per cent, respectively, and the content of digestible DM about 50 per cent (Pio Corrêa & Pena, 1952). Seed is usually well produced but mechanical harvesting is not easy because the pods are situated rather low on the plant; hand picking makes seed expensive.

The plants may vary quite considerably even within the micro-species mentioned; a number of varieties exists and several cultivars have been selected, mainly in the southern states of the USA, Australia and Rhodesia, and there is a further scope for selection.

Phaseolus L.

A number of species formerly known as belonging to *Phaseolus* have been transferred to the genus *Vigna* and some to smaller genera based on *Phaseolus* (*Macroptilium*, and others). The question of distinguishing between the species of *Phaseolus* and *Vigna* is a complex one and the present division is based on ample, though not complete, morphological, geographical and biochemical evidence. Species of *Phaseolus* in the narrow sense occur naturally almost exclusively in the western hemisphere and main morphological feature distinguishing the two genera is the much more spirally twisted keel in *Phaseolus*, and less twisted, maximum up to 360°, in *Vigna*.

Psoralea L.

About 120, predominantly tropical species of little importance for grazing or browsing.

Psoralea eriantha Benth. Bullamon lucerne

Deep-rooted perennial herb with ascending stems usually up to 60 cm high but also said to reach up to 3 m in height. Leaves with three ovate leaflets, usually densely hairy. Flowers bluish, in dense or loose spikes. Pods small, oval, compressed, with one seed.

Occurs naturally in Queensland, Australia, between 20° and 28°S. and under an annual rainfall ranging from 250 to 600 mm; plants from Bullamon Plains attracted attention as a potential pasture species. Although hairy *P. eriantha* is palatable to stock and the nutritive value of the plant is high, the herbage contained 17 per cent CP in the DM, over 1 per cent of Ca and 0.30 per cent P. Tested in small-scale trials the plants yielded 520–570 g DM/10 plants depending on cutting frequency. A high proportion of hard seeds was observed and scarification increased seed germination from 14 to 69 per cent (Kerridge & Skerman, 1968).

Pueraria DC.

There are some 20 species which occur naturally in tropical Asia and Polynesia. Robust climbers with large leaves.

Pueraria lobata (Willd.) Ohwi (*P. thunbergiana* (Sieb. & Zucc.) Benth.). Kudzu. $2n = 24$

Woody climbing and trailing vine more vigorous and robust than *P. phaseoloides* (see p. 392) from which it also differs by larger and usually more lobed leaflets (resembling those of grapes) and by broader pods which are 9–12 mm wide. Originates from the Far East but is now grown throughout the world subtropics and in some warm temperate areas such as the southern states of USA. It is relatively easily naturalized and may become a pest in forests.

The plants grow well under somewhat lower annual rainfall than required by *P. phaseoloides*, and can withstand longer periods of drought and winter frosts but usually lose their leaves during the winters or severe dry seasons. *Pueraria lobata* requires relatively fertile and well-drained soils, neutral or acid, and does not do well on calcareous soils of high pH.

Pueraria lobata is established by seed or vegetatively, by root crowns or cuttings. Seed is sown in wide, 1-metre rows at a rate of 3–4 kg/ha (Toutain, 1973) although Whyte *et al.* (1953) indicate about twice higher rates. There is a considerable percentage of hard seeds and some forms of seed scarification are needed. When established vegetatively, root crowns are planted widely spaced and so are the sprouted cuttings which are usually grown for some time in nurseries before they are transplanted to the field, about 1,000 plants/ha or one plant/10 m^2. The newly established plants grow very slowly, especially on poor soils and fertilizing young plants with phosphates and/or fertilizer K is desirable. Other crops are sometimes grown for a year or two between the young plants. On the second year of growth the plants grow fast and cover the ground.

Pueraria lobata is used as a cover crop and for forage as a grazing, hay or silage plant, the utilization beginning normally on the second year of growth. The plants are sensitive to defoliation and low or frequent grazing or cutting can destroy them, and lenient grazing and cutting high for hay is recommended. Yields of forage are usually lower than those of *P. phaseoloides*, the establishment is extremely slow and whenever the climatic conditions permit *P. phaseoloides* is preferred to *P. lobata* the herbage of which is often of lower quality. However, the available chemical data can be as favourable as that of *P. phaseoloides* and Toutain (1973) gives 9.5 per cent as normal content of DCP, and still more favourable chemical composition is reported by Gill & Negi (1968) from India: 17.3 per cent CP, 2.1 per cent EE, 30.2 per cent CF and 42.8 per cent NFE; the contents of DCP was determined to be 15.0 per cent and of TDN 52.2 per cent.

Seed production is mediocre and seed yields are of the order of 200 kg/ha (Toutain, 1973); in some countries they are still lower or none. Seed germinability is retained for a short period only. There are about 85,000 seeds/kg.

Pueraria phaseoloides (Roxb.) Benth. var. **javanica** (Benth.) Bak. (*P. javanica* Benth.). Tropical kudzu; Puero (Fig. 49). $2n = 22, 24$

Robust climbing or trailing perennial herb with the stems usually covered with rusty-brown hairs. Leaves trifoliolate. Leaflets ovate or rhomboid, entire or lobed, pubescent above and velvety-hirsute beneath, mostly large, 5–12 cm long and up to 11 cm wide. Flowers in long axillary racemes. Corolla up to 2 cm long, mauve, purple or white; standard whitish with a mauve blotch. Pods linear, straight or slightly curved, compressed, up to 11 cm long and 3–5 mm wide, with 10–20 seeds. Seeds dark brown, 3–3.5 mm long and 2 mm thick, oblong or subcylindrical. The cultivated plants apparently belong to var. *javanica* which differs from the wild forms by more vigorous growth, and larger flowers and pods.

Originates from South-East Asia, Malaysia and Indonesia, introduced to other tropical areas and is now widely grown throughout the tropics and often naturalized. It was first grown mainly as a cover crop in tree plantations or green manure but later also as a fodder plant of some importance.

Environment. For optimum growth *P. phaseoloides* requires high temperatures and can be killed by frosts although Barnard (1969) states that in Australia it can persist in areas with light winter frosts. It is not normally grown at altitudes over 1,000 m. It grows well only in sufficiently humid tropics, under 1,200–1,500 mm of annual rainfall but can withstand short dry periods lasting up to 2–3 months, and remains green well into the dry season. It can grow on various soils and tolerates acid soils with pH ranging from 4 to 5.5 and also soils deficient in phosphorus and lime although on such soils the establishment is particularly slow. It responds well to fertilizers and grows best on fertile and comparatively heavy soils rather than on poor sandy soils; it can also tolerate high levels of the ground-water table.

Establishment. The establishment is effected either by seed or vegetatively, by crown divisions or by cuttings of creeping and rooting

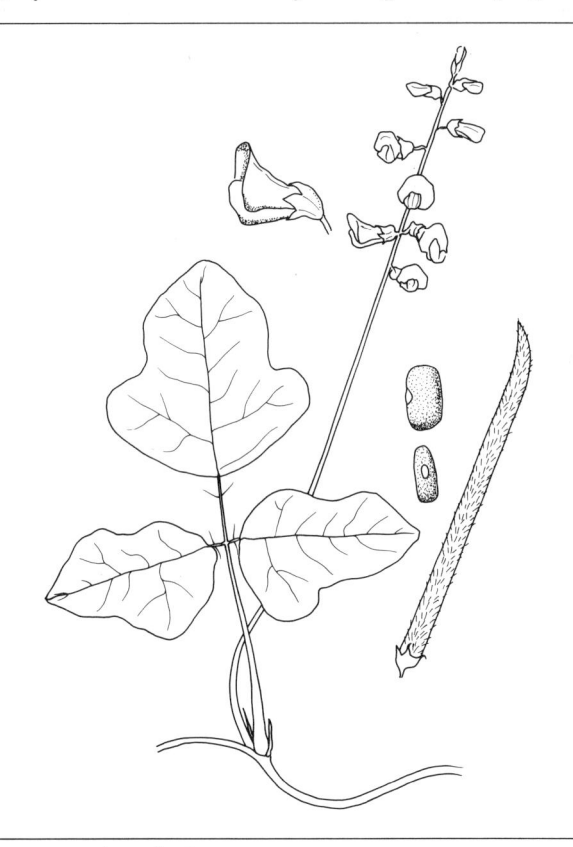

Fig. 49 *Pueraria phaseoloides.*

vines, especially in areas where seed setting is poor. Seed is sown at 3–4 kg/ha in 1-metre or 50 cm rows or in hills placed 1 × 1 m, or at 6 to 17 kg/ha broadcast (Toutain, 1973). There is usually a high percentage of hard seeds, often around 80 per cent but sometimes up to over 95 per cent (Aya, 1973), and dipping in concentrated sulphuric acid for 20 min increased the germination from 22 per cent to 41 per cent in trials by Aya (1973); heating to 40°C increased seed germination to 75–83 per cent in laboratory trials, whereas in field trials heating, combined with soaking in water for 12–24 hours, increased seedling emergence to 41–42 per cent and was superior to sulphuric acid treatment. *Pueraria phaseoloides* can be inoculated by the cowpea-type *Rhizobium*, e.g. by strain CP756 selected in Australia (Obaton, 1974) although it is mildly selective in its *Rhizobium* requirements and not every strain of cowpea *Rhizobium* would produce satisfactory results. Rajaratnam & Guan (1972) observed that nodules appeared on the eighth day after germination and nodule reddening on the twelfth day. The seedlings and the young plants grow slowly in the first 2–4 months, depending on soil fertility, and need protection from weeds in the early stages of development; planting in wide rows or in 1 × 1 m hills allow for early cultivation for weed removal. After 4 or 5 months of growth, or sometimes later, *P. phaseoloides* forms dense entangled masses of vines and leaves and smothers the weeds. When planted vegetatively, by crowns or vines, the plants develop not much faster than when sown by seed and similarly produce a dense growth in 3 to 4 months time. Eventually, in 6 to 9 months after the establishment, the plants develop dense herbage 60 to 80 cm high.

Management, productivity. As a forage plant *P. phaseoloides* is used for grazing, haymaking or for silage. Close or frequent grazing or cutting can exhaust the plants or even eliminate them and lenient grazing or high cutting for hay is recommended. Pure stands or mixtures with grasses can be both used for grazing or haymaking and when grown alone *P. phaseoloides* usually yields reasonably well and Toutain (1973) states that normal yields range from 5 to 10 t DM/ha, but much higher and also lower yields have also been recorded. In Cuba 19.7 t DM/ha were obtained in trials by Febles & Padilla (1970), whereas in humid, tropical Australia a mixture with *Panicum maximum* yielded 3.72 t DM/ha to which 1.16 t was contributed by *P. phaseoloides*. Efficient nitrogen fixation has been noted and Rajaratnam & Guan (1972) estimated that *P. phaseoloides* would fix over 200 kg N/ha over 5 months of growth.

Herbage quality. Toutain (1973) gives the contents of DCP as ranging from 8.5 to 10.0 per cent. Barnard (1969) quotes 19 per cent CP in the herbage during the wet season, whereas in analyses in Colombia (Blasco & Bohórquez, 1968) CP content ranged from 11.8 to 14.8 per cent and CF content from 36.9 to 41.1 per cent; the contents of Ca and P were 0.85 and 0.25 per cent, respectively. Dirven & Ehrencron (1963a) reported the contents of 0.16–0.17 per cent of P in unfertilized plants and 0.23–0.25 per cent when adequately supplied with fertilizer P. Although the plants are hairy they are quite palatable to cattle once the animals get used to them.

Grass mixtures. A number of grass species has been grown or tried, sometimes on a farm scale, in mixtures with *P. phaseoloides*, viz.: *Brachiaria mutica, Panicum maximum, Pennisetum purpureum* cv. Merkeron, *Paspalum dilatatum, Stenotaphrum secundatum, Melinis minutiflora* and *Hyparrhenia rufa* (Whyte *et al.*, 1953), and also *Digitaria decumbens*.

Seed production. In the majority of humid tropical areas such as the West Indies, South America, Sri Lanka, Zanzibar or humid tropics of Australia seed production is moderate and Toutain (1973) indicates seed yields of 50 to 100 kg/ha. Hutton (1970) estimates the yields to be higher – of the order of 300–450 kg/ha. In less humid or subtropical areas, flowering and seed setting are usually very poor and seed yields can be very low and sometimes no seed is produced (Whyte *et al.*, 1953). Plants climbing on trees or other supports usually seed better than those trailing near the ground. There are about 80,000–85,000 seeds/kg.

Rhynchosia Lour.

Climbing, prostrate or sometimes erect herbs or subshrubs. Leaves mostly with three leaflets, usually with conspicuous resinous glands dotted on the lower surface. Flowers mostly yellow, often variously tinged or lined with red. A large genus of about 200 species distributed in the tropics of both hemispheres. Leaves, when crushed, are scented and this is perhaps why a number of species are of low palatability and some not touched by cattle or sheep. Less smelly plants can be well eaten by the animals, as e.g. *R. elegans* A. Rich. in Kenya, *R. phaseoloids* DC. in South and Central America or some varieties of *R. minima*.

Rhynchosia minima (L.) DC. (*Dolichos minimus* L.) $2n = 22$

Climbing or prostrate perennial herb. Leaflets rhomboid, ovate or subcordate, 1–6 cm long and up to 5 cm wide, from almost glabrous to velvety hairy, gland-dotted beneath. Flowers in axillary racemes or heads 2–15 cm long, yellow, with the standard sometimes with red veins or flushed with red, 5–10 mm long. Pods oblong-falcate, up to 2 cm long and 3–5 mm wide. Seeds brown, grey or black, often speckled. A variable species and several varieties are distinguished.

Distributed throughout the world tropics and subtropics, in grasslands or in scattered bush and tolerates seasonal waterlogged for not too long periods. It is drought resistant and can tolerate moderate soil salinity.

Rhynchosia minima contribution to grazing can be substantial but its palatability varies and some varieties are much better eaten than the others, and Whyte *et al.* (1953) state that young herbage is eaten better than that at advanced stages of growth. In India *R. minima* contained 15.1 per cent CP, 1.7 per cent EE, 29.5 per cent CF, 45.9 per cent NFE, 1.28 per cent Ca and 0.28 per cent P (Shukla *et al.*, 1970). In trials with sheep the digestibility of DM was 60 per cent, CP 64 per cent, CF 54 per cent. TDN content was 60.9 per cent and DCP 9.5 per cent. Liveweight gains of experimental animals were high and so was daily intake – 3.5 kg/100 kg body weight. In Kenya, however, *R. minima* was more palatable to cattle than to sheep.

Sesbania Adans.

Annual erect herbs, shrubs or small trees mostly with soft, usually pithy stems and branches. Leaves paripinnate. Flowers of medium size or large, yellow, with the standard often streaked or speckled with purple, or the flowers purplish. Pods long, narrow, dehiscent, with numerous seeds. A medium-size genus of some 50 species distributed throughout the tropics and subtropics, usually in wet places: swamps, seasonal swamps or seasonally waterlogged grasslands. Much used for green manure but of little use for fodder as a number of species are unpalatable and some are toxic. Of the species used for fodder *S. grandiflora* (L.) Poir. and *S. aculeata* (Willd.) Poir. in India and *S. brachycarpa* F. Muell. in Australia can be indicated (Whyte *et al.*, 1953).

Stylosanthes Sw.

About 25 species which occur throughout the world tropics.

Stylosanthes fruticosa (Retz.) Alston (*S. mucronata* Willd.; *S. bojeri* Vogel). $2n = 40$

Much branched perennial herb up to 60 cm high occasionally taller. Stems and leaflets hairy, often glutinose. Leaflets elliptic to lanceolate. Flowers in dense heads, cream-coloured to orange, 5–7 mm long. Pods 4–9 mm long, with one or two joints which are 3.5–4 mm long and 2–2.5 mm wide, with a beak 1.5–2.5 mm long. Resembles *S. guianensis* (see p. 397) from which it differs in flower colour and by a much longer beak of the pod. Distributed in tropical Africa, Arabia, Madagascar, India and Sri Lanka. A drought-resistant plant which in Sudan can grow under an annual rainfall of 300–500 mm. It is grazed by cattle and valued by nomadic pastoralists. *Stylosanthes fruticosa* is a short-day plant: under 8–10 h photoperiods it flowers earlier but yields less herbage than under 14-hour photoperiods. Hard seeds often exceed in number easily germinating seeds, but when scarified 60 per cent germination can be achieved. *Stylosanthes fruticosa* has been tried under cultivation with moderate success in south-eastern Queensland (Skerman, 1970).

Stylosanthes guianensis (Aubl.) Sw. (formerly known as *S. gracilis* Kunth). Stylo; Finestem stylo; Meladinho (Brazil, León & Sgaravatti); Alfafa do Nordeste (Brazil) (Fig. 50). $2n = 20$

Perennial much branched erect or suberect herb 30–120 cm high. Leaves trifoliolate; leaflets 0.5–4 cm long and 0.2–1.5 cm wide, ovate to lanceolate, pointed, slightly succulent, glabrous to pubescent or bristly. Flowering heads dense, 1–4 cm long, 2–40 flowered. Standard yellow, orange-yellow or streaked with red, 4–8 mm long. Pod a single ovoid joint 2.5–3 mm long and 2 mm wide, with a minute inflexed beak. Seeds mostly pale brown but can vary from yellow to almost black in colour, 2.2 mm long and 1.5 mm wide.

Widespread naturally in Central and South America, introduced to other areas and is now grown almost throughout the world tropics and to a lesser extent in the subtropics. Stylo is remarkable in its hardiness and adaptability to various climatic and soil conditions which can

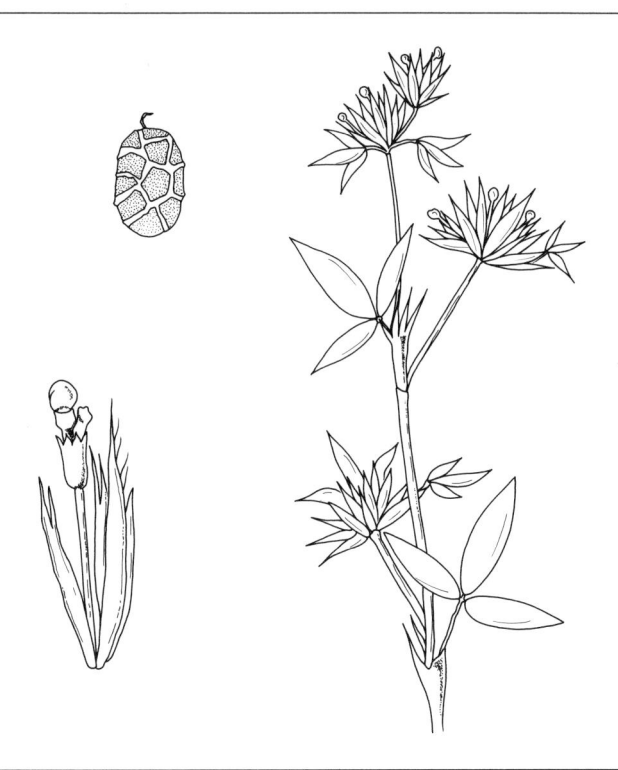

Fig. 50 *Stylosanthes guianensis.*

perhaps be partly explained by a considerable variability of the species. It can be grown on poor soil, is relatively easily established, both under arable cultivation and when oversown to natural grassland where it extends grazing well into the dry season and increases the carrying capacity of pastures. The nutritive value and palatability are reasonable, and the popularity of *S. guianensis* has greatly increased during the last two decades.

Environment. *Stylosanthes guianensis* grows well in open habitats but poor performance was noted in shade. It can grow under a wide range of temperatures and is more tolerant to cool weather than the majority of other tropical cultivated legumes, and can survive frosts. Cool weather reduces the herbage productivity and in Kenya it is not recommended for altitudes above 1,800 m. *Stylosanthes guianensis* can also grow under a wide range of annual rainfall, from 600–700 mm up to 2,500 mm, survives long dry periods, and the proportion of dry-season yields in the total yield of herbage is usually higher than in several other cultivated legumes. Under marginal rainfall its yields in drought years can be greatly reduced. It tolerates poor soil of low P status which can be partly due to endotrophic mycorrhiza found in the roots (Possingham *et al.*, 1971). In pot trials *S. guianersis* was tolerant to flooding (De-Polli *et al.*, 1973).

Establishment. *Stylosanthes guianensis* is normally established from seed. The percentage of hard seed is usually high, and germination of untreated seed is poor. Mechanical scarification and sulphuric acid treatment improve the germination and seedling emergence and Phipps (1973) has also shown that freezing to $-17°C$ for 7 days or immersion in boiling water for 10 sec. increased the germination from 20 to 80 per cent. Seed is sown at 2–3 kg/ha; higher rates can sometimes result in quicker establishment, but seed is expensive and the rates mentioned are usually accepted. Drilling in rows 45–60 cm apart gives satisfactory establishment and allows for inter-row weeding; chemical weeding can also be used and 2.4.5–T about a month before sowing or 2,4–D at the establishment do not harm the seedlings. For the improvement of natural grassland seed can be oversown direct, without any soil tillage or the grass can be disced before sowing. Sometimes discing does not improve the establishment but in trials by Clatworthy & Thomas (1972) in Rhodesia more plants were established from 4.5 kg/seed/ha plus discing than from 18 kg seed without discing. In their trials *S. guianensis* oversown to grass increased the yields of herbage by 50 per cent and of CP by 300 per cent. P fertilizer can improve the establishment. Stylo can also be established vegetatively, from cuttings, but the establishment is slow and expensive and can only be recommended when no seed is available.

Inoculation and nitrogen fixation. Inoculation is not always necessary as it was shown by Chandapillai (1972) in Malaysia where *S. guianensis* nodulated well even if no legumes had recently been grown on the land. However, this may not be true for other varieties and cultivars of stylo, and although cv. Schofield can be effectively inoculated by almost any strain of cowpea *Rhizobium*, cv. Oxley is highly selective and requires a special *Rhizobium* strain; comparable results have been obtained in Brazil for different cultivars. In laboratory trials in Australia, Norris (1964) has shown that the effectiveness of *Rhizobium* can differ in strains isolated at different sites; the best strain in this trial was CP82, which gave 70 mg of fixed N per stylo plant compared with 15–66 mg for the other strains used in the trial. The nodules are small and they begin to appear some 3 weeks after gemination (Oke, 1967b). Nitrogen fixation by *S. guianensis* varies and 240 kg N/ha/year reported by Wendt (1970), is perhaps the highest amount recorded. Amongst other factors nodulation depends on soil pH and the highest N fixation was observed at pH 6.

Association with grasses. In pure stands *S. guianensis* is used mainly for seed production and seldom for herbage for which purpose it is normally grown in grass mixtures. Grasses successfully grown with stylo include: *Digitaria decumbens*, *D. smutsii*, *Chloris gayana*, *Cenchrus ciliaris*, *Melinis minutiflora*, *Setaria anceps*, *Andropogon gayanas*, *Heteropogon contortus*, *Hyparrhenia rufa*, *Panicum maximum*, *Pennisetum polystachion* and stylo was doing reasonably well even with vigorous *P. purpureum*. Grasses are usually grown in alternate rows with stylo, and the mixtures can be established and utilized on soils hardly suitable for mixtures with other legumes. Stylo often, but not always, increases the yields of total herbage over that of pure grasses. In a trial with a *Pennisetum polystachion* mixture R. D. Singh *et al.* (1968) estimated that the contribution of *S. guianensis* was equivalent to 32 kg fertilizer N/ha in terms of DM yields, and 41 kg in terms of CP yields. Haggar's (1971) estimates for a *Chloris gayana* mixture grown in Nigeria were 84 and 187 kg fertilizer N, respectively.

Fertilizing, management. Fertilizer P is needed on poor soils, especially at the early stages of plant growth. Rates of fertilizer normally applied are in the order of 50–100 or 150 kg superphosphate/ha. Responses to fertilizers containing sulphur have also been obtained, but little information is available on the effect of potassium or minor nutrients, although in Queensland the application of about 10 kg/ha copper sulphate is advocated (from Tuley, 1968). It has also been observed that K was effective in the presence of P and Cu. The percentage of *S. guianensis* in grass mixtures is usually increased by fertilizer P, but fertilizer N decreases it and reduces CP content in the total herbage, and also often the yields of CP and even of DM. Liming is usually not recommended, but in Brazil yields of *S. guianensis* were increased by the application of 250 kg Ca/ha (M. B. Jones & De Freitas, 1970) and some response to lime was also obtained elsewhere; however, if soil deficiencies are corrected the application of lime becomes unnecessary. The establishment and the initial growth are slow and utilization of stylo-based pastures can begin some 6 months after sowing, but perhaps earlier in cv. Endeavour and Cook, noted for their relatively vigorous early growth. In established swards early seasonal grazing is based mostly on the grass, which at that time is more palatable and bulky than the legume, whereas late in the season stylo becomes more palatable and can provide the main grazing. Mowing usually encourages the growth of stylo, whereas cattle grazing suppresses it especially if the animals are grazed on a 24-hour basis, but grazing by sheep is less harmful. Cutting below 15 cm is not recommended and the herbage can be cut at 6- to 16-week intervals, depending on local usage and the conditions of growth. Grazing, if deferred, can be done more frequently, with 3- to 6-week intervals. Stylo is utilized mainly for grazing and sometimes for soilage or hay. When hay is being prepared, losses of leaf and reduction in nutritive value can be considerable, and to retain high quality herbage the preparation of stylo meal instead of hay is advocated by some authors. Ensiling is little used although good silage was prepared in Nigeria (A. A. Adegbola from Tuley, 1968). In some parts of Africa, notably in West Africa, stylo alone or in mixture with grass was used experimentally and with some success to replace the short-term bush or grass fallow between the periods of arable cultivation. The value of stylo in fallows is due to its ability to grow under adverse conditions and to its competitive vigour and its resistance to serious pests and diseases which can affect the crops that follow. In Uganda, however, stylo can produce some adverse effect on the following crop of cotton (Tuley, 1968) possibly by root exudates.

Herbage yields. Herbage yields of stylo grown alone range usually from 2.5 to 10 t DM/ha but can reach over 15 t as it was, e.g., observed in small plots in western Samoa. Yields can be equal or sometimes higher than those of the best other pasture legumes such as *Centrosema pubescens, Macroptilium atropurpureum* or *Glycine wightii*. In grass mixtures stylo usually increases total yields of herbage compared with grass alone and always greatly increases the yields of crude protein. In Queensland yields of a *Panicum maximum*/stylo mixture reached 5,780 kg DM/ha, stylo contributing 50 per cent to the yield of DM and over 70 per cent to the yield of CP (Grof & Harding, 1970). In Rhodesia DM yields of natural grass were increased from 2,210 kg/ha to 3,370 kg and of CP from 37 to 151 kg when *S. guianensis* was oversown (Thomas, 1970).

Chemical composition, palatability. CP content in stylo herbage is not particularly high and ranges from 12 to 18 per cent of the DM and this relatively low CP content can perhaps be ascribed to a large proportion of stem. The plants can also contain appreciable amounts of oxalates, and Ndyanabo (1974) determined their content as 1.72 per cent of the DM although the content of water-soluble oxalates was low, 0.15 per cent, about the same percentage of oxalates that was found in *Centrosema pubescens*. Palatability varies, and reports on cattle refusing to eat this legume and, on the other hand, on its high palatability can be found in the literature. Cattle need to be used to stylo, which can be slightly aromatic, and they consume it well after some, often forced, experience; e.g. in Uganda the herbage consumed by cattle grazed on a *Hyparrhenia*/stylo pasture contained 10 per cent of stylo in the first day and 35–40 per cent on the third day (Stobbs, 1969c). Young plants are usually less palatable than more mature ones, and good consumption of mature herbage late in the season has been noted, and wilted herbage or hay can be more readily eaten than fresh herbage (from Tuley, 1968).

Animal production. In mixed swards the presence of stylo increases the liveweight gains of grazing cattle. In Nigeria it increased liveweight gains from 58 kg/head for cattle grazed on grass alone to 71 kg on grass/stylo mixture, the beneficial effect of stylo being particularly evident during the early part of the dry season (Haggar *et al.*, 1971). In Brazil (Buller *et al.*, 1970) a *Digitaria decumbens*/stylo mixture gave 259 kg liveweight gain/ha/year,, pure *D. decumbens* gave 139 kg when unfertilized and 239 kg were obtained when cattle were grazed on *D. decumbens* given 100 kg N/ha. In Uganda, natural grassland oversown with stylo and fertilized with P increased liveweight gains of cattle from 279 kg/ha for the original grassland to 474 kg (Stobbs, 1969e).

Reproduction, variability. *Stylosanthes guianensis* is a short-day plant and does not flower under a day length much over 12 hours. Some varieties flower only under even shorter photoperiods, although herbage yields can be higher under longer than shorter photoperiods (t'Mannetje, 1965). It is an essentially self-pollinated species although cross-pollination occurs frequently. The species is variable, and in Australia, and also in Brazil, its variability was thoroughly studied and superior types suitable for cultivation selected. In Australia four cultivars have been developed and registered; they all apparently belong to the so-called 'fine-stem' stylo (Barnard, 1972) although in other papers 'fine-stem' stylo denotes only some of the four cultivars. The four Australian cultivars are:
Schofield (after J. L. Schofield), a late-flowering standard cultivar of Brazilian origin. Suitable for tropical conditions with an annual rainfall over 1,500 mm. Readily nodulates from commercial cowpea inoculant.
Oxley. Early-flowering cultivar selected from Argentinian and Urug-

uayan introductions. It suits subtropical conditions, tolerates relatively low temperatures and can grow under an annual rainfall of 650–900 mm, and it is also more resistant to heavy grazing and to grass fire than other cultivars. Highly selective to *Rhizobium* strains and requires a special inoculant. This is perhaps the original 'fine-stem' type.

Cook. Early flowering, fast growing and high-yielding robust cultivar. Suits tropical conditions. Originates from Colombia.

Endeavour. Medium-early flowering cultivar noted for its vigorous initial growth. Suits the high-rainfall coastal area of Queensland.

In Brazil cv. **Deodoro** and **Deodoro** 2 have developed and cv. IRI–1022 showed good vigour and gave high yields; this last-named cultivar is however specific in its *Rhizobium* requirements and not easy to inoculate with usually high-efficiency *Rhizobium* strains (Souto *et al.*, 1972).

Seed production. Numerous flowers develop but flowering usually lasts for a considerable period of time as the flowering heads do not emerge simultaneously; moreover the flowers in each head do not all flower at the same time, and mature seed can shed while some flowers are only beginning to open. This and the glutinous secretion on the flower-heads makes harvesting difficult. The difficulty is accentuated by the position of flower heads within the upper part of the sward, mixed with herbage. Methods and equipment for mechanical harvesting have been developed in Australia. Seeds harvested from the same plot can differ in colour, the difference possibly depending on the stage of seed maturity at the time of harvest. Light-coloured seeds germinate easier than the darker seeds. A high proportion of hard seeds has been noted; these can remain in the soil undamaged for a long time, and seedling emergence can occur up to 3 years after the crop has been ploughed in. Seed yields range from 75 to 200 kg/ha, and there are about 250,000 seeds/kg. Basic information from Tuley (1968).

Stylosanthes hamata (L.) Taub.

Much branched perennial herb. Flowers yellow. Pod with one joint; the style, which is as long as the pod segment, persistent, almost orbicular. Occurs naturally in South America, West Indies and Florida, USA. In Jamaica it was found to be high yielding and reasonably palatable (Richards, 1970). In Florida, where it grows on sands, sand dunes and beaches, the plants vary in their agronomic characteristics and provide material for selection; *in vitro* digestibility of DM ranged in different introductions from 60.8 to 66.9 per cent (Brolmann, 1974). *Stylosanthes hamata* has also shown good promise in Australia.

Stylosanthes humilis H.B.K. Townsville stylo; Townsville lucerne. (Fig. 51). $2n = 20$

Stylosanthes humilis was known as Townsville lucerne until about 1968–9 when the Queensland Herbage Plant Liaison Committee

Fig. 51 *Stylosanthes humilis.*

recommended that the name should be changed to Townsville stylo in order to avoid confusion with species of *Medicago*. Townsville is a town in north-eastern Australia, the centre where work with *S. humilis* began.

Erect or subprostrate annual with stems up to 1 m long. Leaves trifoliolate, with narrow-ovate sharply pointed leaflets up to 2 cm long. Flowers yellow to orange, in terminal clusters. Pod 7–10 mm long consisting of two joints, of which the lower contains one seed, whereas the terminal joint, which is 4–6 mm long, is empty and is hook-shaped. The seeds are kidney-shaped, 2.3 mm long, sometimes yellowish but mostly brown or purplish-black. *Stylosanthes sundaica* Taub. is morphologically almost identical with *S. humilis*, except some details of the inflorescence structure, and is sometimes considered a synonym of the latter. It is however an Asiatic species and its distribution is confined to Java and the Lesser Sunda Islands, whereas *S. humilis* is an American species (t'Mannetje, 1968).

Stylosanthes humilis occurs naturally in Mexico, Central America, Venezuela, northern and eastern Brazil and also in the Ivory Coast of Africa.

Stylosanthes humilis is cultivated mainly in Australia where it was first found in 1913 growing in pastures, apparently as an involuntary introduction (Humphreys, 1967); large-scale trials began in Queensland in the 1930s. It is now widely cultivated in northern, and to a lesser extent mid-eastern Queensland, and also in north-western Australia and in the Northern Territory, where it is estimated to cover over 50,000 ha (Woods, 1969). In Australia *S. humilis* has been naturalized and can now be found thoughout eastern Queensland to the very north of the continent. In north-western, northern and north-eastern Australia *S. humilis* is considered to be an outstanding pasture legume, because of its adaptability to a variety of conditions, high protein content, the retention of high-quality herbage for a long period of time and the ease of establishment. It has indeed improved the natural pastures in the areas of its cultivation, almost doubled the stocking capacity of the land and increased animal production. Outside Australia it was partially successful under experimental cultivation carried out in India, Belize, Tanzania, Burma, the Philippines, where it is reported to be a native plant (Farinas, 1970), Florida in USA and in some other countries. *Stylosanthes humilis* has been successfully grown with several grasses: *Cenchrus ciliaris*, *C. setigerus*, *Dichanthium annulatum*, *Sorghum almum*, *Urochloa mosambicensis*, *Digitaria decumbens* and a few others, but its main importance is in the improvement of native Australian pastures, mostly those with the dominance or participation of *Heteropogon contortus*. This self-regenerating annual legume, once sown, seeds freely and establishes itself yearly for a few to several years. It is, however, a weak competitor and when perennial grass is suppressed by heavy grazing it can be overgrown with annual grass weeds.

Environment. *Stylosanthes humilis* required high light intensity for good growth and shading restricts it. In phytotron experiments (Sillar, 1969) the plants grew best at temperatures alternating from 27° to 34°C and were highly responsive to temperature increases if they were below the optimum (Humphreys, 1967). Optimum seed germination was observed at temperatures below those required for plant growth. In Australia *S. humilis* requires a relatively high annual rainfall, from 650–700 mm p.a. upwards, although it can occasionally tolerate drier conditions. Summer-season rainfall is essential and the plants do not grow well in the areas of winter rains. Young plants can be killed by drought, but at later stages of growth droughts can even stimulate growth for some time before reducing it (Fisher, 1970). *Stylosanthes humilis* grows on nearly all types of soil, even on those unsuitable for other forage plants or for cropping, such as shallow, structureless soils, podzolic or solodic soils and badly drained soils where the water table is close to the surface, but it cannot withstand flooding. It tolerates soil acidity up to pH 4 and high levels of labile Al and Mn. In mixed stands it sometimes grows better on poor than on good soils, where grass competition can be too strong.

Establishment. Only a shallow seedbed is needed for the establishment of *S. humilis*, which is achieved by discing the natural sward to loosen the soil and to reduce grass competition at the early stages of establishment. Discing, which can be replaced by the use of a tined cultivator, is preferably done early in the rainy season; in October in the south of Queensland and in November in the north; the seed is broadcast immediately after cultivation. With suitable machinery available, cultivation and sowing can be done in one operation. Sowing before some reliable rains arrive is usually avoided, as rain storms which may occur at the end of the dry season can induce germination, and the seedlings may die if the storm is not followed by more persistent rains. This however does not necessarily happen; Torssell *et al.* (1968) report that a 38-mm rain caused the seedlings to emerge, and most of them then survived 7 weeks of dry weather before the main rains arrived. It is essential to graze heavily, or burn the grass before sowing *S. humilis* in order to reduce grass competition. Complete cultivation to kill the existing vegetation is not necessary on light soil but essential for heavy clays. *Stylosanthes humilis* can be established by direct oversowing without any cultivation, but this method is less reliable and results in thinner stands of the legume and lower yields of herbage. Direct oversowing is usually more reliable under a high than a low rainfall, and in the lower rather than the higher latitudes of Australia. Even under the most suitable conditions, stand densities of *S. humilis* will only be half of those achieved by sowing into disced land. When *S. humilis* is established in mixtures with sown grass, soil cultivation is of course necessary, and if seed is drilled it is essential to plant it not deeper than 0.5 cm; if this cannot be achieved then surface sowing is preferred. For good production the application of P fertilizers is essential, and the amounts of superphosphate needed for high production and the recommended rates vary widely from 120 to 360, or even 600 kg/ha, depending on the soil and the economics of farming. Molybdenum is often added to superphosphate at a rate of 0.03 per cent. On K-deficient soils 120 kg of muriate of potash/ha are applied. Seed is sown together with the fertilizer at a rate of 3–5 kg/ha of standard quality seed; an increase in sowing rates from 3 to 6 kg usually increases herbage production in the year of establishment. It has been estimated (Humphreys, 1967) that about 10 per cent of sown seed develop into adult plant. Heavy grazing of newly established swards reduces grass competition because, when young, *S. humilis* is less palatable to the grazing animals than the grass; heavy grazing is particularly necessary when *S. humilis* is established in undisturbed grassland.

Nodulation, nitrogen fixation. Roots of *S. humilis* nodulate freely from natural *Rhizobium* bacteria found in the soil and no inoculation is usually needed, but if it is required then inoculants of the cowpea group can be used. The content of N in the soil after 7 years under *S. humilis* was reported to be 110 kg/ha higher than before the legume was sown (CSIRO *Plant Introduction Review*, Vol. 3, No. 1, 1966). The quantity of it in the soil however did not progressively increase from year to year, and it is believed that nitrogen fixation is only active in the first year of growth; in subsequent years enough nitrogen was available for plant needs and its presence suppressed further fixation of atmospheric nitrogen.

Management, fertilizing. In the establishment year mixed swards containing *S. humilis* are managed mainly for obtaining maximum seed yields for subsequent regeneration, and heavy stocking at the seed formation stage is avoided. In subsequent years heavy stocking and close grazing are recommended for a longer period. This often results in smothering by annual grass weeds, Chlorthal at 4–6 kg/ha has proved to be an effective herbicide (Wood, 1970) which controls the weeds and thus encourages *S. humilis*. The same herbicide was also applied at establishment when *S. humilis* was sown into weedy land. Roots of *S. humilis* were found to contain endotrophic mycorrhiza which may help the plants to utilize P from soils of low P status (Possingham *et al.*, 1971). Nevertheless P fertilizers are necessary for obtaining high yields of good quality *S. humilis* herbage. Similarly with establishment, fertilizer requirements and recommended rates vary in subsequent years and 60 kg superphosphate/ha/year is the minimum requirement, 120 kg is a normal rate and 600 kg a maximum. Sillar (1969) recommends 360–600 kg/ha in the first 3–5 years and then 120 kg every second year. R. K. Jones (1968) reports that a *S. humilis*/*Urochloa mosambicensis* mixture yielded during the 3 years of trial 12.6 t DM/ha when given 800 kg superphosphate as an initial dressing and only slightly less, 11.8 t, with 370 kg of superphosphate. Fisher (1970) writes that 250 kg superphosphate/ha gave about 75 per cent of the effect produced by 625 kg. At Katherine, northern Australia, on clay soil, the application of 120 kg superphosphate/ha increased DM production by 31 per cent over that obtained from unfertilized stands, and no further response was obtained from 360 kg.

Chemical composition, nutritive value, digestibility. Crude protein content of *S. humilis* herbage varies widely from 9 to 18 per cent of the DM, depending on the growth conditions and the stage of growth, but it usually ranges between 12 and 15 per cent. Phosphate fertilizers almost invariably increase the N content of the herbage. A low concentration of Na, about 0.10 per cent of the DM, which sometimes approaches the critical level of 0.05 per cent, was noted. During the early stages of growth the herbage is less palatable than in flowering or mature plants, and *S. humilis* is readily eaten by the grazing animals as 'standing hay' of mature plants. Organic matter digestibility ranges from 40 to 73 per cent and that of CP from 58 to 75 per cent (Thurbon *et al.*, 1970). Digestibility decreases with age, but the decreases are relatively small.

Herbage production. *Stylosanthes humilis* grown in pure stands, or with grass, forms swards some 40–50 cm in height, but when heavily grazed it can form a dense mat of freely branching shoots. In the establishment year the yields of DM normally vary between 0.6 and 4.8 t/ha and in subsequent years between 1.8 and 2.4 t.

Conservation. Hay is often prepared; the cut plants are crushed, turned with side rakes the next day and baled 3–4 days later. Hay yields are comparable with those of dry matter and they depend on the weather conditions of the year and on soil and fertilizers applied. CP content of the hay and its digestibility are usually reduced compared with those of fresh material at the comparable stage of growth.

Stocking rates and animal production. The introduction of *S. humilis* into local pastures in northern and north-eastern Australia has drastically increased the carrying capacity of grazing land, partly because it increased total yields and the quality of herbage, and partly because it extends grazing well into the dry season. It was recorded e.g. that on a local unimproved pasture cattle gained 87 kg/animal, and 165 kg on a similar pasture oversown with *S. humilis*; the difference was partly due to liveweight losses of 28 kg per animal on unimproved pasture during the dry season of April–November, whereas cattle grazed on the *S. humilis* pasture gained weight during this period. In another trial *Heteropogon contortus* pasture gave 45 kg liveweight increase per animal per year and 29 kg per ha, whereas a similar pasture oversown with *S. humilis* gave 121 and 93 kg, respectively. When fertilized with PK the unimproved pasture gave 100 and 62 kg per animal and per ha, respectively, and the pasture with *S. humilis* 150 and 148 kg (Shaw & t'Mannetje, 1970). In general, liveweight gains of cattle grazed on *S. humilis* pasture can be expected to be about 90 kg per head under relatively unfavourable conditions, and up to 160 kg on good soil and under favourable weather conditions. Stocking rate as high as 0.8–1 ha per steer is often considered to be a safe rate, and this exceeds three- to four-fold the carrying capacity of unimproved pastures. Heavy stocking is normally advocated for *S. humilis* pastures, and there are reports that at one beast to 1.6 ha liveweight increases per land unit were considerably higher than at one beast to 3.2 ha.

Reproduction, variability. In Australia *S. humilis* flowers 81 to 151 days after sowing, earlier at 19°S. and later at 24°S., the time of flowering also depending on the variety (Cameron, 1967). Day length seems to be the main factor controlling the flowering and different maturity types of *S. humilis* grown in controlled environment showed a strong response to short photoperiods and flowered under day lengths not exceeding 13 hours for the early and $11\frac{1}{2}$ hours for the late varieties (Cameron, 1967), e.g. cultivar Lawson flowered under $11-12\frac{1}{2}$ hours photoperiods and cv. Gordon under $10\frac{1}{2}-11\frac{1}{2}$ hours photoperiods. *Stylosanthes humilis* varies considerably in its habit of growth, from erect to semi-prostrate, vigour and the time of flowering. Local varieties suitable for certain conditions have developed naturally in various localities and latitudes in respect to their reactions to day length and to local climate and soils. Ordinary commercial seed is usually a mixture of varieties, most of which are of good vigour. *Stylosanthes humilis* is a self-pollinating plant and the cultivars so far selected in Queensland are based on a single plant each. The three known cultivars **Gordon**, **Lawson** and **Paterson** are of erect types and morphologically similar, except that ripe seeds of Paterson are purplish-black whereas those of Gordon and Lawson are brown. Gordon is a late flowering cultivar, Paterson early flowering and Lawson is intermediate in this respect and they have been selected to suit the 1,100 mm, 630–900 mm and 1,000 mm annual rainfall areas, respectively.

Seed production. Seed of *S. humilis*, as sown, is either in the form of pure hulled seeds (about 420,000 seeds/kg) or in the form of unhulled pods (about 300,000 seeds/kg) containing one seed in the basal joint, the upper joint being empty; this upper joint is in the shape of a hook which is considered an adaptative feature facilitating plant distribution by clinging to grazing animals. This hook makes mechanical sowing difficult because it hinders a free flow of seed and dehooking of the seed is not easy. *Stylosanthes humilis* is a prolific seeder, and when adequately fertilized with P yields of 280–340 kg seed/ha have been obtained, and up to 550 kg under particularly favourable conditions. Stand densities of 850 plants/m^2 gave the highest seed yields: 69 g/m^2, as was shown in trials with plants grown in boxes (Shelton & Humphreys, 1971). Harvesting is done by hand, and more recently also by the sucking machines used for harvesting *Trifolium subterraneum* seed and adapted for *S. humilis*. These machines, which have now been made simple and effective, make harvesting cheaper compared with hand harvesting and reduce the seed price, thus promoting a wider use of *S. humilis*.

Minimum requirements for standard seeds are 97 per cent purity and 40 per cent germination. A large proportion of hard seeds in those freshly harvested was noted; this proportion, which can reach some 90 per cent, is reduced naturally during the storage before the seed is sown; a large proportion of remaining hard seeds can germinate, after a year in the soil, in the second season.

Basic information from Humphreys (1967), Sillar, (1969) and from 'Officers of the Agricultural Branch of the Northern Territory Administration', (1966).

Teramnus P. Br.

A small genus of eight species, mostly perennial twining or trailing herbs resembling perennial species of *Glycine* but differing in having only five well developed anthers, the other five being small and sterile or absent, whereas in *Glycine* all ten anthers are developed normally. Pods linear, straight, or slightly curved or hooked at the apex, and with about eight seeds. Most of the species are distributed throughout the world tropics and two or three show promise for cultivation as fodder or pasture plants, but not sufficiently known so far. *Teramnus labialis* (Linn.f.) Spreng. is perhaps the most suitable for cultivation: it forms dense growth, blends well with grasses and seems to be a good seeder; it is well liked by cattle but less so by sheep. It thrives on wet soils and requires moderate or good annual rainfall, perhaps over 1,000 mm. *Teramnus uncinatus* (L.) Sw. is in general more robust and vigorous and has slightly larger leaflets; the pods have brown or dark hairs, whereas most forms of *T. labialis* have pods with white hairs or are hairless. *Teramnus uncinatus* grows normally in drier areas or in drier habitats than *T. labialis* and is another promising species although Buller *et al.* (1970) report that in their trial *T. uncinatus* yielded only 2.45–3.61 t DM/ha, depending on P and K levels in the soil, less than most other species; moreover it dried up and lost its leaves during the 5 months of cool and dry season. A species worth mentioning is also *T. volubilis* Sw., confined mainly to South America where it provides good grazing in natural pastures.

Trifolium L. Clover.

Annual or perennial herbs. Leaves with three (or five in a few species) leaflets. Stipules fused with petioles to a considerable length. Flowers mostly in dense globose, semiglobose or elongated heads. Pods short, surrounded, at least in the basal part, with the persistent calyx and have 1 to 10 seeds. A large genus of about 300 species distributed mainly in temperate countries, with the greatest concentration of species in the Mediterranean area. Several species are of considerable significance for agriculture, the most important being *T. repens*, *T. pratense*, *T. hybridum*, *T. subterraneum* and *T. alexandrinum*.

Trifolium alexandrinum L. Berseem; Egyptian clover. (Fig. 52).

$2n = 16$

Sparingly pubescent annual 30–80 cm high. Leaflets 1.5–5 cm long, oblong to oblong-lanceolate. Petiole long in the lower leaves and very short in the upper leaves. Stipules lanceolate, the free parts subulate. Flowering heads on short peduncles to almost sessile, ovate, becoming oblong, 15–25 mm long and 15 mm wide. Bracts linear, short. Calyx tube 2–3 mm long, in fruit with triangular-subulate teeth, the lower teeth 3–4 mm long. Corolla twice as long as the calyx, cream coloured to almost white. Pod oblong-ovoid, with one or sometimes two seeds. Seeds yellow to reddish-brown, 2–2.5 mm long.

Fig. 52 *Trifolium alexandrinum.*

Trifolium alexandrinum has been grown in Egypt from ancient times and is apparently of Egyptian origin. It is now widely cultivated and is of considerable importance as a fodder crop, mostly irrigated, in Egypt and in some other, mainly subtropical, near-East countries and in India, and perhaps to a lesser extent in Pakistan. It is also a well-known fodder crop in the Mediterranean area and has also been introduced to South Africa, the southern states of the USA, the Caribbean Islands (Cuba), Argentina and to southern and south-western parts of Australia. It has also been tried, mostly with moderate to good success, in middle-European countries, such as Germany or Switzerland.

Three types of berseem have been distinguished (Kaddah, 1962), mainly on the basis of stem branching which is connected with plant ecology and utilization and typified by common, well-known varieties or cultivars. The Miscawi type at an advanced stage of growth develops short side-branches at the plant base and, when the plant is cut, these branches elongate and produce new growth. The Fahl type develops no such branches and it does not regenerate freely after cutting but its stems branch in the upper half more freely than in the two other types. The Saidi type develops the basal shoots but for a shorter period of time than the Miscari type and it also branches in the upper part of stems but less freely than the Fahl type. Five to six cuts can be taken under irrigation from the Miscari variety, two to three cuts from Saidi berseem and usually only one from the Fahl variety although the yield of a single cut of Fahl can be higher than single cuts of the other two types.

Berseem clover is an annual crop which is valued for its rapid growth under cool conditions of subtropical or tropical winters, good recovery after cutting and several cuts can be taken under irrigation of certain varieties or cultivars. Under dry land conditions only one, or occasionally two, cuts can be taken and the Fahl type cultivars are well adapted for semi-arid areas. The quality of berseem herbage is good and the plant also improves soil nitrogen status and the last regrowth is often ploughed in before planting cotton or other valuable crops. Berseem is essentially a subtropical crop although it is grown in Egypt, India and is under trial in several other tropical countries. In warm countries it is predominantly a winter crop.

Environment. According to Kretschmer (1964) berseem clover can withstand some 2°-3°C below the freezing point and basal parts of adult plants can even survive down to $-7°C$ but some other authors maintain that the plants are highly susceptible to frost. High temperatures during the active growth are not necessary and can even suppress it but they stimulate flowering and seed formation. *Trifolium alexandrinum* responds well to soil moisture and irrigation but does not tolerate waterlogging. It grows on various soils although heavier soils are more suitable than too light, sandy soils. Tolerance to soil alkalinity and salinity is well known; the plants can tolerate up to pH value of 8 and a considerable degree of salinity and small amounts of salts in the soil can benefit plant growth.

Establishment. In subtropical areas berseem is usually planted in the autumn, broadcasted or drilled in rows at sowing rates of 20–30 kg/ha and 10–15 kg seed/ha, respectively; Whyte *et al.* (1953) give even higher sowing rates, 50–60 kg/ha, broadcast, for Egypt and South Africa. When broadcast, discing after sowing is recommended so that seeds are buried to a depth of 1–2 cm. Seed is inoculated with the *Rhizobium trifolii* inoculant and the inoculant in use for *T. repens* can be applied (Kretschmer, 1964). In the fields where *T. alexandrinum* was relatively recently grown no inoculation may be necessary. The application of fertilizer P is recommended before sowing and Narayanan & Dabadghao (1972) advise the application of 65–70 kg P or some 400 kg superphosphate/ha, and Kretschmer (1964) about 750–900 kg of 0–12–12 fertilizer. Narayanan & Dabadghao also recommend small amounts of N together with P and also B and Mo as micronutrients and the application of Cu, Mn and Zn is recommended by Kretschmer in Florida. On irrigated land wet and imbibed seed can be sown into shallow water; the seeds germinate rapidly and young seedlings appear almost as soon as the water recedes.

Management, uses. Berseem can be ready for cutting 50–60 days after sowing and under irrigation, in India, it is given one or two further irrigations before the crop is harvested for the first time. Management of *T. alexandrinum* in its productive stage includes fertilizing with P or P + K applied after every second cut or some 10 days before cutting to make the plants grow fast soon after they have been cut. For irrigated *T. alexandrinum* 12 to 15 irrigations are recommended in India throughout the winter months (rabi season) which may result in five to six cuts of multicut cultivars. Berseem is usually grown in pure stands although, if seed is sown early, rape can be sown as well; rape grows very rapidly and after it has been cut or grazed berseem recovers well and its first crop is not much affected (Narayanan & Dabadghao, 1972). Sometimes berseem is also grown with oats and occasionally with some other crops such as *Euchlaena mexicana* or Sudan grass. The first cut can be taken when the plants are about 30 cm tall, at the time when the basal young shoots of multicut varieties reach about 2–3 cm and cutting to a height of 5–6 cm is preferred to a lower level in order not to damage these young branches. Subsequent cuts can be taken at some 35- to 45-day intervals; the intervals can be longer during the coldest months of winter when the growth can be slow. The speed of growth increases towards the spring, but in late spring or early summer plant growth is adversely affected by high temperatures; berseem is then ploughed and replaced by summer crops such as cotton, sorghum and a variety of other crops. Berseem is usually cut for soilage and fed as green fodder. When it is cut for hay, drying in tripods has an advantage over drying on the ground because it reduces the losses of DM and in Egypt (Danasoury *et al.*, 1971) the first cut dried on tripods produced 35.7 per cent more DM than when dried on the ground. In India, chopping green fodder before drying it for hay in the sun for 2–3 days reduced the losses of leaf (Relwani, 1971). Berseem is also used for grazing, a common practice in the upper Nile area of Egypt (Whyte et al., 1953). For the use as silage berseem herbage is usually mixed with grass; ensilage may, however reduce the digestibility of CP and is seldom used in practice.

Herbage yields. Average annual yields of irrigated multicut varieties can be estimated as 30–45 t fresh fodder/ha in five to six cuts although higher yields, up to 60 or more t/ha, have also been recorded. Dry matter content in the first one or two cuts is low, 10–12 per cent, but increases with the plant age and in subsequent cuts can reach 15–20 per cent. In trials by Kretschmer in Florida, USA, annual yields of oven-dry herbage ranged from 4.7 to 6.0 t/ha (DM yields would be only slightly lower) for the multicut cv. Miscawi and 2.9–3.8 t for the single-cut cv. Fahl; in these trials cv. Miscawi outyielded lucerne and white clover during comparable periods of time. In practice yields of DM are likely to be somewhat lower, in the order of 3–5 t DM/ha for multi-cut varieties. In Pakistan (Shah et al., 1970) lactating buffalo cows fed with lucerne and berseem supplements by equal rations of concentrates gave 8.6 kg and 7.2 kg milk/day, respectively. There are reports that *T. alexandrum* can occasionally cause bloat in cattle.

Nutritive value. Crude protein content in the herbage is generally high and ranges from 17 to 23 per cent (Kretschmer, 1964) but is lower than in lucerne or white clover and an analysis of herbage at the flowering stage showed only 11.6 per cent CP (Dougall, 1962). Crude fibre content is around 20 per cent and is lower than in lucerne and NFE content higher and can reach well over 50 per cent. Digestibility of CP of fresh herbage is reported by Narayanan & Dabadghao (1972) to be 81 per cent, that of CF 60 per cent and of NFE 80 per cent. Digestibility was slightly lower in hay, 70, 49 and 71 per cent, respectively and considerably lower in silage – 31, 52 and 47 per cent.

Pests and diseases. Caterpillars of *Agrotis segetum* and cotton worm (*Prodenia littoralis*) can inflict a considerable damage to young, early sown plants of berseem in India but later sown stands usually escape damage, and stem rot caused by *Rhizoctonia solani* is the main disease. Stands of berseem can be attacked by a few species of dodder (*Cuscuta* spp.), a parasitic weed which under severe infestation can greatly reduce the yields of affected clover. *Cuscuta* usually appears in patches and cutting the affected patches, covering the cut clover with straw, pouring some paraffin and burning is one of the control measures; another measure is not to use berseem seed in which even a few dodder seeds have been found. Another parasitic weed affecting berseem is *Orobanche*, some species of which are parasites on clover roots.

Reproduction, cultivars. *Trifolium alexandrinum* is a predominantly self-pollinated plant. Warm weather in spring stimulates flowering and seed formation in this winter crop and of irrigated multicut berseem the last cut is often preserved for seed production. Seed yields can reach up to 600–700 kg/ha; there are 300,000–450,000 seeds/kg.

The better known multicut cultivars used mainly under irrigation are **Miscawi (Miskawi, Muscawi)**, **Kadrawi (Khadrawi)**, both old Egyptian cultivars, and **Hastler** of USA selection. **Saidi (Saida)** is a cultivar of intermediate type between the multicut and singlecut cultivars; it can be grown under short-term irrigation or on dry land in moderately moist areas. **Fahl** is the best known Egyptian single-cut cultivar suitable for growing in dry areas without irrigation. Other cultivars mentioned in the literature are **Bali (Baali)** introduced to Australia from Israel, **Wafeer**, an Egyptian cultivar, **Imperial** of USA selection, **Tabor** (single cut) and **Carmel** (multicut), both grown in Cuba, and a number of others.

Trifolium pratense L. Red clover.

A short-lived erect or suberect perennial, a species of Euro-Siberian origin now widely cultivated in all temperate countries with sufficiently moist climate but not in arid or semi-arid areas. In the tropics it is grown at high elevations over 1,800 m alt, but mostly above 2,000 m and up to well over 3,000 m, in several countries, including Colombia and Bolivia, and tried in Sri Lanka and Australia. Oversown to natural grassland in Sri Lanka it increased DM yields, from 1.6 t/ha to 2.5 t (Jayawardana & Andrew, 1970). In tropical Australia it did well when oversown to an old sward of *Pennisetum clandestinum* but did not last beyond the year of establishment (Gartner, 1968). In Bolivia and Colombia (Crowder, 1960) it has been grown with good success in mixtures with temperate grasses.

Trifolium repens L. White clover

Creeping perennial which grows naturally in the temperate zone of the Old World, and is now widely cultivated in mixture with grass in temperate countries throughout the world, to a lesser extent in the subtropics and also in high altitude areas of the tropics. It has been grown in East Africa, notably in Kenya, in Rhodesia, India, Sri Lanka, Hawaii, Bolivia, Colombia and subtropical Australia and tried in a number of other tropical countries with various success. The more successful mixtures were with *Setaria anceps* cv. Nandi, *Pennisetum clandestinum*, *Acroceras macrum*, *Brachiaria humidicola*, *Digitaria decumbens*, and at particularly high altitudes with temperate grasses. Of the cultivars grown in the tropics, Louisiana, mainly Sl, is perhaps the best known and most successful, large Ladino, and an Israeli cv. Haifa also of the large Ladino type.

Trifolium rueppellianum Fresen var. **rueppellianum** (*T. subrotundum* A. Rich.). Rueppels clover. $2n = 16$

Erect, suberect or prostrate glabrous annual. Leaflets mostly oblong or obovate up to 24 mm long and 14 mm wide. Flowers purple, in globose heads. Seed oval, 1.5 mm long.

Occurs in Fernando Po, Cameroon, eastern Congo, Sudan, Eritrea and tropical East Africa at altitudes over 1,500 m and an annual rainfall of 650–1,200 mm. Photoperiodically neutral. *Trifolium rueppellianum* nodulates freely and trials in Kenya suggest that it fixes nitrogen effectively. It was tried in mixtures with perennial grasses and contributed considerably to the yields of herbage and its CP content in the first year of growth, but its regeneration in the following years was usually poor because the seedlings, which emerged freely during the first rains, mostly died out during the first spells of dry weather. *Trifolium rueppellianum* was also tried with moderate success in Rhodesia, Mauritius and Australia, mainly subtropical. Analysed at the early flowering stage it contained 23.0 per cent CP, 16.0 per cent CF, 44.2 per cent NFE, 1.24 per cent Ca and 0.19 per cent P but at full flowering CP content was reduced to 21.3 per cent and P to 0.16 per cent, whereas CF content increased to 19.5 per cent (Dougall, 1962). The herbage is well eaten by stock, but mainly at the flowering or later stage; young plants are not willingly grazed. It is a prolific seeder but the pods shatter easily and should be harvested with care.

Trifolium rueppellianum is predominantly self-pollinated but flower tripping, which is normally effected by bees (*Apis mellifera*), is necessary for seed formation and cross-pollination in the field is likely to range from 1 to 10 per cent (Bogdan, 1966b). In East Africa it varies widely in vigour, leafiness and shows good promise for breeding. A very dense upright type promising for cultivation, possibly a mutant, was found in Kenya.

Trifolium mattirolianum Chiov., common in Ethiopia, is almost identical with *T. rueppellianum* except that it requires cross-pollination.

Trifolium semipilosum Fresen. Kenya white clover. $2n = 16$

This species was known for some time in Kenya under the incorrect name of *T. johnstonii*.

Low-growing perennial which can reach a height of over 30 cm when well fertilized and grown with a not too aggressive grass. Stems creeping and freely rooting at the nodes, hairy. Leaflets oblong to obovate, hairy on the outer halves of the lower surface of lateral leaflets. Flowering heads on long hairy peduncles, 2 cm wide, 10–25-flowered, with the pedicels reflected after flowering as in *T. repens*. Corolla white to pale-pink, 8–9 mm long. Pods 5–6 mm long with two to six grey or brown, often mottled seeds 1.5 mm long.

There are two main varieties: var. *semipilosum* and var. *glabrescens* Gillett. Two other formerly distinguished varieties, var. *kilimanjaricum* Bak.f. and var. *microphyllum* Chiov., seem to be identical with var. *semipilosum* (Gillett, 1971). *Trifolium semipilosum* grows naturally at high elevations of Yemen, Ethiopia and east tropical Africa, from 1,400 m upwards and var. *semipilosum* can reach up to 3,000 m alt. Var. *glabresens* is a larger plant than the other variety, with larger and broader leaflets often almost glabrous and with thinly scattered hairs underneath the lateral leaflets, and is of some importance for grazing; it has been tried in cultivation and this variety is considered below as *T. semipilosum*.

Trifolium semipilosum often occurs in Kikuyu grass (*Pennisetum clandestinum*) swards, and if the grass is kept low can constitute up to 15 per cent of the herbage. It is a plant of cool areas, the optimum day/night temperatures being 15°/10°C for the growth of roots and 21°/16°C for leaves and stems. Lower temperatures also increased the number of flowers/head from 33 at the optimum temperature of 21/16°C to 38 at 15°/10°C. The higher temperatures hastened however seed ripening which matured in 19–27 days at 27°/22°C, 28–40 days at 21°/16°C and in 67–78 days at 15°/10°C (Mwakha, 1969). *Trifolium semipilosum* is a photoperiodically neutral plant. It requires over 800 mm annual rainfall and becomes dormant during the dry seasons.

Trifolium semipilosum was introduced into cultivation by D. C. Edwards (Bogdan, 1965), the first cv. **Kabete 4** originating from a mixture of four clones selected for their vigour and persistence. Productive mixtures with *Setaria anceps* cv. Nandi and Rhodes grass (*Chloris gayana*) were grown experimentally but the yields were inconsistent possibly because of erratic nodulation. *Trifolium semipilosum* inoculates naturally in the areas of its distribution but requires specific *Rhizobium* bacteria in other areas, and it responds to fertilizer P and to gypsum. It was also tried with some success for oversowing natural *Hyparrhenia* grassland (Keya *et al.*, 1971) and although yield increases were negligible CP content of the total herbage was increased in the presence of *T. semipilosum* from 6.7 to 9.6 per cent. Introduced to Rhodesia, *T. semipilosum* did well in mixtures with several grasses (Clatworthy, 1970) and was even recommended for farm-scale cultivation at high altitudes. In Rhodesia it was complementary to *Lotononis bainesii* and it was suggested that the two legumes could be grown together. Reasonably good results were also obtained in Australia, mainly south-eastern Queensland, in trials with *Paspalum plicatulum*, Pangola grass (*Digitaria decumbens*) and Kikuyu grass. *Trifolium semipilosum* is well grazed at all stages of growth but there is some possibility of bloat when luxurious growth is grazed in the early morning. The nutritive value is high and the plants analysed in Kenya at the full flowering stage contained 23.7 per cent CP, 2.9 per cent EE, 18.7 per cent CF, 40.8 per cent NFE, 1.78 per cent Ca and 0.32 per cent P (Dougall, 1962). A new cultivar, **Safari**, has been recently developed in Australia.

Trifolium semipilosum is essentially a cross-pollinated plant and if bees (*Apis mellifera*) are present in sufficient numbers seed is well produced. However, because of the low position of flowering heads, which often recline to the ground when overripe, mechanical harvesting is difficult, whereas hand picking is too labour-consuming and therefore expensive. Moreover, in pure stands grown for seed, weed invasion can be severe and difficult to control; in Kenya weed seeds particularly difficult to separate were those of *Commelina subulata* Rich., *C. benghalensis* L., *Nicandra physaloides* (L.) Gaertn. and *Amaranthus hybridus* L. (Bogdan, 1966c).

Trifolium subterraneum L. Subterranean clover; Subclover.

A spreading annual with the flowering heads of three to six fertile flowers on common peduncles which bend down after flowering and bury the seeds in the soil. This species is of Mediterranean origin but introduced to other parts of the world as a winter-growing legume for temperate or sometimes subtropical areas with cool wet winters and hot and dry summers. It is of particular importance in southern Australia where it has contributed a great deal to soil improvement and increased the productivity of pastures. Argentina and Uruguay are subtropical areas where *T. subterraneum* is also grown (Whyte *et al.*, 1953). In the tropics it can be cultivated at high altitudes with cool and wet growing periods not necessarily followed by hot summers, as it has been proved in Kenya, mainly by J. Morrison, but it does not perform well below 1,800 m alt.

Trifolium tembense Fresen $2n = 16$

Erect to prostrate annual. Stems rather thick but hollow and soft. Leaflets elliptic to obovate or rhomboid, under 2 cm long and 10 mm wide; petioles from 5 cm long in the lower leaves to 5–10 mm in upper leaves, fused with the stipules to the most of petiole length. Inflorescence semiglobose of 3 to 16 flowers supported by broad bracts. Calyx broad; corolla purple, standard about 9 mm long. Pods 5–7 mm long with three to eight seeds. Seeds brown.

Distributed in tropical East Africa including Zaire and Ethiopia, from 1,400 m alt upwards, in wet situations, often in shallow water and can form extensive colonies, pure or mixed with creeping species of *Pennisetum*. It is well eaten by stock, but usually not before flowering. Analysed in Kenya at the full flowering stage it contained 23.6 per cent CP, 2.6 per cent EE, 17.8 per cent CF, 42.8 per cent NFE, 1.32 per cent Ca and 0.24 per cent P (Dougall, 1962), the analysis showing a good nutritive value.

Trifolium usambarense Taub. $2n = 16$

Annual or rarely short-lived perennial up to 1 m high, with slender stems. Leaflets 6–22 mm long. Leaf petiole united with the stipules to its full length. Flowering heads oblong, with about 30 flowers. Corolla purple, rarely white. Standard 4–7 mm long. Pods about 3 mm long, usually with two seeds. Seeds dark, mottled.

Occurs in tropical Africa, at altitudes of 900 to 3,000 m, in swampy situations and by streams. *Trifolium usambarense* is not particularly widespread, but when it occurs it usually forms extensive colonies. It showed good promise in trials in Kenya and contained, at the late-flowering stage, 19.9 per cent CP, 2.0 per cent EE, 29.3 per cent CF, 37.9 per cent NFE, 1.33 per cent Ca and 0.23 per cent P (Dougall, 1962).

Vicia L. Vetch.

A genus of about 120 species distributed mainly in temperate areas of the northern hemisphere and, to a lesser extent, in temperate South America; there are a few species found at high altitudes in the tropics and subtropics. Some species, mainly *V. sativa*, *V. villosa* Roth. and *V. benghalensis*, are of importance for agriculture; they are grown for hay, silage or grazing in fodder mixture with spring or winter cereals. Several species have been tried, with some success, in subtropical and tropical areas, mostly at high altitudes.

Vicia benghalensis L. (*V. atropurpurea* Desf.). Purple vetch. $2n = 14$

Annual, biennial or short-lived perennial, rambling or climbing by tendrils. Flowers purple, 2–20 in axillary racemes, pods 8–12 mm wide. Originates from the Mediterranean area and cultivated in warm temperate areas in fodder mixtures with winter cereals. It is also grown in the subtropics and tropics at high altitudes.

Vicia sativa L. Common vetch.

An annual climbing by means of tendrils. Flowers single or paired in leaf axils. Pods 6–8 cm long. Widely grown in temperate countries in fodder mixtures with oats and used for making hay or silage or for grazing. Introduced into Australia, mostly temperate, when it is grown as a winter crop and also to Brazil and other South American countries. In Colombia a narrow-leaved variety, var. *angustifolia* L. (*V. angustifolia* L.), is grown at altitudes from 1,200 to 3,200 m (Crowder, 1960) where it produces 3–5 t DM when cut at the flowering stage.

Vigna Savi

Annual or perennial herbs of various habit, rarely small shrubs. Leaves mostly with three leaflets (always in the species described below). Flowers variously arranged but mostly in short racemes on relatively long peduncles. Calyx two-lipped: lower lip with three lobes or teeth, upper with two teeth. Standard with inflexed auricles and two or four short appendages on the inner side. Keel sometimes twisted up to 360° (in allied *Phaseolus* it is spirally twisted to over 360°). Pods dehiscent, linear or linear-oblong, often cylindrical. Seeds of various shape and can be cylindrical or in the form approaching a cube.

A large genus of about 150 species distributed throughout the world tropics and subtropics. It merges with the genus *Phaseolus* but typical species of both genera are quite distinct and the two genera are treated separately. A number of wild species are palatable and grazed by stock. *Vigna unguiculata* or cowpea is an important crop grown for grain, as a fodder plant or for green manure, and there are some other species under trial as fodder or pasture plants.

Vigna aconitifolia (Jacq.) Maréchal (*Phaseolus aconitifolius* Jacq.). Moth bean; Mat bean; Dewgram. $2n = 22$

Slender annual with short erect main stem and long trailing branches. Leaflets 5–8 cm long; terminal leaflet with five deep acute lobes, lateral leaflets with four lobes. Inflorescences axillary, dense, on 5–10 cm-long peduncles. Flowers yellow, up to 1 cm long. Pods brown, 2.5–5 cm long and 5 mm wide, with four to nine seeds. Seeds rectangular, about 5 mm long, yellow to brown, often mottled with black.

Originates from India and Pakistan where it is widely grown as a pulse crop, for fodder and also for green manure. Introduced to Sri Lanka, China and south-western USA where it is grown for fodder, and to Sudan. *Vigna aconitifolia* is a short-day plant which grows best under high constant temperatures. A drought-resistant species growing well under an annual rainfall of 750 mm (Purseglove, 1968), but can also be grown under an annual rainfall of 500 mm and even 300 mm (Daulay *et al.*, 1970). It grows on a variety of soils, light textured soils being most suitable.

In India *V. aconitifolia* is grown as a rain-fed crop, but sometimes under irrigation, pure or more often in mixture with sorghum and some other grasses, and occasionally with an admixture of other legumes. When grown alone it is sown at 15–20 kg seed/ha and at 2–5 kg in mixtures with grasses, but in the USA seed rates are higher, 30–40 kg/ha broadcast or 5.5–6.5 kg in rows (Whyte *et al.*, 1953). In India *V. aconitifolia* has also been tried in mixture with maize and although it did not increase total yields of stover, yields of CP reached 267 kg/ha compared with 216 kg for maize stover when maize was grown alone (Singh & Chand, 1969). In dry areas of India *V. aconitifolia* has also been tried with drought-tolerant perennial grasses: *Cenchrus ciliaris*, *C. setigerus*, *Dichanthium annulatum* and *Lasiurus hirsutus*; the grasses were grown in wide rows with the legume intersown each year between the rows, increasing markedly total herbage yields (Daulay *et al.*, 1970). Purseglove (1968) quotes the USA yields of green fodder of 15–20 t/ac or 3–4 t/ac of hay which corresponds to about 30–40 t/ha and 6–8 t/ha, respectively. *Vigna aconitifolia* analysed by Johri & Nooruddin (1970) contained 15.5 per cent CP, 2.5 per cent EE, 21.5 per cent CF, 42.2 per cent NFE, 3.03 per cent Ca and 0.19 per cent P. DCP content was 11.1 per cent and that of TDN 48 per cent, and the animals fed on the herbage showed positive balances of N, Ca and P.

The plants are predominantly self-pollinated and give good yields of seed, up to over 1.2 t/ha in the USA, but usually much less. The number of seeds/kg varies from 45,000 to 100,000.

Vigna angularis (Wild.) Ohwi.& Ohashi (*Phaseolus angularis* (Willd.) W. F. Wight). Adzuki bean $2n = 22$

Erect, or slightly twining or prostrate, glabrous or almost glabrous annual up to 75 cm high. Leaflets ovate, 5–9 cm long. Flowers yellow, six to twelve in axillary racemes. Pods cylindrical, slightly constricted between the seeds, 6–12 cm long and 5 mm wide. Seeds oblong, 8 mm long, smooth and shiny, variously coloured and with long white hilum.

A pulse crop originating in the Far East, probably in Japan, and also grown in China and Sarawak; introduced to South America and Zaire (Purseglove, 1968). In Columbia (Crowder, 1960) it is grown from sea level to 1,800 m, but preferably between 300 and 1,500 m. *Vigna angularis* is sown in 75–90 cm rows at 20–30 kg seed/ha to a depth of 1–2 cm. The plants develop rapidly and produce 8–10 t green fodder/ha at the flowering stage, 3–4 months after sowing; higher yields were, however, obtained in trials, and in Brazil (Vieira, 1971) a yield of 24.7 t was recorded. When grown for grain mature herbage is often used for forage or green manure after seed has been harvested. Self-pollinated plants, but cross-pollination not excluded and is frequent. Seed germinability can be retained for over 2 years (Purseglove, 1968).

Vigna frutescens A. Rich. (*V. fragrans* Bak.f.) $2n = 22$

Perennial prostrate, climbing or occasionally erect herb from a woody tuber. Leaflets ovate to rhomboid, entire or more often deeply lobed, 2–9 cm long and 1–5 cm wide. Inflorescences terminal and axillary with several flowers which are lilac to almost white in colour. Pods erect, cylindrical, 6–11 cm long and 4–5 mm wide, with 12–16 seeds. A polymorphic species in which a few varieties have been distinguished. Distributed throughout tropical Africa, mostly in grasslands, and can grow on waterlogged black clays where it quickly recovers after burning and forms early herbage. Readily grazed by stock.

Vigna luteola (Jacq.) Benth. (*V. nilotica* (Del.) Hook.f., *V. repens* (L.) Kuntze (non Bak.)) $2n = 22$

Perennial climbing or spreading herb. Stems 1.2–2.4 m long. Leaflets ovate to ovate-lanceolate, 2.5–10 cm long and 0.5–5 cm wide, with prominent reticulate venation. Inflorescences axillary, up to 5 cm long and on long peduncles bearing a few large flowers. Standard yellow or greenish-yellow, occasionally tinged with red, 1.3–2.5 cm long. Pods linear, 4–8 cm long and 5–6 mm wide, with 6–12 seeds. Seeds brown, speckled with black, 3–6 mm long and 2–3.5 mm thick.

Distributed in tropical Africa including Egypt and Syria, tropical Asia and tropical and subtropical South America. Occurs in swampy or seasonally wet grasslands, lake shores and wet woodlands; in tropical Africa from 600 to 1,900 m alt.

Vigna luteola has been grown for some time in the valley of the Parana river of subtropical Argentina (Burkart, 1952); it was introduced to Australia in 1956 under the name of *V. marina*, later corrected, and tried in various locations of Queensland with a good success; the introduced variety received later the name **Dalrymple** (Barnard, 1969). In Australia *V. luteola* is a summer-growing legume well adapted to tropical conditions of coastal Queensland with its high annual rainfall. It grows well on seasonally wet or waterlogged soils to which it is better adapted than most of the tropical legumes, but shows poor growth on dry soil; Hutton (1971) observed its high tolerance to soil salinity. *Vigna luteola* is frost susceptible although cases of recovery after frosts have been observed (Barnard, 1969). The plants are easily inoculated, the initial growth is rapid and vigorous herbage is produced early in the season. The herbage is highly palatable and requires lenient grazing for reasonable persistence; it can then last from 2 to over 3 years (Barnard, 1969).

Vigna marina (Burm.) Merrill (*V. lutea* (Sw.) A.Gr.; *Phaseolus marinus* Burm.) $2n = 22$

Perennial climbing or rambling herb rather similar to *V. luteola* from which it differs by broader leaflets, wider pods which are 8–9 mm wide and contain only two to six almost globose seeds. It also differs in ecology: *V. marina* grows in the vicinity of sandy sea shores, often just above the high tide mark, in coastal bush, and does not occur at much higher altitudes.

Distributed in tropical America, Asia, Australia and East Africa. In a trial in subtropical Turkmenian SSR (Muradov, 1970) *V. marina* introduced from Australia outyielded *V. unguiculata* and *Lablab purpureus* and produced 65–88 t fresh fodder/ha. Silage prepared from a maize/*V. marina* mixture had a high content of CP. (The identity of the plant under trial perhaps requires confirmation).

Vigna oblongifolia A. Rich. $2n = 22$

Annual or rarely subperennial, suberect, prostrate or climbing herb with slender stems. Leaflets ovate to lanceolate, 1.5–12 cm long and 0.5–3 cm wide. Two to ten flowers are borne on peduncles 2–35 cm long. Flowers yellow, 10–12 mm long. Pods linear-cylindrical, 2.5–6.5 cm long and 3–5 mm wide, straight or curved, with three to nine seeds. Distributed in tropical and subtropical Africa, mainly in its eastern parts. Occurs on swampy ground, also near lake shores and stream banks and as a weed of arable land. *Vigna oblongifolia* is eaten by stock and when grown in observational plots in Kenya it formed a dense growth and contained

23.7 per cent CP at full flowering and 17.2–18.8 per cent at late flowering. The contents of CF ranged from 24 to 28 per cent, NFE from 39 to 48 per cent, Ca 0.96–1.25 per cent and P 0.25–0.37 per cent (Dougall & Bogdan, 1966).

Var. *parviflora* (Bak.) Verdc. (*V. parviflora* Bak.) has smaller flowers and shorter pods; the contents of the main nutrients were found to be about the same as for the main form.

Vigna parkeri Bak. (*V. gracilis* auct.) $2n = 22$

Perennial climbing or procumbent herb with slender stems often rooting at the nodes. Leaflets ovate or ovate-lanceolate 1–9 cm long and 0.8–5 cm wide. Flowers blue or sometimes yellow, mostly in two- to five-flowered inflorescences. Pods linear oblong, compressed, 1–3 cm long and 5 mm wide, with two to five seeds. This plant is better known in the literature as *V. gracilis*. It is distributed in tropical East Africa from Ethiopia to Mozambique and also in Zaire, Angola and Madagascar. A fine legume, often forming natural mixtures with grasses in moist ground and produces excellent grazing.

Vigna radiata (L.) Wilczek (*Phaseolus radiatus* L., *P. aureus* Roxb., *P. mungo* auct. non L.). Mungo; Green gram; Golden gram $2n = 22$ or 24

Erect or climbing annual 20–60 cm high. Stems with brownish bristly hairs. Leaflets elliptic, rhomboid or ovate, 5–16 cm long and 3–12 cm wide, entire or lobed. Flowers in groups of four to several on axillary peduncles. Standard greenish-yellow outside and pinkish inside, 1.2 cm long and 1.6 cm wide. Keel twisted to 180°. Pods spreading or reflexed, linear, cylindrical, 4–9 cm long and 5–6 mm wide, hairy; hairs short, brownish. Seeds globose or subcubical, green to black; seed skin with wavy ridges.

Var. *radiata*, mostly of erect habit, is a cultivated form, a pulse crop of Indian origin now grown in the tropics and subtropics throughout the world.

Var. *sublobata* (Roxb.) Verdc. (*Phaeolus sublobatus* Roxb.) is a wild form of twining or prostrate habit.

The cultivated form is grown to a limited extent as a fodder crop in mixture with grasses, mainly in India, where it has been sown with encouraging results between the rows of perennial drought-tolerant grasses, *Cenchrus setigerus*, *Dichanthium annulatum*, *Lasiurus hirsutus* and *Panicum antidotale*, and satisfactory results were also obtained from mixtures with annual fodder grasses: maize, sorghum and *Euchlaena mexicana*. Good results were obtained from mixtures with *Cenchrus ciliaris* in northern Australia where it was also grown in mixed stands with *Stylosanthes humilis*. In India *V. radiata* has also been grown as a short-term fallow crop: after the pods have been taken the herbage is then grazed or ploughed in as green manure, and a very early maturing **Type 1** was selected for the purpose (Whyte, 1964). Cv. **Celera** is an

Australian introduction from Nicaragua grown in New South Wales where it has been noted for early development, flowering and ripening (Barnard, 1972). Straw of *V. radiata* is also used as fodder. The flowers are self fertile and predominantly self pollinated, the pollination often taking place in closed flowers.

Vigna mungo (L.) Hepper (*Phaseolus mungo* L.) differs from *V. radiata* by erect or suberect pods covered with long hairs and oblong seeds with concave hilum (Purseglove, 1968). Verdcourt (1971) writes however that the two forms 'are scarcely more than variants of one species although they are totally different crops'. A confusion may arise when botanical names are used to denote the crops and some of the information given for *V. radiata* may perhaps be referred to this form. In India *V. mungo* (as *Phaseolus mungo*) is recommended for mixtures with annual *Pennisetum pedicellatum*.

Vigna schimperi Bak. $2n = 22$

A species allied and rather similar to *V. luteola* from which it differs by shorter, almost unbellate inflorescences, narrower pods and by about twice the number of seeds per pod. The distribution is more restricted than that of *V. luteola* and it occurs in tropical eastern and north-eastern Africa – Kenya, Uganda, Tanzania, Zaire and Ethiopia, mainly at high altitudes. In its value as fodder *V. schimperi* is apparently similar to *V. luteola* and when grown in observational plots in Kenya it formed dense herbage of very high CP content – 28.3 per cent. The content of CF was 23.0 per cent, NFE 37.6 per cent, Ca 0.74 per cent and P 0.50 per cent (Dougall & Bogdan, 1966).

Vigna umbellata (Thunb.) Ohwi & Ohashi (*V. calcarata* (Roxb.) Kurz, *Phaseolus calcaratus* Roxb.). Rice bean $2n = 22$

An annual closely related to *V. angularis* (see p. 420) from which it differs by more vigorous and taller growth. 1.5–3 m high, by pods not constricted between the seeds and by concave hilum. It is a pulse crop grown in southern and south-eastern Asia, the Philippines and Mauritius, and also used for silage and hay (Farinas, 1970). Introduced to Latin America and in trials in Brazil it produced 33.7 t fresh fodder/ha outyielding *V. angularis* and *Canavalia ensiformis*. Predominantly a self-pollinated plant.

Vigna unguiculata (L.) Walp. (*V. sinensis* (L.) Hassk., *V. catjang* Walp., *Dolichos unguiculatus* L.). Cowpea $2n = 22$ or 24

Annual or sometimes perennial, bushy, trailing or climbing polymorphic herb. Stems 1–3 m long, glabrous or slightly hairy. Leaflets 1.5–16 cm long and 1–12 cm wide, rhomboid to lanceolate, entire or lobed; lateral leaflets oblique. Inflorescences axillary, with a few to several flowers on peduncles 2–35 cm long. Calyx glabrous. Standard white, greenish, yellow or mauve, round, 1.2–3.3 cm long. Pods erect or deflexed, linear, cylindrical, 5–12 cm long but can be longer, up to 90 cm long in some cultivated forms, 3–11 mm wide. Seeds of different size, shape and colour.

Distributed naturally in tropical Africa as a wild plant and also as a pulse crop. The cultivated forms originate in Africa but have now been spread throughout the tropics, subtropics and warm temperate areas. This species has been variously classified into subspecies and varieties and cultivated forms belong mostly to ssp. *unguiculata* which has long pods and large seeds 6–9 mm long, ssp. *cylindrica* (L.) Van Eseltine (ssp. *catjang* (Bum.) Chiov.) with the pods up to 13 cm long and seeds 5–6 mm long, and ssp. *dekindtiana* (Harms) Verdc. with still smaller seeds and shorter, up to 10 cm long, pods (Verdcourt, 1971). There is also ssp. *sesquipedalis* (L.) Verdc., a form with the pods 30–90 cm long used as vegetables; it is also grown in the USA as a fodder plant in association with sorghum and maize, and used for haymaking or ensiling (Toutain, 1973).

Cowpea is an annual pulse crop also used as a fodder plant for feeding green, making hay, grazing, and for ensiling in mixtures with sorghum or maize. Small farmers often use this crop for grain and for animal feeding after the pods have been harvested. Cowpea is also used as green manure and as green cover in plantation crops. It is a plant of diverse habit and often forms bushy growth although a number of bushy varieties or cultivars produce, at later stages, spreading vines; in other varieties the spreading nature of plant is evident from the early stages of growth.

Cowpea is best suited for moderately humid areas of the tropics and subtropics although some varieties show a considerable drought resistance. The plants cannot withstand frosts and excessive heat can also reduce their growth. In the monsoon areas it is a summer grown plant or a 'kharif' crop in India. Cowpea can grow well on a variety of soils but so called 'cold' or heavy soils of temperate areas are unsuitable.

Cowpea is established by seed and sowing rates for fodder production range from 25–35 kg/ha for wide-row cultivation to about 100 kg when broadcast. In the areas of previously grown cowpea no seed inoculation is necessary, otherwise the 'cowpea' inoculant should be used and the plants inoculate and form root nodules easily; sulphur-containing fertilizer applied in small quantities may improve nodulation. In Brazil (Terada, 1971) mulching improved seed nodulation and increased herbage yields. In the USA the fertilizers used at or before sowing include about 300 kg superphosphate plus 50 kg muriate of potash/ha. Responses to applied P may not however be great and in India Dhan Ram *et al.* (1971) obtained fresh fodder yields of 27.2 t/ha when no P was given and 29.8 and 31.4 t when 40 and 80 kg P_2O_5/ha were applied. No nitrogenous fertilizers are needed and there are reports that in some cases in India N produced an adverse effect on plant growth, but Wheeler

(1950) in USA recommended small amounts of N to be given at planting.

In India (Kohli et al., 1971), African and Indian cultivars began flowering 42–77 days after germination but Barnard (1969) reports from Australia that the best fodder cultivars flower from 90 to 115 days after germination. The plants are cut for fodder usually much later, when the pods have been well formed or even approached ripening. Yields of fresh fodder can increase if the plants are cut twice per season compared with a single cut, but cutting three times produced no further increase in trials by Dhan Ram et al. (1971). Yields of green fodder usually range from 10 to 25 t/ha and of DM from 1 to 5 t, but they can be higher, especially if measured in small-plot trials, and in the Indian Grassland and Fodder Research Institute, Jhansi, 50 best cultivars selected from a collection of over 1,000 introductions yielded from 40 to 50 t green fodder in two cuts.

Crude protein content of cowpea herbage is usually high, ranging mostly from 16 to 25 per cent, and is often favourably compared with that of lucern. A comparison of the contents of the main nutrients in green fodder, hay and silage (Narayanan & Dabadghao, 1972) showed CP contents of 18.4, 21.6 and 14.9 per cent, respectively, indicating to considerable losses of N during the ensiling; the contents of CF were 23, 28 and 27 per cent and of NFE 42.9, 41.3 and 40.1 per cent. The contents of P were 0.84, 0.77 and 0.66 per cent, more than adequate for animal requirements and so were the contents of Ca: 1.86, 1.77 and 2.03 per cent. The digestibility of CP was 76 per cent in green fodder, 68 per cent in hay and 57 per cent in silage; those of CF 60, 47 and 52 per cent, respectively, and of NFE 81, 68 and 57 per cent. The contents of TDN were 67, 57 and 59 per cent and of DCP in the DM 14.1, 14.6 and 8.6 per cent.

In the USA and some other countries cowpea has been grown with good success in mixtures with maize and sorghum, the mixtures being used mainly for preparing silage and those with Sudan grass and sorghum for making hay; the mixtures usually produce high yields and Chundawat (1971) in India observed an increase of 8.3 per cent in the weight of single plants of sorghum grown in mixture with cowpea compared with that of sorghum grown alone. Mixtures with *Euchlaena mexicana*, and with perennial *Panicum antidotale* and *Paspalum dilatatum* have also been tried and found satisfactory.

Of pests and diseases, aphids and weevils damaging ripe seeds have been particularly serious in India and root-knot nematodes in USA and Australia. In Australia considerable damage has also been observed from stem rot caused by *Phytophthora vignae*, *Fusarium* wilt, a *Septoria* leaf spot disease and mildew (Barnard, 1969).

Cowpea is essentially a self-pollinated species; under dry conditions, e.g. in California, it is almost 100 per cent self-pollinated but the proportion of cross-pollination can be high in more humid areas. The flowers open early in the morning, close in late morning and the corolla falls down in the afternoon of the same day (Purseglove, 1968). Seed is well formed but harvesting is difficult because a number of cultivars continue to grow and flower when some pods have already ripened. For obtaining good crops of seed the pods have to be picked by hand or the whole plants should be pulled out, heaped, and when dried the pods have to be picked up by hand or machinery thrashed (Wheeler, 1950). Various harvesting machines have been tried but only with moderate success and both hand-picked and machinery-harvested seeds are expensive. The high price of seed is one of the reasons for a considerable reduction of hectarage under cowpea grown in the United States for hay which was reduced from over 600,000 ha in early 1940s to under 200,000 in late 1940s and to 20,000 ha in 1964 (Wheeler, 1950; Leffel, 1973).

Numerous cultivars have been selected or developed, mostly for the production of grain but also those with bulky herbage for the use as forage, green manure, or green cover. The best cultivars of this nature grown in Australia (Barnard, 1969) are:

Malabar. Plants of spreading or semi-spreading habit with large leaflets, moderately resistant to stem rot but susceptible to root nematodes.

Cristaudo. Bushy when young but spreading at later stages of growth; useful for wet soils.

Reeves. A dense and leafy late-maturing type.

Burnett. Semi-erect type selected mainly for grazing.

Aurora. A high-yielding cultivar of broadly-spreading habit. In India cv. **Russian giant** outyielded the Australian cultivars except cv. Aurora.

Vigna vexillata (L.) Wilczek $2n = 22$

Perennial climbing or trailing herb. Leaflets linear to ovate, 2–16 cm long and 0.5–8 cm wide, pubescent to velvety on both surfaces. Flowers two to six in short axillary racemes, yellow, pink or purple; keel with the beak twisted to 180° and asymmetrical. Pods erect, linear-cylindrical, 4–14 cm long and 2–4 mm wide, hairy, with 10–18 seeds. Seeds buff-coloured to black or reddish-brown and then speckled, 2.5–4.5 mm long and 2–2.5 mm thick. Widely distributed in the world tropics, in grassland, bush, wasteland and other habitats.

Var. *vexillata*, which is common, is very hairy and this makes it almost unpalatable to cattle. There are, however, other, less hairy types, and apparently one of these was tried in south-eastern Queensland, subtropical Australia, with very encouraging results. At the flowering stage the herbage contained 20.3 per cent CP, 26.4 per cent CF, 6.1 per cent EE and 37.2 per cent NFE, and in trials with sheep these nutrients were digested to 80.9, 61.5, 74.1 and 73.4 per cent, respectively. The digestibility of DM was 69.3 per cent, of organic matter 71.6 per cent and the content of TDN 70 per cent, which showed the excellent nutritive value of the variety under trial. Frosted herbage which lost some of its leaves and was partly dry was also of reasonable value although the percentage of CP was reduced to 16.2 per cent and the digestibility of all the nutrients decreased. The intake of DM by sheep was high and reached 60.5 g/kg $W^{0.75}$ for unfrosted herbage and was higher than of the other six tropical legumes under trial, but lower than the intake of lucerne (Milford, 1967).

Zornia J. F. Gmel.

Annual or perennial herbs. Leaves with one or two pairs of leaflets and two well-developed, often large stipules; leaflets and stipules glandulate-punctate. Flowers in terminal or lateral spikes, mostly yellow; they are enclosed in pairs of bracts when in bud. Pods articulate, breaking into segments (joints) at maturity. The 70–80 species of the genus are distributed in the tropics and subtropics, predominantly in America and Africa. Some species are of importance in providing natural grazing and hay, especially in Brazil and the neighbouring countries and also in the drier parts of West Africa. The taxonomy of plants as found in agricultural and pastoral literature is not easy to decipher as the name usually refer to groups of species rather similar in their ecology and general habit but differing in morphology; the name *Z. diphylla* is particularly frequently used.

Zornia diphylla (L.) Pers.

This name is normally applied to a complex of species, both perennial and annual, which include *Z. latifolia* DC., *Z. gracilis* DC., *Z. perforata* Vogel and several others sometimes regarded as varieties of *Z. diphylla*. True *Z. diphylla* occurs in tropical America and possibly also in West Africa. Species of the *Z. diphylla* group prefer sandy soils of relatively humid areas but sometimes also grow under comparatively dry climates. The plants provide fodder for all kind of stock and are preferably used for making hay rather than for grazing.

Zornia glochidiata DC.

An annual widely distributed in Africa, mostly in semi-arid areas, and often mistaken for *Z. diphylla*. Pods up to 20 mm long with four to six segments armed with stiff bristles. In the Sahel zone of West Africa and Sudan it is often associated with *Diheteropogon hagerupii* grass. *Z. glochidiata* is used in its natural habitats for grazing and haymaking.

Zornia pratensis Milne-Redh.

Perennial herb usually prostrate at the base and with numerous upright stems about 30–40 cm high. Pods 1.2–1.7 cm long with five to six bristly segments. Occurs in most parts of tropical Africa usually in grasslands, often on rocky soil, volcanic ash and in general on soils of light texture, and is particularly abundant in the Sahel zone of Senegal, Mali and Sudan where it is much valued and used for grazing and haymaking. This species has also been known under the name *Z. diphylla*.

Appendix

Conversion Table

Weight/Area

kg/ha	kg/ha or lb/ac	lb/ac
1.121	1	0.892
2.242	2	1.784
3.363	3	2.677
4.484	4	3.569
5.605	5	4.461
6.726	6	5.353
7.848	7	6.245
8.969	8	7.138
10.090	9	8.030
11.211	10	8.922
22.421	20	17.844
33.632	30	26.766
44.843	40	35.688
56.054	50	44.609
67.265	60	53.531
78.486	70	62.453
89.696	80	71.374
100.907	90	80.296
112.108	100	89.218

From the *Journal of the British Grassland Society*.

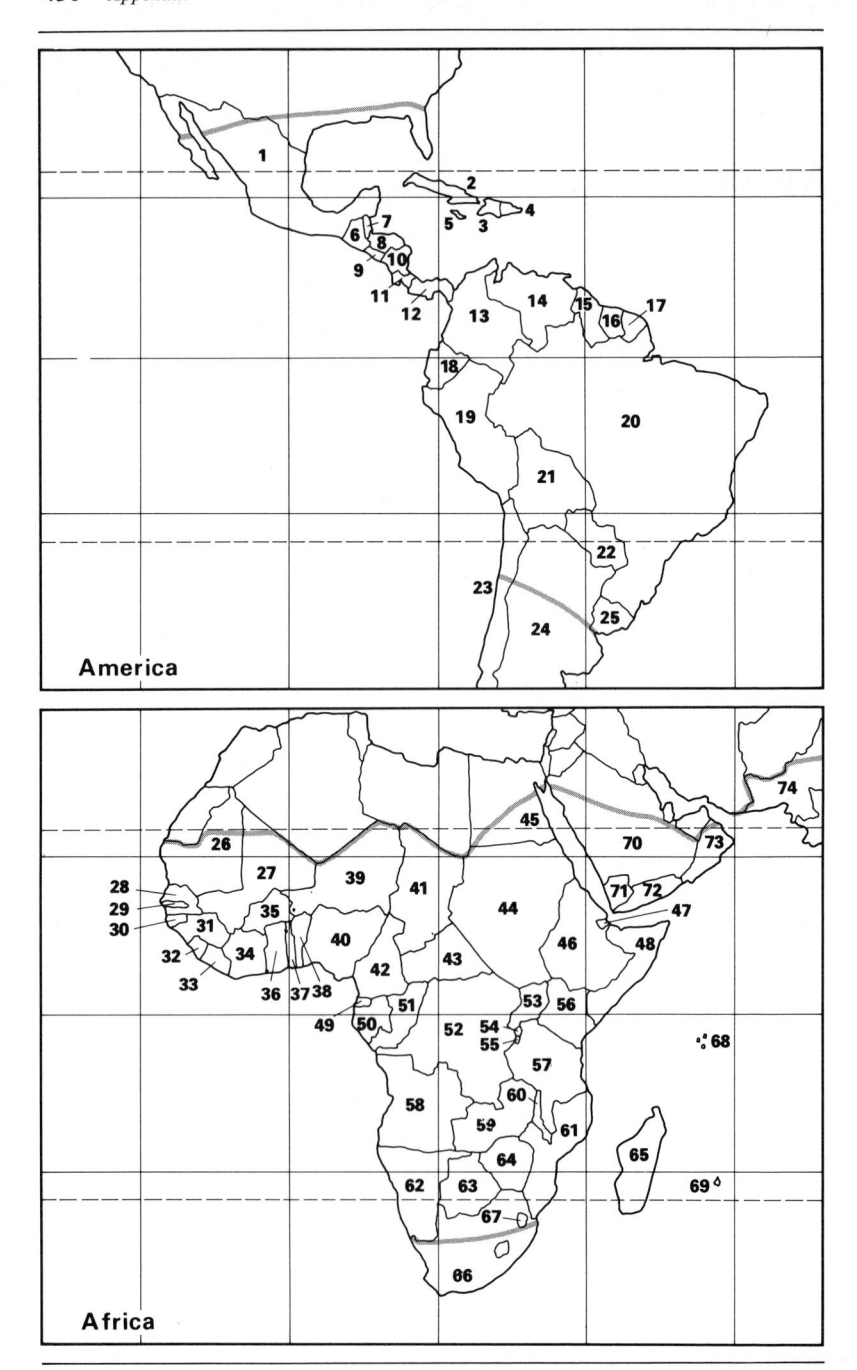

Appendix

America

Africa

Asia & Australia

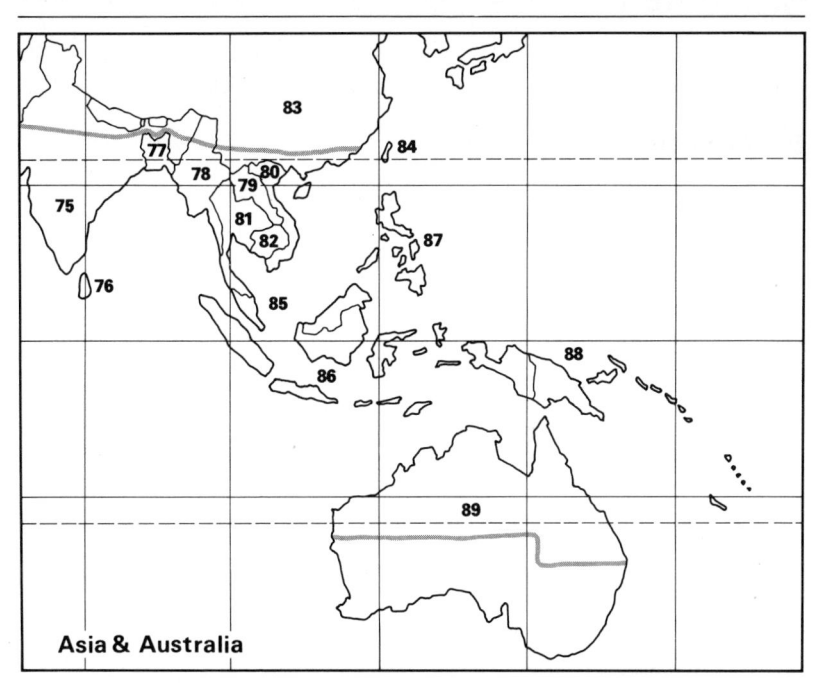

THE TROPICS   Areas covered (in the text)

1 Mexico	**25** Uruguay	**49** Equat. Guinea	**73** Oman
2 Cuba	**26** Mauritania	**50** Gabon	**74** Pakistan
3 Haiti	**27** Mali	**51** Congo Rep.	**75** India
4 Dominican Rep.	**28** Senegal	**52** Zaire	**76** Sri Lanka
5 Jamaica	**29** Gambia	**53** Uganda	**77** Bangladesh
6 Guatemala	**30** Guinea Bissau	**54** Rwanda	**78** Burma
7 Belize	**31** Guinea	**55** Burundi	**79** Laos
8 Honduras	**32** Sierra Leone	**56** Kenya	**80** Vietnam
9 El Salvador	**33** Liberia	**57** Tanzania	**81** Thailand
10 Nicaragua	**34** Ivory Coast	**58** Angola	**82** Khmer Rep.
11 Costa Rica	**35** Upper Volta	**59** Zambia	**83** China
12 Panama	**36** Ghana	**60** Malawi	**84** Taiwan
13 Colombia	**37** Togo	**61** Mozambique	**85** Malaysia
14 Venezuela	**38** Dahomey	**62** Namibia	**86** Indonesia
15 Guyana	**39** Niger	**63** Botswana	**87** Philippines
16 Surinam	**40** Nigeria	**64** Rhodesia	**88** Papua
17 French Guiana	**41** Chad	**65** Madagascar	**89** Australia
18 Ecuador	**42** Cameroun	**66** South Africa	
19 Peru	**43** Cen. Afr. Rep.	**67** Swaziland	
20 Brazil	**44** Sudan	**68** Seychelles	
21 Bolivia	**45** Egypt	**69** Mauritius	
22 Paraguay	**46** Ethiopia	**70** Saudi Arabia	
23 Chile	**47** F.T.A.I.	**71** Yemen	
24 Argentina	**48** Somali Rep.	**72** Dem. Rep. Yemen	

References

* Floras and other important general publications scanned but not cited in the text.

Abiusso, N. G. (1970) 'Analisis quimicos y valor forrajero de plantas indigenas y cultivadas en la Republica Argentina', *Revta Fac., Agron. Univ. nac. La Plata*, **46**, No. 1, 1–14.

Acocks, J. P. H. (1966) 'Non-selective grazing as a means of veld reclamation', *Proc. Grassld Soc. S. Afr.*, **1**, 33–9.

Adegbola, A. A. & Onayinka, B. (1966) 'The production and management of grass/legume mixtures at Agege', *Niger. agric. J.*, **3**, No. 2, 84–91.

Adegbola, A. A., Onayinka, B. O. & Eweje, J. K. (1968) 'The management and improvement of natural grassland in Nigeria', *Niger. agric. J.*, **5**, No. 1, 5–6.

Ahmad, S. N., Tirmazi, S. S. & Khan, A. H. (1968) 'Fodder problems of the central zone of West Pakistan and their possible solution', *Agriculture Pakist.*, **19**, No. 2, 191–5.

Akinola, J. O., Whiteman, P. C. & Wallis, E. S. (1975) 'The agronomy of pigeon pea (*Cajanus cajan*)', *Review Series No. 1/1975. Commonwealth Bureau of Pastures and Field Crops, Hurley.*

Al-Ani, T. A. & Ouda, N. A. (1969) 'Effect of moisture tension, temperature and light on germination of four exotic range grasses', *Ann. arid Zone*, **8**, No. 1, 45–51.

Alarcón, E. M., Lotero, J. C. & Escobar, L. R. (1969) 'Producción de semilla de los pastos angleton, puntero y guinea', *Agricultura trop.*, **25**, 4, 206–14.

* **Amshoff, G. J. H. & Henrard, J. Th.** (1966) 'Gramineae in Flora of Surinam, Leiden, E. J. Brill, Vol. 1, part 1.

Anderson, E. R. (1970) 'Effect of flooding on tropical grasses', *Proc. 11th int. Grassld Congr., Surfers Paradise, 1970*, 591–4.

Anderson, E. R. (1972) 'Emergence of buffel grass (*Cenchrus ciliaris*) from seed after flooding', *Qd J. agric. anim. Sci.*, **29**, No. 3, 167–72.

Andrade, S. O., Peregrino, C. J. B. & Aguiar, A. A. (1971a) 'Estudos sôbre *Brachiaria* sp (Tanner grass). 1. Efeito nocivo para bovinos', *Archos Inst. biol., S. Paulo*, **38**, No. 3, 135–50.

Andrade, S. O., Retz, L. & Marmo, O. (1971b) 'Estudos sôbre *Brachiaria* sp. (Tanner grass). 3. Ocorrências de intoxicações de bovinos durante um ano (1970–1971) e níveis de nitrato em amostras de gramínea', *Archos Inst. biol., S. Paulo*, **38**, No. 4, 239–52.

Andrew, C. S. & Vanden Berg, P. J. (1973) 'The influence of aluminium on phosphate sorption by whole plants and excised roots of some pasture legumes', *Aust. J. agr. Res.*, **24**, No. 3, 341–51.

Andrew, C. S., Johnson, A. D. & Sandland, R. L. (1973a) 'Effect of aluminium on the growth and chemical composition of some tropical and temperate legumes', *Aust. J. agric. Res.*, **24**, No. 3, 325–39.

Andrew, C. S. & Robins, M. F. (1971) 'The effect of phosphorus on the growth, chemical composition, and critical phosphorus percentages of some tropical pasture grasses', *Aust. J. agric. Res.*, **22**, No. 5, 693–706.

Andrew, W. D. & Jayawardana, A. B. P. (1971) 'Kikuyu grass – *Pennisetum clandestinum*

Hochst. ex Chiov. and its value in the montane region of Ceylon'. *Tropical Agriculturist*, **127**, No. 1, 23–42.

Anslow, R. C. (1957) 'Investigation into the potential productivity of "Acacia" (*Leucaena glauca*) in Mauritius', *Revue agric. sucr. Ile Maurice*, **36**, 39–49.

Appelman, H. & Dirven, J. G. P. (1962) 'De invloed van de maaitijd op de chemische samenstelling van verschillende grassoorten', *Surin. Landb.*, **10**, No. 3, 95–102.

Aronovich, S., Sepra, A. & Ribeiro, H. (1970) 'Effect of nitrogen fertilizer and legume upon beef production of pangòlagrass pasture', *Proc. 11th int. Grassld Congr., Surfers Paradise, 1970*, 796–800.

Arroyo, J. A. & Rivera-Brenes, L. (1961) 'Digestibility studies on Napier (Merker) grass (*Pennisetum purpureum*), giant Pangola grass (*Digitaria valida* Stent), and signal grass (*Brachiaria brizantha*)', *J. Agric. Univ. P. Rico*, **45**, No. 3, 151–6.

Arroyo, R. D. & Teunissen, H. (1964) 'Estudio comparativo de produccion de carne en 5 zócates tropicales', *Téc. pec. Méx.*, No. 3, 15–19.

Aya, F. O. (1973) 'Germination inhibitors in the seeds of *Pueraria phaseoloides* (Roxb.) Benth.', *Nigerian Inst. Oil Palm Res.*, **5**, No. 18, 7–12.

Bailey, D. R. (1967) 'Observations on the use of preplant herbicides in pasture establishment in the wet tropics', *Qd J. agric. anim. Sci.*, **24**, No. 1, 31–40.

Barker, S. J. & Kyneur, G. W. (1962) 'Leguminous pasture silage as a production ration for dairy cattle during the northern dry season', *Proc. N. Qd Agrost. Conf.*, 12/4, pp. 6.

Barnard, C. (1969) 'Herbage plant species', *Division of Plant Industry*, CSIRO, Canberra, 1969.

Barnard, C. (1972) *Register of Australian Herbage Plant Cultivars*, CSIRO, Australia, Canberra.

Barnes, D. L. (1960) 'Some studies on nitrogen fertilizing of dryland ley grasses on sandveld', *Rhodesia agric. J.*, **57**, 311–17.

Barnes, D. L. (1968) 'Dryland pastures', *Rhodesia agric. J.*, **65**, No. 1, 6–13.

Barrault, J. (1973) 'La recherche fourragère au Nord-Cameroun. Production et valeur alimentaire de quelques fourrages locaux (Travaux menés par l'IRAT de 1965 a 1971)', *Agron. trop., Paris*, **28**, No. 2, 173–88.

Bartha, R. (1970) *Fodder Plants of the Sahel Zone of Africa*, Weltforum Verlag, München.

Baskin, J. M., Schank, S. C. & West, S. H. (1969) 'Seed dormancy in two species of *Digitaria* from Africa', *Crop Sci.*, **9**, No. 5, 584–6.

Beaty, E. R., Powell, J. D. & Brown, R. H. (1963) 'Effect of nitrogen rate and clipping frequency on yield of Pensacola bahiagrass', *Agron. J.*, **55**, No. 1, 3–4.

Beaty, E. R., Powell, J. D. & Lawrence, R. M. (1970) 'Response of Brunswickgrass (*Paspalum nicorae* Parodi) to N fertilization and intense clipping; *Agron. J.*, **62**, No. 3, 363–5.

Beaty, E. R. Powell, J. D. & Stanley, R. L. (1968a) 'Production and persistence of wild annual peanuts in Bahia and Bermudagrass sods', *J. Range Mgmt*, **21**, No. 5, 331–3.

Beaty, E. R., Stanley, R. L. & Powell, J. D. (1968b) 'Effect of height of cut on yield of Pensacola Bahiagrass', *Agron. J.*, **60**, No. 4, 356–8.

Beetle, A. A. (1974) 'Sour Paspalum – tropical weed or forage?', *J. Range Mgmt*, **27**, No. 5, 347–9.

Begg, J. E. & Burton, G. W. (1971) 'Comparative study of five genotypes of pearl millet under a range of photoperiods and temperatures', *Crop Sci.*, **11**, No. 6, 803–5.

Bennett, H. W., Burson, B. L. & Bashaw, F. C. (1969) 'Intraspecific hybridization in Dallisgrass, *Paspalum dilatatum* Poir', *Crop Sci.*, **9**, 5–6, 807–9.

Bennett, H. W. & Marchbanks, W. W. (1969) 'Seed drying and viability in Dallisgrass', *Agron. J.*, **61**, No. 2, 175–7.

Bhan, S. (1967) 'Effects of methods of sowing and fertility levels on the yield and quality of jowar (*Sorghum vulgare*) and guar (*Cyamopsis tetragonoloba*) for fodder', *Ann. arid Zone*, **6**, No. 2, 153–60.

Birch, E. B. (1967) 'Nitrogen fertilization of weeping love grass (*Eragrostis curvula*) at Dohne', *Proc. Grassld. Soc. S. Afr.* **2**, 39–43.

Birch, H. F., Dougall, H. W. & Hodgson, H. C. (1964) 'The build-up and decline of

ammonia and acidity during the growth of a grass (*Setaria sphacelata*)', *Pl. Soil*, **20**, No. 3, 287–301.
Birch, H. F. & Friend, M. T. (1956) 'The organic-matter and nitrogen status of East African soils', *J. Soil Sci.*, **7**, 156–67.
Black, G. A. (1963) 'Grasses of the genus *Axonopus*', *Adv. Front. Pl. Sci.*, **5**, 1–186.
Blasco, M. L. & Bohórquez, N. A. (1968) 'Pastos en el Amazonas. I. Análisis de algunos componentos químicos', *Agricultura trop.*, **24**, No. 3, 175–7.
Blondon, F. & Lenoble, M. (1973) 'Les exigences pour la mise à fleur de deux lignes de *Sorghum vulgare* Pers. et de *Sorghum sudanense* (Piper) Stapf', *C.r. Séanc. Acad. Agric. Fr.*, **59**, No. 2, 155–62.
Blunt, C. G. & Humphreys, L. R. (1970) 'Phosphate response of mixed swards at Mt. Cotton, south-eastern Queensland', *Aust. J. exp. Agric. Anim. Husb.*, **10**, No. 45, 431–41.
Blydenstein, J., Louis, S., Toledo, J. & Camargo, A. (1969) 'Productivity of tropical pastures. 1. Pangola grass', *J. Br. Grassld Soc.*, **24**, No. 1, 71–5.
Bogdan, A. V. (1959) 'Flowering habits of *Chloris gayana*', *Proc. Linn. Soc.*, **170**, Pt 2, 1957–8, 154–8.
Bogdan, A. V. (1960a) 'The breeding behaviour of molasses grass in Kenya', *E. Afr. agric. For. J.*, **26**, No. 1, 49–50.
Bogdan, A. V. (1960b) 'A molasses grass variety trial', *E. Afr. agric. For. J.*, **26**, No. 2, 132–3.
Bogdan, A. V. (1961a) 'Hybridization in the "*Setaria sphacelata*" complex in Kenya', *Comptes Rendus de la IV Réunion Plénière de l'AETFAT*, Lisboa, 311–17.
Bogdan, A. V. (1961b) 'Intra-variety variation in Rhodes grass (*Chloris gayana* Kunth) in Kenya', *J. Brit. Grassld Soc.*, **16**, No. 3, 238–9.
Bogdan, A. V. (1963a) 'Three interesting introductions from the local grass flora at Kitale, Kenya', *J. Br. Grassld Soc.*, **18**, No. 3, 247–8.
Bogdan, A. V. (1963b) *Chloris gayana* without anthocyanin colouration', *Heredity*, **18**, Pt. 3, 364–8.
Bogdan, A. V. (1964a) 'A study of the depth of germination of tropical grasses. A new approach', *J. Br. Grassld Soc.*, **19**, No. 2, 251–4.
Bogdan, A. V. (1964b) Unpublished data.
Bogdan, A. V. (1965a) 'Cultivated varieties of tropical and subtropical herbage plants in Kenya', *E. Afr. agric. For. J.*, **30**, No. 4, 330–8.
Bogdan, A. V. (1965b) 'Pasture and fodder grasses and legumes for medium and low altitudes', *Kenya Fmr*, 1965, No. 107, 30–5.
Bogdan, A. V. (1966a) 'Seed morphology of some cultivated African grasses', *Proc. Int. Seed Test. Ass.*, **31**, No. 5, 789–99.
Bogdan, A. V. (1966b) 'Pollination and breeding behaviour in the *Trifolium rueppellianum* complex in Kenya', *New Phytol.*, **65**, 417–22.
Bogdan, A. V. (1966c) 'Weeds in herbage seeds in Kenya', *E. Afr. agric. For. J.*, **32**, No. 1, 63–6.
Bogdan, A. V. (1969) 'Rhodes grass', *Herb. Abstr.*, **39**, 1–13.
Bogdan, A. V. (1971) 'Notes on bunt disease of setaria grass', *Kenya Fmr*, No. 9, 33.
Bogdan, A. V. & Pratt, D. J. (1967) *Reseeding Denuded Pastoral Land in Kenya*, Republic of Kenya. Ministry of Agriculture and Animal Husbandry, Nairobi, Govt. Printer, pp. 1–46.
Bolkhovskikh, Z. V., Grif, V. G., Zakharyeva, O. I. & Matvejeva, T. C. (1969) '*Chromosome numbers of flowering plants*', [Ru] Acad. Sci. USSR, Leningrad.
Bolton, J. L., Goplen, B. P. & Baenziger, H. (1972) 'World distribution and historical developments', in *C. H. Hanson (ed), Alfalfa science and technology*, Amer. Soc., Madison, USA, pp. 1–34.
Bondale, K. V. (1969) 'Alysicarpus – fodder legumes', *Indian Dairyman*, **21**, No. 8, 230–1 and 233.
Boonman, J. G. (1971a) 'Experimental studies on seed production of tropical grasses in Kenya. 1. General introduction and analysis of problems', *Neth. J. agric. Sci.*, **19**, 23–36.

Boonman, J. G. (1971b) 'Experimental studies on seed production of tropical grasses in Kenya. 2. Tillering and heading in seed crops of eight grasses', *Neth. J. agric. Sci.*, **19**, 237–49.
Boonman, J. G. (1972a) 'Experimental studies on seed production of tropical grasses in Kenya. 4. The effect of fertilizers and planting density on *Chloris gayana* cv. Mbarara', *Neth. J. agric. Sci.*, **20**, 218–24.
Boonman, J. G. (1972b) 'Experimental studies on seed production of tropical grasses in Kenya. 6. The effect of harvest date on seed yields in varieties of *Setaria sphacelata*, *Chloris gayana* and *Panicum coloratum*', *Neth. J. agric. Sci.*, **21**, 3–11.
Boonman, J. G. (1973) *On the Seed Production of Tropical Grasses in Kenya*, Center of Agricultural Publishing and Documentation, Wageningen.
Boonman, J. G. & Van Wijk, A. J. P. (1973) 'Experimental studies on seed production of tropical grasses in Kenya. 7. The breeding for improved seed and herbage productivity', *Neth. J. Agric. Sci.*, 1973, **21**, 12–23.
Booysen, P. de V., Tainton, N. M. & Scott, J. D. (1963) 'Shoot-apex development in grasses and its importance in grassland management', *Herbage Abstr.*, **33**, No. 4, 209–12.
Bor, N. L. (1960) *The Grasses of Burma, Ceylon, India and Pakistan (excluding Bambuseae)*, Pergamon, Oxford and New York.
Borget, M. (1966) 'Rendements et caractéristiques de cinq Graminées fourragères sur sables côtiers a Cayenne (Guyane Française)', *Agron. trop. Paris*, **21**, No. 2, 250–9.
Borget, M. (1968) 'Les recherches fourragères a l'IRAT/Cameroun (bilan à la mi-1968)', *Agron. trop. Paris*, **23**, No. 11, 1231–41.
Bose, B. B. (1965) '*Pennisetum* overscores other fodder crops', *Indian Fmg*, **15**, No. 3, 9.
Boudet, G. (1970) 'Management of savannah woodland in West Africa', *Proc. 11th int. Grassld Congr., Surfers Paradise, 1970*.
Bowden, B. N. (1963) 'Studies on *Andropogon gayanus* Kunth. 1. The use of *Andropogon gayanus* in agriculture', *Emp. J. exp. Agric.*, **31**, No. 123, 267–73.
Bowden, B. N. (1964) 'Studies on *Andropogon gayanus* Kunth. 3. An outline of its biology', *J. Ecol.*, **52**, No. 2, 255–71.
Boyd, F. T. & Perry, V. G. (1970) 'The effect of sting nematodes on establishment, yield, and growth of forage grasses on Florida sandy soils', *Proc. Soil Crop Sci. Soc. Fla 1970*, **29**, 288–300.
Bredon, R. M. & Horrell, C. R. (1961) 'The chemical composition and nutritive value of some common grasses in Uganda. 1. General pattern of behaviour of grasses', *Trop. Agr., Trin.*, **38**, 297–304.
Brolmann, J. B. (1974) 'Growth studies in some new *Stylosanthes hamata* selections, *Port Fierce ARC Res. Report RL-1974-6, Floida University*, pp. 2.
Brown, W. V. & Emery, W. H. P. (1957) 'Apomixis in the Gramineae: *Themeda triandra* and *Bothriochloa ischaemum*', *Bot. Gaz.*, **118**, No. 4.
Brown, W. V. & Emery, W. H. P. (1958) 'Apomixis in the Gramineae. 2. Panicoideae', *Amer. J. Bot.*, **45**, 253–63.
De Bruyn, J. A. & McIlrath, W. J. (1966) 'Effect of boron, manganese, copper and zinc upon the growth of *Setaria sphacelata*', *Jl S. Afr. Bot.*, **32**, Pt 4, 313–24.
Bryan, W. W. (1961) '*Lotononis bainesii* Baker – a legume for sub-tropical pastures', *Aust. J. exp. Agric. Anim. Husb.*, **1**, No. 1, 4–10.
Bryan, W. W. (1967) 'Botanical changes following application of fertilizer and seed to rundown paspalum, Kikuyu and mat grass pastures on a scrub soil at Maleny, South-East Queensland', *Trop. Grasslds*, **1**, No. 2, 167–70.
Bryan, W. W. (1968) 'Grazing trials on the Wallum of south-eastern Queensland. 1. A comparison of four pastures', *Aust. J. exp. Agric. Anim. Husb.*, **8**, No. 34, 512–20.
Bryan, W. W. (1969) *Desmodium intortum* and *Desmodium uncinatum*. *Herb. Abs.*, **39**, 183–191.
Bryan, W. W. (1970) 'Changes in botanical composition in some subtropical sown pastures', *Proc. 11th int. Grassld Congr., Surfers Paradise*, 1970, 636–9.
Bryan, W. W., Sharpe, J. P. & Haydock, K. P. (1971) 'Some factors affecting the growth of lotononis (*Lotononis bainesii*)', *Aust. J. exp. Agric. Anum. Husb.*, **11**, No. 48, 29–34.

Bryant, W. G. (1961) 'Studies on Buffel grass. The effect of varying periods of presowing contact with superphosphate', *J. Soil Conserv. NSW*, **17**, No. 2, 123–5.
Bryant, W. G. (1967a) 'Interim assessment of introduced plants. No. 1. *Panicum coloratum* L.', *Plant Introd. Rev.*, **3**, No. 3, 18–33.
Bryant, W. G. (1967b) 'Plant testing at Scone Research Station – A note on morphological variability within the species *Panicum coloratum* L.', *J. Soil Conserv. Serv. N.S.W.*, **23**, No. 4, 290–302.
Brzostowski, H. W. (1961) 'Establishment of *Cenchrus ciliaris* from caryopses', *E. Afr. agric. For. J.*, **26**, No. 4, 242–4.
Brzostowski, H. W. & Owen, M. A. (1964) 'Botanical changes in the sown pasture', *Trop. Agric., Trin.*, **41**, No. 3, 231–42.
Brzostowski, H. W. & Owen, M. A. (1966) 'Production and germination capacity of buffel grass (*Cenchrus ciliaris*) seed', *Trop. Agric., Trin.*, **43**, No. 1, 1–10.
Buckle, J. A. (1972) 'Temperature effect on maize emergence', *Rhodesia agric. J.*, **69**, No. 5, 90.
Buenaventura, P. R. (1962) 'Respuesta del pasto elefante (*Pennisetum purpureum* Sh.) a la aplicacion de fertilizantes nitrogenados', *Acta agron.*, *Palmira*, **12**, Nos. 1–2, 1–15.
Buller, R. E., Aronovich, S., Quinn, L. R. & Bisschoff, W. V. A. (1970) 'Performance of tropical legumes in the upland savannah of Central Brazil', *Proc. 11th int. Grassld Congr., Surfers Paradise, 1970*, 143–6.
Bumpus, E. D. (1958) 'Ley pasture plants and ley management', *Rep. Grassl. Res. Stn, Kitale, Kenya 1958 Pt 2*, 3–6.
Burbridge, N. T. (1966–70) *Australian Grasses*, Vol. 1 – 1966, vol. 2 – 1968, vol. 3 – 1970. Angus and Robertson, Sydney–London–Melbourne.
Burkart, A. (1952) *Las leguminosas Argentinas silvestres y cultivadas*, 2nd ed., Acme Agency, Soc. de Resp. Ltda, Buenos Aires.
Burkart, A. (Ed) (1969) 'Flora illustrada de Entre Rios (Argentina). Parte 2. Gramíneas, *Collection cientifica de J.N.T.A. Tomo VI, 2, Buenos Aires.*
Burson, B. L. & Bennett, H. W. (1970) 'Cytology, method of reproduction and fertility of Brunswickgrass, *Paspalum nicorae* Parodi.', *Crop. Sci.*, **10**, No. 2, 184–7.
Burton, G. W. (1942) 'A cytological study of some species in the tribe Paniceae', *Am. J. Bot.*, **29**, 355–9.
Burton, G. W. (1947) 'Breeding Bermuda grass for the southeastern United States', *J. Am. Soc. Agron.*, **39**, 551–69.
Burton, G. W. (1962) 'Registration of varieties of Bermudagrass', *Crop. Sci.*, **2**, No. 4, 352–3.
Burton, G. W. (1967) 'A search for the origin of Pensacola Bahia grass', *Econ. Bot.*, **21**, No. 4, 379–82.
Burton, G. W. (1969) 'Breaking dormancy in seeds of pearl millet, *Pennisetum typhoides*', *Crop Sci.*, **9**, No. 5, 659–64.
Burton, G. W., Hart, R. H. & Lowrey, R. S. (1967) 'Improving forage quality in Bermudagrass by breeding', *Crop Sci.*, **7**, 329–32.
Burton, G. W., Jackson, J. E. & Knox, F. E. (1959) 'The influence of light reduction upon the production, persistence and chemical composition of Coastal Bermudagrass, *Cynodon dactylon*,' *Agron. J.*, **51**, No 9, 537–42.
Butterworth, M. H. (1967) 'The digestibility of tropical grasses', *Nutr. Abstr. Rev.*, **37**, No. 2, 349–68.
Cabanis, T., Chabouis, L. & Chabouis, F. (1970) *Végétaux et groupements végétaux de Madagascar et des Mascareignes*, Vol. 1–4, Tananarive, 1970.
Cameron, D. F. (1967) 'Flowering in Townsville lucerne (*Stylosanthes humilis*). 1. Studies in controlled environments. 2. The effect of latitude and time of sowing on the flowering time of single plants', *Aust. J. exp. Agric. Anim. Husb.*, **7**, No. 29, 489–94 and 495–500.
Carneiro, A. M., Carvalho, S. R., De Souto, S. M. & Cesar, T. I. (1972) 'Competição entre veriedades e híbridos de *Sorghum vulgae* de épocas, espeçamento e densidade de plantio na produção', *Pesq. Agropec. Brasil., Zootech.*, **7**, 47–51.
Caro-Costas, R., Abruña, F. & Vicente-Chandler, J. (1972) 'Comparison of heavily

fertilized Pangola and Star grass pastures in terms of beef production and carrying capacity in the humid mountain region of Puerto Rico', *J. Agric. Univ. P. Rico.*, **56**, No. 2, 104–9.

Caro-Costas, R., Abruña, F. & Vicente-Chandler, J. (1973) 'Comparison of heavily fertilized Pangola and Star grass pastures under humid tropical conditions', *Agron. J.*, **65**, No. 1, 132–3.

Caro-Costas, R. & Vicente-Chandler, J. (1961) 'Cutting height strongly affects yields of tropical grasses', *Agron. J.*, **53**, No. 1, 59–60.

Caro-Costas, R. & Vicente-Chandler, J. (1969) 'Milk production with all-grass rations from steep intensively managed tropical pastures', *J. Agric. Univ. P. Rico*, **53**, No. 4, 251–8.

Carrera, M. C. & Ferrer, F. M. (1963) 'Produccion de carne de ganado cebú, con seis especies de zácates tropicales', *Agric. téc., Méx.*, **2**, No. 2, 81–6.

Carvalho, M. M., de, Mozzer, O. L., Emrich, E. S. & Contijo, V. de P. M. (1972) 'Competição de variedades e híbridos de capim-elefante (*Pennisetum purpureum*) em um solo hidromórfico em Sete Lagoas, Minas Gerais', *Pesq. Agropec. Brasil., Zoot.*, **7**, 39–44.

Casamayor, R. (1970) 'Pre-emergent herbicides in elephant grass (*Pennisetum purpureum* Schum.)', *Revta. cub. Cienc. agric.*, **4**, No. 1, 79–83.

Cassidy, G. J. (1971) 'Response of a mat grass – Paspalum sward to fertilizer application', *Trop. Grasslds*, **5**, No. 1, 11–22.

Cassidy, N. G. (1972) 'Observations on nutrient deficiencies in Kikuyu grass (*Pennisetum clandestinum*)', *Qd J. agric. anim. Sci.*, **29**, No. 1, 51–7.

Catchpoole, V. R. (1968) 'Effects of season, maturity and rate of nitrogen fertilizer on ensilage of *Setaria sphacelata*', *Aust. J. exp. Agric. Anim. Husb.*, **8**, No. 34, 569–73.

Catchpoole, V. R. (1969) 'Preliminary studies on curing and storing Nandi setaria hay', *Trop. Grasslds*, **3**, No. 1, 65–74.

Catchpoole, V. R. (1970) 'Laboratory ensilage of three tropical pasture legumes – *Phaseolus atropurpureus*, *Desmodium intortum* and *Lotononis bainesii*', *Aust. J. exp. Agric. Anim. Husb.*, **10**, No. 46, 568–76.

Catchpoole, V. R. (1972) 'Laboratory ensilage of *Sorghum almum* cv. Crooble', *Trop. Grasslds*, **6**, No. 3, 171–6.

Catchpoole, V. R. & Henzell, E. F. (1971) 'Silage and silage-making from tropical herbage species', *Herb. Abs.*, **41**, No. 3, 213–21.

Chadhokar, P. A. & Humphreys, L. R. (1970) 'Effects of time of nitrogen deficiency on seed production of *Paspalum plicatulum* Michx.', *Proc. 11th int. Grassld Congr., Surfers Paradise, 1970*, 315–19.

Chadhokar, P. A. & Humphreys, L. R. (1973) 'Effect of tiller age and time of nitrogen stress on seed production of *Paspalum plicatulum*', *J. agric. Sci., Camb.*, **81**, No. 2, 219–29.

Chakravarty, A. K. (1971) 'Karad – a hardy perennial grass for pasture of semi-arid zones', *Indian Fmg*, **21**, No. 1, 32–3, 38.

Chakravarty, A. K., Ratan, R. & Murari, K. (1970) 'Variation in morphological and physiological characters in bunch-grass (*Cenchrus ciliaris* L.) and selection of high-yielding, nutritious types', *Indian J. agric. Sci.*, **40**, No. 10, 912–16.

Chakravarty, A. K. & Verma, C. M. (1972) 'Study on the pasture establishment technique. 4. Effect of different spacings and weedings on establishment and forage production of *Cenchrus ciliaris* Linn., *Lasiurus sindicus* Henr. and *Panicum antidotale* Retz under arid conditions', *Ann. Arid Zone*, **11**, Nos 1 and 2, 60–6.

Champ. B. R., Sillar, D. I. & Lavery, H. J. (1961) 'Seed-harvesting ant control in Cloncurry district', *Qd. J. agric. Sci.*, **18**, No. 2, 257–60.

Chandapillai, M. M. (1972) 'Studies on the nodulation of *Stylosanthes guyanensis* Aubl. 1. Effect of added organic matter in four types of Malaysian soil', *Trop. Agric., Trin.*, **49**, No. 3, 205–13.

Chandra, K. (1964) 'The chemical composition and nutritive value of musal hay (*Iscielema* [*Iseilema*] *laxum*) at the pre-flowering stage', *Indian vet. J.*, **41**, No. 3, 216–21.

Chapman, H. L., Haines, C. E. & Kidder, R. W. (1960) 'Apparent digestibility of nutrients in silages, pasture forages and feeds produced in the Everglades', *Rep. Fla agric. Exp. Stas 1960*, 252.

Chatterjee, B. N., Premchand & Singh, R. D. (1969) 'Herbage growth in spear grass (*Heteropogon contortus*) swards mixed with legumes in Indian tropics', *Ranchi Univ. J. agric. Res.*, **4**, 5–7.

Chatterjee, B. N. & Singh, R. D. (1967) 'Herbage growth analysis of Deenanath grass (variety T15) in comparison to jowar cultivars', *Indian Agric.*, **11**, No. 1, 62–8.

Chatterton, N. J., Carlson, G. E., Hungerford, W. E. & Lee, D. R. (1972) 'Effect of tillering and cool nights on photosynthesis and chloroplast starch in Pangola', *Crop. Sci.*, **12**, No. 2, 206–8.

Chaudhry, M. H., Bhatti, M. S. & Sheikh, N. A. (1969) 'Fertilizer effect on the yield potential in bajra Napier grass hybrid', *W. Pakist. J. agric. Res.*, **7**, No. 5, 22–7.

Chaudhuri, A. P. & Prasad, B. (1969) 'Maize-teosinte hybrid for fodder', *Indian J. agric. Sci.*, **39**, No. 6, 467–72.

Chesney, H. A. D. (1969a) 'Fertilizer studies with Pangola grass (*Digitaria decumbens* Stent.) on Tiwiwid fine sand in Guyana. 1. Effect of fertilizer nitrogen, phosphorus and potassium on dry matter production', *Agric. Res., Guyana*, **3**, 131–5.

Chesney, H. A. D. (1969b) 'Fertilizer studies with Pangola grass (*Digitaria decumbens* Stent.) on Tiwiwid fine sand, Guyana. 2. Effect of magnesium and trace elements on dry matter production', *Agric. Res., Guyana*, **3**, 136–8.

Chesney, H. A. D. (1972) 'Response of *Digitaria setivalva* to nitrogen, phosphorus, potassium, magnesium and calcium on Ebini sandy loam, Guyana. 1. Effect on yield, tissue composition and nutrient uptake', *Trop. Agric., Trin.*, **49**, No. 2, 115–24.

Chippindall, L. K. A. (1955) 'A guide to the identification of grasses of South Africa', in *The Grasses and Pastures of South Africa*, Central Newsagency, Parov, Cape Province, South Africa, 1–527.

Chundawat, G. S. (1971) 'Note on growth of sorghum as affected by methods of sowing, crop mixtures and phosphate levels', *Indian J. agric. Res.*, **5**, No. 3, 212–14.

Clatworthy, J. N. (1968) 'Results of grazing Paraguay and beehive paspalums', *Rhodesia agric. J.*, **65**, No. 2, 36.

Clatworthy, J. N. (1970) 'A comparison of legume and fertilizer nitrogen in Rhodesia', *Proc. 11th int. Grassld Congr., Surfers Paradise, 1970*, 408–11.

Clatworthy, J. N. & Thomas, P. I. (1972) 'Establishment of *Stylosanthes guyanensis* in Marandellas sandveld', *Proc. Grassld Soc. S. Afr.*, **7**, 76–83.

Clayton, W. D. (1961) 'Proposal to conserve the generic name *Sorghum* Moench (Gramineae) versus *Sorghum* Adans. (Gramineae)', *Taxon*, **10**, No. 8, 242–3.

Clayton, W. D. (1966) 'Studies in the Gramineae', *Kew Bull.*, **20**, No. 1, 73–6.

Clayton, W. D. (1969) 'A revision of the genus *Hyparrhenia*', *Kew Bull.*, Additional Series, **2**, 1–196.

Clayton, W. D. (1970) 'Gramineae (Part 1)', in *Flora of Tropical East Africa*, Crown Agents for Overseas Governments and Administrations, London.

Clayton, W. D. & Harlan, J. R. (1970) 'The genus *Cynodon* L. C. Rich. in tropical Africa', *Kew Bull.*, **24**, No. 1, 185–9.

Clayton, W. D., Phillips, S. M. & Renvoize, S. A. (1974) 'Gramineae (Part 2)', in *Flora of Tropical East Africa*, Crown Agents for Overseas Governments and Administrations, London.

Coaldrake, J. E. & Russell, M. J. (1969) 'Establishment and persistence of some legumes and grasses after ash seeding on newly burnt brigalow land', *Trop. Grasslds*, **3**, No. 1, 49–55.

Coaldrake, J. E. & Smith, C. A. (1967) 'Estimates of annual production from pastures on brigalow land in the Fitzroy Basin, Queensland', *J. Aust. Inst. agric. Sci.*, **33**, No. 1, 52–4.

Colman, R. L. & Holder, J. M. (1968) 'Effect of stocking rate on butterfat production of dairy cows grazing kikuyu grass pastures fertilized with nitrogen', *Proceedings Australian Society of Animal Production*, **7**, 129.

Combellas, J., Centeno, A. & Mazzani, B. (1971) 'Aprovechamiento de la parte aerea del

mani. 1. Rendimiento, composicion química y digestibilidad *in vitro*', *Agronomía trop.*, **21**, No. 6, 533–7.
Combellas, J. & Gonzáles, J. (1973) 'Rendimiento y valor nutritivo de forrajes tropicales. 4, Pasto Aleman (*Echinochloa polystachya* (H.B.K.) Hitchc.)', *Agron. trop.*, *Maracay*, **23**, No. 3, 269–75.
Combes, D. & Pernès, J. (1970) 'Variations dans les nombres chromosomiques du *Panicum maximum* Jacq. en relation avec le mode de reproduction', *C. R. Acad. sc. Paris*, **270**, 782–5.
Cooper, J. P. (1970) 'Potential production and energy conversion in temperate and tropical grasses', *Herb. Abstr.*, **40**, 1–15.
Cooper, J. P. & Tainton, N. M. (1968) 'Light and temperature requirements for the growth of tropical and temperate grasses', *Herb. Abstr.*, **38**, No. 3, 167–76.
Crampton, E. W. & Harris, L. E. (1969) '*Applied animal nutrition*', W. H. Freeman & Co., San Francisco, 2nd ed.
Crespo, G. (1972a) 'Effects of three levels of urea and two systems of application on the yield and nitrogen content of Pangola grass', *Revta cub. Cienc. agric.*, **6**, No. 2, 235–44.
Crespo, G. (1972b) 'Influence of foliar spraying of urea on the composition yield of Pangola during the dry season', *Revta cub. Cienc. agric.*, **6**, No. 2, 245–9.
Crowder, L. V. (1960) 'Gramíneas y leguminosas forrajeras en Colombia', *Min. Agric. Colombia, Bogotá, Bol. Tec.*, No. 8.
Crowder, L. V., Chaverra, H. & Lotero, J. (1970) 'Productive improved grasses in Colombia', *Proc. 11th int. Grassld Congr., Surfers Paradise, 1970*, 147–9.
Crush, J. R. (1974) 'Plant growth responses to vesicular-arbuscular mycorrhiza. 7. Growth and nodulation of some herbage legumes', *New Phytol.*, **73**, No. 4, 743–52.
Cunha, E., Cabello, P. & Chicco, C. F. (1971) 'Composicion química y digestibilidad *in vitro* del *Trachypogon* sp.', *Agronomia trop.*, **21**, No. 3, 183–93.
Cummins, D. G. & Smith, D. H. (1973) 'Effect of *Cercospora* leaf spot of peanuts on forage yield and quality and on seed yield', *Agron. J.*, **85**, No. 6, 919–92.
Dabadghao, P. M. & Marwaha, S. P. (1962) 'Relative palatability studies on important indigenous grass species of Western Rajasthan', *Indian J. Agron.*, **6**, No. 4, 323–7.
Dabadghao, P. M. & Shankarnarayan, K. A. (1970) 'Studies of *Iseilema*, *Sehima* and *Heteropogon* communities of the *Sehima-Dichanthium* zone', *Proc. 11th int. Grassld Congr., Surfers Paradise, 1970*, 36–8.
Dabadghao, P. M. & Shankarnarayan, K. A. (1973) *The grass cover of India*, New Delhi, Indian Council of Agricultural Research.
Daftardar, S. Y. & Zende, G. K. (1968) 'Periodical changes in the protein contents of Gajraj grass', *Poona agric. Coll. Mag.*, **58**, Nos 2–3, 110–16.
Dalziel, J. M. (1948) *The Useful Plants of West Tropical Africa*, London.
Danasoury, M. S., El-Nouby, H. M. & Makky, A. (1971) 'Yield and losses of dry matter and nutrients in berseem hay (Egyptian clover) cured by ground and tripod methods', *Agric. Res. Rev.*, Cairo, **49**, No. 4, 121–30.
Daubenmire, R. (1972) 'Ecology of *Hyparrhenia rufa* (Nees) in derived savanna in northwestern Costa Rica', *J. appl. Ecol.*, **9**, No. 1, 11–23.
Daulay, H. S., Chakravarty, A. K. & Bhati, G. N. (1968) 'Study on the pasture establishment technique. 3. Effect of intercropping with different legumes on the growth and forage production of dhaman (*Cenchrus ciliaris*) and sewan (*Lasiurus sindicus*) pastures in the establishment year', *Ann. arid Zone*, **7**, No. 2, 265–9.
Daulay, H. S., Chakravarty, A. K. & Bhati, G. N. (1970) 'Intercropping of grasses and laegumes', *Indian Fmg*, **19**, No. 10, 12–14, 44.
Davies, J. G. & Edye, L. A. (1959) '*Sorghum almum* Parodi – a valuable summer-growing perennial grass', *J. Aust. Inst. agric. Sci.*, **25**, No. 2, 117–27.
Deinum, B. & Dirven, J. G. P. (1967) 'Een oriënterende proef omtrent de invloed van licht en temperatuur op opbrengst en chemische samenstelling van *Brachiaria ruziziensis* Germain et Evrard', *Surin. Landb.*, **15**, No. 1, 5–10.
Delgadillo, G. & Rossiter, J. (1972) 'Plantas forrajeras para el tropico Boliviano. Glycine (*Glycine javanica*)', *Bol. téc., Minist. Asuntos Camp. Agric., Bolivia*, No. 12.

De-Polli, H., Vargas, M. A. T., Franco, A. A. & Döbereiner, J. (1973) 'Efeitos da inundação na nodulação e desenvolvimento de leguminosas forrageiras tropicais', *Pesq. Agropec. Brasil.*, *Zootecnica*, **8**, No. 2, 27–34.

Derbal, Z., Pagot, J. & Lahore, J. (1959) 'Résumé synthétique des recherches faites au Centre Fédéral de Rescherchers Zootechniques de l'Afrique Occidentale Française de 1950 à 1957 sur les pâturages tropicaux de la Zone Soudanienne', in *Etude des pâturages Tropicaux de la Zone Soudanienne*, Paris.

Dhan Ram, Tomer, P. S. & Tripathi, H. P. (1971) 'Effect of number of cuttings and levels of phosphorus and nitrogen on summer cowpea forage', *Haryana agric. Univ. J. Res.*, **1**, No. 3, 39–43.

Diatloff, A. (1968) 'Nodulation and nitrogen fixation in some *Desmodium* spp.', *Qd. J. agric. anim. Sci.*, **25**, 165–7.

Diatloff, A. (1973) '*Leucaena* needs inoculation', *Qd. agric. J.*, **99**, No. 12, 642–4.

Diatloff, A., & Ferguson, J. E. (1970) 'Nodule number, time to nodulation and its effectiveness in eleven accessions of *Glycine wightii*', *Trop. Grasslds*, **4**, No. 4, 223–8.

Díaz, H. B. & Lagomarsino, E. F. (1969) 'Suelos salino-sodicos y sodicos. Su utilizacion con especies forrajeras', *Revta agron. Noroeste Argent.*, **6**, Nos. 3–4, 221–35.

Dirven, J. G. P. (1962) 'De voederwaarde van bladeren en stengels bij tropische grassen', *Surin. Landb.*, No. 5, 199–202.

Dirven, J. G. P. (1963) 'Snijmais in de tropen', *Surin. Landb.*, **11**, No. 1, 31–4.

Dirven, J. G. P. (1971) 'De chemische samenstelling van enige grassoorten uit de gematigde en tropische gebieden, geteeld in Suriname', *Surin. Landb.*, No. 1, 5–13.

Dirven, J. G. P., Dulder, I. G. H. & Hermelijn, W. C. (1960) 'De braakvegetative op rijstvelden in Nickerie', *Surin. Landb.*, **8**, No. 1, 1–7.

Dirven, J. P. G. & Ehrencron, V. K. R. (1963a) 'Bemestingsproef bij koedzoe (*Pueraria phaseoloides* (Roxb.) Benth.)', *Surinam. Landb.*, **11**, No. 2, 39–45.

Dirven, J. G. P. & Ehrencron, V. K. R. (1963b) 'Dry matter percentages of grasses in the humid tropics', *Surinam. Landb.*, **11**, No. 3, 88–93.

Dirven, J. G. P. & Van Hoof, H. A. (1960) 'A destructive virus disease of Pangola grass', *Tijdschr. PlZiekt*, **66**, No. 6, 344–9.

Döbereiner, J. (1966) '*Azotobacter paspali* sp. n., uma bactéria fixadora de nitrogênio na rizosfera de *Paspalum*', *Pesq. Agropec. Brasil*, **1**, 357–65.

Döbereiner, J. (1970) 'Further research on *Azotobacter paspali* and its variety specific occurence in the rhizosphere of *Paspalum notatum* Flügge', *Zentralblatt für Bacteriologie und Parasitenkunde*, **II**, 124, 224–30.

Döbereiner, J., Day, J. M. & Dart, P. J. (1972) 'Nitrogenase activity and oxygen sensitivity of the *Paspalum notatum* – *Azotobacter paspali* association, *J. gen. Microbiol.*, **71**, No. 1, 103–16.

Doggett, H. (1970) *Sorghum, Tropical Agriculture Series*, Longmans.

Donnelly, E. D. & Hoveland, C. S. (1966) 'Interspecific reseeding *Vicia* hybrids for use on summer perennial grass sods in south-eastern USA', *Proc. 10th int. Grassld Congr.*, Helsinki, July 1966, 679–83.

Dougall, H. W. (1960) 'Average nutritive values of Kenya feeding stuffs for ruminants', *E. Afr. agric. For. J.*, **26**, No. 2, 119–28.

Dougall, H. W. (1962) 'The chemical composition of some species and varieties of *Trifolium*', *E. Afr. agric. For. J.*, **27**, No. 3, 142–4.

Dougall, H. W. & Birch, H. F. (1966) 'pH of grasses in relation to genera', *Nature, Lond.*, **210**, No. 5038, 844.

Dougall, H. W. & Bogdan, A. V. (1958) 'The chemical composition of the grasses of Kenya. 1', *E. Afr. agric. J.*, **25**, 17–23.

Dougall, H. W. & Bogdan, A. V. (1960) 'The chemical composition of the grasses of Kenya. 2', *E. Afr. agric. J.*, **25**, No. 4, 241–4.

Dougall, H. W. & Bogdan, A. V. (1965) 'The chemical composition of the grasses of Kenya. 3', *E. Afr. agric. For. J.*, **30**, No. 4, 314–19.

Dougall, H. W. & Bogdan, A. V. (1966) 'The chemical composition of some leguminous plants grown in the herbage nursery at Kitale, Kenya', *E. Afr. agric.*

For. J., **32**, No. 1, 45–9.
Dougall, H. W., Drysdale, V. M. & Glover, P. E. (1964) 'The chemical composition of Kenya browse and pasture herbage', *E. Afr. Wildlife J.*, **2**, 86–121.
Dougall, H. W. & Glover, P. E. (1964) 'On the chemical composition of *Themeda triandra* and *Cynodon dactylon*', *E. Afr. Wildlife J.*, **2**, 67–70.
Dunavin, L. S. (1970) 'Gahi-1 pearl millet and two sorghum × Sudangrass hybrids as pasture for yearling beef cattle', *Agron. J.*, **62**, No. 3, 375–7.
Durango, M. O. & Padilla, V. H. (1972) 'Evalucion de quatro gramíneas tropicales para produccion de leche', *Acta agron., Palmira*, **22**, No. 3/4, 163–83.
Edwards, D. C. (1937) 'Three ecotypes of *Pennisetum clandestinum* Hochst. Kikuyu grass', *Emp. J. exp. Agric.*, **5**, 371–6.
Edwards, D. C. (1940) 'The reaction of Kikuyu grass (*Pennisetum clandestinum*) herbage to management', *Emp. J. exp. Agric.*, **8**, No. 30, 101–10.
Edwards, D. C. (1948) 'Kavirondo perennial sorghum: an improved type R2 S42', *E. Afr. agr. J.*, **13**, 202.
Edwards, D. C. & Bogdan, A. V. (1951) *Important Grassland Plants of Kenya*, Nairobi, Pitman & Sons.
Edwards, P. J. & Visser, J. H. (1967) 'Columbus grass as a cultivated pasture crop', *Fmg. S. Afr.*, **43**, No. 6, 11.
Edye, L. A., Williams, W. T. & Pritchard, A. J. (1970) 'A numerical analysis of variation patterns in Australian introductions of *Glycine wightii* (*G. javanica*)', *Aust. J. agric. Res.*, **21**, 57–69.
Eggeling, W. J. (1947) *An Annotated List of the Grasses of Uganda*, Entebbe, 1947.
Ehara, K., Maeno, N. & Yamada, T. (1966) 'Physiological and ecological studies on the regrowth of herbage plants. 4. The evidence of utilization of food reserves during the early stages of regrowth in bahiagrass (*Paspalum notatum* Flügge) with $C^{14}O_2$ [in Japanese], *J. Jap. Soc. Grassld Sci.*, **12**, No. 1, 1–13.
Ehara, K. & Tanaka, S. (1961) 'Effect of temperature on the growth behaviour and chemical composition of the warm- and cool-season grasses' [in Japanese], *Jap. Proc. Crop Sci, Soc. Japan*, **29**, No. 2, 304–6.
Ehara, K. & Tanaka, S. (1972) 'Studies on the ecological and growth characteristics of warm season native grasses. 2. Growth pattern and responses of several warm-season native species to temperature, nitrogen and cutting' [in Japanese], *Sci. Bull. Fac. Agric. Kyushu Univ.*, **26**, No. 1/4, 423–8.
Elliott, R. C. & Fokkema, K. (1960) 'Digestion trials on Rhodesian feedstuffs', *Rhod. agric. J.*, **57**, No. 3, 252–6.
Engels, E. A., Schalkwyk, A. van & Hugo, J. M. (1969) 'The determination of the nutritive value potential of natural pastures by means of oesophageal fistula and feacal indicator technique', *Agroanimalia*, **1**, No. 3/4, 119–22.
Evans, L. T. & Knox, R. B. (1969) 'Environmental control of reproduction in *Themeda australis*', *Aust. J. Bot.*, **17**, No. 3, 375–89.
Evans, T. R. (1969) 'Beef production from nitrogen fertilized Pangola grass (*Digitaria decumbens*) on the coastal lowlands of southern Queensland', *Aust. J. exp. Agric. Anim. Husb.*, **9**, No. 38, 282–6.
Evers, G. W., Holt, E. C. & Bashaw, E. C. (1969) 'Seed production characteristics and photoperiodic responses in buffelgrass, *Cenchrus ciliaris* L.', *Crop Sci.*, **9**, No. 3, 309–10.
Fagan, E. B. & Vargas, P. O. (1971) 'The influence of adult *Prosapia distanti* on the forage quality of Kikuyu grass in Costa Rica', *Turrialba*, **21**, No. 2, 181–3.
Farinas, E. C. (1970) 'Pasture legumes and grasses and other forage plants in the National Forage Park, Philippines (1958–68)', *Proc. 11th int. Grassld Congr. Surfers Paradise, 1970*, 224–6.
Febles, G. & Padilla, C. (1970) 'The effect of inoculation and foliar urea on kudza (*Pueraria phaseoloides*) and pigeon pea (*Cajanus cajan*)', *Revta cub. Cienc. agric.*, **4**, No. 2, 149–51.
Febles, G. & Padilla, C. (1972) 'Effect of grazing on associations of gramineae and tropical legumes', *Revta Cub. Cienc. agric.*, **6**, No. 3, 385–90.
Fernandes, M. I. B. de M., Barreto, I. L. & Salzano, F. M. (1968) 'Cytogenetic, ecologic and

morphologic studies in Brazilian forms of *Paspalum dilatatum*', *Can. J. Genet. Cytol.*, **10**, No. 1, 131–8.

Fernando, G. W. E. (1961) 'Preliminary studies on the associated growth of grass and legumes', *Trop. agriculturist*, **117**, No. 3, 167–79.

Fisher, M. J. (1970) 'The effect of phosphorus and water stress on Townsville lucerne (*Stylosanthes humilis* H.B.K.)', *Proc. 11th int. Grassld Congr., Surfers Paradise, 1970*, 481–3.

Fisher, W. D., Bashaw, E. C. & Holt, E. C. (1954) 'Evidence for apomixis in *Pennisetum ciliare* and *Cenchrus setigerus*', *Agron. J.*, **46**, 401–4.

* *Flora of Tropical Africa* (*1917–1934*), Vol. 9, 'Gramineae', Ed. D. Prain.
* *Flora of Tropical Africa* (*1937*), Vol. 10, part 1, 'Gramineae' (cont.), Ed. A. W. Hill.

Fosberg, F. R. (1968) 'Critical notes on Pacific island plants, 2, *Micronesica*, **4**, No. 2, 255–9.

Fredenslung, A. & Cassady, J. (1969) 'Seeding grasses on denuded Kenya bushland', *Kenya Fmr*, No. 159, 7 and 30.

Gamboa, G. L. M. & Guerrero, S. D. N. (1969) ['Scarification of Bahia grass (*Paspalum notatum*) to hasten its germination'], *Agricultura téc. Méx.*, **2**, No. 10, 445–9 (quoted from *Herb. Abstr.*, 1971, 1888).

Gangstad, E. O. (1963) 'Columbus grass for grazing', *Bull. 14 Hoblitzelle agric. Lab., Texas Res. Fdn, Renner*.

Gangstad, E. O. (1967) 'Variation and hybridization in *Sorghum almum* Parodi', *Turrialba*, **17**, No. 2, 191–6.

García-Barriga, H. (1960) 'Una nueva graminea Colombiana de importancia economica', *Caldasia*, **8**, No. 39, 431–4.

* Gardner, C. A. (1952) *Flora of Western Australia*, Vol. 1, 'Gramineae', Govt. Printer, Perth.

Gartner, J. A. (1968) '*Trifolium* species at Millaa Millaa, North Queensland', *CIRO Pl. Introd. Rev.*, **5**, No. 1, 49–55.

Garza, T. R., Arroyo, D. & Pérez, S. A. (1970) 'Produccion de carne con los zacates Pangola y Jaragua en el trópico Aw', *Téc. pecuaria Méx.*, No. 14, 20–4.

Gibson, T. A. & Humphreys, L. R. (1973) 'The influence of nitrogen nutrition of *Desmodium uncinatum* on seed production', *Aust. J. agric. Res.*, **24**, No. 5, 667–76.

Gill, A. S., Pandey, R. K., Maurya, R. K., Mukhtar Singh & Abichandani, C. T. (1972) 'Response of NPK on the fodder yield of hybrid maize'. *Indian J. agric. Res.*, **6**, No. 2, 159–62.

Gill, G. S., Batra, P. C. & Singh, M. (1967) 'Effect of high doses of nitrogen on the forage of Sudangrass (*Sorghum vulgare* var. *sudanensis*)', *J. Res. Punjab Agric. Univ.*, **4**, No. 2, 179–84.

Gill, R. S. & Negi, S. S. (1968) 'Nutritive values of *Phalaris tuberosa* and *Pueraria thunbergiana* (kudzu) evaluated on ram lambs as sole feeds and in combination', *J. Res. Punjab agric. Univ.*, **5**, No. 3, 30–5.

Gillett, J. B. (1971) '*Indigofereae*', in *Flora of Tropical East Africa*, 'Leguminosae', part 3, & 4, 212–330, 1013–41.

* Gilliland, H. B. (1971) *A Revised Flora of Malaya*, Vol. 3, 'Grasses of Malaya', Singapore.

Gillis, W. T. & Stearn, W. T. (1974) 'Typification of the names of the species of *Leucaena* and *Lysiloma* in the Bahamas', *Taxon*, **23**, No. 1, 185–91.

Giménez Ferrer, A. (1970) 'Presencia de la cochinilla de los pastos en Paraguay', *Idia*, No. 276, 64–6.

Gledhill, D. (1966) 'Cytotaxonomic revision of the *Axonopus compressus* (Sw.) Beauv. complex', *Bol. Soc. Broteriana*, 40 (2nd ser.), 125–47.

Glover, J. & Forsgate, J. (1964) 'Transpiration from short grass', *Quart. J. Royal Met. Soc.*, **90**, 320.

Gomide, J. A., Noller, C. H., Mott, G. O., Conrad, J. H. & Hill, D. L. (1969a) 'Effect of plant age and nitrogen fertilization on the chemical composition and *in vitro* cellulose digestibility of tropical grasses', *Agron. J.*, **61**, No. 1, 116–20.

Gomide, J. A., Noller, C. H., Mott, G. O., Conrad, J. H. & Hill, D. L. (1969b) 'Mineral composition of six tropical grasses as influenced by plant age and nitrogen fertilization', *Agron. J.*, **61**, No. 1, 120–3.
Gonzales, V., Brewbaker, J. L. & Hamill, D. E. (1967) '*Leucaena* cytogenetics in relation to the breeding of low mimosine lines', *Crop. Sci.*, **7**, 140–3.
* **Gooding, E. G. B., Loveless, A. R. & Proctor, G. R.** (1965) *Flora of Barbados*, Overseas Res. Publications No. 7, H.M. Stationary Office, London.
Gosnell, J. M. (1963) 'Vleiland development in West Kenya. 1. Pasture introduction', *E. Afr. agric. For. J.*, **29**, No. 2, 99–105.
Gosnell, J. M. & Weiss, E. A. (1965) 'Vleiland development in west. Kenya. 2. Pasture fertilizer and grazing trials', *E. Afr. agric. For. J.*, **30**, No. 3, 169–76.
Goswami, A. K., Gupta, B. K. & Sharma, K. P. (1970) 'The chemical composition of bajra fodder', *J. Res. Punjab agric. Univ.*, **7**, No. 1, 58–61.
* **Gould, F. W.** (1968) *Grass Systematics*. McGraw-Hill Book Co., New York, St Louis, San Francisco, Toronto, London, Sydney.
Graham, N. McC. (1964) 'Utilization by fattening sheep of the energy, and nitrogen in fresh herbage and in hay made from it,' *Aust. J. agric. Res.*, **15**, No. 6, 974–81.
Granier, P. & Chatillon, G. (1972) '*Desmodium intortum*. Utilisation dans l'alimentation des vaches laitières', *Revue Elev. Méd. vét. Pays trop.*, **25**, No. 3, 425–31.
Granier, P & Razafindratsita, R. (1970) 'Contribution à l'étude de la culture dérobée des fourrages en rizière dans le région de Tananarive', *Revue Elev. Méd. vét. Pays trop.*, **23**, No. 1, 101–8.
Gray, S. G. (1968) 'A review of research on *Leucaena leucocephala*', *Trop. Grasslands*, **2**, 19–30.
Grof, B. (1961) 'Two pasture grasses show promise', *Qd. agric. J.*, **87**, 741–2.
Grof, B. (1968) 'Viability of seed of *Brachiaria decumbens*' *Qd J. agric. anim. Sci.*, **25**, No. 3, 149–52.
Grof, B. (1969a) 'Viability of Para grass (*Brachiaria mutica*) seed and the effect of fertilizer nitrogen on seed yields', *Qd J. agric. anim. Sci.*, **26**, No. 2, 271–6.
Grof, B. (1969b) 'Elephant grass for warmer and wetter lands', *Qd. agric. J.*, **95**, No. 4, 227–34.
Grof, B. (1969c) 'Notes on selections from hybrid derivatives of elephant grass (*Pennisetum purpureum* Schum.)', *Qd J. agric. anim., Sci.*, **26**, No. 1, 49–53.
Grof, B. & Harding, W. A. T. (1970) 'Dry matter yields and animal production of Guinea grass (*Panicum maximum*) on the humid tropical coast of north Queensland', *Trop. Grasslds*, **4**, No. 1, 85–95.
Guerrero, R., Fassbender, H. W. & Blydenstein, J. (1970) 'Fertilización del pasto elefante (*Pennisetum purpureum*) en Turrialba, Costa Rica. 1. Efecto de dosis crecientes de nitrógeno', *Turrialba*, **20**, No. 1, 53–8.
Gupta, P. K. (1969–70) 'Apomixis in *Bothriochloa pertusa* (L.) A. Camus. *Port. Acta biol.* (*A*), **11**, Nos 3–4, 279–87.
Gupta, R. K. & Saxena, S. K. (1970) 'Some ecological aspects of improvement and management of sewan (*Lasiurus sindicus*) rangelands', *Annals of Arid Zone*, **9**, No. 3, 193–208.
Hacker, J. B. (1966) 'Cytological investigations in the *Setaria sphacelata* complex', *Aust. J. agric. Res.*, **17**, 297–301.
Hacker, J. B., Forde, B. J. & Gow, J. M. (1974) 'Simulated frosting of tropical grasses', *Aust. J. agric. Res.*, **25**, No. 1, 45–57.
Hacker, J. B. & Jones, R. J. (1969) 'The *Setaria sphacelata* complex – a review', *Trop. Grasslds*, **3**, No. 1, 13–34.
Hacker, J. B. & Jones, R. J. (1971) 'The effect of nitrogen fertilizer and row spacing on seed production in *Setaria sphacelata*', *Trop. Grasslds*, **5**, No. 2, 61–73.
Haggar, R. J. (1966) 'The production of seed from *Andropogon gayanus*', *Proc. int. Seed Test. Ass.*, **31**, No. 2, 251–9.
Haggar, R. J. (1969) 'Use of companion crops in grassland establishment in Nigeria', *Exp. Agric.*, **5**, No. 1, 47–52.

Haggar, R. J. (1971) 'The production and managment of *Stylosanthes gracilis* at Shika, Nigeria. 1. In sown pastures', *J. agric. Sci. Camb.*, **77**, No. 3, 427–36.

Haggar, R. J. & Couper, D. C. (1972) 'Effects of plant population and fertilizer nitrogen on growth and components of yield of maize grown for silage in Nigeria', *Exp. Agric.*, **8**, No. 3, 251–63.

Haggar, R. J., Leeuw, P. N. de & Agishi, E. (1971) 'The production and management of *Stylosanthes gracilis* at Shika, Nigeria. 2. In savanna grassland', *J. agric. Sci. Camb.*, **77**, No. 3, 437–44.

Haines, C. E., Chapman, H. L., Allen, R. J. & Kidder, R. W. (1961) 'Comparison of major perennial pasture forages of the Everglades', *Rep. Fla agric. Exp. Stas*, 272.

Haines, C. E., Chapman, H. L., Allen, R. J. & Kidder, R. W. (1965) 'Roselawn St Augustinegrass as a perennial pasture forage for organic soils of south Florida', *Bull. 689, Fla agric. Exp. Stn.*

Hamilton, R. I., Donaldson, L. E. & Lambourne, L. J. (1971) '*Leucaena leucocephala* as a feed for dairy cows; direct effect on reproduction and residual effect on the calf and lactation', *Aust. J. agric. Res.*, **22**, No. 4, 681–92.

Hamilton, R. I., Fraser, J. & Armitt, J. D. (1969) 'Preliminary assessment of tropical pasture species for taint in milk', *Aust. J. Dairy Technol.*, **24**, No. 2, 62–5.

Hamilton, R. I., Lambourne, L. J., Roe, R. & Minson, D. J. (1970) 'Quality of tropical grasses for milk production', *Proc. 11th int. Grassld Congr., Surfers Paradise, 1970*, 860–4.

Hanna, W. W. (1973) 'Effect of seed treatment and planting depth on germination and seedling emergence in *Aeschynomene americana* L., *Crop Sci.*, **13**, No. 1, 123–4.

Harding, W. A. T. & Cameron, D. G. (1972) 'New pasture legumes for the wet tropics', *Qd. agric. J.*, **98**, No. 8, 394–406.

Harker, K. W. (1962) 'A fertilizer trial on *Paspalum notatum* pasture. 1. The effects on yield', *E. Afr. agric. For. J.*, **27**, No. 4, 201–3.

Harlan, J. R. (1970) '*Cynodon* species and their value for grazing and hay', *Herbage Abst.*, **40**, No. 3, 233–8.

Harlan, J. R. & de Wet, J. M. J. (1963) 'Role of apomixis in the evolution of *Bothriochloa-Dichanthium* complex', *Crop Sci.*, **3**, 314.

Harlan, J. R. & de Wet, J. M. J. (1972) 'A simplified classification of cultivated sorghums', *Crop Sci.*, **12**, No. 2, 172–6.

Harlan, J. R., de Wet, J. M. J. & Rawal, K. M. (1970) 'Geographical distribution of the species of *Cynodon* L. C. Rich. (Gramineae)', *E. Afr. agric. For. J*, **36**, No. 2, 220–6.

Harrington, G. N. (1969) 'Liveweight production from *Themeda triandra* grassland, under two management systems and three stocking rates', *1968 Annual report of the Animal Health Research Centre. Department of Veterinary Services and Animal Industry, Uganda*, 62–4.

Harrington, G. N. & Thornton, D. D. (1969) 'A comparison of controlled grazing and manual hoeing as a means of reducing the incidence of *Cymbopogon afronardus* Stapf in Ankole pastures, Uganda', *E. Afr. agric. For. J.*, **35**, No. 2, 154–9.

Harrison, E. (1942) Digestibility trials on green fodders, *Trop. Agric., Trinidad*, **19**, No. 8, 147–50.

Hart, R. H. & Burton, G. W. (1966) 'Prostrate vs. common dallisgrass under different clipping frequencies and fertility levels', *Agron. J.*, **58**, No. 5, 521–2.

Hartley, W. & Williams, R. J. (1956) 'Centres of distribution of cultivated pasture species and their significance for plant introduction', *Proc. 7th int. Grassld Congr., Palmerston-North*, 190–201.

Hatch, M. D. (1972) 'Photosynthesis and the C_4-pathway. Division of Plant Industry', CSIRO, *Annual Report 1971*, Canberra, 19–26.

*Havard-Duclos, B. (1967) '*Les plantes fourrageres tropicales*', Maisonneuve & Larouss, Paris.

Hawkins, G. E., Smith, L. A., Grimes, H., Patterson, R. M., Little, J. A. & Rollo, C. A. (1969) 'Managing Johnsongrass for dairy cows. Relative efficiency of several methods of utilizing forage determined in Alabama research', *Bull. 389 Ala. agric. Res. Stn.*

References 445

Haylett, D. G. (1970) 'Maize following heavily fertilized pasture', *Agroplantae*, **2**, No. 4, 113–20.
Healey, J. S. (1969) 'Japanese millet may cause photosensitivity', *Agric. Gaz. N.S.W.*, **80**, No. 1, 40.
Hearn, C. J. & Holt, E. C. (1969) 'Variability in components of seed production in *Panicum coloratum* L.', *Crop. Sc.*, **9**, No. 1, 38–40.
Heinrichs, D. H., Bingefors, S., Crowder, L. V. & Langer, R. H. M. (1972) 'Highlight of research around the world', in *C. H. Hanson (ed.), Alfalfa Science and Technology*, Amer. Soc. Agron., Madison, USA, 737–80.
Hendy, K. (1972) 'The response of a Pangola grass pasture near Darwin to the wet season application of nitrogen', *Trop. Grasslds*, **6**, No. 1, 25–32.
Henrard, J. T. (1950) '*Monograph of the genus Digitaria*', Univ. Press, Leiden.
Henty, E. E. (1969) 'A manual of the grasses of New Guinea', *Botany Bulletin No. 1*, Department of Forests, Lae, New Guinea.
Henzell, E. F. (1962) 'Nitrogen fixation and transfer by some tropical and temperate pasture legumes', *Aust. J. exp. Agric. anim. Husb.*, **2**, 132–40.
Henzell, E. F. (1963) 'Nitrogen fertilizer responses of pasture grasses in south-eastern Queensland', *Aust. J. exp. Agric. Anim. Husb.*, **3**, 290–9.
Hermann, F. J. (1954) 'A synopsis of the genus *Arachis*', *Agriculture Monograph*, No. 19, USDA, Washington DC.
Hermann, F. J. (1962) 'A revision of the genus *Glycine* and its immediate allies', *U.S. Dept. Agric. Tech. Bull.*, No. 1268.
Hernández, O. A. & Abiusso, N. G. (1969) 'Efecto de distintes intensidades de utilización en sorgo forrejero, sobre el rendimiento de pasto, materia seca, proteinas y carbohidratos solubles', *Revta Invest. agropec., Ser. 2*, **6**, No. 7, 131–44.
Herrera, P., Lotero, C. J. & Crowder, L. V. (1966) ['Cutting frequency with tropical forage legumes'], *Agricultural trop.*, **22**, No. 9, 473–83. (from *Herb. Abst. 1967, 577*).
Herridge, D. F. & Roughley, R. J. (1974) 'Survival of some slow-growing *Rhizobium* on inoculated legume seed', *Pl. Soil*, **40**, No. 2, 441–4.
Hill, G. D. (1969) 'Performance of forage sorghum hybrids and Katherine pearl millet at Bubia', *Papua New Guin. agric. J.*, **21**, No. 1, 1–6.
Hill, G. D. (1970) 'Effect of environment on the growth of *Leucaena leucocephala*', *J. Aust. Inst. agric. Sci.*, **36**, No. 4, 301.
Hill, G. D. (1971) '*Leucaena leucocephala* for pastures in the tropics', *Herbage Abst.*, **41**, 111–19.
Hitchcock, A. S. (1950) *Manual of the Grasses of the United States*, *U.S. Dep. Agr. Misc. Publ. No. 200*, 2nd ed, rev. Agnes Chase, Washington.
Hogan, W. H., Brooks, O. L., Beaty, E. R. & McCreery, R. A. (1962) 'Effect of pelleting Coastal Bermudagrass on livestock gains', *Agron. J.*, **54**, No. 3, 193–5.
Hogg, P. G. & Collins, J. C. (1965) 'Clover and Coastal Bermudagrass', *Miss. Fm. Res.*, **28**, No. 5, 5.
Holm, J. (1972) 'The yields of some tropical fodder plants from northern Thailand', *Thai. J. agric. Sci.*, **5**, 227–36.
Holt, E. C. & Lancaster, J. A. (1968) 'Yield and stand survival of Coastal Bermudagrass as influenced by management practices'. *Agron. J.*, **60**, No. 1, 7–11.
Hopkinson, J. M. & Loch, D. S. (1973) 'Improvement of seed yields of siratro (*Macroptilium atropurpureum*). 1. Production and loss of seed in the crop, *Trop. Grasslds*, **7**, No. 3, 255–68.
Horowitz, M. **(1973)** 'Spatial growth of *Sorghum halepense* (L.) Pers., *Weed Res.*, **13**, No. 2, 200–8.
Hosaka, E. Y. & Ripperton, J. C. (1948) 'Promising pasture species, *Univ. Hawaii agric., exp. Sta. Rep*.
Hubbard, C. E. & Vaughan, R. E. (1940) *The grasses of Mauritius and Rodriguez*, London, Crown Agents.
Hubbard, W. A. (1960) '*Sorghum almum*', *Forage Notes*, **6**, No. 1, 18–19.
Humphreys, L. R. (1967) 'Townsville lucerne: history and prospect', *J. Aust. Inst. agric.*

Sci., **33**, No. 1, 3–13.
Humphreys, L. R. & Davidson, D. E. (1967) 'Some aspects of pasture seed production,' *Trop. Grasslds*, **1**, No. 1, 84–7.
Hunkar, A. E. S. (1969) 'Grasland onderzoek', *Surin. Landb.*, **17**, No. 1, 37–40.
Hutchinson, J. (1964) '*The genera of flowering plants (Angiospermae), Dicotyledones*, Vol. 1, Clarendon Press, Oxford.
Hutchison, D. J. & Bashaw, E. C. (1964) 'Cytology and reproduction of *Panicum coloratum* and related species', *Crop Sci.*, **4**, 151–3.
Hutton, E. M. (1960) Flowering and pollination in *Indigofera spicata*, *Phaseolus lathyroides*, *Desmodium uncinatum*, and some other tropical pasture legumes, *Emp. J. exp. Agric.*, **28**, 235–43.
Hutton, E. M. (1962) 'Siratro – a tropical pasture legume bred from *Phaseolus atropurpureus*', *Aust. J. exp. Agric. anim. Husb.*, **2**, No. 5, 117–25.
Hutton, E. M. (1970) 'Tropical pastures', *Adv. Agron.*, **22**, 1–73.
Hutton, E. M. (1971) 'Variation in salt response between tropical pasture legumes', *SABRAO Newsletter*, **3**, No. 2, 75–81.
Hutton, E. M. & Gray, S. G. (1959) 'Problems in adopting *Leucaena glauca* as a forage for the Australian tropics', *Emp. J. exp. Agric.*, **27**, 187–96.
Ingle, M. & Rogers, B. J. (1961) 'The growth of a midwestern strain of *Sorghum halepense* under controlled conditions', *Amer. J. Bot.*, **48**, No. 5, 392–6.
International Code of Botanical Nomenclature (1972) Utrecht.
International Code of Nomenclature for Cultivated Plants (1969) Utrecht.
I'Ons, J. H. (1969) 'Cultivated pastures', *Swaziland Dept. Agric. Annual Rep. Res. Division 1967–8. Malkerns*, 80–5.
*   **Jacques-Felix, H.** (1962) 'Les Graminées (Poaceae) d'Afrique tropicale. Généralities, classification, description des genres', Inst. Rech. Agron. Trop. Cult. Vivrieres, *Bull. Sc.*, No. 8, Paris.
Jacques-Félix, H. (1968) 'Evolution de la végétation au Cameroun sous l'influence de l'homme', *J. Agric. trop. Bot. appl.*, **15**, Ns 9–11, 350–6.
Javier, E. Q. (1970) 'The flowering habits and mode of reproduction of Guinea grass (*Panicum maximum* Jacq.), *Proc. 11th int. Grassld Congr., Surfers Paradise, 1970*, 284–9.
Jayawardana, A. B. P. & Andrew, W. D. (1970) 'Surface sowing a simple and safe technique for pasture estabishment in the wet upper montane zone of Ceylon', *Trop. Agriculturist*, **126**, No. 4, 143–58.
Jeffery, H. (1971) 'Nutritive value of *Pennisetum clandestinum* based pastures in a subtropical environment', *Aust. J. exp. Agric. Anim. Husb.*, **11**, (49), 173–7.
Jodhpur, P. (1965) Pusa Giant Napier – an Indian fodder grass, *CSIRO Pl. Introd. Rev.*, **2**, No. 3, 24–5.
Johnson, J. C., Lowrey, R. S., Monson, W. G. & Burton, G. W. (1968) 'Influence of the dwarf characteristics on composition and feeding value of near-isogenic pearl millets', *J. Dairy Sci.*, **51**, No. 9, 1423–5.
Johnson, S. C. & Brown, W. V. (1973) 'Grass leaf ultrastructural variations', *Amer. J. Bot.*, **60**, No. 8, 727–35.
Johnston, M. E. H. & Miller, J. G. (1964) 'Fumigation of seed with methyl bromide', *Proc. int. Seed Test. Ass.*, **29**, No. 3, 451–62.
Johri, C. B., Kulshrestha, S. K. & Saxena, J. S. (1971) 'Chemical composition and nutritive value of green soyabean and soyabean straw', *Indian vet. J.*, **48**, No. 9, 938–40.
Johri, P. N. & Nooruddin (1970) 'Studies on the digestibility and nutritive value of bhirni (*Phaseolus aconitifolius*) green', *Indian vet. J.*, **47**, No. 4, 344–7.
Johri, P. N., Srivastava, J. P. & Sinha, S. K. (1969) 'Studies on the digestibility and nutritive value of *Pennisetum pedicellatum* (Dinanath) grass at flowering stage', *Indian J. Dairy Sci.*, **22**, No. 1, 1–4.
Jones, M. B. & Freitas, L. M. M. de (1970) 'Respostas de quatro leguminosas tropicais a fósforo potássio e calcário num latossolo vermelho-amarelho de campo cerrado', *Pesq. Agropec. Brasil.*, **5**, 91–9.

Jones, M. B., Quagliato, J. & De Freitas, L. M. M. (1970) 'Respostas de alfafa e algumas leguminosas tropicais a aplicações de nutrientes minerais, em três solos de campo cerrado', *Pesq. Agropec. Brasil.*, **5**, 209–14.
Jones, R. I. (1967) 'Comparative effects of differential defoliation on grass plants in pure and mixed stands', *S. Afr. J. agric. Sci.*, **10**, No. 2, 429–44.
Jones, R. J. (1970 'The effect of nitrogen fertilizer applied in spring and autumn on the production and botanical composition of two sub-tropical grass–legume mixtures', *Trop. Grasslds*, **4**, No. 1, 97–109.
Jones, R. J. (1973a) 'Tropical legumes – their growth and response to management variables in a subtropical environment. *J. Aust. Inst. agric. Sci.*, **39**, 192–3.
Jones, R. J. (1973b) 'The effect of frequency and severity of cutting on yield and persistence of *Desmodium intortum* cv. Greenleaf in a subtropical environment', *Aust. J. exp. Agric. Anim. Husb.*, **13**, No. 61, 171–7.
Jones, R. J. (1974) 'The relation of animal and pasture production to stocking rate on legume based and nitrogen fertilized subtropical pastures', *Proc. Aust. Soc. Anim. Prod.*, **10**, 340–3.
Jones, R. J., Davies, J. G. & Waite, R. B. (1967) 'The contribution of some tropical legumes to pasture yields of dry matter and nitrogen at Samford, south-eastern Queensland', *Aust. J. exp. Agric. Anim. Husb.*, **7**, No. 24, 57–65.
Jones, R. J., Davies, J. G. & Waite, R. B. (1969) 'The competitive and yielding ability of some subtropical pasture species sown alone and in mixtures under intermittent grazing at Samford, south-eastern Queensland, *Aust. J. exp. Agric. Anim. Husb.*, **9**, No. 37, 181–91.
Jones, R. J. & Ford, C. W. (1972) 'The soluble oxalate content of some tropical pasture grasses grown in south-east Queensland', *Trop. Agric., Trin.*, **6**, No. 3, 201–4.
Jones, R. J. & Pritchard, A. J. (1971) 'The method of reproduction in Rhodes grass (*Chloris gayana* Kunth)', *Trop. Agric., Trin.*, **48**, No. 4, 301–7.
Jones, R. J., Seawright, A. A. & Little, D. A. (1970) 'Oxalate poisoning in animal grazing the tropical grass *Setaria sphacelata*', *J. Aust. Inst. agric. Sci.*, **36**, No. 1, 41–3.
Jones, R. K. (1968) 'Initial and residual effects of superphosphate on a Townsville lucerne pasture in north-eastern Queensland', *Aust. J. expt. Agric. Animal Husb.*, **8**, No. 34, 521–7.
Jones, R. M. (1969) 'Mortality of some tropical grasses and legumes following frosting in the first winter after sowing', *Trop. Grasslds*, **3**, No. 1, 57–63.
Jordan, S. M. (1957) 'Reclamation and pasture management in the semi-arid areas of Kitui District, Kenya', *E. Afr. agric. J.*, **25**, 18–22.
Joshi, A. B., Patil, B. D. & Manchanda, P. L. (1959) 'Chromosome numbers in some grasses', *Curr. Sci.*, **28**, 454–5.
Jozwik, F. X. (1970) 'Response of Mitchell grasses (*Astrebla* F. Muell.) to photoperiod and temperature', *Aust. J. agric. Res.*, **21**, No. 3, 395–405.
Kaddah, M. T. (1962) 'Tolerance of berseem clover to salt', *Agron. J.* **54**, No. 5, 421–5.
Kass, D. L., Drosdoff, M. & Alexander, M. (1971) 'Nitrogen fixation by *Azotobacter paspali* in association with Bahiagrass (*Paspalum notatum*)', *Proc. Soil Sci. Soc. Am.*, **35**, No. 2, 286–9.
Katiyar, R. C. & Ranjhan, S. K. (1969) 'Chemical composition and nutritive value of Kikuyu grass (*Pennisetum clandestinum*) for sheep', *Indian J. Dairy Sci.*, **22**, No. 1, 42–5.
Katiyar, R. C., Ranjhan, S. K. & Shukla, K. S. (1970) 'Yield and nutritive value of *Clitoria ternatea* – a wild perennial legume – for sheep', *Indian J. Dairy Sci.*, **23**, No. 2, 79–81.
Kawamura, A. & Yamasaki, S. (1972) ['Studies on the soil adaptability of warm-season grasses. 1. Effect of phosphorus level in mineral soils on the early growth of Bahia grass'], *Bull. Shikoku Agric. Exp. Sta.*, No. 24, 109–22 [in Japanese].
Kenya Report (1970) National Agricultural Research Station, Kitale, Kenya, Pasture Research Section. *Annual Report* for 1970, Part 2.
Kerridge, P. C. & Skerman, P. J. (1968) 'The distribution and growth characteristics of the native legume *Psoralea eriantha* i western Queensland', *Trop. Grasslds*, **2**, No. 1, 41–50.
Keya, N. C. O., Olsen, F. J. & Holliday, R. (1971) 'Oversowing improved pasture legumes

in natural grasslands of the medium altitudes of western Kenya', *E. Afr. agric. For. J.*, **37**, No. 2, 148–55.

Keya, N. C. O., Olsen, F. J. & Holliday, R. (1972) 'Comparison of seed-beds for oversowing a *Chloris gayana* Kunth/*Desmodium uncinatum* Jacq. mixture in *Hyparrhenia* grassland', *E. Afr. agric. For. J.*, **37**, No. 4, 286–93.

Khan, A. M. & Syed, E. A. (1970) 'Investigations on methods of propagation on Napier grass – bajra hybrid', *W. Pakistan J. agric. Res.*, **8**, No. 2, 152–6.

Khan, C. M. A. (1970) 'Effect of clipping intensities on forage yield of *Cenchrus ciliaris* (Linn.) in Thal, Pakistan', *Pakist. J. For.*, **20**, No. 1, 75–87.

Kidder, R. W., Beardsley, D. W. & Erwin, T. C. (1961) 'Photosensitization in cattle grazing frosted common Bermudagrass', *Bull. 630, Florida agric. Exp. Sta.*

Kirk, W. G., Easley, J. F., Shirley, R. L. & Hodges, E. M. (1972) 'Effect of pregnancy and lactation on liver vitamin A of beef cows grazing Pangolagrass', *J. Range Mgmt*, **25**, No. 2, 114–16.

Knight, W. E. (1955) 'The influence of photoperiod and temperature on growth, flowering, and seed production of dallisgrass (*Paspalum dilatatum* Poir.)', *J. Am. Soc. Agron.*, **47**, 555–9.

Knight, W. E. & Bennett, H. W. (1953) 'Preliminary report on the effect of photoperiod and temperature on the flowering and growth of several southern grasses', *Agron. J.*, **45**, 268–9.

Knipe, O. D. (1967) 'Influence of temperature on the germination of some range grasses', *J. Range Mgmt*, **20**, No. 5, 208–9.

Kohli, K. S., Singh, C. B., Singh, A., Mehra, K. L. & Magoon, M. L. (1971) 'Variability of quantitative characters in a world collection of cowpea; interregional comparisons', *Genetica Agraria*, **25**, No. 3/4, 231–42.

Kretschmer, A. E. (1964) 'Berseem clover, a new winter annual for Florida', *Agr. exp. Stations, Florida Univ., Circ. S.–163*, pp. 16.

Kretschmer, A. E. (1970) 'Production of annual and perennial tropical legumes in mixtures with pangolagrass and other grasses in Florida', *Proc. 11th int. Grassld Congr., Surfers Paradise, 1970*, 149–53.

Kretschmer, A. E. (1972) 'Siratro (*Phaseolus atropurpureus* D.C.), a summer-growing perennial pasture legume for central and south Florida', *Circ. Fla Agric. Exp. Sta.*, No. S–124.

Krishnaswamy, N. (1962) *Bajra. Pennisetum typhoides* S. & H. Indian Council Agric. Res., New Delhi.

Kyneur, G. W. & Tow, P. G. (1958) 'Rhodes grass – lucerne pasture at Kairi', *Qd agric. J.*, **84**, 398–406, and 453–60.

Ladeira, N. P., Sykes, D. J., Daker, A. & Gomide, J. A. (1966) 'Estudos sôbre produção e irrigação dos capins Pangola, sempre-verde e gordura, durante o ano de 1965', *Rev. Ceres*, **13**, No. 74, 105–16.

Lahiri, A. N. & Kharabanda, B. C. (1961) 'Dimorphic seeds in some arid zone grasses and the significance of growth differences in their seedlings', *Sci. and Cult.*, **27**, No. 9, 448–50.

Lahiri, A. N. & Kharabanda, B. C. (1962–3) 'Germination studies on arid zone plants. 2. Germination inhibitors in the spikelet glumes of *Lasiurus sindicus, Cenchrus ciliaris* and *Cenchrus setigerus*. *Ann. Arid Zone, Jodhpur*', **1**, No. 1–2, 114–26.

Larin, I. V. (Ed.) (1950) *Fodder and Pasture Plants of the USSR* [in Russian], Vol. 1, Moscow–Leningrad.

Lazarides, M. (1970) *The Grasses of Central Australia*. Canberra, Australian National University Press.

Lazarides, M., Norman, M. J. T. & Perry, R. A. (1965) 'Wet-season development pattern of some native grasses at Katherine, N.T.', *Tech. Pap. 26 Div. Ld Res. reg. Surv. C.S.I.R.O. Aust.*

Leffel, R. C. (1973) 'Other legumes', in *Forages. The Science of Grassland Agriculture*, Iowa State Univ. Press, Ames, USA, 3rd edn, 208–20.

Leigh, J. H. & Davidson, R. L. (1968) '*Eragrostis curvula* (Schrad.) Nees and some other

African lovegrasses', *CSIRO Pl. Introd. Rev.*, **5**, No. 1, 21–44.
León, J. & Sgaravatti, E. (1971) '*Tropical pastures; grasses and legumes. Provisional Catalogue of genetic materials for introduction and exchange*', FAO, Rome.
Leslie, J. K. (1965) 'Factors responsible for failures in the establishment of summer grasses on the black earth of the Darling Downs, Queensland', *Qd J. agric. anim. Sci.*, **22**, 17–38.
Lester, D. C. & Carter, O. G. (1970) 'The influence of temperature upon the effect of gibberellic acid on the growth of *Paspalum dilatatum*', *Proc. 11th int. Grassld Congr., Surfers Paradise, 1970*, 615–18.
Lewin, P. & Melotti, L. (1965–6) 'Estudo dos teores de lignina e de outros componentes quimicos nos capins jaraguá e gordura', *Bolm. Ind. anim.*, **23**, 169–75.
Lima, F. P., Martinelli, D. & Werner, J. C. (1968) 'Produção de carne de bovinos em pastagenes de Gramíneas em região de terras roxas (latosol roxo)', *Bolm. Ind. anim.*, **25**, 129–37.
Lima, C. R. & Souto, S. M. (1972) 'Valor nutritivo do feno proveniente de diferentes estádios de crescimento da cultura de soja perenne (*Glycine javanica*)', *Pesq. Agropec. Brasil. Zoot.*, **7**, 59–62.
Lima, C. R., Souto, S. M., Garcia, J. M. R. & Araujo, M. R. (1972) 'Valores nutritivos de feno de Siratro (*Phaseolus atropurpureus*) em diferentes estadios de crescimento', *Pesq. Agropec. Brasil., Zootech.*, **7**, 63–6.
Little, S., Vicente, J. & Abruna, F. (1959) 'Yield and protein content of irrigated Napier grass, Guinea grass and Pangola grass as affected by nitrogen fertilization', *Agron. J.*, **51**, No. 2, 111–13.
Lloyd, D. L. & Scateni, W. (1968) 'Makarikari grasses for heavy soils', *Qd agric. J.*, **94**, No. 12, 721–4.
Long, M. I. E., Thornton, D. D. & Marshall, B. (1969) 'Nutritive value of grasses in Ankole and the Queen Elizabeth National Park, Uganda. 2. Crude protein, crude fibre and soil nitrogen', *Trop. Agric., Trin.*, **46**, No. 1, 31–42.
Lotero, C. J., Ramirez, O. R. & Crowder, L. V. (1960) 'Estudio preliminar de la asociación del pasto parà con leguminosas', *Agric. trop., Bogota*, **16**, No. 7, 450–5.
Lovadini, L. A. C. (1971) 'Método de plantio para soja perenne (*Glycine wightii* Verdc.)', *Bragantia*, **30**, No. 1, XVII–XIX.
Lucci, C. de S., Boin, C. & Lobao, A. de O. (1968) 'Estudo comparativo das silagenes de Napier, de milho e de sorgo, como unicos volumosos para vacas em lactação', *Bolm. Ind. anim.*, **25**, 161–73.
Mackay, J. H. E. (1973) 'Register of Australian herbage plant cultivars. B. Legumes. 2. Desmodium. c. *Desmodium heterophyllum* DC (Hetero) cv. Johnstone (Reg. No. B–2c–1)', *J. Aust. Inst. agric. Sci.*, **39**, No. 2, 147–8.
Mahudeswaran, K. (1973) 'A note on the fodder quality of ragi (*Eleusine coracana*) straw', *Madras agric. J.*, **60**, No. 1, 69.
Maire, R. (1952) *Flore de l'Afrique du Nord*, Vol. 1, Paris, Paul Lechevalier.
Majumdar, B. R. & Roy, S. R. (1968) 'Thin Napier (*Pennisetum polystachyon* Schult), a promising tropical grass as silage and pasture', *Indian Dairyman*, **20**, No. 12, 355–8.
't Mannetje, L. (1965) 'The effect of photoperiod on flowering, growth habit, and dry matter production in four species of the genus *Stylosanthes* Sw.', *Aust. J. agric. Res.*, **16**, 767–71.
't Mannetje, L. (1968) '*Stylosanthes sundaica* Taub.', *CSIRO Plant Introduction Review*, **5**, No. 23.
't Mannetje, L. (1972) 'The effects of some management practices on pasture production', *Trop. Grasslds*, **6**, No. 3, 260–3.
Mappledoram, B. D. & Theron, E. P. (1972) 'Notes on the adaptability of several tropical legumes to different environments in Natal', *Proc. Grassld Soc. S. Africa*, **7**, 84–6.
Marshall, B. & Bredon, R. M. (1967) 'The nutritive value of *Themeda triandra*', *E. Afr. agric. For. J.*, **32**, No. 4, 375–9.
Matos, E. de & Torre, R. de la (1971) 'Trials with five populations of *Clitoria ternatea* L.', *Revta. cub. Cienc. agric.*, **4**, 217–21.
Mattos, H. B. de (1970–1) 'Efeito de escarificação em sementes de *Phaseolus atropurpureus*

cv. Siratro', *Bolm Ind. anim.*, **27/28**, 379–82.
Mazzani, B., Allievi, J., Centeno, A. & Combellas, J. (1972) 'Aprovechamiente de las partes aéreas del mani. 3. Producción y caracteristicas de frutas y semillas cosechades en diferentes épocas', *Agronomia trop.*, **22**, No. 4, 375–89.
McCormick, W. C., Maechant, W. H. & Southwell, B. L. (1967) 'Coastal Bermudagrass and Pensacola Bahiagrass hays for wintering beef calves', *Res. Bull. 19 Ga agric. Exp. Stn.*
McDonald, P., Edwards, R. A. & Greenhalgh, J. F. D. (1973) *Animal Nutrition*, Edinburgh, Oliver & Boyd, 2nd edn.
McLean, D. & Grof, B. (1968) 'Effect of seed treatment on *Brachiaria mutica* and *B. ruziziensis*', *Qd J. agric. anim. Sci.*, **25**, Nos 1–2, 81–3.
Mears, P. T. (1970) 'Kikuyu (*Pennisetum clandestinum*) as a pasture grass – a review', *Tropical Grasslands*, **4**, No. 2, 139–52.
Melotti, L. (1969) 'Determinação do valor nutritivo dos capins gordura (*Melinis minutflora* Pal. de Beauv.) e angolinha do rio (*Eriochloa polystachya* H.B.K. – Hitchc.), através de ensaio de digestibilidade (aparente) con carneiros', *Bolm Ind. anim.*, **26**, 285–94.
Melotti, L., Boin, C. & Lobao, A. O. (1969) 'Determinação do valor nutritivo da soja perenne (*Glycine javanica*) como forragem verde e na forma de feno, através de ensaio de digestibilidade (aparente) com ovinos', *Bolm Ind. anim.*, **26**, 295–302.
Melotti, L. & Velloso, L. (1970–1) 'Determinação do valor nutritivo de feno de soja (*Glycine max* (L.) Merr.) var. Santa Maria através de ensaio de digestibilidade (aparente) com carneiros', *Bolm Ind. anim.*, **27/28**, 197–205.
Metcalfe, D. S. (1973) 'Forage statistics', in *Forages*, Iowa State Univ. Press, Ames, 3rd ed, 1973, 64–79.
Middleton, C. H. (1970) 'Some effects of grass–legume sowing rates on tropical species establishment and production', *Proc. 11th int. Grassld Congr., Surfers Paradise, 1970*, 119–23.
Middleton, C. H. (1973) 'Effects of sowing rate on yield and composition of a Siratro–Nandi setaria pasture', *Qd J. agric. anim. Sci.*, **30**, No. 1, 45–52.
Milford, R. (1960) 'Nutritional values for 17 subtropical grasses', *Aust. J. agric. Res.*, **11**, 138–48.
Milford, R. (1967) 'Nutritive values and chemical composition of seven tropical legumes and lucerne grown in subtropical south-eastern Queensland', *Aust. J. exp. Agric. Anim. Husb.*, **7**, No. 29, 540–5.
Milford, R. & Minson, D. J. (1966) 'Intake of tropical pasture species', *Proc. 9th int. Grassld Congr., Brazil, 1965*, 815–22.
Millar, R. P. (1967) 'Oestrogenic activity in Central African highveld grasses', *Rhod. Zam. & Malawi Jnl. agric. Res.*, **5**, No. 2, 179–83.
Miller, T. B. & Rains, A. B. (1963) 'The nutritive value and agronomic aspects of some fodders in Northern Nigeria. 1. Fresh herbage', *J. Brit. Grassld. Soc.*, **18**, No. 2, 158–67.
Miller, T. B., Rains, A. B. & Thorpe, R. J. (1963) 'The nutritive value and agronomic aspects of some fodders in northern Nigeria. 2. Silages', *J. Brit. Grassld. Soc.*, **18**, No. 3, 223–9.
Mills, P. F. L., Rodel, M. G. W. & Boultwood, J. N. (1973) 'Effects of applied nitrogen on herbage production of four grasses receiving supplementary irrigation', *Rhodesia agric. J.*, **70**, No. 5, 131–3.
Mishra, M. L. & Chatterjee, B. N. (1968) 'Seed production in the forage grasses *Pennisetum polystachyon* and *Andropogon gayanus* in the Indian tropics', *Trop. Grasslds*, **2**, No. 1, 51–6.
Moffett, M. L. (1973) 'Seed transmission of *Pseudomonas phaseolicola* (halo blight) in *Macroptilium atropurpureum* cv. Siratro', *Trop. Grasslds*, **7**, No. 2, 195–9.
Montgomery, C. R., Ellzey, H. D., Allen, M. & Rusoff, L. L. (1972) 'Effect of age and season on the quality of bahiagrass', *Louisiana Agr.*, **15**, No. 3, 8–9, 11.
Moore, W. H. & Hilmon, J. B. (1969) 'Jointvetch: native legume in a new role for deer and cattle', *Res. Note SE–114 U.S. Dep. Agric. For. Serv.*, pp. 5.

Morrison, J. (1966) 'The effects of nitrogen, phosphorus and cultivation on the productivity of Kikuyu grass at high altitudes in Kenya', *E. Afr. Agric. For. J.*, **31**, 29–7.
Mosse, B. (1972) 'Effect of different *Endogone* strains on the growth of *Paspalum notatum*', *Nature, Lond.*, **239**, (5369), 221–3.
Motooka, P. S., Plucknett, D. L., Saiki, D. F. & Younge, O. R. (1967) 'Pasture establishment in tropical brushlands by aerial herbicide and seeding treatments on Kauai', *Tech. Prog. Rep. 165 Hawaii agric. Exp. Stn.*
Motta, M. S. (1953) '*Panicum maximum*', *Emp. J. exp. Agric.*, **21**, No. 81, 33–41.
Mukherjee, A. & Chatterji, U. N. (1970) 'Photoblastism in some of the desert grass seeds', *Ann. arid Zone*, **9**, No. 2, 104–13.
Mullick, P. & Chatterji, U. N. (1967) 'Eco-physiological studies on seed germination: germination experiments with the seeds of *Clitoria ternatea* Linn.', *Trop. Ecol.*, **8**, 116–25.
Muradov, K. M. (1970) Legumes as silage crops newly cultivated in the Turkmenia [in Russian] in *Fifth sympodium on new silage plants*. Leningrad, Bot. Inst.
Murata, Y., Iyama, J. & Honma, T. (1965) 'Studies on the photosynthesis of forage crops. 4. Influence of air-temperature upon the photosynthesis and respiration of alfalfa and several southern-type forage crops', *Proc. Crop. Sci. Soc. Japan*, **34**, 154–8.
Murphy, L. S. (1966) 'Nitrate accumulation in forage crops', *Diss Abstr.*, **27**, No. 6, 1686B–7B (from *Herbage Abst.*, 1968, 236).
Murtagh, G. J. & Dougherty, A. B. (1968) 'Relative yields of lablab and velvet bean', *Trop. Grasslds*, **2**, No. 1, 57–63.
Mwakha, E. (1969) 'Observations on the effect of temperature on the growth of *Trifolium semipilosum* Fres.', *E. Afr. agric. For. J.*, **34**, No. 3, 289–92.
Mwakha, E. (1970) 'Observations on the growth of Bungoma grass (*Entolasia imbricata* Stapf)', *C.S.I.R.O. Plant Introduction Review*, **7**, No. 1, 24–9.
Mwakha, E. (1971) 'Germination depth of Bungoma grass *Entolasia imbricata* Stapf', *E. Afr. agric. For. J.*, **37**, No. 1, 26–8.
Narayan, K. N. (1955) 'Cytogenetic studies of apomixis in *Pennisetum* (*Pennisetum clandestinum* Hochst.)', *Indian Academy of Science*, **41**, Section B, 196.
Narayanan, T. R. & Dabadghao, P. M. (1972) *Forage Crops of India*, Indian Council of Agricultural Research, New Delhi.
Nath, K., Malik, N. S. & Singh, O. N. (1971) 'Chemical composition and nutritive value of *Indigofera enneaphylla* and *I. cordifolia* as sheep feeds', *Aust. J. exp. Agric. Anim. Husb.*, **11**, No. 49, 178–80.
Nath, J. & Swaminathan, M. S. (1957) 'Chromosome numbers of some grasses', *Indian J. Genet.*, **17**, 102.
Naveh, Z. & Anderson, G. D. (1965) 'Some preliminary yields from promising grass/legume mixtures at Tengeru, Northern Tanzania', *Sols Afr.*, **10**, Nos 2–3, 493–508.
Ndyanabo, W. K. (1974) 'Oxalate content of some commonly grazed pasture forages of Lango and Acholi districts of Uganda', *E. Afr. agric. For. J.*, **39**, No. 3, 210–14.
Nel, J. W., Grunow, J. O., Hugo, W. J., Pienaar, A. J. & Voss, H. C. (1964) 'The feeding value of *Eragrostis curvula* as hay and pasture for sheep', *Tech. Comm. 23*, Dep. Agric. Tech. Servs, Pretoria.
Neme, N. A. (1958) 'Soja perenne. Leguminosa para forragem e conservação do solo', *Agronômico, Campinas*, **10**, No. 9–10, 20–3.
Nestel, B. L. & Creek, M. J. (1962) 'Pangola grass', *Herb. Abstr.*, **32**, No. 4, 265–71.
Ng, T. T. (1972) 'Comparative responses of some tropical grasses to fertilizer nitrogen in Sarawak, E. Malaysia', *Trop. Grasslds*, **6**, No. 3, 229–36.
Norris, D. O. (1958) 'A red strain of *Rhizobium* from *Lotononis bainesii*', *Aust. J. agric. Res.*, **9**, 629–32.
Norris, D. O. (1964) 'Legume bacteriology', in *Some concepts and methods in sub-tropical pasture research. Bull. 47 Commonwealth Bureau of Pastures and Field Crops, CAB, Hurley*, 120–17.
Norris, D. O. (1967) 'The intelligent use of inoculants and lime pelleting for tropical

legumes', *Trop. Grasslds*, **1**, 107–21.
Norris, D. O. (1971) 'Seed pelleting to improve nodulation of tropical and subtropical legumes. 3. A field evaluation of inoculant survival under lime and rock phosphate pellet on *Dolichos lablab*', *Aust. J. exp. Agric. Anim. Husb.*, **11**, No. 53, 677–83.
Norris, D. O. (1972) 'Seed pelleting to improve nodulation of tropical and subtropical legumes. 4. The effects of various mineral dusts on nodulation of *Desmodium uncinatum*', *Aust. J. exp. Agric. Anim. Husb.*, **12**, No. 55, 152–8.
Norris, D. O. (1973a) 'Seed pelleting to improve nodulation of tropical and subtropical legumes. 5. The contrasting response to lime pelleting of two *Rhizobium* strains on *Leucaena leucocephala*', *Aust. J. exp. Agric. Anim. Husb.*, **13**, No. 60, 98–101.
Norris, D. O. (1973b) 'Seed pelleting to improve nodulation of tropical and subtropical legumes. 6. The effects of dilute sticker, and of bauxite pelleting on nodulation of six legumes', *Aust. J. exp. Agric. Anim. Husb.*, **13**, No. 65, 700–4.
Northwood, P. J. & Macartney, J. C. (1969) 'Pasture establishment and renovation by direct seeding', *E. Afr. agric. For. J.*, **35**, No. 2, 185–9.
Nourrissat, P. (1966) 'Les introductions de plantes fourragères au CRA de Bambey', *L'Agronomie tropicale, Serie 2*, **21**, No. 9, 1013–35.
Novoa, L. G. & Rodríguez-Carrasquel, S. (1972) 'Estudio del comportamiento de los pastos Pará (*Brachiaria mutica* Stapf) y Alemán (*Echinochloa polystachya* (H.B.K.) Hitche.)', *Agronomía trop.*, **22**, No. 6, 643–55.
Nowar, M. S., Marai, I. F. M. & Yamani, K. A. (1973) 'Biochemical studies on silage. 2. Effect of wilting on the course of fermentation and the quality of silage made from corn cut at the beginning of flowering', *Beitr. trop. & subtrop. Landwirt. & Tropenvet.–Med.*, **11**, No. 3, 153–9.
Oakes, A. J. (1968a) *Leucaena leucocephala* – description, culture, ultilization, *Adv. Front. Pl. Sci.*, **20**, 1–114.
Oakes, A. J. (1968b) 'Replacing hurricane grass in pastures of the dry tropics', *Trop. Agric., Trin.*, **45**, No. 3, 235–41.
Oakes, A. J. (1969a) 'Pangolagrass (*Digitaria decumbens* Stent)', *Crop Sci.*, **9**, No. 6, 835.
Oakes, A. J. (1969b) 'Pasture grasses in the U.S. Virgin Islands', *Turrialba*, **19**, No. 3, 359–67.
Obaton, M. (1974) 'Légumineuses tropicales: Problèmes particuliers posés par la symbiose fixatrice d'azote et l'inoculation des semences', *Agron. trop., Paris*, 1128–39.
Odu, C. T. I., Fayemi, A. A. & Ogunwale, J. A. (1971) 'Effect of pH on the growth, nodulation and nitrogen fixation of *Centrosema pubescens* and *Stylosanthes gracilis*', *J. Sci. Fd Agric.*, **22**, No. 2, 50–9.
Oke, O. L. (1967a) 'Nitrogen fixing capacity of *Calopogonium* and *Pueraria*', *Trop. Sci.*, **9**, No. 2, 90–3.
Oke, O. L. (1967b) 'Nitrogen fixing capacity of some Nigerian legumes', *Exp. Agric.*, **3**, 315–21.
Oliveira, B. A. D., de, Faria, P. R. de S., Souto, S. M., Carneiro, A. M., Döbereiner, J. & Aronovich, S. (1973) 'Identificação de gramíneas tropicais com via fotossintética "C 4" pela anatomia foliar', *Pesq. Agropec. Bras., Sér. Agron.*, **8**, 267–71.
Olsen, F. J. (1973) 'Effects of cutting management of a *Desmodium intortum* (Mill.), Urb./*Setaria spacelata* (Schumach.) Stapf mixture', *Agron. J.*, **65**, No. 5, 714–16.
Osbourn, D. F., Terry, R. A., Cammell, S. B. & Outen, G. E. (1971) 'The effect of leucoanthocyanins in sainfoin (*Onobrychis viciifolia* Scop.) on the availability of protein to sheep and upon the determination of the acid detergent fibre and lignin factors', *Proc. Nutr. Soc.*, **30**, 13–14A.
Oschita, M., Andrade, S. O. & Bueno, P. (1972) 'Intoxicação de búfalos alimentados com *Brachiaria* sp. (Tanner grass)', *Archos Inst. biol., S. Paulo*, **39**, No. 3, 209–11.
Owen, M. A. & Brzostowski, H. W. (1966) 'A grass cover for the upland soils of the Kongwa plain, Tanganyika', *Trop. Agric., Trin.*, **43**, No. 4, 303–14.
Pal, M. & Pandey, S. L. (1969) 'Straw mulch brings water economy and higher yields of summer fodders', *Indian Fmg*, **19**, No. 8, 25–7.
Panday, R. K., Singh, R. P. & Singh, M. (1969) 'Weed control in the fodder crops of

teosinte and maize', *Indian J. Weed Sci.*, **1**, No. 2, 95–102.
Pandeya, S. C. & Jayan, P. K. (1970) 'Population differences in buffel grass (*Cenchrus ciliaris*) at Ahmedabad, India: productivity under various agronomic conditions', *Proc. 11th int. Grassld Congr., Surfers Paradise, 1970* 239–44.
Park, R. J. & Minson, D. J. (1972) 'Flavour differences in meat from lambs grazed on tropical legumes', *J. agric. Sci., UK*, **79**, No. 3, 473–8.
Parodi, L. R. (1943) 'Una nueva especie de "Sorghum" cultivada en la Argentina', *Rev. Argent. Agron.*, **10**, 361–72.
Parodi, L. R. (1946) 'Las especies de *Sorghum* cultivadas en la Argentina', *Rev. Argent. Agron.*, **13**, 1–35.
Parsons, J. J. (1972) 'Spread of African pasture grasses to the American tropics', *J. Range Mgmt*, **25**, No. 1, 12–17.
Patil, B. D. & Ghosh, R. (1962) 'The grass that shrugs off drought and cold', *Indian Fmg*, **12**, No. 2, 11–12.
Paula, R. R., Gomide, J. A. & Sykes, D. J. (1967) 'Influencia de diferentes sistemas de corte sôbre o capim-gordura (*Melinis minutiflora* Beauv.)', *Revta Ceres*, **14**, No. 80, 157–86.
*****Payne, W. J. A.** (1969) 'Problems of the nutrition of ruminants in the tropics', in *Nutrition of Animals of Agricultural Importance*, Part 2, 849–82, Pergamon Press.
Pereira, H. C., Hosegood, P. H. & Thomas, D. B. (1961) 'The productivity of tropical semi-arid thorn-scrub country under intensive management', *Emp. J. exp. Agric.*, **29**, No. 115, 269–86.
Pereira, R. M. A., Sykes, D. J., Gomide, J. A. & Vidigal, G. T. (1966) 'Competição de 10 Gramíneas para capineiras, no cerrado, em 1965', *Rev. Ceres*, **13**, No. 74, 141–53.
Pérez Infante, F. (1970) 'Effect of cutting interval and N fertilizer on the productivity of eight grasses', *Revta cub. Cienc. agric.*, **4**, No. 2, 137–48.
Pernès, J., Combes, D. & Rene-Chaume, R. (1970) 'Différenciation des populations naturelles du *Panicum maximum* Jacq. en Côte-d'Ivoire par acquisition de modifications transmissibles, les unes par graines apomictiques, d'autres par multiplication vég-étative', *C.R. Acad. Sc. Paris*, **270**, 1992–95.
Philpotts, H. (1969) 'Rongai *Dolichos lablab*, a drought-hardy forage legume for the northern wheat belt of New South Wales', *Agric. Gaz. N.S.W.*, **80**, 541–3.
Phipps, R. H. (1973) 'Methods of increasing the germination percentage of some tropical legumes', *Trop. Agric., Trin.*, **50**, No. 4.
Picard, D. & Fillonneau, C. (1972) 'Mise en évidence d'une période critique pour la fauche chez les Graminées. L'exemple de *Panicum maximum*', *Fourrages*, No. 52, 71–80.
Pio Corrêa, M. (1926 and 1931) *Diccionário das plantas úteis do Brasil e das exóticas cultivadas*, Vol. 1 and Vol. 2, Imprensa Nacional, Rio de Janeiro.
Pio Corrêa, M. & De Azeredo Pena, L. (1952) *Diccionário das plantas úteis do Brasil e das exóticas cultivadas*, Vol. 3, Ministerio da Agricultura, Rio de Janeiro.
Pittier, H. (1944) 'Leguminosas de Venezuela. 1. Papilionáceas', *Bol. Tec. No. 5 Min. Agric. Cria Venezuela*, Caracas.
Playne, M. J. (1969) 'The effect of dicalcium phosphate supplement on the intake and digestibility of Townsville lucerne and spear grass by sheep', *Aust. J. exp. Agric. Anim. Husb.*, **9**, No. 37, 192–5.
Plucknett, D. L. (1970) 'Productivity of tropical pastures in Hawaii', *Proc. 11th int. Grassld Congr., Surfers Paradise, 1970*, A38–A49.
Polhill, R. M. (1971) '*Crotalaria*' in Flora of Tropical East Africa, Leguminosae, Part 4, 817–994.
Possingham, J. V., Groot Obbink, J. & Jones, R. K. (1971) 'Tropical legumes and vesicular arbuscular mycorrhiza', *J. Aust. Inst. agric. Sci.*, **37**, No. 2, 160–1.
Poultney, R. G. (1963) 'A comparison of direct seeding and undersowing on the establishment of grass and the effect on the cover crop', *E. Afr. agric. For. J.*, **29**, No. 1, 26–30.
Prasad, L. K. & Mukerji, S. K. (1961) 'Studies on the effect of clipping at different stages of plant growth on seed production of annual grass *Pennisetum pedicellatum* Trin.',

Indian J. Agron., **5**, No. 3, 157–65.
Pratt, D. J. (1963) 'Reseeding denuded land in Baringo district, Kenya', *E. Afr. agric. For. J.*, **29**, No. 1, 78–91.
Pratt, D. J. (1964) 'Reseeding denuded land in Baringo district, Kenya. 2. Techniques for dry alluvial sites', *E. Afr. agric. For. J.*, **29**, No. 3, 243–60.
Pratt, D. J. & Knight, J. (1964) 'Reseeding denuded land in Baringo district, Kenya. 3. Techniques for capped red loam soils', *E. Afr. agric. For. J.*, **30**, No. 2, 117–25.
Prine, G. M. (1964) 'Forage possibilities in the genus *Arachis*', *Proc. Soil Crop Sci. Soc. Fla*, **24**, 187–96.
Prine, G. M. (1973) 'Perennial peanuts for forage', *Proc. Soil Crop Sci. Soc. Fla*, **32**, 33–5.
Pritchard, A. J. (1971) 'The hybrid between *Pennisetum typhoides* and *P. purpureum* as a potential forage crop in south-eastern Queensland', *Trop. Grasslds*, **5**, No. 1, 35–9.
Prodonoff, E. (1967) 'The determination and maintenance of seed quality', *Trop. Grasslds*, **1**, No. 1, 91–8.
Purseglove, J. W. (1968) *Tropical Crops.* 'Dicotyledons', Longman, London.
Purseglove, J. W. (1972) *Tropical Crops.* Monocotyledons. Longman, London.
Rai, R. R., Singh, G. S. & Singh, S. N. (1966) 'Studies on Para grass (*Panicum barbinode* or *Brachiaria mutica* Stapf). 2. Effect of hoeing, organic and inorganic nitrogen on the chemical composition and yield of Para grass', *B. V. J. agric. Sci. Res.*, **8**, Nos 1–2, 26–35.
Rains, A. B. & Foster, W. H. (1956–7) 'Effect of cutting *Andropogon gayanus* (gamba) at different heights', *Rep. Dep. Agric. N. Nigeria* Pt. 2, 163–4.
Rajaratnam, J. A. & Ang Poo Guan (1972) 'Nitrogen fixation by *Pueraria phaseoloides* in Malaysia', *Malay. agric. J.*, **1**, No. 2, 92–7.
Ramaswamy, K. R., Raman, V. S. & Menon, P. M. (1969) 'An analysis of morphological variation in relation to chromosomal forms in the *Cenchrus* complex', *J. Indian Bot. Soc.*, **48**, Nos 1–2, 102–11.
Ramia, M. (1967) 'Tipos de sabanas en los llanos de Venezuela', *Boln Soc. venez. Cienc. nat.*, **27**, No. 112, 264–88.
Randolph, L. E. (1970) 'Variation among *Tripsacum* populations of Mexico and Guatemala', *Brittonia*, **22**, No. 4, 305–37.
Rao, B. V. V. & Sadasivaiah, T. (1968) 'Studies on nitrogen mobilisation through phosphate fertilizing of a legume in the Bangalore red soil', *Mysore J. agric. Sci.*, **2**, No. 4, 251–6.
Rao, K. A., Kandlikar, S. S., Sarma, V. S. & Anantharaman, P. V. (1969) 'Note on the fodder value of Asiriya Mwitunde, a promising variety of groundnut', *Sci. Cult.*, **35**, No. 9, 472–3.
Rao, R. S. (1970) 'Studies on the flora of Kutch, Gujarat State (India) and their utility in the economic development of the semi-arid region', *Ann. Arid Zone*, **9**, No. 2, 125–42.
Ravens, P. H. (1960) 'The correct name for rescue grass', *Brittonia*, **12**, No. 3, 219–21.
Reddy, M. R. (1968) 'Sunn-hemp hay can cut down concentrate needs of cattle', *Indian Fmg*, **18**, No. 6, 45–6.
* **Reeder, J. R.** (1948) 'The Gramineae – Panicoideae of New Guinea', *J. Arnold. Arbor.*, **29**, 257–392.
Reichert, F. & Trelles, R. A. (1923) 'Introducción e investigación química de las forrajeras', *Las plantas forrajeras indígenas y cultivadas de la República Argentina*, Buenos Aires.
Relwani, L. L. (1968a) 'Teosinte (*Euchlaena mexicana* Schrad.) a fodder crop for warm and humid tropics', *Indian Dairyman*, **20**, No. 2, 61–6.
Relwani, L. L. (1968b) 'Sudan grass – an excellent summer forage crop', *Indian J. Agron.*, **13**, No. 3, 205–9.
Relwani, L. L. (1971) 'Make berseem hay the easier way', *Indian Fmg.*, **21**, No. 3, 10–12, 15.
Relwani, L. L. & Bagga, R. K. (1969) 'Blue panic (*Panicum antidotale* Retz.)', *Indian Dairyman*, **21**, No. 6, 179–81.
Relwani, L. L. & Kumar, A. (1970) 'Jowar (*Sorghum vulgare*), a high yielding, heat and drought resistant fodder crop for the tropics', *Indian Dairyman*, **22**, No. 4, 93–8.
Rensburg, H. J. van (1969) 'Selection of productive strains of *Chloris gayana* in Zambia',

Zambia Min. Rur. Dev. Lusaka, pp. 13.
Renvoize, S. A. (1974) Private communications.
Rethman, N. F. G. & Malherbe, C. E. (1970) 'The influence of fertilization on the production and digestibility of natural veld', *Agric. Sci. S. Afr. (1), Agroplantae*, **2**, No. 1, 43.
Revilla, M. V. A. (1966) 'Una virosis destructiva del Pasto Pangola en el Peru', *Boln Sitac. exp. agric. La Molina*, 12.
Rhind, D. (1945) *The Grasses of Burma*. Calcutta.
Rhodesia Report (1966-7) Matopos Research Station, *Annual Report, 1966-7*.
Richards, J. A. (1970) 'Productivity of tropical pastures in the Caribbean', *Proc. 11th int. Grassld Congr., Surfers Paradise, 1970*, A49-A56.
Richardson, W. D. (1963) 'Observations on the vegetation and ecology of the Apiro Savannas, Trinidad', *J. Ecol.*, **51**, 295-313.
Rivera-Brenes, L., Cestero, H. & Sierra, A. (1962) 'Napier grass (*Pennisetum purpureum*) *versus* sugar cane as forage crops in Puerto Rico', *J. agr. Univ. Puerto Rico*, **46**, No. 4, 307-12.
Rivera-Brenes, L., Torres Más, J. & Arroyo, J. A. (1961) 'Response of Guinea, Pangola, and Coastal Bermuda grasses to different nitrogen fertilization levels under irrigation in the Lajas Valley of Puerto Rico', *J. Agric. Univ. P.R.*, **45**, No. 3, 123-46.
Riveros, F. & Wilson, G. L. (1970) 'Responses of a *Setaria sphacelata-Desmodium intortum* mixture to height and frequency of cutting', *Proc. 11th int. Grassld Congre., Surfers Paradise, 1970*.
Riveros, R. G. (1960) 'Comportamiento del pasto Pangola (*Digitaria decumbens* Stent) en mezcla con leguminosas', *Acta agron., Palmitra*, **10**, No. 1, 101-29.
Roberts, O. T. (1970a) 'A review of pasture species in Fiji. 1. Grasses', *Trop. Grasslds*, **4**, No. 2, 129-37.
Roberts, O. T. (1970b) 'A review of pasture species in Fiji. 2. Legumes', *Trop. Grasslds*, **4**, No. 4, 213-22.
Rocha, G. L. da, Cintra, B., Freire, A. & Montagnini, M. I. (1965) 'Estudo da variação do teor germinativo de sementes de capim gordura (*Melinis minutiflora*) armazenadas a baixas temperaturas e em ambiente normal', *Proc. 9th int. Grassld Congr., San Paulo*, 531-4.
Rodel, M. G. W. (1970) 'Herbage yields of five grasses and their ability to withstand intensive grazing', *Proc. 11th int. Grassld Congr., Surfers Paradise, 1970*, 618-21.
Rodel, M. G. W. (1972) 'Effect of different grasses on the incidence of neonatal goitre and skeletal deformities in autumn born lambs', *Rhodesia Agr. J.*, **69**, No. 3, 59-60.
Rodel, M. G. W. & Boultwood, J. N. (1971) 'Herbage yields of thirty grasses fertilized heavily with nitrogen', *Proc. Grassld Soc. S. Afr.*, **6**, 129-33.
Rodríguez, S. C. & Blanco, E. (1970) 'Composicion quimica de hojas y tallos de 21 cultivares de elefante (*Pennisetum purpureum* Schumacher)', *Agron. Trop. (Maracay)*, **20**, No. 6, 383-96.
Rosengurtt, R., Arrillaga de Maffei, R. R. & Izaguirre de Artucio, P. (1970) *Gramineas Uruguayas*, Montevideo, Universidad de la Republica.
Rossiter, J. & Delgadillo, G. (1971) 'Buffel (*Cenchrus ciliaris*)', *Bol. Tec. No. 7. Min. Asunt. Camp. Agric. Bolivia*.
Rotar, P. P., Park, S. J., Bromdep, A. & Urata, U. (1967) 'Crossing and flowering behaviour in Spanish clover, *Desmodium sandwicense* E. Mey. and other *Desmodium* species', *Tech. Prog. Rep. 164 Hawaii agric. Exp. Stn.*
Sant, H. R. (1964) 'Ecological studies with special reference to reproductive capacity in relation to grazing *Dactyloctenium aegyptium* Beauv.', *Proc. natn. Inst. Sci. India (B)*, **30**, No. 1, 30-46.
Santhirasegaram, K. & Ferdinandez, D. E. F. (1967) 'Yield and competitive relationship between two species of *Brachiaria*', *Trop. Agric. Trin.*, **44**, No. 3, 229-34.
Sauer, J. (1964) 'Revision of *Canavalia*', *Brittonia*, **16**, No. 2, 106-81.
Saxena, J. S., Kulshrestha, S. K. & Johri, C. B. (1971) 'Studies on exotic legume fodder, siratro (*Phaseolus atropurpureus*) cultivation and nutritive value for sheep', *Indian vet.*

J., **48**, No. 8, 849–53.
Scarbrook, C. E. (1970) 'Regression of nitrogen uptake on nitrogen added from four sources applied to grass', *Agron. J.*, **62**, No. 5, 618–20.
Schank, S. C., Edwardson, J. R., Christie, R. G. & Overman, M. A. (1972) 'Pangola stunt virus studied in Pangolagrass and Digitaria hybrids', *Euphytica*, **21**, No. 2, 344–51.
Schank, S. C., Klock, M. A., Moore, J. E. (1973) 'Laboratory evaluation of quality in subtropical grasses: 2. Genetic variation among *Hemarthrias* in *in vitro* digestion and stem morphology', *Agron. J.*, **65**, No. 2, 256–8.
*****Schmid, M.** (1958) 'Flore agrostologique de l'Indochine', *Agron. Tropicale*, Nos. 1–6, 1958.
Schroder, V. N. (1966) 'Photosynthesis and respiration of four pasture grasses as affected by moisture conditions and salinity', *Proc. 10th Int. Grassld Congr., Helsinki,* July, 1966, 181–4.
Schroder, V. N. (1971) 'Soil temperature effect on shoot and root growth of Pangolagrass, slenderstem digitgrass, Coastal Bermudagrass and Pensacola Bahiagrass', *Proc. Soil Crop Sci. Soc. Fla*, 1970 (publ. 1971), **30**, 241–5.
Schuster, M. F. & Boling, J. C. (1971) 'Biological control of Rhodesgrass scale in Texas by *Neodusmetia sangwani* (Rao): effectiveness and colonization studies', *Publication, Texas Agric. Exp. Stn*, No. B–1104, 15 pp.
Semb, G. & Garberg, P. K. (1969) 'Some effects of planting date and nitrogen fertilizer in maize', *E. Afr. agric. For. J.*, **34**, No. 3, 371–9.
Semple, A. T. (1964) '*Desmodium barbatum* (L.) Benth. from natural tropical pastures of Central and South America,' *Turrialba*, **14**, No. 4, 205.
Sen, K. M. & Mabey, G. L. (1966) 'The chemical composition of some indigenous grasses of coastal savanna of Ghana at different stages of growth', *Proc. 9th int. Grassld Congr., Sao Paulo, Brazil, 1965,* 763–71.
Sen, K. C. & Ray, S. N. (1964) 'Nutritive values of Indian cattle feeds and the feeding of animals', *I.C.A.R. Bull*, No. 25, New Delhi.
Senaratna, S. D. J. E. (1956) 'The grasses of Ceylon', *Paradenal Manual No. 8*, Gvt. Printer, Paradeniya.
Sepra, A. (1971) 'A influência do meio na permeabilidade das sementes de *Centrosema pubescens*,' *Pesq. Agropec. Brasil. Agron.*, **6**, 151–3.
Seton, D. H. C. (1962) 'Problems of pasture establishment and growth on the Burdekin', *Proc. N. Qd Agrost. Conf. 1962*, 2/5, pp. 5.
Shah, S. K., Sial, M. A., Hussain, T. (1970) 'A comparative study of the nutritive value of alfalfa and Egyptian clover for lactating buffaloes', *J. agric. Res., Pakistan*, **8**, No. 4, 391–5.
Shaw, N. H., Elich, T. W., Haydock, K. P. & Waite, R. B. (1965) 'A comparison of seventeen introductions of *Paspalum* species and naturalized *P. dilatatum* under cutting at Samford, southeastern Queensland', *Aust. J. exp. Agric. Anim. Husb.*, **5**, No. 19, 423–32.
Shaw, N. H. & 't Mannetje, L. (1970) 'Studies on a spear grass pasture in central coastal Queensland – the effect of fertilizers, stocking rate, and oversowing with *Stylosanthes humilis* on beef production and botanical composition', *Trop. Grasslds*, **4**, No. 1, 43–56.
Shelton, H. M. & Humphreys, L. R. (1971) 'Effect of variation in density and phosphate supply on seed production of *Stylosanthes humilis*', *J. agric. Sci., Camb.*, **76**, No. 3, 325–8.
Sherrod, L. B., Tamimi, Y. N. & Ishiaki, S. M. (1968) 'Effects of stage of maturity upon yield, composition, and nutritive value of whole plant corn and forage sorghum', *Tech. Bull. 72 Hawaii agric. Exp. Stn*.
Sheth, A. A., Yu, L. & Edwardson, J. (1956) 'Sterility in Pangolagrass (*Digitaria decumbens* Stent.)', *Agron. J.*, **48**, 505–7.
Shukla, K. S., Ranjhan, S. K. & Katiyar, R. C. (1970) '*Rhynchosia minima* as a feed for sheep', *Indian J. Dairy Sci.*, **23**, No. 2, 82–4.
Siebert, B. D., Newman, D. M. R. & Nelson, D. J. (1968) 'The chemical composition of some arid zone pasture species', *Trop. Grasslds*, **2**, No. 1, 31–40.

Silcock, R. G. (1971) 'Drying temperature and its effect on viability of *Setaria sphacelata* seed', *Trop. Grasslds*, **5**, No. 2., 75–80.
Sillar, D. I. (1969) 'Townsville lucerne in Queensland', *Qd agric. J.*, **95**, No. 1, 2–11.
Simpson, J. R. & Fretes, R. (1972) 'An economic evaluation of buffelgrass in Paraguay', *J. Range Mgmt*, **25**, No. 4, 261–6.
Singh, A. & Patil, B. D. (1970) 'Studies on induced polyploids of forages. 1. Some aspects of growth, morphology and seeding habit of tetraploid Siratro (*Phaseolus atropurpureus* DC.)', *Indian J. Sci. & Ind.*, (A), **4**, No. 1, 43–8.
Singh, A. P. & Yadav, M. S. (1971) 'Internode pattern in *Pennisetum pedicellatum* Trin.', *J. Ind. Bot. Soc.*, **50**, No. 3, 285–8.
Singh, G. S. (1967) 'Studies on the Bundelkhand pastures of Utar Pradesh. 1. The quality and quantity of the sai (*Sehima nervosum*) grass', *Indian J. Dairy Sci.*, **20**, 118–21.
Singh, H. K. (1969) 'Natal grass', *Indian Fmg*, **19**, No. 3, 23.
Singh, R. D. & Chand, P. (1969) 'Intercropping of maize with forage legumes', *Indian J. Agron.*, **14**, No. 1, 67–70.
Singh, R. D., Chatterjee, B. N. & Das, N. C. (1968) 'Herbage production with fertilizer nitrogen and legumes', *J. Indian Soc. Soil Sci.*, **16**, No. 4, 331–6.
Skerman, P. J. (1970) '*Stylosanthes mucronata* Willd., an important natural perennial legume in Eastern Africa', *Proc. 11th int. Grassld Congr., Surfers Paradise, 1970*, 196–8.
Smith, C. J. (1970) 'Seed dormancy in Sabi panicum', *Proc. int. Seed Test. Ass.*, **36**, No. 1, 81–97.
Snowden, J. D. (1936) *The cultivated races of Sorghum*. Bentham-Moxon Fund, London.
Snowden, J. D. (1961) 'The classification and nomenclature of *Sorghum*, section *Sorghum*', *Sorghum Newsletter*, 1961, **4**, 67–74.
Snyder, L. A., Hernandez, A. R. & Warmke, H. E. (1955) 'The mechanism of apomixis in *Pennisetum ciliare*', *Bot. Gaz.*, **116**, 209.
Sotomayor-Ríos, A., Matienzo, A. A. & Vélez-Fortuño, J. (1973) 'Evaluation of seven forage grasses at two cutting stages', *J. agric. Univ. Puerto Rico*, **57**, No. 3, 173–85.
Sotomayor-Ríos, A., Vélez-Fortuño, J. & Spain, G. (1971) 'Forage yields and plant character correlations in 30 *Digitaria* selections', *J. Agric. Univ. P. Rico*, **55**, No. 1, 63–9.
Souto, S. M., Cóser, A. C. & Döbereiner, J. (1972) 'Especificidade de uma variedade nativa de "Alfafa do Nordeste" (*Stylosanthes gracilis*) na simbiose com *Rhizobium* sp.', *Pesq. Agropec. Brasil. Zootecnica*, **7**, 1–5.
Srivastava, V. C. (1969) 'Effect of nitrogen levels, seeding rates and cutting on yield and protein content of sorghum forage', *Madras agric. J.*, **56**, No. 3, 99–103.
Stephens, D. (1967) 'Effect of fertilizers on grazed and cut elephant grass leys at Kawanda Research Station, Uganda', *E. Afr. agric. For. J.*, **32**, No. 4, 383–92.
Stewart, D. R. M. & Stewart, J. (1970) 'Food preference data by feacal analysis for African Plains ungulates', *Zool. Afr.*, **5**, No. 1, 115–29.
Stewart, D. R. M. & Stewart, J. (1971) 'Comparative food preferences of five East African ungulates at different seasons', in *The Scientific Management of Animal and Plant Communities for Conservation*, Oxford, UK, Blackwell Scientific Publications, 351–66.
Stobbs, T. H. (1969 a,b,c) 'The effect of grazing management upon pasture productivity in Uganda', (a) '1. Stocking rate', (b) '2. Grazing frequency', (c) '4. Selective grazing', *Trop. Agric., Trin.*, **46**, No. 3, 187–94, No. 3, 195–200, No. 4, 303–9.
Stobbs, T. H. (1969d) 'The influence of inorganic fertilizers upon the adaptation, persistency and production of grass and grass/legume swards in eastern Uganda', *E. Afr. agric. For. J.*, **35**, No. 2, 112–17.
Stobbs, T. H. (1969e) 'Animal production from *Hyparrhenia* grassland oversown with *Stylosanthes gracilis*', *E. Afr. agric. For. J.*, **35**, No. 2, 128–34.
Stobbs, T. H. (1971) 'Production and composition of milk from cows grazing siratro (*Phaseolus atropurpureus*) and greenleaf desmodium (*Desmodium intortum*)', *Aust. J. exp. Agric. Anim. Husb.*, **11**, No. 50, 268–73.
Strange, R. (1961) 'Effect of legumes and fertilizers on yields of temporary leys', *E. Afr. agric. For. J.*, **26**, No. 4, 231–4.
Streetman, L. J. (1963) 'Reproduction of the lovegrasses, the genus *Eragrostis* 1. *E.*

chloromelas Steud., *E. curvula* (Schrad.) Nees, *E. Lehmanniana* Nees and *E. superba* Peyr. *Wrightia'*, **3**, No. 3, 41–51.

Strickland, R. W. (1970) 'Seed production and testing problems in tropical and subtropical pasture species', *Proc. int. Seed Test. Ass.*, **36**, No. 1, 189–99.

Sundararaj, D. D. & Nagarajan, M. (1963) 'Plant introduction – *Desmanthus virgatus* Willd. (hedge lucerne). A new fodder cum hedge plant for Madras State', *Madras agric. J.*, **50**, No. 7, 279–82.

Suttie, J. M. (1970) 'The effect of single superphosphate on the yield and chemical composition of a grazed grass/legume pasture at Kitale, Kenya', *E. Afr. agric. For. J.*, **35**, No. 4, 259–63.

Suttie, J. M. & Moore, C. E. M. (1966) '*Desmodium uncinatum*', *Kenya Fmr*, No. 116, 18.

Suttie, J. M. & Ogada, J. (1967) 'The production of *Desmodium uncinatum* seed with special reference to mechanised harvesting', *Kenya Fmr*, No. 131, 22 and 36.

* **Swallen, J. R.** (1943–4) 'Gramineae', in *Flora of Panama* by R. E. Woodson (Jr) and R. W. Schery, *Ann. Mo. bot. Gdn*, No. 3, 30–1, 104–280.

Taerum, R. (1970) 'Comparative shoot and root growth studies on 6 grasses in Kenya', *E. Afr. agric. For. J.*, **36**, No. 1, 94–113.

Tainton, N. M. & Booysen, P. de V. (1965) 'Growth and development in perennial veld grasses. 1. *Themeda triandra* tillers under various systems of defoliation', *S. Afr. J. agric. Sci.*, **8**, No. 1, 93–110.

Takahashi, M. & Ripperton, J. C. (1949) 'Koa haole (*Leucaena glauca*) its establishment, culture and utilization as a forage crop', *Bull. 100, Hawaii agric. Exper. Stn*.

Tamimi, Y. N., Sherrod, L. B., Ishizaki, S. M. & Izuno, T. (1968) 'The effect of levels of nitrogen, phosphorus, and potassium fertilization upon beef production on Kikuyugrass', *Tech. Bull. 76 Hawaii agric. Exp. Station*, pp. 12.

Tang, C. N. & Lin, P. W. (1970) ['Study on the nutrition of tropical pasture legume on lateritic soil'], *Taiwan Livestock Research*, **3**, No. 1, 98–105. From English abstract in *Soils and Fertilizers in Taiwan*, **78**.

Tardin, A. C., Calles, C. H. & Gomide, J. A. (1971) [Vegetative growth of Guatemala grass.] *Experientiae*, **12**, No. 1, 1–31. (quoted from *Herb. Abstr.* 1972, 1063).

Terada, S. (1971) 'Effect of method of applying organic matter on soil environment and plant growth in the Amazon region' (in Japanese), *Jap. J. trop. Agric.*, **15**, No. 1, 11–19.

Tergas, L. E., Blue, W. G. & Moore, J. E. (1971) 'Nutritive value of fertilized Jaragua grass (*Hyparrhenia rufa* (Nees), Stapf) in the wet-dry Pacific region of Costa Rica', *Trop. Agric., Trin.*, **48**, No. 1 1–8.

Tetteh, A. (1972) 'Comparative dry matter yield patterns of grass/legume mixtures and their pure stands', *Ghana. J. Agric. Sci.*, **5**, No. 3, 195–9.

Teunissen, H., Arroyo, R. D. & Garza, T. R. (1966) 'Estudio comparativo de producción de carne en 5 zacates tropicales. 2', *Téc. pec. Méx.*, No. 8, 38–45.

Teunissen, H. & Villarreal, Q. O. (1966) 'Algunos aspectos de la punta de caña de azúcar como forraje para el ganado', *Téc. pec. Méx.*, No. 8, 53–5.

Thaiphanich, N. (1968) 'Effects of different rates of nitrogen on yield and quality of Napier (*Pennisetum purpureum*) and hybrid Napier (*Pennisetum purpureophoides*) grasses'. Thesis Kassertsart Univ. Thailand. From *Herb. Abstr.*, 1972, abst. 715.

Theron, E. P. & Booysen, P. de V. (1966) 'Palatability in grasses', *Proc. Grassld Soc. S. Afr.*, **1**, 111–20.

Thomas, P. I. (1970) 'Increased forage from stylo on sandveld', *Rhodesia agric. J.*, **67**, No. 2, 38–9.

Thurbon, P., Byford, I. & Winks, L. (1970) 'Evaluation of hays of *Dolichos lablab*, cv. Rongai, a sorghum/Sudan grass hybrid, cv. Zulu and Townsville lucerne (*Stylosanthes humilis* H.B.K.) on the basis of organic matter and crude protein digestibility', *Proc. 11th int. Grassld Congr., Surfers Paradise, 1970*, 743–7.

Tilley, J. M. A. & Terry, R. A. (1963) 'A two-stage technique for the *in vitro* digestion of forage crops', *J. Br. Grassl. Soc.*, **18**, No. 2, 104–11.

Tiwari, S. R. (1966) 'Cattle feed in India', *Wld Crops*, **18**, No. 2, 59–61.

Tomer, P. S. & Singh, R. R. (1969) 'Manurial practices affecting yield potential of winter

forages grown pure and mixed', *Indian J. agric. Sci.*, **38**, No. 6, 971–7.
Topps, J. J. & Manson, J. L. (1967) 'The value of maize stover after early harvesting of grain', *Rhod., Zam. & Malwai J. agric. Res.*, **5**, No. 2, 191–2.
Torssell, B. W. R., Begg, J. E., Rose, C. W. & Byrne, G. F. (1968) 'Stand morphology of Townsville lucerne (*Stylosanthes humilis*). Seasonal growth and root development', *Aust. J. exp. Agric. Anim. Husb.*, **8**, No. 34, 533–43.
Tothill, J. C. (1967) 'Variability pattern in *Heteropogon contortus*', *CSIRO Plant Intr. Review*, **3**, No. 3, 34–8.
Tothill, J. C. & Hacker, J. B. (1973) *The Grasses of Southeast Queensland*, University of Queensland Press.
Toutain, B. (1973) 'Principales plantes fourragères tropicales cultivées', *Inst. d'Elevage Méd. Vét. Pays Trop. Note de Synthese*, No. 3.
Tucker, V. C., O'Grady, P., Smith, R. A. D. & Byford, I. (1972) 'Effect on milk yield and composition of using hay and silage conservation for dairy cows grazing Glycine-green panic pastures', *Australian J. Dairy Techn.*, **27**, No. 4, 144–8.
Tuley, P. (1968) '*Stylosanthes gracilis*', *Herb. Abstr.*, **38**, No. 2, 87–94.
Tutin, T. G. (1958) 'Classification of the legumes', in *Nutrition Of The Legumes*, (ed. E. G. Hallsworth), London, Butterworth Sci. Publ., 3–14.
Valenza, J. (1970) 'Survey of different types of natural pasture land in the Senegal Republic', *Proc. 11th int. Grassld Congr., Surfers Paradise, 1970*, 78–82.
Vallis, I. & Jones, R. J. (1973) 'Net mineralization of nitrogen in leaves and leaf litter of *Desmodium intortum* and *Phaseolus atropurpureus* mixed with soil', *Soil Biol. Biochem.*, **5**, No. 4, 391–8.
Vásquez, R. (1965) 'Effects of irrigation and nitrogen level on the yields of Guinea grass, Para grass and Guinea grass-kudzu and Para grass-kudzu mixtures in Lajas Valley', *J. agric. Univ. P. Rico*, **49**, No. 4, 389–412.
Velásquez, J. A. & Gonzáles, J. E. (1972) 'El valor nutritivo de la paja de maní (*Arachis hypogaea*)', *Agronomia trop.*, **22**, No. 3, 287–90.
Verboom, W. C. (1966) '*Brachiaria dura*, a promising new forage grass', *J. Range. Mgmt*, **19**, No. 2, 91–3.
Verboom, W. G. (1968) 'Grassland successions and associations in Pahang, Central Malaya', *Trop. Agric., Trin.*, **45**, No. 1, 47–59.
Verdcourt, B. (1966) 'A proposal concerning *Glycine* L.', *Taxon*, **15**, No. 1, 34–6.
Verdcourt, B. (1971) 'Phaseoleae', in *Flora of Tropical East Africa. Leguminosae*, part 4, 503–807.
Verhoeven, G. (1958) 'Tropical legume seed can be harvested commercially', *Qd. agric. J.*, **84**, No. 2, 77–82.
Verma, C. M. & Chakravarty, A. K. (1969) 'Study on the pasture establishment technique. 4. Comparative efficiency of seedlings and rooted slips as a transplanting material for *Lasiurus sindicus* pasture', *Ann. arid Zone*, **8**, No. 1, 52–7.
Vicente-Chandler, J., Silva, S., Rodríguez, J. & Abruña, F. (1972) 'Effect of two heights and three intervals of grazing on the productivity of a heavily fertilized Pangola grass pasture', *J. Agric. Univ. P. Rico*, **56**, No. 2, 110–14.
Vicente-Chandler, R., Silva, S. & Figarella, J. (1959) 'The effect of nitrogen fertilization and frequency of cutting on the yield and composition of three tropical grasses', *Agron. J.*, **51**, No. 4, 202–6.
Vieira, C. (1971) 'Nota sôbre o comportamento de variedades de *Phaseolus calcaratus* Roxb., em Viçosa, Minas Gerais', *Rev. Ceres*, **18**, No. 98, 303–7.
Virkki, N. & Purcell, C. M. (1967) 'Observations on the behaviour, genetics and cytology of two South African *Digitaria valida* Stent accessions in Puerto Rico', *J. Agric. Univ. P. Rico*, **51**, No. 4, 269–85.
Visser, J. H. (1965) 'Root exudates of *Eragrostis curvula* as an ecological factor', *Proc. 9th int. Grassld Congr., São Paulo*, 453–5.
Voigt, P. W. (1971) 'Discovery of sexuality in *Eragrostis curvula* (Schrad.) Nees', *Crop Sci.*, **11**, No. 3, 424–5.
Vonesch, E. E. & Riverós, M. H. C. K. de (1967–8) 'Composición y digestibilidad de

forrajeras de la Provincia de Buenos Aries', *Revta Fac. Agron. Vet. Univ. B. Aires*, **17**, No. 1, 49–67.

Voorthuizen, E. G. van (1971) 'A quality evaluation of four widely distributed native grasses in Tanzania', *E. Afr. agric. For. J.*, **36**, No. 4, 384–91.

Walker, B (1969) 'Effect of nitrogen fertilizer and forage legumes on a *Cenchrus ciliaris* pasture in western Tanzania', *E. Afr. agric. For. J.*, **35**, No. 1, 2–5.

*****Walter, H.** (1971) *Ecology of Tropical and Subtropical Vegetation*, Oliver & Boyd, Edinburgh.

Wang, C.-C. (1961) 'Growth, flowering and forage production of some grasses and legumes in response to different photoperiods', *J. agric. Assoc., China*, No. 36, 27–52.

Wang, C.-C., Hsu, C.-S. & Tsay, R.-C. (1969) 'Evaluation of the competitive ability of grasses and legumes from the competition with weeds', *J. agric. Ass. China*, No. 66, 50–6.

Ware-Austin, W. D. (1963) 'Napier grass for milk production in the Trans Nzoia', *E. Afr. agric. For. J.*, **28**, No. 4, 223–7.

Warmke, H. E. (1954) 'Apomixis in *Panicum maximum*', *Amer. J. Bot.*, **41**, 5–11.

Warmke, H. E., Freyre, R. H. & Morris, M. P. (1952) 'Studies on palatability of some tropical legumes', *Agron. J.*, **44**, No. 10, 517–20.

Webster, C. C. & Wilson, P. N. (1966) *Agriculture in the tropics*, Longmans, London.

Wendt, W. B. (1970) 'Responses of pasture species in eastern Uganda to phosphorus, sulphur and potassium', *E. Afr. agric. For. J.*, **36**, No. 2, 211–19.

Wendt, W. B. (1971) 'Effects of nodulation and fertilizers on *Desmodium intortum* at Serere, Uganda', *E. Afr. agric. For. J.*, **36**, No. 4, 317–21.

Werner, J. C., Pereira, L. F., Martinelli, D. & Cintra, B. (1965–6) 'Estudo de três diferentes alturas de corte em capim elefante Napier', *Bolm. Ind. anim.*, **23**, 161–8.

Wesley-Smith, R. N. (1973) 'Para grass in the Northern Territory – parentage and propagation', *Trop. Grasslds*, **7**, No. 2, 249–50.

Weston, E. J. (1962) 'The Northern Downs and Western Gulf Region', *Proc. N. Qd Agrost. Conf. 1962*, 4/3.

Weston, E. J. (1969) 'Grazing preferences of sheep and nutritive value of plant components in a Mitchell grass association in north-western Queensland', *Qd J. agric. anim. Sci.*, **26**, No. 4, 639–50.

De Wet, J. M. J. & Harlan, J. R. (1970a) 'Apomixis, polyploidy, and speciation in *Dichanthium*', *Evolution*, **24**, No. 2, 270–7.

De Wet, J. M. J. & Harlan, J. R. (1970b) '*Bothriochloa intermedia* – a taxonomic dilemma', *Taxon*, **19**, 339–40.

Wheeler, L. J. & Hedges, D. A. (1971) 'Summer forage crops'. *Pastoral Res. Laboratory. Div. Animal Physiology, CSIRO*, Armidale, **11**.

Wheeler, W. A. (1950) *Forage and Pasture Crops*, D. Van Nostrad Co., Inc., New York, Toronto, London.

*****Whitehead, D. C.** (1966) Nutrient minerals in grassland herbage. *Mimeographed Publication No. 1/1966. Commonwealth Bureau of Pastures and Field Crops*, Hurley.

Whiteman, P. C. (1968) 'Effect of temperature on the vegetative growth of six tropical legume species', *Aust. J. exp. Agric. Anim. Husb.*, **8**, 528–32.

Whiteman, P. C. (1972) 'The effects of inoculation and nitrogen application on seedling growth and nodulation of *Glycine wightii* and *Phaseolus atropurpureus* in the field', *Trop. Grasslds*, **6**, No. 1, 11–16.

Whiteman, P. C. & Gillard, P. (1971) 'Species of *Urochloa* as pasture plants', *Herb. Abstr.*, **41**, 351–7.

Whitney, A. S. (1966) 'Nitrogen fixation by three forage legumes and the utilization of legume-fixed nitrogen by their associated grasses', *Diss. Abstr.*, 1966, **27**, No. 6, 1688B.

Whitney, A. S. (1970) 'Effect of harvesting interval, height of cut, and nitrogen fertilization on the performance of *Desmodium intortum* mixtures in Hawaii', *Proc. 11th int. Grassld Congr., Surfers Paradise, 1970*, 632–6.

Whitney, A. S. & Green, R. E. (1969) 'Legume contribution to yields and composition of *Desmodium* spp. – Pangolagrass mixtures, *Agron. J.*, **61**, No. 5, 741–6.

Whitney, A. S., Kaneheiro, Y. & Sherman, G. D. (1967) 'Nitrogen relationships of three tropical forage legumes in pure stands and in grass mixtures', *Agron. J.*, **59**, No. 1, 47–50.
Whyte, R. O. (1964) *The Grassland and Fodder Resources of India*, Indian Council of Agricultural Research, New Delhi.
Whyte, R. O., Moir, T. R. G. & Cooper, J. P. (1959) 'Grasses in agriculture', *FAO Agricultural Studies*, No. 42, FAO, Rome.
Whyte, R. O., Nilsson-Leissner, G. & Trumble, H. C. (1953) 'Legumes in Agriculture', *FAO Agricultural Studies*, No. 21, Rome.
Wild, H. (1965) 'The flora of the Great Dyke of Southern Rhodesia with special reference to the serpentine soils', *Kirkia*, **5**, Pt. 1, 49–86.
Willard, E. E. & Schuster, J. L. (1973) 'Chemical composition of six Southern Great Plains grasses as related to season and precipitation', *J. Range Mgmt*, **26**, No. 1, 37–8.
Williams, J. T. & Farias, R. M. (1972) 'Utilization and taxonomy of the desert grass *Panicum turgidum*', *Econ. Bot.*, **26**, No. 1, 13–20.
* **Willis, J. C.** (1973) *A Dictionary of the Flowering Plants and Ferns*, 8th ed, rev. H. K. Airy Shaw, Univ. Press, Cambridge.
Wilson, G. P. M. (1970) 'Method and practicability of Kikuyu grass seed production', *Proc. 11th int. Grassld Congr, Surfers Paradise, 1970*, 312–15.
Wilson, J. R. & Norris, D. O. (1970) 'Some effects of salinity on *Glycine javanica* and its Rhizobium symbiosis', *Proc. 11th int. grassld. Congr., Surfers Paradise, 1970*, 455–8.
Wood, I. M. W. 1070) 'Herbicides for the control of grass weeds in pastures of Townsville stylo (*Stylosanthes humilis*)', *Aust. J. exp. Agric. Anim. Husb.*, **10**, No. 47, 790–4.
Woods, L. E. (1969) 'A survey of Townsville stylo (Townsville lucerne) pastures established in the Northern Territory up to 1969', *Trop. Grasslds*, **3**, No. 2, 91–8.
Wutoh, J. G., Hutton, E. M. & Pritchard, A. J. (1968a) 'Inheritance of flowering time, yield and stolon development in *Glycine javanica*', *Aust. J. exp. Agric. Anim. Husb.*, **8**, No. 32, 317–22.
Wutoh, J. G., Hutton, E. M. & Pritchard, A. J. (1968b) 'The effects of photoperiod and temperature on flowering in *Glycine javanica*', *Aust. J. exp. Agric. Anim. Husb.*, **8**, No. 34, 544–7.
Yadav, N. K. & Vyas, S. R. (1971) 'Note on the response of root-nodule rhizobia to saline, alkaline and acid conditions', *Indian J. agric. Sci.*, **41**, No. 12, 1123–5.
Yamada, T., Matsuo, S. & Tamura, K. (1972) 'Dissemination of pasture plants by livestock. 3. Recovery of some pasture plant seeds passed through the digestive tract of beef cattle and emergence of seedlings from seeds recovered from the feaces', *J. Jap. Soc. Grassl. Sci.*, **18**, No. 1, 16–27 [in Japanese, quoted from *Herb. Abst.*, 1973, 1068].
Yates, J. J., Edye, L. A., Davies, J. G. & Haydock, K. P. (1964) 'Animal production from a *Sorghum almum* pasture in south-east Queensland', *Aust. J. exp. Agric. Anim. Husb.*, **4**, No. 15, 326–35.
Young, J. G. & Chippendale, F. (1970) 'Beef cattle performance of pastures on heath plains in southeast Queensland', *Proc. 11th int. Grassld Congr., Surfers Paradise, 1970*, 849–52.
Younge, O. R., Plucknett, D. L. & Rotar, P. P. (1964) 'Culture and yield performance of *Desmodium intortum* and *D. canum* in Hawaii', *Tech. Bull. 59 Hawaii agric. Exp. Stn.*
Zúñiga, M. P., Sykes, D. J. & Gomide, J. A. (1965) 'Produção de onze variedades de gramíneas para capineiras, em Viçosa, MG – Resultadas preliminares', *Revta Ceres*, **12**, No. 71, 315–31.
Zúñiga, M. P., Sykes, D. J. & Gomide, J. A. (1967) 'Competição de treze gramíneas forrageiras para corte, com e sem adubação, em Viçosa, Minas Gerais', *Revta Ceres*, **13**, No. 77, 324–43.

Index of botanical names

(Figures in **bold type** denote main entries with descriptions, uses, etc.)

Grasses

Acroceras macrum, 2, 4, **31–2**, 377, 414
Agrostis
  alba, 24
    var. *schimperana*, 32
  producta, **32**
  schimperana, **32**
Alloteropsis
  cimicina, 33
  semialata, **33**
Amphilophis
  insculpta, 50
  pertusa, 51
  radicans, 51
Andropogon
  amplectens, 127
  annulatus, 106
  apricus var. *africanus*, 40
  aristatus, 107
  barbinodis, 49
  caricosus, 108
  gayanus, 9, **34–9**, 334, 399
    var. *bisquamulatus*, xi, 34, 36, 38
    var. *gayanus*, 34
    var. *squamulatus*, 34, 39
    var. *tridentatus*, 34
  grandiflorus, 127
  insculptus, 50
  laguroides, 50
  nutans, 261
  pertusus, 51
  pilosus, 160
  plumosus, 290
  pseudapricus, **40**, 127
  saccharoides, 52
    ssp. *laguroides*, 50
  sorghum, 269

Anthistiria
  australis, 286
  glauca, 286
  imberbis, 286
Aristida
  adscensionis, **40–1**, 42, 100
  browniana, **41**, 140
  contorta, **41**
  curvata, 40
  ingrata,
    var. *jerichoensis*, 41
  jerichoensis, **41–2**
  kenyensis, 100
  muelleri, 41
  mutabilis, **42**
  pruinosa, **42**
  stipoides, 41
  submucronata, 40
Astrebla
  elymoides, 43
  lappacea, **43**
  pectinata, **43**
  squarrosa, 43
Avena sativa, **44**
Axonopus affinis, **44–5**, 47
  arenosus, 45
  brevipedunculatus, 45
  canescens, **45**
  compressus, 44, **45–7**, 196, 226
  flexuosus, 45
  micay, **47**
  purpusii, **47**
  scoparius, 45, **47–8**

Beckera polystachya, 261
Beckeropsis uniseta, xi, **48–9**

Bothriochloa
 barbinodis, **49**
 ewartiana, 49
 insculpta, **50**, 51
  var. *insculpta*, 50
  var. *vegetior*, 50
 intermedia, 49
 ischaemum, 49
 laguroides, **50-1**, 52
 pertusa, **51**, 115
 radicans, **51-2**
 saccharoides, 49, 50, **52**
Bouteloua
 americana, 53
 curtipendula, **53**
 heterostega, **53**
 megapotamica, **53**
 repens, **53**
Brachiaria
 brizantha, xi, **54**, 58, 64, 320, 345
 decumbens, **54-7**, 64, 113, 119, 343, 344
 dictyoneura, 57, 58
 distachya, 321
 dura, **57**
 humidicola, **57-8**, 334, 382, 414
 miliiformis, **58**
 mutica, xi, 46, **59-62**, 63, 119, 129, 300, 334, 345, 390, 395
 plantaginea, **62-3**
 radicans, **63-4**
 rugulosa, 63
 ruziziensis, 26, **64-5**, 256
Bromus
 catharticus, 65
 fibrosus, 4
 inermis, 65
 marginatus, 65
 unioloides, **65-6**
 wildenowii, 65

Capillipedium parviflorum, 49
Cenchrus
 barbatus, 66
 biflorus, **66**
 ciliaris, 9, 22, 24, **66-74**, 75, 76, 100, 229, 334, 340, 361, 382, 384, 399, 404, 419, 422
 glaucus, 73
 pennisetiformis, **74**
 prieurii, **74-5**
 setigerus, 66, 74, **75-6**, 340, 404, 419, 422
 tribuloides, 66
Chloris acicularis, **76**

 capensis, 148
 distichophylla, 147
 elegans, 87
 gayana, xi, 5, 10, 22, 23, 24, 26, 29, 70, **77-86**, 199, 253, 254, 256, 320, 329, 334, 335, 352, 361, 377, 380, 382, 383, 385, 399, 416
 myriostachya, 86
 petraea, 148
 polydactyla, **86**
  forma *stolonifera*, 86
 roxburghiana, **86-7**
 virgata, **87**
  var. *elegans*, 87
Chrysopogon
 acicularis, **88**
 aucheri, **88-9**
  var. *pluvinatus*, 89
  var. *quinqueplumis*, 88-9
 fallax, **89**
 fulvus, **89-90**
 lancearius, 90
 latifolius, 90
 montanus, 89, 90
 orientalis, 90
Coix lacrima-jobi, **90-1**
Cymbopogon afronardus, 55, 290
Cynodon aethiopicus, 91, **99-102**, 103
 arcuatus, **91**
 barberi, **91-2**
 dactylon, 3, 9, 11, 91, **92-8**, 99, 100, 101, 102, 127, 208, 209, 212, 322
  var. *aridus*, 93
  var. *coursii*, 94
  var. *dactylon*, 93, 94
  var. *elegans*, 9, 93
  var. *sarmentosus*, 98
  race *seleucidus*, 93
 nlemfuensis, 10, 29, 91, **98-102**, 103, 113, 119, 296, 377
  var. *nlemfuensis*, 98, 99
  var. *robustus*, 94, 98
 plectostachyus, 9, 91, 100, **102-3**
 polevansii, 92
 transvaalensis, 94

Dactylis glomerata, **103-4**
Dactyloctenium
 aegyptium, **104-5**, 106
 bogdanii, **105**
 giganteum, 105
 radulans, **106**
Danthonia
 lappacea, 43
 pectinata, 43

Dichanthium
  annulatum, 49, 70, 75–6, **106–7**, 108, 109, 174, 414, 419, 422
  aristatum, **107–8**
  caricosum, 106, 107, **108**, 109, 382
  fecundum, **108–9**
  nodosum, 107
  sericeum, **109**
Digitaria
  adscensionis, 111
  californica, **111**
  decumbens, 57, 98, 101, 109, **111–22**, 124, 158, 253, 321, 322, 329, 334, 342, 343, 344, 345, 348, 349, 352, 361, 362, 365, 377, 382, 385, 395, 399, 401, 404, 414, 416
  didactyla, **122**, 126, 266
  eriantha, 109, **122**, 126
  gazensis, **123**
  longiflora, 111
  macroblephara, 9, 109, 110, **123**, 125
  milanjiana, 110, 111, **123–4**
  nodosa, 109, 110, 111, **124**
  pentzii, 109, 112, 123, **124–5**, 126
    var. *stolonifera*, 124
  sanguinalis, 111
  scalarum, 9, **125**, 228
  setivalva, 109, **125**
  smutsii, 109, **125–6**, 399
  swazilandensis, **126**
  ternatea, 111
  valida, 109, 124, **126**
  velutina, 85
Digraphis arundinacea, 244
Diheteropogon
  amplectens, **127**
    var. *amplectens*, 127
    var. *catangensis*, 127
  grandiflorus, **127**
  hagerupii, **127–8**, 427

Echinochloa
  colona, **128**
  crus-galli, 128
  crus-pavonis, 128
  frumentacea, **128**
  haploclada, **129**
  polystachya, **129–30**
    var. *polystachya*, 129
    var. *spectabilis*, 129
  pyramidalis, **130**
  stagnina, **130–1**
Eleusine
  compressa, **131**
  coracana, **131–2**

  flagellifera, 131
  indica, 8, 131, **132–3**
    ssp. *africana*, 131, 132
    ssp. *indica*, 131, 132
  multiflora, **133**
  tristachya, **133**
Elyonurus hirsutus, 163
Enneapogon polyphyllus, **133–4**
Enteropogon
  acicularis, 76
  macrostachyus, **134**
  rupestris, 134
  sechellensis, 134
  somalensis, 134
Entolasia imbricata, **134–5**
Eragrostis
  abyssinica, 142
  caespitosa, 24, **135**
  chloromelas, 136, 138
  cilianensis, **136**
  curvula, 23, **136–9**, 345
  eriopoda, **139–40**
  lehmanniana, **140**
  lugens, **140**
  major, 136
  polysticha, 140
  robusta, 136
  superba, **140–1**
    var. *contracta*, 140
  tef, **141–2**, 390
  tenuifolia, 8
Eriochloa
  meyeriana, **142**
  montevidensis, 143
  nubica, 143
  polystachya, 142, 143
  procera, 143
  punctata, **142–3**
    var. *montevidensis*, 143
Euchlaena mexicana, 27, **143–7**, 297, 412, 422, 425
Eulalia fulva, 109
Eustachys
  distichophylla, **147**
  paspaloides, **147–8**
  petraea, **148**

Festuca
  arundinacea, **148**
  pratensis, 148

Hemarthria
  altissima, **149**
  compressa, 149
Heteropogon contortus, 89, **150–2**, 288, 325, 334, 382, 399, 404, 407

Holcus
  plumosus, 275
  sorghum, 269
Hymenachne
  amplexicaule, **152–3**, 296
  pseudo-interrupta, **153**
Hyparrhenia
  anamesa, 154
  collina, **153–4**
  cymbaria, **154**
  dissoluta, 159
  filipendula, **154**
  hirta, 2, **154**
  lintonii, 155
  nyassae, **155**
  papillipes, 8, **155**
  pilgerana, **155**
  rufa, xi, 9, 27, 153, **155–9**, 167, 185, 334, 390, 395, 399
    var. *rufa*, 156
    var. *siamensis*, 156
  ruprechtii, 159
Hyperthelia dissoluta, **159**

Imperata cylindrica, 236, 333
Ischaemum
  afrum, 160
  aristatum ssp. *imberbe*,
    var. *imbricatum*, 159
    var. *indicum*, 160
  brachyatherum, 160
  goebelii, **159–60**
  imbricatum, 159
  indicum, **160**, 334, 382
  pilosum, **160**
  rugosum, **160**
  timorense, 159, **161**
Iseilema
  laxum, **161–2**
  prostratum, **162**
Ixophorus unisetus, 162–3

Lasiurus
  hirsutus, 70, **163–4**, 340, 419, 422
  sindicus, 163
Latipes senegalensis, 165
Leersia
  abyssinica, 164
  capensis, 164
  denudata, 164
  hexandra, **164**
Leptochloa obtusiflora, 165
Leptothrium senegalense, **165–6**
Lolium
  multiflorum, 166
  perenne, 14, 24, **166**

Melinis minutiflora, 8, 9, 22, 23, 27, 38, 115, **167–72**, 199, 256, 329, 334, 361, 395, 399

Panicum amplexicaule, 152
  antidotale, **173–5**, 320, 422, 425
  barbinode, 59
  brizanthum, xi
  bulbosum, **175**
  californicum, 111
  coloratum, 173, **175–81**, 334, 349, 352
    var. *coloratum*, **176–8**, 179, 181
    var. *makarikariense*, 176, 177, **178–81**
  compressum, 131
  desertorum, 193
  elephantipes, **181**, 296
  geminatum, 194
  hochstetteri, 192
  kabulabula, 176
  makarikariense, 176
  maximum, 5, 9, 11, 22, 24, 26, 27, 38, 74, 115, 173, 175, **181–91**, 256, 267, 292, 334, 345, 352, 353, 361, 371, 373, 382, 385, 390, 395, 399, 400
    var. *gonylodes*, 175
    var. *pubiglume*, 182
    var. *trichoglume*, 182, 187, 188, 191
  meyerianum, 142
  muticum, 59
  paludivagum, 194
  plantagineum, 62
  purpurascens, xi, 59
  repens, **191–2**
  transvenulosum, 192
  trichocladum, **192**
  turgidum, **193**
Paspalidium
  desertorum, **193–4**
  geminatum, 193, **194**
  paludivagum, **194**
Paspalum
  almum, **195**
  commersonii, **195–6**, 308, 352, 377, 385
  conjugatum, 119, **196–7**
    var. *parviflorum*, 196
    var. *pubescens*, 196
  dilatatum, 2, 5, 9, 26, 195, **197–202**, 207, 210, 214, 226, 321, 345, 346, 352, 377, 385, 395, 425
    var. *dilatatum*, 198
    var. *flavescens*, 198, 201
  distichum, **202–3**, 215
  fasciculatum, **203**
  guenoarum, **203–4**

var. *guenoarum*, 204
var. *rojasii*, 204
maritimum, **204**
nicorae, **204–5**
notatum, 10, 29, 45, 195, **205–12**, 213, 215, 322, 334, 343, 352, 365, 377, 382, 394
var. *latiflorum*, 205, 210, 212
var. *notatum*, 205, 207
var. *saure*, 206, 207, 212
orbiculare, **213**
paspaloides, 215
pauciciliatum, 197, 198, 202
plicatulum, 204, 210, 211, **213–14**, 349, 352, 377, 382, 416
var, *arenarium*, 204
var. *glabrum*, 214
var. *plicatulum*, 214
pulchellum, **214**
rojasii, 204
saure, 206
scrobiculatum,
var. *commersonii*, 195
uruguayense, 205
urvillei, **214–15**
vaginatum, 210, **215**
vaseyanum, 214
virgatum, **215**
wettsteinii, **215–16**
Pennisetum
americanum, **216–22**, 230, 242
cenchroides, 66
ciliare, 66, 73
clandestinum, ix, 5, 9, 22, 45, 199, 216, **222–9**, 334, 345, 352, 361, 414, 416
glaucum, 216, 242
merkeri, 233
orientale, 221
pedicellatum, 8, 36, **229–31**, 423
polystachion, **231–3**, 243, 334, 399
prieurii, 74
purpureum, 9, 11, 38, 216, **233–41**, 242, 243, 271, 292, 334, 342, 345, 352, 353, 371, 373, 395, 399
ssp. *benthamii*, 233, 240
ssp. *flexispica*, 233
var. *merkeri*, 233
ssp. *purpureum*, 233
purpureum × *P. americanum*, **242–3**
purpureum × *P. typhoides*, 242
schimperi, 216, 224, 233, 288
spicatum, 216
squamulatum, **243**
subangustum, **243–4**
typhoides, 216, 242

typhoideum, 216
unisetum, 48
Phalaris
aquatica, 244
var. *stenoptera*, 244
arundinacea, 244
stenoptera, 244
tuberosa, 244
Phleum pratense, 25
Poa tef, 142
Pollinia fulva, 89

Rhynchelytrum
repens, **244–5**
roseum, 244
villosum, 245
Rottboellia
altissima, 149
compressa,
var. *fasciculata*, 149
exaltata, **246**
selloana, **246**

Saccharum
benghalense, **246–7**
munja, 246
officinarum, 246, **247**
sinense, **247–8**
spontaneum, **248**
Sehima nervosum, **248–9**
Setaria
anceps, ix, 5, 9, 25, 26, 64, 81, 177, 186, **249–59**, 260, 310, 334, 343, 344, 345, 346, 349, 352, 361, 365, 381, 383, 384, 385, 399, 414, 416
geniculata, **259–60**
italica, 390
longiseta, **260**
sphacelata, 249, 250, 251, 257
splendida, **260–1**
trinervia, 257
Snowdenia polystachya, **261**
Sorghastrum
nutans, **261–2**
ssp. *albescens*, 262
ssp. *nutans*, 262
ssp. *pellitum*, 262
rigidifolium, 262
Sorghum
aethiopicum, 263, **264**
almum, 230, 243, **264–9**, 404
var. *almum*, 264
var. *parvispiculatum*, 264
var. *typicum*, 264
arundinaceum, 263, **269**, 279, 282
var. *kavirondense*, 282

var. *virgatum*, 276
australiense, 275
bicolor, 263, **269–73**, 278, 279, 280, 281, 282
　ssp. *arundinaceum*, 263, 282
　　race *virgatum*, 276
　ssp. *bicolor*, 263, 269
　　race bicolor, 276
　var. *roxburghii*, 275
halepense, 2, 263, 264, 266, 268, **273–4**, 275, 276, 279, 281
× *S. roxburghii*, **275**
nutans, 261
plumosum, 89, **275**
propinquum, 263, **275**
randolphianum ×, 264
rigidifolium, 262
sudanense, 263, 264, 274, **275–80**
× *S. arundinaceum*, **280**
× *S. bicolor*, **280–1**
× *S. halepense*, **281**
verticilliflorum, 263, **282**
virgatum, 276, **282**
vulgare, 269
Sporobolus
　cordofanus, **283**, 284
　helvolus, 9, 111, **283**
　isoclados, 10, **283–4**
　longibrachiatus, 284
　marginatus, 283
　nervosus, **284**
　pyramidalis, 282
Stenotaphrum
　dimidiatum, **284**
　secundatum, **284–5**, 395

Themeda
　australis, 89, **286**, 287
　triandra, 161, **286–90**
Trachypogon
　montufari, 290
　polymorphus, 290
　rufus, 155
　spicatus, **290–91**
Trichachne californica, 111
Tricholaena
　repens, 244
　rosea, 244
Tripsacum
　dactyloides, 2, 144, **291**
　fasciculatum, 292
　latifolium, **291–2**, 294
　laxum, 11, 291, **292–4**, 345, 352, 353, 371
　pilosum, 294
Urochloa
　bolbodes, **294–5**
　mosambicensis, **295–6**, 404, 406
　oligothricha, 294
　panicoides, 296
　pullulans, 295
　stolonifera, 296
　uniseta, 162

Vilfa isoclados, 283
Vossia cuspidata, **296**

Zea mays, **296–301**,
　var. *indentata*, 297
　var. *indurata*, 297

Legumes

Aeschynomene
　americana, **318**, 319
　indica, **319**
Alysicarpus glumaceus, **319**
　reticulatus, 320
　rugosus, **319–20**, 321
　　ssp. *perennirufus*, **319–20**
　　ssp. *reticulatus*, 320
　　ssp. *rugosus*, 320
　vaginalis, 54, 118, **320–1**
Arachis
　diogoi, 322
　glabrata, 97, **321–2**
　　var. *hagenbeckii*, 321

hagenbeckii, 322
hypogaea, 305, 321, **322–5**
marginata, 321
monticola, 209
prostrata, 321
villosa, 321
Atylosia scarabaeoides, 107, 232, **325**

Cajanus
　cajan, 60, 81, 305, **325–8**
　　var. *bicolor*, 326
　　var. *fulvus*, 326
　indicus, 325
Calopogonium

mucunoides, 118, 232, **328–9**
orthocarpum, 185
Canavalia
  bonariensis, 330
  ensiformis, 305, 329, **330**, 423
  gladiata, 329, **330**
  plagiosperma, 330
  virosa, 330
Centrosema
  brasilianum, **331**
  plumieri, **331**
  pubescens, 37, 54, 58, 60, 70, 80, 101, 117, 151, 157, 158, 174, 185, 186, 200, 232, 238, 239, 305, 315, **331–5**, 348, 400, 401
Clitoria ternatea, 37, 80–1, 174, 232, 305, **335–6**
Crotalaria
  anagyroides, 337
  brevidens, 337
  incana, **337**
  intermedia, **337–8**
  juncea, 336, **338**
  lanceolata, 337
  ochroleuca, 337
Cyamopsis
  psoraleoides, 338
  tetragonoloba, 70, 75, 107, 163, 174, 271, 307, 315, **338–40**

Desmanthus virgatus, 302, **340**
Desmodium
  adscendens, **341**
  barbatum, **341–2**
  canum, 118, **342**, 354
  discolor, **342–3**
  heterocarpon, 118, **343**
  heterophyllum, 56, **343–4**
  incanum, 342
  intortum, xi, 64, 70, 117, 118, 185, 186, 196, 200, 216, 225, 238, 252, 253, 294, 317, 341, 342, 343, **344–8**, 349, 352, 354, 377
  leiocarpum, 81, **348**
  limense, 349
  pabulare, **348–9**
  pilosiusculum, 349
  purpureum, 349
  sandwicense, 118, 178, 315, 316, **349**
  tortuosum, **349–50**
  triflorum, **350**
  uncinatum, 81, 118, 170, 177, 185, 186, 213, 216, 225, 238, 252, 294, 308, 315, 316, 341, 342, 344, 348, 349, **350–4**, 377
Dolichos

axillaris, 386
biflorus, 386
lablab, 367
minimus, 395
purpureus, 367
unguiculatus, 423
uniflorus, 386

Glycine
  hispida, 355
  javanica, 357, 358
  max, 278, 305, **355–7**
  petitiana, 357
  soja, 355
  wightii, 70, 81, 118, 177, 185, 216, 225, 238, 252, 266, 305, 316, 352, 354–5, **357–64**, 400
    ssp. *petitiana*, 358
    var. *maernsii*, 358, 359
    var. *petitiana*, 358
    ssp. *pseudojavanica*, 358
    ssp. *wightii*, 358
    var. *longicauda*, 358, 359
    var. *wightii*, 358

Indigofera
  cordifolia, **365**
  endecaphylla, 366
  enneaphylla, **365**
  hendecaphylla, 366
  hirsuta, 305, 364, **365**
  medicaginea, 365
  retroflexa, 364
  schimperi, 364
  spicata, 185, 308, 312, 316, 364, **366–7**
  subulata, 364
  tettensis, 364
  tirta, 364

Lablab
  niger, 367
  purpureus, 60, 70, 200, 305, 307, 312, 315, **367–9**, 421
    ssp. *benghalensis*, 367
    ssp. *purpureus*, 367
    ssp. *uncinatus*, 368
  uncinatus, 368
  vulgaris, 367
Lespedeza sericea, 137
Leucaena
  glauca, 369
  latisiliqua, 369
  leucocephala, 185, 188, 302, 305, 312, **369–74**
Lotononis
  angolensis, **374**

bainesii, 58, 101, 117, 123, 209, 213, 253, 305, 309, 310, 315, 317, **374-9**
laxum, 378
Lotus
  corniculatus, 379
    var. *eremanthus*, 379
  maernsii, 379
  uliginosus, 118, 379

Macroptilium
  atropurpureum, xi, 58, 70, 81, 117, 179, 185, 186, 200, 209, 213, 216, 252, 253, 305, 310, 315, 316, 333, 347, 360, **379-85**, 406
  lathyroides, 118, 200, 252, 305, 315, 377, 383, **385-6**
Macrotyloma
  axillare, **386**
  uniflorum, **386-7**
    var. *uniflorum*, 387
Medicago
  denticulata, 388
  laciniata, **387-8**
  orbicularis, 388
  polymorpha, 388
  sativa, 70, 81, 146, 387, **388**
  scutellata, 388
  truncatula, 388
Melilotus
  alba, 230, 389
  indica, **388-9**
  officinalis, 389
  parviflora, 146
Mimosa
  leucocephala, 369
  pudica, 60
  virgata, 340
Mucuna
  pruriens, **389-90**
    var. *utilis*, 389
  utilis, 389

Phaseolus
  aconitifolius, 419
  angularis, 420
  atropurpureus, xi, 379
  aureus, 422
  calcaratus, 423
  lathyroides, 385
  marinus, 421
  mungo, 230, 422, 423
  radiatus, 422
  sublobatus, 422
Psoralea eriantha, **391**
Pueraria
  javanica, 392
  lobata, **391-2**
  phaseoloides, 54, 60, 118, 162, 170, 185, 238, 266, 305, 329, 391, **392-5**
    var. *javanica*, **392-5**
  thunbergiana, 391

Rhynchosia
  elegans, 395
  minima, 313, **395-6**
  phaseoloides, 395

Sesbania
  aculeata, 396
  brachycarpa, 396
  grandiflora, 396
Stizolobium
  aterrimum, 389
  cochinchinensis, 389
  deeringianum, 60, 185, 238, 330, 389, 390
  niveum, 381
  pruriens var. *utile*, 389
  utile, 389
Stylosanthes
  bojeri, 396
  fruticosa, **396-7**
  gracilis, xi, 397
  guianensis, xi, 37, 54, 64, 70, 81, 101, 118, 151, 157, 158, 170, 174, 185, 186, 213, 232, 253, 305, 306, 362, 396, **397-402**
  hamata, **402**
  humilis, 70, 73, 75, 81, 118, 151, 152, 170, 232, 266, 295, 310, 312, **402-9**, 422
  mucronata, 396
  sundaica, 464

Teramnus
  labialis, 118, 409
  uncinatus, 409
  volubilis, 409
Trifolium
  alexandrinum, 44, 146, 307, 312, 409, **410-14**
  hybridum, 409
  incarnatum, 96
  johnstonii, 415
  mattirolianum, 414
  pratense, 409, **414**
  repens, 58, 81, 97, 101, 117, 123, 126, 166, 178, 196, 200, 201, 209, 213, 252, 253, 256, 377, 409, 412, **414**

rueppellianum, 252, 312, 315, **414–15**
 var. *rueppellianum*, 414–15
semipilosum, 170, 224, 225, 252, 305, 316, **415–17**
 var. *glabrescens*, 415
 var. *kilimanjaricum*, 415
 var. *microphyllum*, 415
 var. *semipilosum*, 415
subrotundum, 414
subterraneum, 408, 409, **417**
tembense, **417**
usambarense, 415, **417–18**

Vicia
 angustifolia, 418
 atropurpurea, 418
 benghalensis, **418**
 cordata, 209
 dasycarpa, 118
 sativa, 44, 209, 225, **418**
  var. *angustifolia*, 418
 villosa, 96, 418
Vigna
 aconitifolia, 7, 70, 75, 107, 146, 163, 271, **419–20**
 angularis, **430**, 423
 calcarata, 423
 catjang, 423
 fragrans, 420
 frutescens, **420**
 gracilis, 422
 lutea, 421
 luteola, **420–1**, 423
 marina, **421**
 mungo, 423
 nilotica, 420
 oblongifolia, **421–2**
  var. *parviflora*, 422
 parkeri, **422**
 parviflora, 422
 radiata, 75, 107, 146, 163, 174, 230, **422–3**
  var. *radiata*, 422
  var. *sublobata*, 422
 repens, 420
 schimperi, **423**
 sinensis, 423
 umbellata, **423**
 unguiculata, 146, 174, 200, 271, 278, 368, 421, **423–6**
  ssp. *cylindrica*, 424
  ssp. *deckindtiana*, 424
  ssp. *sesquipedalis*, 424
  ssp. *unguiculata*, 424
 vexillata, **426–7**
  var. *vexillata*, 426

Zornia
 diphylla, **427**, 428
 glochidiata, 127, **427**
 gracilis, 427
 latifolia, 427
 perforata, 427
 pratensis, **427–8**

Index of common names

Grasses

Abyssinian grass, 261
Adlay, 90
African couch grass, 125
African foxtail, 66
African millet, 131
Alabang, 107
Alyce clover 319, 320
Admirable, 59
Angleton grass, 107, 108
Angola grass, 59
Angolinha, 142
Anjan, Anjan grass, 66, 75
Annual kyasuwa grass, 229
Antelope grass, 130
Apang, 106
Arroz bravo, 164
Avena, 44
Avoine, 44

Babala-Napier hybrid, 242
Bahia grass, 205
Bajra, 216
Baksha, 149
Bana grass, 242
Barit, 164
Barley Mitchell grass, 43
Barwari, 173
Beck grass, 48
Bermuda grass, 92
Bharra grass, 246
Bher, 160
Birdwood grass, 75
Bitter grass, 196
Black kolukattai, 75
Black speargrass, 150
Blue couch, 122

Blue panic, 173
Bourgou, 130
Broadleaf carpet grass, 45
Broadleaf paspalum, 215
Brunswick grass, 204
Buffel grass, 66
Bulb panicum, 175
Bulo, 131
Bulrush millet, 216
Bunched kerosene grass, 41
Bungoma grass, 134
Button grass, 106

Calinguero, 167
Canutillo, 152, 181
Capim amaroso, 196
Capim angola, 59
Capim bóbó, 52
Capim cebola, 147
Capim colonião, 181
Capim columbia, 47
Capim gamalote, 149
Capim imperial, 47
Carib grass, 142
Carpet grass, 44, 45
Carrizo chico, 152
Cebadilla, 65
Chari, 269
Chiendent, 92
Cholam, 269
Cocksfoot, 103
Cocorobo, 147
Cola de zorro, 50
Coloured guinea grass, 175
Columbus grass, 264
Congo grass, 64

Index

Congo signal grass, 64
Corn, 296
Coronivia grass, 57
Creeping signal grass, 57
Criolla, 65
Crowfoot grass, 104
Curly windmill grass, 76
Cushion lovegrass, 135

Dallis grass, 197
Daremo, 88
Deenabandhu grass, 229
Deenanath grass, 229
Dhaman, 66
Dhub, 92
Dog's tooth grass, 92
Donkey grass, 192
Doob, 92
Du-chasi, 193

Eastern gama grass, 291
Egipto, 59
Elefante, 233
Elephant grass, 233

Faragua, 155
Fataque, 181
Finger millet, 131
Forquinha, 205
Fowlfoot grass, 132

Gajraj, 242
Gamalote, 181, 203
Gamba grass, 34
Gengibrillo, 205
Ghamur, 173
Giant elephant grass, 242
Giant Pangola grass, 126
Giant panic, 173
Giant setaria, 260
Giant star grass, 99, 102
Gift, 173
Gondirimi, 132
Goosegrass, 132
Gordura, 167
Goria, 89
Grama de antena, 196
Grama dulce, 205
Grama grass, 52
Gramilla, 202
Gramilla blanca, 202
Gramilla canita, 149
Grey lovegrass, 136
Guaratara, 47
Guatemala grass, 292
Guinea grass, 181

Gumai, 273
Guria, 89

Hariali, 92
Herbe de Guinée, 181
Herbe du Brézil, 167
Herbe elephant, 233
Hierba blanca Honduras, 162
Hierba de India, 181
Hindi grass, 106
Hoodgrasses, 153
Horo, 290
Hurricane grass, 51

Ikoka, 192
Indian corn, 296
Indian grass, 261
Italian ryegrass, 166

Janeiro, 142
Japanese barnyard millet, 128
Jaragua, 155
Jharua grass, 128
Job's tears, 90
Johnson grass, 273
Jonna, 269
Jowar, 269
Jungle rice, 128

Kala-sat, 216
Kangaroo grass, 286
Karad, 106
Kennedy, ruzi, 64
Kikuyu (kikuyo) grass, 222
Kleingrass, 175
Kolukattai, 66
Koracan, 131
Kyasuwa, 231

Lagrimas de San Pedro, 90
Leafy nineawn, 133
Limpograss, 149
Little para, 295
Loekoentoegras, 161

Maize, 296
Makarikari grass, 175
Makchari, 143
Malohilla, 142
Malohillo, 59
Marmalade grass, 62
Marvel grass, 106
Masai lovegrass, 140
Massambara, 273
Matgrass, 44
Mauritius grass, 59

Melado, 167
Micay, 47
Mission grass, 231
Mitchell grasses, 42
Mollasses grass, 167
Morkuba, 193
Mulga grass, 41
Musal, 161
Mwele, 216

Nadi blue grass, 108
Naivasha star grass, 102
Nakuru grass, 65
Napier grass, 233
Napier's fodder, 233
Napierzinho, 249
Narrowleaf carpet grass, 44
Natal grass, 48, 244
N'golo, 243
Nile grass, 31
Noble cane, 247

Oat, 44
Orchard grass, 103

Paja de agua, 181
Palisade grass, 54
Pangola grass, 111
Pangola gigante, 126
Panisharu, 149
Para grass, 59
Paraguillas, 147
Pasto alemán, 129
Pasto amargo, 196
Pasto Argentina, 92
Pasto Bermuda, 92
Pasto bora, 147
Pasto borla, 86
Pasto clavel, 149
Pasto colon, 264
Pasto Dalis, 197
Pasto dulce, 202
Pasto gigante, 233
Pasto guatemala, 292
Pasto Guinea, 181
Pasto Hatico, 162
Pasto Honduras, 162
Pasto horqueta, 205
Pasto Johnson, 273
Pasto lloron, 136
Pasto micay, 47
Pasto miel, 197
Pasto mosquito, 140
Pasto Rodes, 77
Pasto rojas, 203
Pasto San Augustin, 284

Pasto Sudan, 275
Pata de gallina, 202
Pearl millet, 216
Pemba grass, 284
Penhalonga grass, 59
Perennial ryegrass, 166
Plicatulum, 213
Pongola grass, 111
Prairie grass, 65
Privilegio, 181
Prodigioso, 291
Pusa giant Napier, 242
Pyaung, 269

Queensland blue grass, 109

Ragi, 131
Rapoka grass, 132
Red oatgrass, 286
Redtop Natal grass, 244
Rescue grass, 65
Rhodes grass, 77
Rooigras, 286
Ruzi grass, 64

Sadabahar, 34
Sain grass, 248
St. Augustine grass, 284
Salt water couch, 215
Sanwa millet, 128
Scrobic paspalum, 195
Sea-shore paspalum, 215
Sempreverde, 181
Setaria grass, 249
Sewan, 163
Shama millet, 128
Sheda, 106
Side-oats grama, 53
Signal grass, 54
Sorgho d'Alep, 273
Sorghum, 269
Sorgo, 269
Sorgo amargo, 203
Sorgo negro, 264
Sorghum, 269
Sour grass, 196
Sour paspalum, 196
Spear grass, 150
Star grass, 92, 98, 99
Sudan grass, 275
Sugar cane, 247
Surinam grass, 54
Swazigrass, 126
Sweet pitted grass, 50, 51

Tall fescue, 148

Tangle grass, 150
Tanner grass, 63
Teff, T'ef, 141
Telebun, 131
Teosinte, 143
Thangari, 125
Thin Napier grass, 231
Thoman, 193
Tongolonakanga, 33
Torpedo grass, 191
Tumam, 193

Uba cane, 247

Vasey grass, 214
Venezuela grass, 203

Veyale, 155

Weeping lovegrass, 136
Wheat Mitchell grass, 43
Whitetop, 133
Wimbi, 131
Wintergreen paspalum, 203
Woolly finger grass, 122, 123, 124, 125
Wynne grass, 167

Yaragua, 155, 167
Yayale, 155

Zacate amargo, 44, 45
Zacate Bermuda, 92
Zacaton, 181

Legumes

Adzuki bean, 420
Alfafa do Nordeste, 397
Alfalfa, 388
Alyce clover, 319, 320
Arhar, 325

Berseem, 410
Big trefoil, 379
Birdsfoot trefoil, 379
Bonavista bean, 367
Bullamon lucerne, 391
Burr medic, 387
Butterfly pea, 381

Cacahuete, 322
Calopo, 328
Capparicho, 342
Centro, 331
Clover, 409
Cluster bean, 338
Common vetch, 418
Cordofan pea, 335
Cowpea, 423
Cutleaf medic, 387

Dewgram, 419
Dwarf koa, 340

Egyptian clover, 410

Feijao de boi, 348
Finestem stylo, 397
Florida beggar weed, 349

Glycine, 357

Golden gram, 422
Green gram, 422
Greenleaf desmodium, 344
Groundnut, 322
Guar, 338

Hairy indigo, 365
Hetero, 343
Horsegram, 386
Hyacinth bean, 367
Indian sweet clover, 388
Indigo, 364
Ipil-ipil, 369

Jack bean, 330
Jitirana, 331
Joint vetch, 318

Kaimi, Kaimi clover, 342
Kenya white clover, 415
Koa haole, 369
Kudzu, 391
Kuru vine, 344

Lablab bean, 367
Lotononis, 374
Lubia, 367
Lucerne, 388

Mani, 322
Marmelada de caballo, 342
Mat bean, 419
Meladinho, 397
Miles lotononis, 374
Moth bean, 419

Mungo, 422

One-leaf clover, 320

Peanut, 322
Phasey bean, 385
Pigeon pea, 325
Pois d'Angola, 325
Puero, 392

Rabo de iguana, 328
Red clover, 414
Red gram, 325
Rice bean, 423
Rueppell's clover, 414

Senji, 388
Silverleaf desmodium, 350
Siratro, 379
Soja, 355
Soja perene, 357

Soyabean, 355
Spanish clover, 343, 350
Stylo, 397
Subclover, 417
Subterranean clover, 417
Sunn hemp, 338
Sword bean, 330

Tick clover, 350
Townsville lucerne, 402
Townsville stylo, 402
Trailing indigo, 366
Trebol Alicia, 320
Tropical kudzu, 392
Tur, 325

Velvet bean, 389
Vetch, 418

White clover, 414
White popinac, 369